ADVANCES IN

GEOPHYSICS

VOLUME 25

Contributors to This Volume

Robert E. Dickinson

G. S. Golitsyn

Arnold Gruber

Syukuro Manabe

George Ohring

Abraham H. Oort

José P. Peixóto

Barry Saltzman

G. J. Shutts

Joseph Smagorinsky

Insignia of the Academy of Sciences of Lisbon

Advances in

GEOPHYSICS

VOLUME 25

Theory of Climate

Proceedings of a Symposium Commemorating the Two-Hundredth
Anniversary of the Academy of Sciences of Lisbon
October 12–14, 1981, Lisbon, Portugal

Edited by

BARRY SALTZMAN

Department of Geology and Geophysics
Yale University
New Haven, Connecticut

1983

ACADEMIC PRESS

A Subsidiary of Harcourt Brace Jovanovich, Publishers

New York   London
Paris   San Diego   San Francisco   São Paulo   Sydney   Tokyo   Toronto

11115284

ACADEMIC PRESS, INC.
111 Fifth Avenue, New York, New York 10003

United Kingdom Edition published by
ACADEMIC PRESS, INC. (LONDON) LTD.
24/28 Oval Road, London NW1 7DX

LIBRARY OF CONGRESS CATALOG CARD NUMBER: 52- 12266

ISBN 0-12-018825-2

PRINTED IN THE UNITED STATES OF AMERICA

83 84 85 86   9 8 7 6 5 4 3 2 1

CONTENTS

Parameterization of Traveling Weather Systems in a Simple Model of Large-Scale Atmospheric Flow

G. J. SHUTTS

Climatic Systems Analysis

BARRY SALTZMAN

Part III.   Radiative, Surficial, and Dynamical Properties of the Earth–Atmosphere System

Satellite Radiation Observations and Climate Theory

GEORGE OHRING AND ARNOLD GRUBER

Land Surface Processes and Climate—Surface Albedos and
Energy Balance

ROBERT E. DICKINSON

Global Angular Momentum and Energy Balance Requirements
from Observations

ABRAHAM H. OORT AND JOSÉ P. PEIXÓTO

CONTRIBUTORS

Numbers in parentheses indicate the pages on which the authors' contributions begin.

ROBERT E. DICKINSON, *National Center for Atmospheric Research, Boulder, Colorado 80307* (305)

G. S. GOLITSYN, *Institute of Atmospheric Physics, Academy of Sciences of the USSR, Moscow, USSR* (85)

ARNOLD GRUBER, *Earth Sciences Laboratory, National Earth Satellite Service, NOAA, Washington, D.C. 20233* (237)

SYUKURO MANABE, *Geophysical Fluid Dynamics Laboratory/NOAA, Princeton University, Princeton, New Jersey 08540* (39)

GEORGE OHRING,[1] *Department of Geophysics and Planetary Sciences, Tel-Aviv University, Tel-Aviv, Israel, and National Earth Satellite Service, NOAA, Suitland, Maryland* (237)

ABRAHAM H. OORT, *Geophysical Fluid Dynamics Laboratory/NOAA, Princeton University, Princeton, New Jersey 08540* (355)

JOSÉ P. PEIXÓTO, *Geophysical Institute, University of Lisbon, Lisbon, Portugal* (355)

BARRY SALTZMAN, *Department of Geology and Geophysics, Yale University, New Haven, Connecticut 06511* (173)

G. J. SHUTTS,[2] *Atmospheric Physics Group, Department of Physics, Imperial College of Science and Technology, London, United Kingdom* (117)

JOSEPH SMAGORINSKY, *Geophysical Fluid Dynamics Laboratory/NOAA, Princeton University, Princeton, New Jersey 08540* (3)

[1] Present address: Laboratory for Atmospheric Sciences, Climate and Radiation Branch, Code 915, NASA Goddard Space Flight Center, Greenbelt, Maryland 20771.

[2] Present address: Meteorological Office, Bracknell, Berkshire RG12 2SZ, United Kingdom.

Conferees attending the *Symposium on the Theory of Climate,* commemorating the bicentennial of the Academy of Sciences of Lisbon, October 12–14, 1981. From left to right: Erik Eliasen (representing the Danish Academy of Sciences), G. J. Shutts, Barry Saltzman, Robert E. Dickinson, W. Lawrence Gates, George Ohring, Joseph Smagorinsky (rear), Syukuro Manabe (front), Edward N. Lorenz, G. S. Golitsyn, Abraham H. Oort, and José P. Peixóto.

FOREWORD

The challenge of achieving a deductive understanding of the long-term behavior of the earth's fluid environment—i.e., a theory of climate—is now recognized to be one of the most important and difficult ever posed, embracing all domains of science. Aside from its purely scientific content, there is a wide awareness of the enormous social ramifications of this subject which universally affects mankind across all national boundaries.

It is altogether fitting, therefore, that in celebration of its two-hundredth anniversary the Academy of Sciences of Lisbon should choose climate theory as the subject for one of the symposia in its commemorative series "Frontiers of Knowledge." We are delighted to be able to present the contributions to this symposium as a special volume of *Advances in Geophysics*.

In all, 10 papers were presented. It was suggested by the conveners that the authors not attempt to review the whole field related to their work, but rather present a more personal account of their "school of thought"— i.e., a synthesis of the body of work that they and their colleagues have produced over a period of time—culminating in some new results. As a whole, the papers conform admirably to this suggestion, representing an excellent mix of much needed review and synthesis with a good deal of new and original material that is testimony to the rapid pace of ongoing work in climate theory.

The papers are divided into three main groups: I. History and Application of General Circulation Models, emphasizing the development and use of large-scale numerical models of the atmosphere to obtain asymptotic equilibrium solutions for the "fast response" parts of the climatic system (Smagorinsky and Manabe); II. Statistical-Dynamical Models, emphasizing simpler, more heavily parameterized models with potential for forming a basis for a time-dependent theory of very long term climatic change involving the slow response parts (Golitsyn, Shutts, and Saltzman); and III. Radiative, Surficial, and Dynamical Properties of the Earth–Atmosphere System, emphasizing observational and diagnostic aspects of global climate (Ohring and Gruber, Dickinson, and Oort and Peixóto). These groups are by no means mutually exclusive; each of the papers contains material overlapping that in other groups. For example, the important role of CO_2 in climate is discussed not only in a major paper by Syukuro Manabe, but also in lesser detail and from other viewpoints in papers by George Ohring and Arnold Gruber, G. S. Golitsyn, and Barry Saltzman.

Only two of the papers delivered at the symposium are not included in this volume. These are the excellent discussions of "The Predictability of the Atmosphere and Its Surroundings" by Edward N. Lorenz and "The Numerical Modeling of Climate" by W. Lawrence Gates. However, both Lorenz and Gates plan to prepare articles covering the material of their talks, with substantial new material, for a forthcoming volume of this serial.

It was with great sadness that the conveners and conferees learned of the death of Jule Charney, whose participation in the symposium had been keenly anticipated. All participants of the symposium agreed that his lifelong work in dynamical meteorology and climate theory provided underpinnings for much that was said at the symposium. Some of Charney's influence on this subject is described in Joseph Smagorinsky's contribution giving a historical account of the beginnings of numerical weather prediction and general circulation modeling.

Speaking as one of the participants in this symposium, and I believe also in behalf of the whole group, I wish to express our special thanks to José Peixóto, President of the Academy of Sciences of Lisbon, who was the driving force behind all the proceedings. These thanks are not only for his major scientific contribution and his introductions and background discussions on all subjects scientific, historical, and cultural, but also for the unsurpassable hospitality he and his associates showed us during our stay in Portugal.

It is the hope of the Publisher that this volume will be a lasting, scientifically significant remembrance of this bicentennial celebration of one of the world's oldest and most renowned scientific institutions, whose insignia we are proud to display as the frontispiece to this volume.

BARRY SALTZMAN

ADVANCES IN

GEOPHYSICS

VOLUME 25

Part I

HISTORY AND APPLICATION OF GENERAL CIRCULATION MODELS

THE BEGINNINGS OF NUMERICAL WEATHER PREDICTION AND GENERAL CIRCULATION MODELING: EARLY RECOLLECTIONS

JOSEPH SMAGORINSKY

Geophysical Fluid Dynamics Laboratory/NOAA
Princeton University
Princeton, New Jersey

1. INTRODUCTORY REMARKS

I should say at the outset that my intention here is not to deliver a comprehensive history of the important formative circumstances of the late 1940s and the following decade. The invitation was to make a personal presentation of events in which I had been involved and of which I had first-hand knowledge. I have tried to assemble, from memory and from personal documents, my impressions of the time. My account will therefore be quite selective, but I hope it will be viewed as a useful, if not insightful, contribution to the history by a witness and participant. I therefore apologize in advance for the many lapses in completeness which, no doubt, will be detected by the many others who had been involved in that fascinating period.

The appropriateness of including the era of numerical weather prediction in a symposium on "Advances in the Theory of Climate" is, in retrospect, quite obvious. It was the development of the scientific base and technical methodology needed for the modeling for prediction that paved the way later on for modeling the processes responsible for the general circulation and thereafter for the simulation of climate. In turn, it was the general circulation models that provided the vehicle in the 1960s and 1970s for extending numerical weather prediction beyond a few days.

Hardly any of the events and circumstances touched upon in this account could have happened without Jule Charney. They might have occurred eventually and probably in some other form, but Charney's genius

3

and driving force were singularly responsible for their happening when and how they did. Charney passed from our midst in June 1981. It is my honor to dedicate this work to his memory.

2. SOME PERSONAL ANTECEDENTS

My interest in meteorology began in my midteens (in the late 1930s) when I thought that weather prediction was somehow accomplished deterministically by the application of physical principles. Quite consistently, I also thought this was true for the design of ship hulls, and in fact at that time my first interest was in naval architecture. But financial and family considerations dominated my career decision, and I entered a university course of study in meteorology. I, of course, quickly learned that my basic assumption was quite incorrect. Weather forecasting was quite subjective, but based on powerful conceptual procedures—the construction of the isobaric weather map and an identification of the air mass and frontal systems. The predicted time–space evolution of the synoptic map was based on the experience of having observed and classified many such evolutions. The forecast of wind, temperature, and precipitation was based on empirical models of how these meteorological parameters would be associated with the predicted pressure field and its attendant air mass and frontal systems.

World War II interrupted my formal university education and I entered a military meteorology training course at the Massachusetts Institute of Technology (MIT) where, in 1943, I came into contact with the eminent dynamic meteorologist, Professor Bernhard Haurwitz. When I asked him why physical principles had not been applied to the practical problem of weather prediction, he quickly pointed out the futility of using the tendency equation to predict surface pressure changes. The actual winds were not sufficiently accurate and the geostrophic approximation would give nonsensical measures of the horizontal divergence. When queried further, Haurwitz did recall the work of L. F. Richardson during and just after World War I, but, as I remember, did not attach great importance to its implications.

I was resigned to frustration and disappointment which remained dormant until the end of the decade. I returned to civilian life, resumed my university education, and went on to complete a master's degree with emphasis on dynamic meteorology. My first position was as a research meteorologist at the U.S. Weather Bureau under Dr. Harry Wexler. In 1949, I heard a lecture by Jule Charney which changed my life. His systematic analysis of the scale properties of large-scale atmospheric motions and his presentation of a rational approach to deriving a geostrophically consistent set of prediction equations, reawakened my hopes for a hydrodynamic framework for prediction. I did not, of course, know how far Charney's ideas would carry in shaping, and indeed revolutionizing, the physical and dynamical basis for weather prediction. In fact, we now know that the basic methodology would eventually find its way into the study of a much broader part of the spectrum of phenomena than midtropospheric Rossby waves. With the modern high-speed electronic computer, then under development by von Neumann and his colleagues at the Institute for Advanced Study in Princeton, it would eventually be possible to study synoptic-scale baroclinic processes, the dynamics of convection and mesoscale phenomena, the general circulation, climate, and even the ocean circulation.

In one day, my visions were completely transformed. Little did I know that I would be privileged to participate in a scientific revolution that, when I first made my career choice, I had mistakenly thought had already happened at the time.

3. The Institute for Advanced Study 1949–1953

The formation of the Meteorology Group at the Institute for Advanced Study (IAS) in Princeton and its first numerical forecasts on the Electronic Numerical Integrator and Computer (ENIAC) were key events in the early history of numerical weather prediction. These events were eloquently described in authoritative detail in a lecture[1] in memory of Professor Victor P. Starr at MIT in 1979 by Professor George W. Platzman, who himself was instrumental during that period. Here, I will only try to supplement his account with additional documented contemporary impressions, keeping duplication at a minimum. At this point one should note that remarkably parallel developments were taking place in the Soviet Union during the 1940s and 1950s. But because scientific communications with the West did not begin to fully develop until the late 1950s, much of the Soviet work was largely unknown until an excellent comparative survey of research through 1959 was published by Phillips, Blumen, and Coté in 1960.[2]

Based on John von Neumann's radically new logical ideas for a stored program computer using Williams's cathode ray tube technology as a

[1] Platzman, G. W. The ENIAC computations of 1950—gateway to numerical weather prediction. *Bull. Am. Meteorol. Soc.* **60**(4), 302–312 (1970).

[2] Phillips, N. A., Blumen, W., and Coté, O. Numerical weather prediction in the Soviet Union. *Bull. Am. Meteorol. Soc.* **41**(11), 599–617 (1960).

storage device, an Electronic Computer Project was established in 1946 at the IAS. Only von Neumann's great reputation and persuasive power were able to overcome the opposition of the faculty to so mundane an enterprise. The circumstances surrounding this event are well documented in a book by H. H. Goldstine,[3] who was one of the prime movers on the project. As he points out, a threefold thrust was intended: engineering, numerical mathematics, and some important and large-scale applications. For the latter, von Neumann selected numerical meteorology. This was based on his knowledge of Richardson's earlier work and also on encouragement by Carl-Gustav Rossby of the University of Chicago and Harry Wexler of the U.S. Weather Bureau. It was recorded at the time:[4]

> A project whose ultimate effects on weather forecasting may be revolutionary has been quietly under way during the past year in the academic surroundings of the Institute for Advanced Study, Princeton, New Jersey. . . . In August 1946, a conference of meteorologists met in Princeton to discuss the project. . . . Since last summer, work has gone forward in promising fashion, though it is still far too early to expect immediate, tangible results. . . . The immediate aims of this group are the selection and mathematical formulation of meteorological problems to be solved by the electronic computer . . . the most interesting feature of the project is the effort being made to link the theory behind atmospheric processes with future weather.

After failing to persuade Rossby to come to the Institute to lead the effort, von Neumann invited one of Rossby's young proteges from the University of Chicago, Albert Cahn, Jr., who was then succeeded by Philip D. Thompson.

Charney, who had been a graduate student of J. Holmboe's at the University of California at Los Angeles, came to Rossby's attention when he briefly served as a research associate at the University of Chicago in 1946–1947 on his way to a postdoctoral appointment at the University of Oslo. During that academic year, Rossby, with a distinguished group of collaborators, produced a famous synoptic, theoretical, and experimental paper on the interaction of long waves with the zonal circulation.[5] Although Charney was at Chicago for only part of the duration of that project, he impressed Rossby to the point where Charney was invited to lead the IAS Meteorology Group upon his return from Oslo in 1948. It was in Oslo that he wrote his scale paper.[6]

Charney immediately invited Arnt Eliassen to join him. Eliassen had by

[3] Goldstine, H. H. "The Computer from Pascal to von Neumann." Princeton Univ. Press, Princeton, New Jersey, 1972.

[4] "Electronic Computer Project," Weather Bureau Topics and Personnel, July 1947.

[5] Staff Members of the Department of Meteorology of the University of Chicago (J. G. Charney, G. P. Cressman, D. Fultz, L. Hess, A. D. Nyberg, E. V. Palmen, H. Riehl, C. G. Rossby, Z. Sekera, V. P. Starr, and T.-C. Yeh). On the general circulation of the atmosphere in middle latitudes. *Bull. Am. Meteorol. Soc.* **28**, 255–280 (1947).

[6] Charney, J. G. On the scale of atmospheric motions. *Geofys. Publ.* **17**(2), 1–17 (1948).

that time completed his definitive paper on a consistent formulation of the hydrostatically conditioned equations in pressure coordinates.[7] That was the beginning of the famous Meteorology Group. Charney was also joined by a young mathematician, Gilbert A. Hunt. It was this triumvirate that, in January 1949, reported on a "Program for Numerical Weather Prediction" in New York that had captivated me. Hunt, soon after, returned to his first love and is now a distinguished Professor of Mathematics at Princeton University.

The beginning of the collaboration of Charney and Eliassen in Oslo produced two key papers after they reunited in Princeton. The first, by Charney himself,[8] was a comprehensive rationale which laid the foundation for dynamical prediction. It justified the use of the geostrophic approximation to filter small-scale high-frequency noise from the vorticity equation, discussed the propagation of signal and its implications on data requirements, introduced the notion of the equivalent-barotropic atmosphere to reduce the forecast problem to a two-dimensional one, and finally, showed how Green's functions could be used to make a linear one-dimensional prediction for an arbitrary initial geopotential distribution at midtroposphere. A companion paper, submitted a few days later by Charney together with Eliassen,[9] gave the results of one-dimensional predictions (along a latitude band) and also applied these techniques to the study of topographically produced quasi-stationary perturbations.

In those early days, Charney's group for the most part consisted of two to four meteorologists on visits for about one year. The main exception was Norman A. Phillips, who arrived in 1951 after completing his Ph.D. at the University of Chicago and moved to MIT with Charney in 1956.

In 1949, I was invited as an occasional visitor, from my base in Washington, D.C., to assist the group in extending its one-dimensional linear barotropic calculations. On behalf of the Weather Bureau, I also was asked to become familiar with the theoretical aspects of a more realistic model. As a result of a month-long visit in the spring of 1949, I recorded in a report:[10]

> Essentially, the new method is a much refined form of the vorticity theorem enunciated by Rossby in the late 1930's. Although this model is, as Rossby's, a barotropic fluid in one-dimensional motion which only considers small perturbations, it can take into account [equivalent-barotropic] divergence, the mean finite lateral width of a disturbance, friction, topography, an arbitrary initial pressure disturbance, and the

[7] Eliassen, A. The quasi-static equations of motion with pressure as independent variable. *Geofys. Publ.* **17**(3), 1–44 (1949).

[8] Charney, J. G. On a physical basis for numerical prediction of large-scale motions in the atmosphere. *J. Meteorol.* **6**, 371–385 (1949).

[9] Charney, J. G., and Eliassen, A. A numerical method for predicting the perturbations of the middle latitude westerlies. *Tellus* **1**(2), 38–54 (1949).

[10] Memorandum, Smagorinsky to Chief of Bureau [F. W. Reichelderfer], June 30, 1949.

boundary conditions which arise from considering circular latitude lines. To construct this model, it was necessary to introduce a number of arbitrary parameters in order to describe more fully actual atmospheric motions. The parameters involve (1) a measure of the finite lateral extent of the disturbances and (2) a second approximation on the assumption of a constant basic zonal current. These can best be evaluated by repeated application of the forecast formula to many varied situations for different seasons, performing, more or less, a controlled experiment. It should be remarked that these parameters have some physical meaning, and this is utilized in testing.

The Meteorology Group had only made a few test forecasts. With the aid of Mrs. Margaret Smagorinsky,[11] well over one hundred 24-hour winter forecasts at one latitude were made and analyzed. The forecasts verified fairly well, and may be considered competitively with subjective forecasts. This is very encouraging since the latter type of forecast has only limited physical basis, while the Charney–Eliassen method is based wholly on dynamic considerations. However, because of the simplicity of the model, it is recommended that this method be used only as a supplementary tool, recognizing where its shortcomings lie and how they will affect the forecast. It is found that one can usually state *a priori* how well the objective method will verify, but it is thought that improvement in verification will come from additional detailed analyses of the forecasts.

A small number of trial 5-day forecasts at 500 mb were made. These verified very poorly. However, theoretical examination showed that the proper choice of the arbitrary parameters mentioned above becomes extremely critical in the forecast and that they are a major source of discrepancy. The breakdown can best be observed by successive forecasts from the same profile for 1, 2, . . . , 6, 7 days. One of the most important and successful tools of the Extended Forecast Group [of the Weather Bureau] is an empirically corrected form of Rossby's original formula. It is hoped that proper employment of the more refined technique will enhance the usefulness of the vorticity concept.

Dr. Charney and Mr. Eliassen are not considering extending this model to two dimensions, since hand computations would become formidable. Instead, their plans are to construct a two-dimensional barotropic model which also permits non-linear motions. The solution, of course, can only be found by the electronic computer. Studies of how the present simple model fails will aid Charney and Eliassen in their attempt on the non-linear problem by giving them a clue as to which simplifications are most detrimental.

A short while later, I attended a conference at the University of Chicago in which Charney, Eliassen, and the Staff of the Department of Meteorology participated. The purpose was to assess the basic theory for the numerical forecasting technique, and to examine the results of some trial forecasts. I reported:[12]

> The role of forced stationary perturbations was reviewed, and it was agreed that the effects of topography [and] friction . . . were taken into account as well as possible with this simple atmospheric model.

[11] It was not unusual for wives to be professionally involved. Adele Goldstine, Klara von Neumann and, for a short period, Margaret Smagorinsky all programmed for the IAS computer—in absolute octal, of course.

[12] Memorandum, Smagorinsky to Chief of Bureau, July 18, 1949.

The fact that the entire energy balance cannot be described by a single-layer barotropic model led to the tentative conclusion that this method would fail for long period forecasts even with the two-dimensional non-linear model. . . .

Professor Rossby gave high praise to the work of Charney and Eliassen and expressed the view that this represented one of the most significant turning points in the history of theoretical and synoptic meteorology.

Subsequently, Eliassen[13] repeated his misgivings: "Personally, I must confess that I don't expect too much from the application to 5-day mean maps but experience will, of course, be the best judge."

I was also applying the influence functions to the 36-hr, 700-mbar forecast problem, at the three latitudes, 35° N, 48° N, and 55° N for the longitudes 50° W to 120° W. Upon comparing the results with the Weather Bureau Analysis Center (WBAN) operational forecasts, Wexler commented:[14] "It is seen that the two techniques give quite similar verification which should be considered quite a victory for the numerical, objective procedure, based upon the results of dynamic meteorology." Actually, the dynamical forecasts were slightly inferior.

Meanwhile, the Princeton group was already moving onward rapidly to deal with the barotropic finite amplitude problem.

In the early stages, Charney wrote:[15] "We had so many difficulties with the hand computations for the 2-dim. finite amplitude motion that I decided to abandon the project, especially in consideration of the fact that we are planning to do the same thing on the ENIAC beginning December 1."

The ENIAC at the U.S. Army Aberdeen Proving Ground in Maryland had been enlisted for the nonlinear barotropic forecast integrations. Actually, the Aberdeen operation was delayed until March 1950. There was some uncertainty in the choice of the particular case to be studied. I had suggested a blocking situation in December 1949. However, Charney ultimately settled on January 5, 30, and 31 and February 13, 1949. Some of the considerations were:[16]

How will this model explain the subsequent motions on a map

- (a) that is essentially barotropic;
- (b) which is predominantly baroclinic;
- (c) which displays pronounced blocking of the jet stream.

These tests are designed so that one may discover which properties of baroclinic motions are most essential in devising a more realistic atmospheric model. In addition

---

[13] Letter, Eliassen to Smagorinsky, July 31, 1949.

[14] Memorandum, Wexler to Chief of Bureau, January 12, 1950.

[15] Letter, Charney to Smagorinsky, November 3, 1949.

[16] Memorandum, Smagorinsky to Chief, Weather Bureau, February 7, 1950.

it is planned to test some hypothetical situations which exhibit such "pure" disturbances as:

    (a) an isolated vortex in a field of no relative vorticity,
    (b) an isolated vortex in a field of zonal motion which possesses vorticity,
    (c) periodically distributed vortices in a field of zonal motion which is dynamically
        unstable.

Because of the limited capacity of the ENIAC it is also necessary to decide upon the
most suitable map projection, and the appropriate time and space scales.

In reporting on the outcome of the ENIAC expedition, I noted:[17]

Unfortunately, the lack of time made it impossible to make as many tests as would
have been desirable. Before one could actively test the hypothesis, a number of fundamental questions had to be answered:

    1. What are the upper and lower limits of the grid spacing from which observed
       contour heights are selected?
    2. What is the most feasible time interval to be used in going into the future?

Tentatively, the results of a number of test forecasts indicate a grid size of about 5°
longitude at 45° latitude and a time interval of two or three hours. On this basis, two
forecasts were made from synoptic situations that were characteristically barotropic,
and a third was partially completed. The first was a 24-hour forecast from the 0300 map
of January 5, 1949. The computed forecast was not particularly good. Some of the
failure could be attributed to the small size of a closed low over the United States
which fell in a blank spot of the grid used and so the data picked off did not describe
sufficiently the flow about the low. However, the computation did predict the westward motion of the western cell of the Bermuda high. Another 24 hour forecast, from
the 0300 map of January 31, 1949, gave excellent results. It called for such changes of
the atmospheric motion as the intensification of a blocking high, the filling of trough
and the pivoting of an elongated low pressure area.

This relatively crude theoretical model of the atmosphere succeeded in serving as
the basis for the prediction of some very pronounced modifications of the planetary
flow pattern and demonstrated that for the cases and forecast period chosen the
atmospheric processes were essentially barotropic. However, this is not meant to
minimize the role played by baroclinic phenomena, which, through the transformation
of potential to kinetic energy, are able to generate new disturbances. The small amount
of experience thus gained indicates that the barotropic model is adequate for forecasting up to 36 or 48 hours. It is planned that succeeding models will incorporate baroclinic mechanisms which take into account the effects of variations in the vertical
structure of the atmosphere.

The fact that the memory capacity of the ENIAC was being overtaxed was already
evident. This together with the relative slowness of the machine (36 hours for a 24 hour
forecast) deems it impractical to use the ENIAC on a baroclinic model, since it would
require many times the memory capacity and correspondingly greater speed. For this
purpose the Princeton machine will probably be used. The date of availability of the
[IAS computer] is still a moot question, with optimistic estimates ranging from June to
October of this year.

---

[17] Memorandum, Smagorinsky to Chief, Weather Bureau, April 14, 1950.

Some of the results of this month's work are immediately applicable to subjective forecasting techniques. The experiments showed that a knowledge of the field of maximum absolute vorticity transport gave an excellent indication of the regions of extreme instantaneous pressure tendency. The absolute vorticity transport corresponding to the latest available synoptic map can be calculated by desk computer. Making use of the relatively conservative property of these instantaneous tendencies, one can use them as a supplementary tool to the one-dimensional numerical forecasts. This tool will be experimented with in the Short Range Forecast Development Section [of the U.S. Weather Bureau].

In conclusion the writer wishes to express his appreciation for being given the opportunity of being associated with this historic development in the science of meteorology. We are at the beginning of a new era in weather forecasting—an era that will be based on the use of high speed automatic computers. For best use of the computers, it is essential that our aerological data procurement over the oceans be greatly increased.

The formal scientific report of this first ENIAC expedition was published by the principals in *Tellus*.[18] John C. Freeman of the Meteorology Group and George Platzman were the other participants (Fig. 1). In following up, I wrote about future plans:[19]

The Weather Bureau will participate in a number of experiments for some preliminary tests in anticipation of the use of the [IAS computer] for the solution of this [baroclinic] problem. Experience with the earlier tests at Aberdeen indicated that many questions of fundamental nature can be investigated before use is made of a high speed computer.

In the tests at Aberdeen, it is thought that one of the failures of the barotropic model to explain a pronounced development will be remedied by taking into account the baroclinicity which obviously was present. This situation, a low which first deepened at higher levels in eastern Canada on January 30, 1949, will be used for the first test.

Examination of the data for that date once again brought to focus the great lack of sufficient and reliable data to high levels. It is hoped that the future will show this unfortunate situation remedied.

In the summer of 1950, I began an extended stay at the Institute; it was to last until the spring of 1953. Norman Phillips was already working on his two-layer model for his doctoral dissertation at the University of Chicago. In his 1951 paper[20] he showed that one can construct a model of two superimposed barotropic fluids which indirectly can be related to the real baroclinic continuum. With this model he calculated sea level tendencies and midtropospheric vertical motions for the famous 1950 Thanksgiving Day storm over eastern North America. The results were quite encouraging. This was to be a test case of baroclinic development again 2

---

[18] Charney, J. G., Fjørtoft, R., and von Neumann, J. Numerical integration of the barotropic vorticity equation. *Tellus* **2**, 237–254 (1950).

[19] Memorandum, Smagorinsky to Chief, Weather Bureau, May 31, 1950.

[20] Phillips, N. A. A simple three dimensional model for the study of large-scale extratropical flow patterns. *J. Atmos. Meteorol.* **8**, 381–394 (1951).

FIG. 1. Some of the participants in the first ENIAC expedition, March 1950. Left to right, standing: R. Fjørtoft, J. G. Charney, J. C. Freeman, J. Smagorinsky (J. von Neumann and G. W. Platzman, absent); front: programmer assistants of the Aberdeen Proving Ground. The ENIAC is in the background.

years later. Phillips was invited to join the IAS group in 1951. Meanwhile, both Eady[21] in the United Kingdom and Eliassen[22] in Norway had designed their "2½-dimensional" models. It was pointed out by Eliassen that the three models were mathematically equivalent to each other with the proper interpretation of the dependent variables and constant parame-

[21] Eady, E. T. Note on weather computing and the so-called 2½-dimensional model. *Tellus* **4**, 157–167 (1952).

[22] Eliassen, A. Simplified dynamic models of the atmosphere, designed for the purpose of numerical prediction. *Tellus* **4**, 145–156 (1952).

ters. Of course, the greater problem remained, that is, establishing the equivalence of these models to the real atmosphere. The generalization was undertaken by Charney and Phillips.[23] The limit for large n of their $n + \frac{1}{2}$ level formulation was "not the most general three-dimensional model, but is one that can be obtained from the next general model through ignoring certain effects due to the spatial variations of static stability and absolute vorticity." They then used the $2\frac{1}{2}$-dimensional version to make the first finite interval baroclinic forecasts on the IAS computer which was completed in early 1952.[24] A series of six 12- and 24-hr forecasts were produced; the case was the Thanksgiving Day Storm of 1950. Based on the ENIAC experience with sparse oceanic data, a limited area covering eastern United States and southern Canada was selected for the integration.

In the Meteorology Group's (Fig. 2) report for 1952,[25] which, of course, covered these new results, Charney exposed his contextual philosophy as well as described some of the details:

The problem of primary interest at present is the prediction of changes of atmospheric flow over a period of 24 to 48 hours. The prediction of the field of motion is a necessary, though not a sufficient, pre-requisite for predicting cloudiness and precipitation. The philosophy guiding the approach to this problem has been to construct a hierarchy of atmospheric models of increasing complexity, the features of each successive model being determined by an analysis of the shortcomings of the previous model.

The primitive equations of motion reflect the fact that the atmosphere is capable of sustaining a wide spectrum of disturbances. For the purpose of short-range weather prediction, only those disturbances of planetary dimensions with periods of 3 to 7 days are of importance. These motions may be characterized as quasi-hydrostatic and quasi-geostrophic. . . .

The work during the calendar year 1952 was geared to the use of the Institute computer which was completed early in the year. It was decided that the most logical procedure was to test the various models on a single sequence of weather events. For this purpose the storm of November 25, 1950 over the eastern United States was admirably suited. This storm was one of the most rapid and intense developments ever to have been recorded by a modern observational network. Since its development involved large conversions of potential to kinetic energy and since it was well documented it appears to be an excellent laboratory in which to apply the various models. . . .

[23] Charney, J. G., and Phillips, N. A. Numerical integration of the quasi-geostrophic equations for barotropic and simple baroclinic flows. *J. Meteorol.* **10,** 71–99 (1953).

[24] Versions of this computer were constructed at several locations in the United States and carried such names as MANIAC (Mathematical Analyzer, Numerical Integrator and Computer) at Los Alamos, New Mexico and JOHNNIAC (named for John von Neumann) at Rand Corporation in Santa Monica, California. MANIAC is sometimes erroneously used for the IAS computer.

[25] The Institute for Advanced Study, the Meteorology Project, Summary of work under Contract N-6-ori-139(1), NR 082-008 during the Calendar Year 1952.

Fɪɢ. 2. Some of the members of the IAS Meteorology Group in 1952. Left to right: J. G. Charney, N. A. Phillips, G. Lewis, N. Gilbarg, G. W. Platzman (behind the camera: J. Smagorinsky). The IAS Computer is in the background.

The [baroclinic] model consists essentially of two barotropic layers and requires initial data at the 700 mb and 300 mb levels. It was necessary in order to avoid computational instability to proceed in half-hour time steps. The total computation time for a 24-hour forecast was approximately $2\frac{1}{2}$ hours at full speed. However, the machine usually operated at half speed.

No account can be taken of the horizontal variations of the static-stability in the two-layer model. A diagnosis of the nature of the two-layer model's shortcomings indicates that this artificial constraint may be of importance. In order to remove it, a model with information at a minimum of three levels is required. Such a model is now in the process of preparation for the machine. [Lorenz showed in 1960[26] that an energetically consistent two-level model with variable static stability can be constructed.]

Much thought has been given to the construction of a full three-dimensional model which will adequately describe the vertical variability of the atmospheric motion. The theoretical and programmatical problems are manifold. Since non-adiabatic effects must at present be neglected for lack of adequate knowledge, the motion is regarded as adiabatic. The potential temperature is then a conservative quantity and may be used as the vertical coordinate in a semi-Lagrangian coordinate system. In this system the equation of motion has a beautifully simple form and is well-adapted to numerical

[26] Lorenz, E. N. Energy and numerical weather prediction. *Tellus* **12**(4), 361–373 (1960).

integration. The complete integration of this equation has now been programmed for the I.A.S. machine. The actual coding and computation now awaits the completion of a magnetic drum auxiliary memory, which is needed in a computation of so large a magnitude as this.

We plan eventually to consider the influence of non-adiabatic effects, large-scale orography, and friction at the earth's surface. Also, since the geostrophic approximation does not always appear to be valid, investigations are being made of higher order geostrophic approximations.

Even with the present crude models, cursory comparison with subjective prognoses made by experienced forecasters indicates at least comparable accuracy. Moreover, whereas subjective methods have not shown significant improvements in the past 20 years, the present approach may be refined in a logical manner. It is therefore expected that more realistic atmospheric models will yield predictions becoming progressively and significantly better.

The activities of the Meteorology Group at Princeton have created much interest in numerical weather prediction throughout the world. Research in this direction is concurrently being conducted in England, Norway, Sweden, Denmark, Germany and Japan. In this country, the Weather Bureau and the weather services of the Navy and Air Force have expressed their desire to investigate the possibility of preparing numerical forecasts on an operational basis.

In a subsequent article,[27] Charney commented that the predictions of the November 1950 storm with the $2\frac{1}{2}$-dimensional model "were moderately accurate during the period immediately preceding the storm but deteriorated markedly in accuracy after its onset." He did not feel that the cyclogenesis had been predicted. He felt that the constraint of horizontal invariance of the static-stability precluded important effects of the low-level thermal asymmetries associated with a front, reflecting the earlier conclusions in the 1952 Project report. Indeed, the paper goes on to show the results of an integration for this case with a three-level quasi-geostrophic model. The full intensity of cyclogenesis was predicted (Figs. 3 and 4), as it was in still another, less intense, but more typical case. He showed that by changing the three levels from 200, 500, and 850 mbar to 400, 700, and 900 mbar, the forecast was improved because of a better representation of the baroclinic structure in the lower troposphere. From this, Charney concluded that nongeostrophic and nonadiabatic effects were not essential for the cyclogenetic instability process.

During much of this interval of baroclinic modeling and testing at IAS, I was only peripherally involved. My major occupation was to explore the nature of the quasi-stationary components of the atmosphere. This was in keeping with Charney's broad perspective of looking into forced as well as transient modes of the general circulation. Charney and Eliassen's[28]

[27] Charney, J. G. Numerical prediction of cyclogenesis. *Proc. Natl. Acad. Sci. U.S.A.* **40**(2), 99–110 (1954).

[28] Charney, J. G., and Eliassen, A. A numerical method for predicting the perturbations of the middle latitude westerlies. *Tellus* **1**(2), 38–54 (1949).

16 JOSEPH SMAGORINSKY

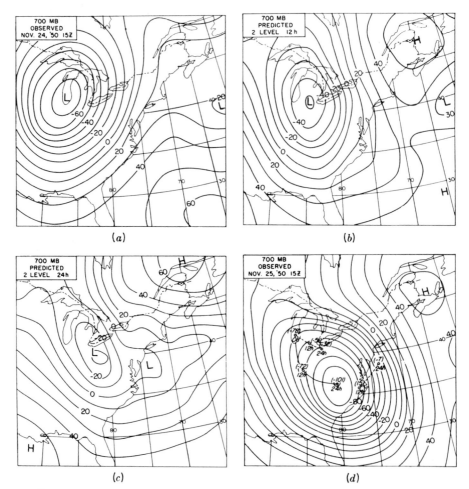

FIG. 3. (a) Observed 700-mbar height contours at 15 Z November 24, 1950. The contours are labeled as deviations from the standard height of 9879 ft in units of 10 ft. (b) Two-level 700-mbar prediction for 03 Z November 25. (c) Two-level 700-mbar prediction for 15 Z November 25. (d) Observed 700-mbar chart for 15 Z November 25. The small circles connected by solid lines indicate the successive positions of the observed low center, and those connected by dashed lines, the predicted positions. The height difference at the center is printed above, and the time below each circle. From J. G. Charney, *Proc. Natl. Acad. Sci. U.S.A.* **40**(2), 99–110 (1954).

linear barotropic results of orographically produced perturbations and the subsequent barotropic calculations by Bolin[29] had left unexplained the relative role of the ocean–continent distribution in forcing quasi-station-

[29] Bolin, B. On the influence of the earth's orography on the general character of the westerlies. *Tellus* **2**, 184–195 (1950).

The problem of initiating such a program can be divided into two parts, an educational problem and a technological problem. There is an educational problem because there are practically no people available at the present time capable of supervising and operating such a program. Synoptic meteorologists who are capable of understanding the physical reasoning behind the numerical forecast are needed to evaluate the forecasts, for example. Mathematicians are needed to formulate the numerical aspects of the computations. During the first several years of the program the meteorological and mathematical aspects probably cannot be separated and personnel familiar with both aspects are needed. An intense educational program could conceivably produce enough people in about three years.

In response to an inquiry by Petterssen on the size of the forecast area, von Neumann offered:

We considered the United States the largest area at the present time with adequate data. This is probably not sufficient for the best possible 36-hour forecasts. I believe that my request for more extensive data had best follow practical demonstration of the usefulness of numerical forecasting. The memory limitations of the machine are also of importance, placing an upper limit to the data which can be handled by the machine.

The area from Japan east to eastern Europe is about four times the area we used here at Princeton. Therefore, an appreciably larger machine would be needed to forecast for such an area, especially when more complicated atmospheric models are used. This technical problem might be solved, let's say, within about five years after the program under consideration is started.

Wexler reported:

In talking to various forecasters they have given me the impression that they expect numerical forecasting methods to yield improvement in forecasts beyond 36 hours, since they believe that present methods yield sufficiently accurate 24- or 36-hour forecasts. I don't necessarily share this opinion but I think this group should be informed of its existence.

There was a generally expressed opinion that it was too early to encourage any expectations of improvement in forecasts for longer periods. Charney added:

From our experience I would say that the barotropic forecasts are perhaps not as good as the best conventional forecasts, but the indications are that baroclinic forecasts will be much better.

Some discussion followed on a proposal by Philip Thompson (who was not present) that it would be more economical to use a two-dimensional "equivalent baroclinic" model (sometimes referred to as a "thermotropic" model) which might produce predictions just as accurate as those from a three-dimensional model. Thompson had just completed some work on such a model, and a paper was to be published shortly.[35] At this

[35] Thompson, P. D. On the theory of large-scale disturbances in a two-dimensional baroclinic equivalent of the atmosphere. *Q. J. R. Meteorol. Soc.* **79**, 3–38 (1953).

point, it was a moot question because Charney's three-level results were not yet available and Thompson had only a theoretical framework. Charney's opinion was, "I do not believe that a very simple model will be sufficient for operational forecasts." To which von Neumann added, "the problem of deciding at what level to begin forecasts can be decided only by experience which we hope to have within another year."

The general consensus of this historic meeting was to gain momentum rapidly. It reflected a deep commitment by the operational forecasting agencies of the United States, both individually and collectively, to rapidly build their internal competence and to accelerate their activities in numerical weather prediction.

In early 1953, I returned to the U.S. Weather Bureau to head a modest effort to begin to introduce the fruits of numerical methods into the research environment. Mr. Louis P. Cartensen was assigned to spend part of his time with me. He already had been involved earlier, in 1949, in the linear barotropic forecast calculations. Contacts with the IAS group were being maintained. In 1952 and 1953 the IAS visitors increased substantially: Ernst Hovmöller and Roy Berggren of Sweden; Eric T. Eady and Bruce Gilchrist of the United Kingdom; Kanzaburo Gambo of Japan; Frederick G. Shuman, George P. Cressman, and Jacob F. Blackburn of the United States; and Ragnar Fjørtoft returned.

One of the earlier concerns was the potential utility of the baroclinic models for forecasting precipitation. I had noted:[36]

> It is possible in principle to predict the 3 dimensional large-scale field of motions with the existing U.S. data density. To date, vertical motions of 500 mb have been predicted by means of high speed computing devices. More vertical detail in the vertical motion will be forthcoming as physical models of greater complexity are devised. However application to present operational needs without the aid of a computer are not entirely hopeless. It seems that certain approximative schemes, in particular the one suggested by Fjørtoft for barotropic models, may lend themselves for application to a $2\frac{1}{2}$ dimensional model. If this can be done successfully then it would be possible to predict the horizontal flow at 700 and 300 mb and the vertical velocity at 500 mb for 24 and possibly 36 hours by means of graphical techniques. The technical aspects of this problem are being explored presently.

In the absence of a computer, the Fjørtoft graphical method was the only hope available to gain experience with the new theoretical frameworks. I was to return to this problem of precipitation prediction in a few years.

The research agenda for the small Weather Bureau research group, which also would include Charles L. Bristor, was intended to include studies in objective analysis "to determine the most suitable method with

---

[36] Memorandum, Smagorinsky to Wexler, March 4, 1953.

respect to dynamical, numerical and operational requirements.'' The previous work by Hans A. Panofsky was to be the starting point.

The early frustration with inadequate data, especially over the oceans, for establishing initial conditions, seemed to be a critical weak link in what we needed to exploit the numerical models. My desperation is expressed in a memorandum[37] which grasped for a possible solution.

One basic method of cutting costs is to automatize present sounding methods so as to sharply reduce the number of operating personnel. Some possible ways of doing this suggest themselves, but each will take a great deal of development. The virtue of such an approach is that an automatic sounding technique would be quite adaptable for oceanic observations. Anything would be less expensive than weather ships.

One can even go a step further and suggest looking for an entirely new method for taking soundings in addition to the process being automatic. Presently, the instrument must pass *through* the atmosphere. What about methods for indirect measurement? Isn't the density stratification on the ocean determined indirectly by sonic devices? Can one take advantage of the fact that radar is somewhat sensitive to variation in atmospheric density?

It seems that development work in meteorological observations should be a continuing process not only for small modifications on existing methods but also further into the future on Buck Rogerish innovations. Had this been so in the past, we now probably would not be working with instruments which are basically 20 and 30 years behind the times. It must be said that rawins are an outstanding exception.

In retrospect, a solution did not begin to emerge until the weather satellite was proposed. Actually in the 1950s the number of weather ships decreased somewhat.

In an earlier memorandum,[38] I had the opportunity to comment on future World Meteorological Organization (WMO) data requirements:

The recommendation that the minimum aerological density over the oceans be 1000 km between stations would not satisfy the needs for numerical prediction. Experience indicates that an absolute minimum for any region to be treated numerically is approximately 500 km, roughly the density of Canada. Of course it is more desirable to have a density corresponding to that of the U.S.—a separation of about 300 km between soundings. The above recommended minimum of 500 km assumes that there are at least as many winds as soundings.

It appears that if upper air predictions, subjective or numerical, are worth their salt they should be reliable for at least 12 hours. Based on this assumption, it would seem that 6-hourly observations would be superfluous. 8 hourlies or even 12 would appear to be adequate. An exception would be the needs for research purposes, and perhaps one can provide for 6 hourlies for periods of a few days.

Now that surface and upper air data are used together in formulating a forecast, it seems paradoxical that the two observation times do not coincide. Although there are no doubt practical reasons for trying to avoid the crowding of the observer's schedule, it would be extremely desirable that *synoptic* surface and aerological data be available.

[37] Memorandum, Smagorinsky to Wexler, May 26, 1953.
[38] Memorandum, Smagorinsky to Tannehill, March 5, 1953.

This is especially true since all upper air height calculations have their origin based on the surface pressure, so that any objective analysis procedure would need the surface pressure in making vertical and horizontal consistency checks of the data.

These density specifications are not too different from those settled on for the Global Weather Experiment of 1979. However, I was still thinking in terms of simultaneous observations.

In December 1953, I had the opportunity to speak at a national computer conference.[39] I commented that "preliminary experience indicates that a general objective analysis prepared on a four-dimensional distribution of data with widely varying density will require as many logical as arithmetic operations. This problem is thus ideally suited for high speed digital computers." This implied asynoptic assimilation but without a specific proposal. I do recall an early realization that objective analysis could in principle deal with data arbitrarily distributed in three-dimensional space. As a consequence, significant level data could be used instead of mandatory level data, with the attendant communications economy.

The Weather Bureau was also thinking ahead in terms of training the people needed. George Platzman was offering a concentrated 10-week course at the University of Chicago during the summer of 1953. The Bureau sent Shuman, Carstensen, Bristor, and two others. Correspondence from the Chicago "students" contained such messages as: "The [Weather Bureau] library is getting unhappy about some of these publications we brought along. . . . These publications are and have been in constant use by all five of us guys" and "living expenses are low enough so that $5 per day pretty well covers them."

The possibility of operational utilization of numerical methods was much discussed. As a result of a visit to the IAS in late May 1953, I reported:[40]

At the time of my visit to the IAS, Colonel George F. Taylor, Air Weather Service, was also there to discuss operational numerical weather prediction. Although he was not voicing an official opinion, Colonel Tayler said that some of the responsible weather people at ARDC [Air Research and Development Command] in Baltimore feel that a joint operational group should be formed immediately. He also indicated that many of the important technical problems connected with such a venture are fully within the Weather Bureau's domain. However, he was cognizant of the fiscal situation and thought the only reasonable arrangement was to have the military bear the bulk of the financial burden. In a discussion with Professor von Neumann and Dr. Charney it was agreed that because a machine could not be obtained before 6 to 18 months, even with high priorities, a joint meeting at least progress to the point where

[39] Smagorinsky, J. Data processing requirements for numerical weather prediction. *Proc. East. Comput. Conf., Washington, D.C., Dec. 1953* pp. 22–30.
[40] Memorandum, Smagorinsky to Chief of Bureau, May 29, 1953.

an order can be placed for a machine. Colonel Taylor agreed that it is extremely desirable that any working committee which is formed by the JMC [the Joint Meteorological Committee of the Joint Chiefs of Staff] or ACC/MET [Air Coordinating Committee/Meteorology] be given the authority to act rather than merely to recommend.

My recommendation was that the Weather Bureau pursue its efforts toward the formation of a joint Weather Bureau–Air Force–Navy operational numerical forecasting group so that a commitment could be made at the earliest possible date for a high-speed calculator of the IBM 701 type. This was a somewhat faster commercial version of the IAS machine but with IBM card equipment for input–output.

Developments were occurring swiftly in the dialogue on the establishment of a Joint Operational Numerical Weather Prediction Unit (JNWPU). In a memorandum, Wexler wrote:[41]

> On June 4 Dr. Joseph Smagorinsky, Dr. George Cressman (AWS [Air Weather Service]) and Major Thomas Lewis (AWS) met at Andrews Air Force Base with representatives of IBM in order to ascertain information regarding the availability of and costs of IBM Computer 701. Present indications are that because of order cancellations a machine could be available between January and June 1954. However, IBM would like a letter of intent as soon as possible. The annual rental fee for the 701 is between $175,000 and $300,000 depending on the auxiliary equipment and the number of operating hours. . . .
>
> On June 9 [1953] Dr. Smagorinsky presented a proposed agenda of problems to be considered by the *ad hoc* committee[42] and with minor modification the following was adopted:
>
> 1. Functions and Organizational Structure
> 2. Personnel Problems (Stability, Training, Selection)
> 3. a. Machine Availability—letter of intent to IBM
>   b. Physical Location of Unit
> 4. a. Initial Cost Estimate
>   b. Continuing Cost Estimates
>   c. Joint-Financing Arrangement
>
> As far as Weather Bureau representation on the *ad hoc* committee is concerned, it is recommended that Dr. Smagorinsky be appointed a member.

Within 2 months, plans solidified to the point where the Bureau was making specific budgetary provisions.[43]

RECOMMENDATION

> The Weather Bureau make available $32,000 in FY 1954 and $39,000 per annum thereafter to give one-third support for the Joint NWP Unit as proposed by JMC.

---

[41] Memorandum, Wexler to Chief of Bureau, June 11, 1953.

[42] The initial membership of the JMC ad hoc Committee on Numerical Weather Prediction was Commander D. F. Rex (Chairman), Majors W. H. Best and T. H. Lewis, and H. Wexler, with R. A. Allen, P. D. Thompson, and J. Smagorinsky as participants.

[43] Memorandum, Wexler to Chief of Bureau, August 12, 1953.

FACTS BEARING ON CASE

The *ad hoc* Committee on NWP appointed by JMC on June 23, 1953 at the request of the Weather Bureau (see attached supporting paper) has formulated a plan for the establishment of a Joint NWP Unit to begin operations July 1, 1954. Some findings and recommendations which will soon be submitted to JMC are summarized below:

1. Initially, daily prognosis of the 3-dimensional atmospheric flow field will be made 7 days per week.

2. A Staff of 34 will be required to carry out the functions planned for the first year of which 13 will be professional meteorologists. By July 1, 1954, the Weather Bureau expects to have trained in NWP methods six of its meteorologists, the Air Force six, and the Navy two.

3. For incorporation of NWP results in current analyses and prognoses it was agreed that the NWP Unit should be located adjacent to WBAN [Weather Bureau–Air Force–Navy] Analysis Center. With increased experience it is expected many of the functions now performed by WBAN Analysis Center will be absorbed by the NWP Unit.

4. The Committee has agreed to recommend that administration of the Unit be assigned to the Weather Bureau.

5. Initial expenditures of $94,500 will be required four to six months prior to July 1954; continuing operating costs for FY 1955 will be $415,019 of which $199,559 will be for machine rental.

After submitting its final report, which formed the basis for an organizational plan, the ad hoc Committee on NWP was dissolved on September 11, 1953 and on September 17, the JMC created a new ad hoc Group for the Establishment of a Joint Numerical Weather Prediction Unit consisting of Rex (Chairman), Lewis, and Wexler. This new group was empowered to select a director, whom it would assist in implementing the organizational plan. On September 22, 1953, Dr. George P. Cressman was nominated for the directorship. On October 9, 1954, Herman Goldstine and I were commissioned to conduct comparative tests on the two large-scale computers available at the time, the IBM Type 701 and the ERA (Engineering Research Associates) Model 1103.[44] The 701 was recommended, and then accepted in January 1954 by a Technical Advisory Group chaired by von Neumann.

By June 30, 1954, the three participating agencies had identified their personnel contributions to JNWPU: seven each from the Weather Bureau and Air Force and three from the Navy. The IBM 701 had been ordered for delivery by March 1, 1955 and a site was selected at Suitland, Maryland. At its 15th meeting on July 1, 1954, the ad hoc Group for Establishment of a JNWPU unanimously recommended its own dissolution and also the formation of a standing steering committee, which nevertheless

---

[44] Goldstine, H. H. "The Computer from Pascal to von Neumann," p. 329. Princeton Univ. Press, Princeton, New Jersey, 1972.

was named the ad hoc Committee on Numerical Weather Prediction, chaired by Wexler.

The latter half of 1954 was a busy period of assembling the new group, preparing the computer program for the IAS three-level (900, 700, and 400 mbar) quasi-geostrophic model, and setting up an operational system for receiving and processing data and for disseminating results. The IBM 701 was delivered early in 1955 and the first 36-hr forecast was produced from 1500Z April 18, 1955 initial conditions (Fig. 5).

A report of the first year's experience was published in 1957.[45]

5. The Advent of the General Circulation Modeling Era

Meanwhile, Norman Phillips had completed, in mid-1955, his monumental general circulation experiment.[46] As he pointed out in his paper, it was a natural extension of the work of Charney on numerical prediction, but Phillips's modesty could not obscure his own important contributions to NWP. The enabling innovation by Phillips was to construct an energetically complete and self-sufficient two-level quasi-geostrophic model which could sustain a stable integration for the order of a month of simulated time. Despite the simplicity of the formulation of energy sources and sinks, the results were remarkable in their ability to reproduce the salient features of the general circulation. A new era had been opened.

Von Neumann quickly recognized the great significance of Phillips's paper and immediately moved along two simultaneous lines.

One was to call a conference on "The Application of Numerical Integration Techniques to the Problem of General Circulation" in Princeton during October 26–28, 1955. Of course, the centerpiece was Phillips's results, but many others presented papers on related research. Of particular interest were von Neumann's published remarks on climate forecasting.[47]

The discussion centered on many questions raised in the papers. I mention only a few that have special historical significance. There was extended discussion on the streakiness developed in the flow during the latter stages of Phillips's integration. This "noodling" was the result of convolutions of the vortex lines, presumably the nonlinear result of truncation error. It already was a property of quasi-geostrophic flows. The

[45] Staff Members JNWPU. One year of operational numerical weather prediction. *Bull. Am. Meteorol. Soc.* **38,** Part I, 263–268; Part II, 315–328 (1957).

[46] Phillips, N. A. The general circulation of the atmosphere: A numerical experiment. *Q. J. R. Meteorol. Soc.* **82,** 123–164 (1956).

[47] Pfeffer, R. L., ed. "Dynamics of Climate." Pergamon, Oxford, 1960.

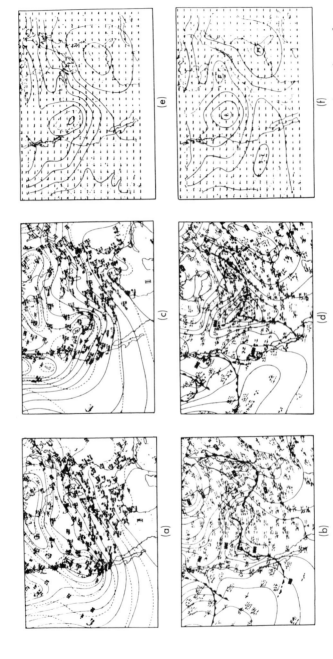

Fig. 5. The first operational three-level numerical forecast by JNWPU from 15 Z April 18, 1955. Shown are observed maps from April 18 at (a) 500 mbar (15 Z) and (b) sea level (1230 Z) and also for April 19 (c) and (d). Shown also are the 24-hr forecast results at (e) 700 mbar and (f) 900 mbar. Actually a 36-hr forecast was made for 900, 700, and 400 mbar and vertical motions produced at 800 and 550 mbar.

small-scale eddy viscosity should properly represent the energy (and enstrophy) exchange between the explicitly represented modes and the truncated smaller (subgrid) scales so as to preserve the spectrum in the vicinity of the limit of computational resolution. Charney recalled von Neumann and Richtmyer's experience with shock-wave flows[48] in which they found that a nonlinear viscosity served to preserve the scale of the shock wave during the course of its propagation. In their paper they said "the dissipation is introduced for purely mathematical reasons." The oft-used terminology that the viscosity is "artificial," in my view, gives a misleading connotation. Surely, an eddy viscosity in any form is an artifice to replace a reality missing from the finitistic representation, that is, the real spectral communication between the explicity resolved flow and the molecular range where a physical viscosity does its work. Charney and Phillips recommended that I apply such a nonlinear viscosity to my primitive equation general circulation model later on. I am often credited with the original idea, but it belongs to others; I only used it and rationalized it.

Another area of discussion centered around the prediction of precipitation. I had, at that time, been in the midst of diagnostic calculations of precipitation from a vertical velocity field[49] and also was trying to show the effect of released latent heat in positively feeding back to amplify the vertical motion and reduce the horizontal scale.[50] I was also working on the formulation of a water vapor predictive framework to be incorporated in a numerical model.[51] Other researchers in Japan and the United States were also working on the problem at the time, making important contributions.[52-55]

---

[48] von Neumann, J., and Richtmyer, R. D. A method for the numerical calculation of hydrodynamic shocks. *J. Appl. Phys.* **21**, 232–237 (1950).

[49] Smagorinsky, J., and Collins, G. O. On the numerical prediction of precipitation. *Mon. Weather Rev.* **83**(3), 53–68 (1955).

[50] Smagorinsky, J. On the inclusion of moist adiabatic processes in numerical prediction models. *Ber. Dtsch. Wetterdienstes* **38**, 82–90 (1957).

[51] Smagorinsky, J. On the dynamical prediction of large-scale condensation by numerical methods. *Am. Geophys. Union, Mongr.* No. 5, pp. 71–78 (1960).

[52] Kombayasi, M., Miyakoda, K., Aihara, M., Manabe, S., and Katow, K. The quantitative forecast of precipitation with a numerical prediction method. *J. Meteorol. Soc. Jpn.* **33**(5), 205–216 (1955).

[53] Miyakoda, K. Forecasting formula for precipitation and the problem of conveyance of water vapor. *J. Meteorol. Soc. Jpn.* **34**(4), 212–225 (1956).

[54] Estoque, M. A. "An Approach to Quantitative Precipitation Forecasting," Sci. Rep. No. 7, Contract AF19(604)-1293 between University of Chicago and G.R.D., AFCRC, 1956.

[55] Aubert, E. J. On the release of latent heat as a factor in large scale atmospheric motions. *J. Meteorol. Soc.* **14**(6), 527–542 (1957).

V. N. I d   Draft
July 29, 1955

DYNAMICS OF THE GENERAL CIRCULATION   *title suggested by Hw at lunch to day with v.n. + J. Charney*

1. In 1947, a project was started in Princeton by the U.S. Navy, and
U.S. Air Force, for ~~theoretical and~~ *theoretical and* computational investigations in
meteorology, with particular regard to ~~wards~~ the development of methods
of numerical weather forecasting. After a few years of experimenting,
the project ~~consisted of~~ *concentrated on* exploring the validity and the use of the
differential equation methods ~~of~~ *developed by* Dr. J. Charney, for numerical forecasting.
For this purpose, the U.S. Army Ordnance Corps ENIAC computing machine
was used in 1950 ~~in~~ *and in 1951*, and the Institute for Advanced Study's own computing
machine from 1952 onward. Subsequently, use was also made of the IBM 701
machine in New York City. With the help of these computing tools, it was
found that forecasts over periods like 24 (and up to *48* ~~40~~) hours are possible,
and give significant improvements over the normal, subjective method of
forecasting. Certain experiments demonstrated that even phenomena of ~~cyclonic~~
cyclogenesis ~~genesis~~ could be predicted. A considerable number of sample forecasts were
made, which permitted the above mentioned evaluation of the ~~use~~ *validity* of the
method. A large number of variants were also explored, particularly with
respect to ~~eliminate successfully~~ *eliminating successively* the major mathematical approximations that
the original method contained. It must be noted, however, that the method,
and also all its variants which exist at the present, are still affected
with considerable simplifications of a physical nature. Thus, the effects
of radiation have only been taken into consideration in exceptional cases,
the same is true for the effects of *geography and* topography, while humidity and
precipitation have not been considered at all. ~~But significant~~ *That significant* results could,
nevertheless, be obtained, is due to the relatively short *time-* span of the
forecasts. Indeed, over 24 or 48 hours ~~of~~ the above mentioned effects do
not yet come into play decisively.

FIG. 6. The first page of a draft proposal by J. von Neumann to establish a project on the
dynamics of the general circulation—eventually to become the Geophysical Fluid Dynamics
Laboratory. The written changes in the text are von Neumann's and notes in the upper right-
hand corner are by Harry Wexler.

The specific problem of discerning marginal cloudiness which does not yield precipitation was identified as a difficult and as yet unresolved problem. It still is unsolved! I recall that sometime in the early 1950s, von Neumann, I, and several others were standing outside of the Electronic Computer Project Building in Princeton, and Johnny looked up at a partially cloudy sky and said, "Do you think we will ever be able to predict that?" In an attempt to answer that question, I had shown in my 1960 paper that an empirical correlation can be found between the large-scale fields of relative humidity and cloud amount for three layers in the troposphere.

Finally, I mention without further comment that the conference discussion also touched on the CO_2 cycle, but without exciting much concern.

This meeting did much to coalesce thinking on problems and opportunities that lay ahead, from all perspectives: observational, theoretical, and experimental.

The other initiative by von Neumann was stimulated by his realization that the exploitation of Phillips's breakthrough would require a new, large, separate, and dedicated undertaking. He followed a path similar to the one he took 2 years earlier in connection with establishing the JNWPU. Von Neumann drafted a proposal to the Weather Bureau, Air Force, and Navy justifying a joint project on the dynamics of the general circulation. Because of its historical interest, it is reproduced here in its entirety together with a photograph (Fig. 6) of the first page of an earlier draft with von Neumann's and Wexler's handwritten changes and comments.

PROPOSAL FOR A PROJECT ON THE
DYNAMICS OF THE GENERAL CIRCULATION

1. In 1947 a project was started in Princeton by the U.S. Navy and U.S. Air Force for theoretical and computational investigations in meteorology, with particular regard to the development of methods of numerical weather forecasting. After a few years of experimenting, the project concentrated on exploring the validity and the use of the differential equation methods developed by Dr. J. Charney for numerical forecasting. For this purpose, the U.S. Army Ordnance Corps ENIAC computing machine was used in 1950 and 1951, and the Institute for Advanced Study's own computing machine from 1952 onward. Subsequently, use was also made of the IBY [sic] 701 machine in New York City. With the help of these computing tools, it was found that forecasts over periods from 24 to 48 hours are possible, and give significant improvements over the normal, subjective method of forecasting. Certain experiments demonstrated that even phenomena of cyclogenesis could be predicted. A considerable number of sample forecasts was made, which permitted the above-mentioned evaluation of the validity of the method. A large number of variants was also explored, particularly with respect to eliminating successively the major mathematical approximations that the original method contained. It must be noted, however, that the method, and also all its variants which exist at the present, are still affected with considerable simplifications of a

physical nature. Thus, the effects of radiation have only been taken into consideration in exceptional cases, the same is true for the effects of geography and topography, while humidity and precipitation have not been considered at all. That significant results could, nevertheless, be obtained is due to the relatively short-time span of the forecasts. Indeed, over 24 or 48 hours the above-mentioned effects do not yet come into play decisively.

On the basis of the results cited it was determined by the sponsoring agencies that a routine 24–36 hour numerical forecasting service has become possible, and should be set up on a permanent basis. This was done by a joint organization of U.S. Air Force, U.S. Navy, and Weather Bureau (JNWPU—Joint Numerical Weather Prediction Unit) which is being operated by the U.S. Weather Bureau at Suitland, Md. It has now been making daily forecasts for over 3 months, and with very good success.

2. The logical next step after this is to pass to longer-range forecasts and, more generally speaking, to a determination of the ordinary general circulation of the terrestrial atmosphere. Indeed, determining the ordinary circulation pattern may be viewed as a forecast over an infinite period of time, since it predicts what atmospheric conditions will generally prevail when they have become, due to the lapse of very long time intervals, causally and statistically independent of whatever initial conditions may have existed.

There is reason to believe that the above-mentioned "infinite" forecast, i.e., deriving the general circulation, is less difficult than intermediate length forecasts, say, to 30 or 90 days. This is just a reflection of the fact that extreme cases are usually easier to treat than intermediate ones, since in extreme cases only a part of factors plays a role, dominating all others, while in intermediate cases, all factors become of comparable importance. It should be added that both the "infinite" and the "intermediate" forecasts have to be performed for the entire earth, or at least for an entire hemisphere. Indeed, the spread of meteorological effects is such that, already after 2 to 3 weeks, every part of the terrestrial atmosphere will have interacted with every other—except for the relative weakness of the interaction between the Northern and Southern Hemispheres. Thus, in both cases, a hemispheric forecast is the minimum that can be envisaged.

In view of the above, it seems logical to investigate now the "infinite" forecast, i.e., the general circulation. It is hoped that this will subsequently lead to a better understanding of the factors involved in the "intermediate" forecasts (compare above). Thus, the "intermediate" forecasts should enter into the program at a somewhat later stage.[56]

3. With regard to calculating the general circulation in the Northern Hemisphere, quite significant progress was made in Princeton. Several calculations were made in which the Northern Hemisphere—or rather a quadrant of it—was treated in a highly simplified way. The simplifications were as follows: The quadrant of the hemisphere was treated as a "flat" area, thus distorting the geometry, primarily in the Arctic, considerably. (It was treated with a "periodic" east–west boundary condition, i.e., the calculation deals with "planetary waves" of wave number 4 (or 8, 12 and so on).) Instead of using a coriolis parameter with its proper meridional variability, the "Rossby plane" was used, i.e., the coriolis parameter was given its mean value, and treated as a constant; however, in all places where the exact theory makes reference to

[56] And that is how it did happen. Intermediate-range forecasting did not begin until the middle and late 1960s after experience had been accumulated with general circulation equilibrium experiments.

the meridional derivative of the coriolis parameter, the (positive) mean value of that quantity was used.

The solar radiation impinging upon the earth was considered without its seasonal or diurnal variations. Indeed, it was treated as a heat source with a linear meridional variation. This model was treated on a horizontal 16 × 16 lattice, with two vertical strata. Starting with an atmosphere at rest, the integration was carried out over 30 days.

The effects of humidity, and of geography and topography, were disregarded. The calculations on this model were started with an atmosphere at rest, and at a uniform temperature. The developing motions and adjustments were calculated over a period of 30 "real" days. The circulation pattern which developed was first the one that one usually obtains by verbal discussion: Northward flow of heated air aloft, and southward flow of cooled air below, with easterly winds on the lower, and westerly winds on the upper level. This (not real!) flow was observed to pass its turbulent stability limit after 5 "real" days. At this point, its breakdown was induced by adding (computational) "noise" to the motion. Hereupon, in the course of the next 25 days a cyclone and an anticyclone, of familiar type, developed with westerlies in the lower level was about 30 miles per hour, and on the high level the maximum westerly velocity reached 200 miles per hour. The temperature difference between the tropics and the Arctic was, as it should be, about twice what it is in reality. (This doubling should correspond to the fact, that in reality half the heat transported north is latent heat of humidity, hence, when this contribution is neglected, the temperature increment that is needed to take care of all the requirements, will be double of what it is in reality.)

Thus, even this very primitive model disclosed the main features of the general circulation, in a rather detailed way, which no verbal, or less elaborate computational, analyses have ever been able to do. Several calculations of this type were made, that gave concordant results, and also disclosed the limitations of the method used. The above described calculation (repeated for checking) required 30 computing hours on the Princeton machine.

4. It seems clear that these general circulation calculations should now be expanded and improved. Even applying only the obvious mathematical and geometrical improvements will greatly increase the size of each calculation. As a minimum program, the entire Northern Hemisphere should be considered; its curvature and the meridional variation of the coriolis parameter should be properly treated; and the meridional variation of the solar energy input, with or without its seasonal or diurnal variations, should be introduced into the calculation. In addition to this, we know that the optimum grid size is about twice as fine (in linear dimension) than what was used, and that one should properly consider 3 or 4 vertical levels (rather than 2, compare above). All of this, with various secondary complications that it induces, is likely to increase the size of the calculation, allowing for reasonable improvements, by at least a factor of thirty. This would mean a problem time of about 900 hours on the Princeton machine, or if the problem is checked in a less-time-consuming way than by repetition, 450 hours.

Comparing the Princeton machine with the IBM 701, it appears likely that the latter will be about 5 times faster on this problem. (The intrinsic speed of the IBM 701 is only twice that of the Princeton machine, but various memory limitations of that machine probably increase this factor for the problem under consideration, to something like 5.) Thus, on the IBM 701, presumably about 90 hours would be needed per problem, allowing for the above indicated refinements. This means that the time on the IBM 704 would probably be about 45 hours, and on the NORC, perhaps 22 hours.

Since a research program of this type requires large scale experimentation, with computing methods, with variations of parameter, and physical approximations of various kinds, there is no doubt that in any rational program, a large number of such problems will have to be solved. Therefore, even the best time mentioned above (22 hours on the NORC) would not be too fast, i.e., even under these conditions, computing would probably take more time than analyzing and planning. This is increasingly true for the IBM 704 and the IBM 701. Consequently, the use of the IBM 701 or, if feasible, of one of the faster machines would be, in principle, amply justified.

5. It is, therefore, proposed to set up a project which has available to it at least a machine of the IBM 701 type. Since the first improvements and refinements on the problem are sufficiently understood today, to be put immediately into the phase of mathematical planning and coding, it would be important to think in terms of a machine which can be made available soon. The only machine of this speed class which is immediately available is the IBM 701. While this machine exists today in about 20 copies, only a few of them have easy access. At this moment, neither Princeton nor New York offer such a possibility, whereas one exists in Washington, at the Suitland establishment of the U.S. Weather Bureau (the JNWPU referred to earlier). It would, therefore, be very profitable to initiate measures immediately which make it possible to use this machine for the calculations mentioned above.

The obvious vehicle for this work would be a project organized around the Suitland machine, and with the advice and collaboration of those who directed the Princeton project, and the above circulation calculations—J. Charney, N. Phillips, and J. von Neumann—readily available.

It is proposed that such a project be set up at the U.S. Weather Bureau to be located at Suitland, with adequate personnel, physical space and facilities, and with about one shift of the Suitland IBM 701 machine available. It is proposed that within the Weather Bureau organization, Dr. H. Wexler, who has considerable familiarity with this work, be made the project officer. It is contemplated that in scientific and policy matters he would be guided by the decisions of a committee to consist of J. Charney, J. von Neumann, and himself.

6. The progress of this project can now be mapped out for about two years. During the first year, the general organization of personnel and facilities should take place, the setting up of computing methods in the sense of the "minimum improved" general circulation problem, as outlined above, and the carrying out of a sufficient sample of calculations on this basis. In the second year, the obvious physical improvements should be gradually introduced into the treatment. As such, one would consider in order of increasing difficulty the introduction of the following factors:

(a) Purely kinematic effects of geography and topography;
(b) Acquisition of humidity in the atmosphere by evaporation. This necessitates the (geographical) consideration of position of the oceans. It also requires the introduction of, presently reasonably well understood, semi-empirical rules regarding the dependence of the rate of evaporation on the local atmospheric and oceanic temperatures, atmospheric stability, and wind velocity.
(c) Some, as yet imperfect semi-empirical rules about the delay-relationships of over-saturation, cloud formation, and precipitation. Also, some semi-empirical rules about the absorption of solar radiation by clouds.
(d) The very difficult problem of the effects of atmospheric humidity on the solar irradiation of the earth and on the long wave radiation from the earth and atmosphere.

It is worth repeating that (d) is an extremely difficult problem, which will probably only be reached at the end of the two year period, and on which progress will only be made at still later stages, and then only in combination with a great deal of theoretical and experimental work, some of which is now under way. (c) is, in principle, even more difficult, but in this case, acceptable practical approximations can probably be made. (a) is quite simple; (b) while not very simple, is nevertheless based on things that we understand reasonably well at present. Personnel and budget for the project are envisaged as follows:

1	Meteorologist-in-Charge	GS-14	$10,750
2	Meteorologists	GS-13	18,840
1	Senior Programmer	GS-12	8,000
1	Synoptic Meteorologist	GS-11	7,035
1	Programmer	GS-11	7,035
2	Programmers	GS-9	11,690
2	Electronic Computer Operators	GS-7	9,860
2	Meteorological Aids	GS-5	8,150
1	Clerk-Typist	GS-4	3,670

	Personnel	$85,030
Travel		4,000
Consultants		5,000
Computing Machine Time		162,000
Other equipment and office furniture		6,000
	Other	$177,000
	GRAND TOTAL	$262,030 per annum

Total for 9 months 1 October 1955 to
30 June 1956 is $196,521
or shared by three   65,507 each

It should be noted that the above figures apply to the first year only. They should be reconsidered at the end of that year, and the budget of the second year determined on the basis of the experiences gained in the first year. It is expected that the latter will not differ very significantly from the budget of the first year on an annual basis, but that it will probably be somewhat higher.

At the end of the first year we may also find that a faster machine than the IBM 701 is becoming available.

In addition to the above, the consultations with J. Charney, N. Phillips, and J. von Neumann (without compensation) will be needed.

The proposal, dated August 1, 1955, was more or less accepted the following month as a joint Weather Bureau–Air Force–Navy venture. I was asked to lead the new General Circulation Research Section,[57] and

[57] This group subsequently changed its name several times: General Circulation Research Laboratory (1959) and Geophysical Fluid Dynamics Laboratory (1963).

reported for duty on October 23, 1955. By the end of the year there were five of us.

In a recent biography of von Neumann[58] it was asserted:

> One of von Neumann's interests was in weather modification and he participated in a panel on "possible effects of atomic and thermonuclear explosions in modifying weather." Von Neumann's most interesting conclusion was that the most likely way to affect the weather and climate is the possible modification of the albedo of the earth. Thinking had moved toward the question of how might we change the weather at will. Von Neumann thought that the evidence so far was that nuclear explosions had only negligible effects on the weather, but that more theoretical and computer studies are needed, like the ones he and Jules [sic] Charney had initiated at Princeton.

One wonders whether this was a motivation in his proposal to form this new project. If so, it was not readily apparent to me at the time.

In the spring of 1955, von Neumann left the Institute to take up one of the posts of Commissioner of Atomic Energy in Washington, D.C. Up until the time he fell ill in 1956, von Neumann kept in close contact with me concerning the work of our group. He died in February 1957 at the age of 53. As a result of von Neumann's departure, the Institute's latent indifference to the Computer Project ultimately resulted in Charney's and Phillips's move to MIT in 1956. A consequence was that the Air Force and Navy withdrew further support from our group in Suitland, Maryland and the Weather Bureau assumed full responsibility, thanks to Francis W. Reichelderfer and Harry Wexler.

However, in the brief interval of our close cooperation with the IAS group, they were instrumental in getting us started on a fruitful line of research. We already were busy with the precipitation problem. In the case of general circulation modeling, it seemed the next logical step beyond Phillips's model was to allow nongeostrophic modes which could be of great significance in how the tropics operated in, and interacted with, the general circulation. Some new work by Arnt Eliassen[59] at UCLA seemed the logical starting point. He did not use the full primitive equations, but allowed only internal gravity waves by constraining the surface pressure tendency to vanish. The domain was a zonal channel on a spherical earth, with one boundary at the equator. A nonlinear lateral viscosity of the von Neumann–Richtmyer type was formulated with the help of Charney and Phillips. Other aspects of the model were quite similar to that of Phillips. The two-level model required that the static stability be

[58] Heims, S. J. "John von Neumann and Norbert Wiener, from Mathematics to the Technologies of Life and Death." MIT Press, Cambridge, Massachusetts, 1980.

[59] Eliassen, A. "A Procedure for Numerical Integration of the Primitive Equations of the Two-Parameter Model of the Atmosphere," Sci. Rep. No. 4 on Contract AF19(604)-1286. Dept. of Meteorology, University of California at Los Angeles, 1956.

entered as an externally specified parameter which could be adjusted so as to crudely take into account the mean effect of released latent heat.

The model and integration scheme were described at an international NWP conference in Stockholm in June 1957.[60] Stable integrations for an extended period of 60 days were achieved soon after, and our very first results were exhibited by Harry Wexler at the 5th General Assembly at CSAGI in Moscow in July–August 1958. By December 1958, at a symposium of the American Association for the Advancement of Science in Washington, D.C., I was able to show the model's ability to sustain an index cycle with the attendant fluctuation in the energy and momentum fluxes. This property was already suggested by Phillips's results.

The long lapse between this stage and final publication in 1963[61] was the result of a personal desire to first perform thorough analyses of the non-geostrophic modes and of the energetics. In retrospect, it was a mark of immaturity that I decided not to publish the results in several intermediate stages but rather at the end as a comprehensive work.

As an important historical aside, it should be said that Hinkelmann published in 1959[62] the results of a numerical experiment with the primitive equations from which sound waves and external gravity waves also were filtered. Friction, nonadiabatic effects, and orography were neglected. This was a five-level model which was stably integrated for 3 days from idealized initial conditions. Hinkelmann must have started this work about the same time as we did.

In late 1958, encouraged by our success with the two-level model, we began to design a nine-level primitive equation hemispheric model. In a private discussion, Charney expressed skepticism on the value of so many levels. This model would admit external gravity waves and extend high enough into the stratosphere to account for significant energetic coupling with the troposphere. The model would have a general radiative algorithm, predict water vapor transport and condensation, and incorporate a convective parameterization and an explicit boundary layer. This model was to be the prototype for much of the laboratory's work in future years. In 1959 we began to pass our experience on to JNWPU. They started their own line of research and although they felt ready to have

[60] Smagorinsky, J. On the numerical integration of the primitive equations of motion for baroclinic flow in a closed region. *Mon. Weather Rev.* **86**(12), 457–466 (1958).

[61] Smagorinsky, J. General circulation experiments with the primitive equations. I. The basic experiment. *Mon. Weather Rev.* **91**(3), 99–164 (1963).

[62] Hinkelmann, K. Ein Numerisches Experiment mit den Primitiven Gleichungen. *In* "The Atmosphere and the Sea in Motion—The Rossby Memorial Volume" (B. Bolin, ed.), pp. 486–500. Rockefeller Institute Press, New York, 1959.

launched a primitive equation operation in 1961, inadequate computing power delayed such an operation until 1966.[63]

Arrangements were made for a new computer to be delivered in 1962. It was the IBM 7030, or "Stretch," which was to be about 40 times faster than the IBM 701.

In October 1959, Syukuro Manabe joined our group. He was to become my close collaborator in this massive enterprise, eventually becoming the leader of our growing general circulation modeling group. In this new modeling venture, we were to be generously assisted by research scientists working for IBM. In 1960, Yale Mintz invited me to lecture about our two-level results at UCLA, and I also talked in detail about our new model and some of our earlier experience with it.

We had decided to test a three-level version of the new model in an octagonal domain with real initial conditions derived from Hinkelmann's analysis for January 22, 1959, with the initial divergence set at zero. The model was stripped of condensation, mountains, and friction. Twenty-four forecast results were shown in November 1960 at the International Symposium on Numerical Weather Prediction in Tokyo. The errors were comparable in magnitude and distribution to those of a forecast by Hinkelmann. Our results were never published. This is one of several examples in my career of a paper that should have been published, but was not. But conversely, I can think of some that should not have been published, but were!

It was in 1960 that I decided that we should consider getting involved with the oceans, for two reasons. First, the techniques we were developing seemed transferable even though a theoretical framework for the oceans comparable to that of the atmosphere was lacking. The other reason was that it was clear that long-term evolutions of the atmosphere and its climatic properties could not be understood without understanding the interaction with the oceans. It was for this reason that Kirk Bryan joined our group in March 1961.

We see then that in the late 1950s, the field of climatology was rapidly on its way to being transformed from a branch of descriptive geography to one of quantitative physical science.

6. Epilogue

I am approaching the limit of the scope that I intended for this account. Although it may seem to be a rather arbitrary stopping point, subsequent

---

[63] Shuman, F. The research and development program at the National Meteorological Center, NOAA (an internal unpublished report), 1972.

developments, from 1960 onward, both in numerical weather prediction and in general circulation and climate modeling, have been exponential in their growth and significance. I doubt whether my own knowledge could do justice to the breadth of achievements. Many of my views in the early 1960s were expressed in a Symons Memorial Lecture.[64]

Today, we still encounter problems that we thought had already been solved in the 1950s. The continuing central problem is that of systematically building a framework of understanding. Jule Charney's original strategy of constructing a hierarchy of models is still quite sound. But as models become more complex, it is difficult, with highly nonlinear and interactive processes, to say why we obtain a given result. There have been many disturbing examples of a result being apparently correct but for the wrong reason. Series of well-designed experiments must be employed to delineate cause and effect. For this purpose, thorough diagnostic techniques must continue to be developed and applied. One must also be prepared to go backward, hierarchically speaking, in order to isolate essential processes responsible for results observed from more comprehensive models.

There now are a tremendous number of scientists throughout the world engaged in modeling research. In contrast, at the International Conference in Stockholm in 1957, most of the world's expertise occupied about 40 seats.

[64] Smagorinsky, J. Some aspects of the general circulation. *Q. J. R. Meteorol. Soc.* **90,** 1–14 (1964).

CARBON DIOXIDE AND CLIMATIC CHANGE

Syukuro Manabe

Geophysical Fluid Dynamics Laboratory/NOAA
Princeton University
Princeton, New Jersey

1. Introduction

According to the observations at Mauna Loa and other stations, the atmospheric CO_2 concentration is about 335 ppm at present and is increasing continuously. In the latter half of the next century the CO_2 concentration may reach 600 ppm, which is twice as much as the preindustrial concentration of 290 ppm (Geophysics Study Committee, 1977).

It has been suggested that such an increase in CO_2 concentration may lead to the general warming of climate (see, for example, Callendar, 1938). At the Geophysical Fluid Dynamics Laboratory of the NOAA, the study of climate change resulting from the future increase of atmospheric CO_2 concentration has been the subject of a long-term research project during the past 15 years. This article reviews some of the results from this research project.

39

2. HISTORICAL BACKGROUND

In the latter half of the nineteenth century, Tyndall (1863) and Arrhenius (1896) suggested that a climate change may be induced by a change of CO_2 concentration in the atmosphere. These works were followed by the studies of Callendar (1938), Plass (1956), Kondratiev and Niilisk (1960), Kaplan (1960), and Möller (1963). In these studies, the CO_2-induced warming was evaluated from a condition of radiative heat budget at the earth's surface. I shall discuss briefly these studies before discussing the results from mathematical models of climate.

Although the atmospheric CO_2 absorbs solar radiation at near-infrared wavelength, the magnitude of the absorbed energy is very small. On the other hand, it strongly absorbs and emits terrestrial radiation at the wavelength of ~ 12–18 μm. In his study, Callendar shows that additional CO_2 in the atmosphere increases the downward flux of terrestrial radiation. To satisfy the condition of the surface heat balance, this CO_2-induced change of radiative flux should be compensated by an increase in the upward terrestrial radiation resulting from an increased surface temperature, if other components of surface heat balance remain unchanged. From consideration of the surface heat balance described above, Callendar estimated the CO_2-induced rise of surface temperature.

Most of the studies mentioned above employ similar approaches for the estimation of the CO_2-induced warming of the earth's surface, although there are some differences among them. For example, Kaplan takes into consideration the effect of cloud cover on the CO_2-induced change in the downward flux of terrestrial radiation. Kondratiev and Niilisk incorporate in their computation the effect of overlapping between an absorption band of water vapor and that of CO_2.

Möller attempted to improve these estimates by taking into consideration the effect of CO_2-induced change of water vapor in the atmosphere. He noted that the climatological distribution of relative humidity in the troposphere changes little with season despite a large change of air temperature. For example, an increase in air temperature is accompanied by an increase in absolute humidity of air, keeping the relative humidity unchanged. Therefore, the CO_2-induced change in the temperature of the earth's surface is accompanied by not only a change of temperature but also a change of absolute humidity of the overlying air which, in turn, causes a change in the downward flux of terrestrial radiation at the earth's surface. The change in absolute humidity also affects the absorption of solar radiation and, accordingly, the amount of solar radiation reaching the earth's surface. Möller estimated the warming of the earth's surface resulting from these increases of both terrestrial and solar radiation. For this purpose, he employed the following assumption:

$$d[S(W(T_s)) - E(T_s, W(T_s), C)] = 0 \qquad (2.1)$$

where S is the net downward solar radiation; E is the net upward terrestrial radiation at the earth's surface; W is the total liquid equivalent of water vapor; C is the CO_2 concentration of air; T_s is the temperature of the earth's surface. E is a function of only T_s and W because the vertical distributions of relative humidity and static stability are assumed to be preserved despite the CO_2-induced temperature change. Carrying out the differentiation on the left-hand side of Eq. (2.1), one gets

$$\frac{\partial S}{\partial W}\frac{\partial W}{\partial T_s}dT_s - \frac{\partial E}{\partial T_s}dT_s - \frac{\partial E}{\partial W}\frac{dW}{dT_s}dT_s - \frac{\partial E}{\partial C}dC = 0 \qquad (2.2)$$

From this equation, one can obtain the following relation which gives the temperature change dT_s resulting from the change in CO_2 concentration dC:

$$dT_s = \frac{\partial E/\partial C}{(\partial S/\partial W)(dW/dT_s) - \partial E/\partial T_s - (\partial E/\partial W)(dW/dT_s)}dC \qquad (2.3)$$

From this formula, Möller obtained rather surprising results: an increase in the water vapor content of the atmosphere with rising temperature causes a self-amplification effect which results in a large temperature change. When the air temperature is around 15°C, the doubling of CO_2 content results in a large temperature increase of 10°C. For other temperatures, the results may be completely different. Möller obtained such a wide variety of results because the denominator on the right-hand side of Eq. (2.3) is small and changes in sign and magnitude, depending upon the surface temperature T_s. In other words, the net upward radiation at the earth's surface hardly increases with increasing surface temperature because of the dependence of absolute humidity upon the atmospheric temperature. On the other hand, the net upward radiation increases significantly with increasing surface temperature when it is assumed that the absolute humidity of air is unchanged by an increase of atmospheric CO_2 concentration. Therefore, the earlier studies of Callendar and others yielded less arbitrary results.

The study of Möller described above exposed the basic difficulty of the surface radiation balance approach, which does not take into consideration the CO_2-induced changes in other components of the surface heat balance. It is expected that the change in the atmospheric CO_2 concentration alters not only the net radiative flux, but also the boundary-layer exchanges of sensible and latent heat between the earth's surface and the atmosphere. Accordingly, the following balance requirement [instead of Eq. (2.1)] should hold at the earth's surface:

$$d(S - E - H) = 0 \qquad (2.4)$$

where H is the sum of sensible and latent heat fluxes from the earth's surface into the atmosphere. In order to evaluate the change in H, it is necessary to consider not only the heat balance of the earth's surface, but also that of the atmosphere. Instead, Möller assumed that $dH = 0$. This is why his approach produced a result which is extremely sensitive to the surface temperature.

Recently, Newell and Dopplick (1979) attempted to estimate the CO_2-induced warming of the earth's surface by employing a surface energy balance method which incorporates not only the CO_2-induced change of radiative flux E but also that of H. Neglecting the CO_2-induced change in atmospheric water vapor W and solar flux S, one gets the following relationship from Eq. (2.4):

$$dT_s = - \frac{\partial E/\partial C}{\partial E/\partial T_s + \partial H/\partial T_s} dC \tag{2.5}$$

In order to estimate the CO_2-induced warming from Eq. (2.5), it is necessary to compute $\partial H/\partial T_s$. For this purpose, they made an unrealistic assumption that the change of surface temperature is not accompanied by the changes of temperature and absolute humidity of the overlying air. This assumption is responsible for the strong dependence of H upon surface temperature, i.e., the large value of $\partial H/\partial T_s$. (Note that it was assumed that the boundary-layer fluxes of sensible and latent heat are proportional to the air–surface difference of temperature and absolute humidity, respectively.) Thus, they obtained a very small value of $\partial T_s/\partial C$ from Eq. (2.5). In short, surface temperature can hardly change in response to a CO_2 increase when it is assumed that the temperature and absolute humidity of the overlying air do not change.

To evaluate H and its dependence on other variables, it is obviously necessary to consider not only the heat balance of the earth's surface but also that of the atmosphere. In the following section, the CO_2-induced change of the atmospheric temperature is discussed based upon the results from radiative, convective models of the atmospheric–surface system.

3. Radiative, Convective Equilibrium

In order to evaluate the CO_2-induced change of atmospheric temperature, Manabe and Wetherald (1967) used a "radiative, convective model of the atmosphere." This one-dimensional model consists of a system of the equations which represent the effects of radiative transfer and vertical convective mixing upon the heat balance of the atmosphere. As discussed below, the model may be used for the determination of the vertical distribution of the global mean temperature of the atmosphere in thermal equilibrium. By comparing the two thermal equilibria for the normal and above-normal concentrations of atmospheric CO_2, Manabe and Wetherald estimated the change of the global mean atmospheric temperature which occurs in response to an increase in the CO_2 concentration of air. It turned out that their study was useful for getting a preliminary insight into the physical mechanisms responsible for the CO_2-induced temperature change in the atmosphere. This section begins with the definition of radiative, convective equilibrium followed by a discussion of the CO_2-induced change of the atmospheric temperature.

3.1. Approach toward a Thermal Equilibrium

An atmosphere in radiative, convective equilibrium satisfies the following conditions:

(1) The net radiative flux at the top of the atmosphere is zero.
(2) The lapse rate of the atmosphere does not exceed a certain critical lapse rate because of the stabilizing effect of convection and other dynamical processes.
(3) In a layer where the lapse rate is subcritical (convectively stable), the condition of local radiative equilibrium is satisfied, i.e., the net radiative flux is zero.
(4) The heat capacity of the earth's surface is zero. This implies that, at the earth's surface, the net downward flux of radiation is equal to the upward convective heat flux.

In order to determine the temperature distribution of the atmosphere in radiative, convective equilibrium, Manabe and Wetherald employed "the time integration method" in which the state of radiative, convective equilibrium was approached asymptotically through the numerical time integration of the following equation:

$$C_p \frac{\partial T}{\partial t} = Q_R + Q_c \qquad (3.1)$$

where C_p is the specific heat of air; T is temperature; Q_R and Q_c are heating rate per unit mass due to radiation and convection, respectively. For the computation of heating (or cooling) due to solar and terrestrial radiation, the effects of water vapor, carbon dioxide, ozone, and cloud cover were taken into consideration. The effect of convection was incorporated into the model by use of the very simple procedure of "convective adjustment." This procedure involves the adjustment of the lapse rate to a critical value whenever it becomes supercritical during the course of a time integration. The adjustment was made such that the sum of internal and potential energy was unaltered. For their study, the critical lapse rate was chosen to be 6.5°C/km in reference to the static stability of the actual atmosphere. Based upon the assumption that the earth's surface has no heat capacity, the net downward radiative flux received by the earth's surface was immediately returned to the lowest layer of the atmosphere, inducing the convective adjustment in the overlying layer. Following the suggestion of Möller, it was also assumed that the model atmosphere maintains a given distribution of relative humidity. As air temperature changed during the course of a time integration, the distribution of absolute humidity was continuously updated such that relative humidity remains unchanged. On the other hand, it was assumed that the uniform mixing ratio of CO_2 and the vertical distributions of cloud cover and ozone did not change with time. In addition, the globally averaged, annual mean insolation was prescribed at the top of the model atmosphere throughout the course of a numerical time integration.

Figure 1 illustrates how the temperature of the model atmosphere approaches the equilibrium value starting from the initial condition of a cold

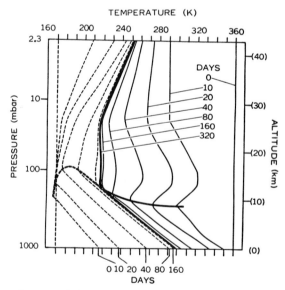

FIG. 1. Approaches toward the state of radiative, convective equilibrium. The solid and dashed lines show the approaches from warm and cold isothermal atmospheres, respectively. (From Manabe and Wetherald, 1967.)

(or warm) isothermal atmosphere. Toward the end of each time integration, the model atmosphere attains the realistic structure consisting of a convective troposphere and a stable stratosphere. The net incoming solar radiation becomes almost exactly equal to the outgoing terrestrial radiative at the top of the atmosphere, indicating that the atmosphere–surface system of the model is in radiative equilibrium as a whole. In the stable model stratosphere, the radiative heating (or cooling) is equal to zero, satisfying the condition of local radiative equilibrium. On the other hand, the model troposphere is not in radiative equilibrium. The total radiative heat loss from the model troposphere is equal to the net radiative heat gain by the earth's surface, implying the convective transfer of heat from the earth's surface to the atmosphere of the model.

3.2. Equilibrium Response

To evaluate the sensitivity of the model atmosphere to changes in atmospheric CO_2 concentration, a set of numerical time integrations was performed with the radiative, convective model of the atmosphere described above. Figure 2 illustrates the vertical distributions of the equilibrium temperature of the model atmosphere with the normal, half the normal, and twice the normal concentration of CO_2. This figure indicates that, in response to the doubling of the atmospheric CO_2, the temperature of the model troposphere increases by about 2.3°C whereas that of the middle stratosphere decreases by several degrees. In addition, it reveals that the magnitude of the warming resulting from the doubling of CO_2 concentration is approximately equal to the magnitude of the cooling from the halving of CO_2 concentration. This result suggests that CO_2-induced temperature change is not linearly proportional to the change in CO_2 concentration. Instead, it is proportional to the change in the logarithm of CO_2 concentration (see also the results of Rasool and Schneider, 1971; Augustson and Ramanathan, 1977).

The physical process of the warming due to an increase in CO_2 concentration has traditionally been explained by citing the greenhouse effect as an analogy. However, this is not a satisfactory analogy for the warming of the model troposphere. The warming process may be understood by considering the CO_2-induced change in the emissivity of the model atmosphere. In response to an increase in CO_2 concentration, the infrared opacity of the model atmosphere increases, raising the altitude of the effective source of the infrared emission into the space. Since the tropospheric temperature decreases with increasing altitude, this results in the lowering of the effective emission temperature for the outgoing radiation

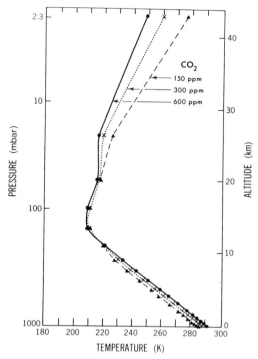

FIG. 2. Vertical distributions of temperature in radiative, convective equilibrium for various values of atmospheric CO_2 concentration, i.e., 150, 300, 600 ppm by volume. (From Manabe and Wetherald, 1967.)

and reduces the intensity of upward radiation at the top of the model atmosphere. In order to satisfy the condition that the outgoing terrestrial radiation be equal to the net incoming solar radiation at the top of the model atmosphere, it is therefore necessary to raise the temperature of the model troposphere, preventing the reduction of the outgoing terrestrial radiation.

In response to this warming, the absolute humidity of the model troposphere increases. This causes further increase in the infrared opacity of the atmosphere and raises the altitude of the effective source of the outgoing radiation. Thus, the temperature of the model troposphere must increase in order to maintain the radiation balance of the earth–atmosphere system. In addition, this increase of absolute humidity increases the fraction of solar radiation absorbed by the model troposphere and further enhances the CO_2-induced warming. Because of these positive feedback effects, the radiative, convective equilibrium of a model with a fixed distribution of relative humidity is twice as sensitive to a doubling of the atmospheric CO_2 concentration as that of another model with a fixed distribution of absolute humidity (Manabe and Wetherald, 1967). In short, the change of absolute humidity resulting from the change in air temperature significantly enhances the sensitivity of climate.

The above discussion implies that the earth's surface and the troposphere of the model are strongly coupled. The CO_2-induced warming of the coupled system results from the reduction in the radiative cooling of the system. The CO_2-induced change in the turbulent heat flux from the earth's surface to the overlying troposphere is essentially determined by the change in radiative heat deficit of the troposphere, which is equal to the change in the net radiative heat energy absorbed by the earth's surface.

As pointed out already, the temperature of the model stratosphere reduces in response to an increase of the CO_2 concentration in the atmosphere. In order to comprehend this result, it is necessary to recognize that the radiative heat budget of the model stratosphere is essentially maintained as a balance between the heating due to the absorption of the solar ultraviolet radiation by ozone and the net radiative cooling due to carbon dioxide. The increase of the emissivity of air resulting from an increase in CO_2 concentration enhances this cooling due to CO_2 and is responsible for the reduction of the temperature in the model stratosphere.

It is significant that the CO_2-induced change in the temperature of the model atmosphere is proportional to the change in the logarithm of atmospheric CO_2 concentration rather than the concentration itself. This is because the emissivity of CO_2 is approximately proportional to the logarithm of the CO_2 concentration.

Table I tabulates the changes of surface temperature due to the doubling of the atmospheric CO_2 concentration. The data were obtained from

TABLE I. INCREASE OF SURFACE AIR TEMPERATURE OF THE ATMOSPHERE IN RADIATIVE, CONVECTIVE EQUILIBRIUM RESULTING FROM THE DOUBLING OF CO_2 CONCENTRATION OF AIR

Reference	ΔT (°C)
Manabe and Wetherald (1967)	2.3
Manabe (1971)	1.9
Ramanathan (1976)	1.5
Wang et al. (1976)	1.6[a]
Augustsson and Ramanathan (1977)	2.0

[a] The estimates of Wang et al. (1976) are inferred from the result of numerical experiments in which CO_2 concentration is increased by a factor of 1.25.

radiative, convective models of the atmosphere developed by various authors. The magnitudes of warming contained in this table range from 1.5° to 2.3°C and are not very different from one another.

One must keep in mind, however, that the radiative, convective models used to obtain the results in Table I do not include various feedback mechanisms which can alter the sensitivity of climate. For example, the study of Augustson and Ramanathan quoted in Table I demonstrates that the sensitivity of climate can change significantly when the assumption of fixed cloud cover is abandoned. A recent study of Hansen et al. (1981) with a radiative, convective model indicates that the sensitivity of a model climate depends significantly upon the choice of a parameterization of the moist convective process.

Obviously, it is necessary to use a more comprehensive model of climate for the further discussion of the CO_2–climate sensitivity problem. Nevertheless, there is little doubt that a radiative, convective model is a very useful tool for getting a preliminary insight into this problem and identifying various factors which affect the sensitivity of climate.

4. DISTRIBUTION OF THE GLOBAL CLIMATE CHANGE

In the preceding section, the CO_2-induced change of global mean temperature is discussed based upon the results from one-dimensional models of the atmosphere in radiative, convective equilibrium. In order to discuss the latitudinal or geographical distribution of the CO_2-induced climate change, however, it is necessary to use a three-dimensional model of climate in which the effects of the atmospheric circulation and other physical process are explicitly taken into account. Furthermore, such a three-dimensional model is indispensable for the comprehensive assessment of the influences of various feedback mechanisms upon the sensitivity of climate.

Almost 25 years have passed since the pioneering attempts of Phillips (1956) and Smagorinsky (1963) to simulate the atmospheric general circulation by the use of three-dimensional, dynamical models of the atmosphere. Owing to the improvement of climate models and rapid advance of computer technology, it has become possible to simulate many of the large-scale characteristics of atmospheric circulation and climate by these models.

Encouraged by the similarity between the model climate and the actual climate, Manabe and Wetherald (1975) made the first attempt to study the CO_2-induced climate change by the use of a general circulation model of the atmosphere. For economy of computer time, their model had a limited computational domain with an idealized geography and had no seasonal variation of insolation. Nevertheless, this study, together with a companion study by Manabe and Wetherald (1980), yielded a preliminary insight into how the latitudinal distribution of the CO_2-induced climate change may be determined. These studies were followed by the investigations of Manabe and Stouffer (1979, 1980) and Manabe *et al.* (1981) which employed a global model of the joint atmosphere–mixed-layer ocean system with seasonal variation of insolation. This section discusses the latest results from this global model. Some of the conclusions from the earlier studies mentioned above will be incorporated in this discussion.

4.1. General Circulation Model

A general circulation model of climate is a prognostic system of equations representing the physical and dynamical processes which control climate. As was done to obtain a radiative, convective equilibrium solution of the atmosphere, a statistically stationary climate is approached asymptotically through the long-term integration of the general circulation model.

The general circulation model of climate used by Manabe and Stouffer consists of three major components: (1) a general circulation model of the atmosphere, (2) a heat- and water-balance model of the continental surface, and (3) a static, mixed-layer model of the ocean. Figure 3 contains the box diagram illustrating the basic structure of the model. The structure of each component is briefly described below.

The general circulation model of the atmosphere predicts the changes of the vertical component of vorticity, horizontal divergence, tempera-

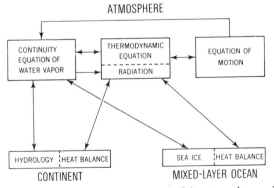

FIG. 3. Box diagram illustrating the basic structure of the atmosphere–mixed-layer ocean model of Manabe and Stouffer (1980).

ture, moisture, and surface pressure based upon the equation of motion, the thermodynamical equation, and the continuity equations of air mass and water vapor. The horizontal distributions of these predicted variables are represented by a finite number of spherical harmonics. The model has a global computational domain with realistic geography.

For the computation of solar and terrestrial radiation, the distributions of ozone and cloud cover are prescribed beforehand and the uniform concentration of carbon dioxide is set differently for each experiment, whereas the distribution of water vapor is determined from the prognostic system of water vapor.

Condensation of water vapor is predicted whenever supersaturation is indicated in the computation of the continuity equation of water vapor. Snowfall is predicted when air temperature near the earth's surface falls below the freezing temperature. Otherwise, rainfall is predicted.

The temperature of the continental surface is determined so that it satisfies the requirement of the heat balance. The changes of soil moisture and snow depths are obtained from the budget computations of water and snow, respectively.

The ocean model is a vertically isothermal and static water layer of uniform thickness with provision for a sea ice layer. The thickness of 68 m is chosen to ensure that the heat storage associated with the seasonal variation of observed sea surface temperature is correctly modeled. The rate of temperature change of the mixed-layer ocean is computed based upon the budget among surface heat fluxes. For this computation, the contributions of the horizontal heat transport by ocean currents and heat exchange between the mixed layer and the deeper layer of ocean are not taken into consideration. In the presence of sea ice, the temperature of the underlying water of the mixed-layer ocean is at the freezing point and the heat flux through the ice is balanced by the latent heat of freezing and melting at the bottom of the ice. This process, together with the melting at the upper surface of the ice, sublimation, and snowfall, determines the change of ice thickness. The albedo of sea ice and continental snow is assumed to vary between 50 and 80% depending on latitude and its thickness. Smaller values are assigned for thin sea ice, thin snow, or melting surface of sea ice.

4.2. Simulated Climate

In order to determine with confidence the CO_2-induced change of climate based upon the results from a numerical experiment with a mathematical model, it is necessary that the model be capable of reproducing the seasonal and geographical variation of climate. This section contains a

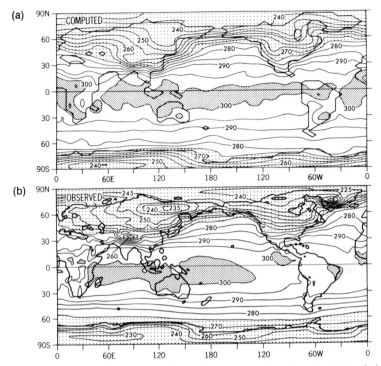

FIG. 4. Geographical distribution of monthly mean surface air temperature (kelvins) in February: (a) computed distribution from the $1 \times CO_2$ experiment (Manabe and Stouffer, 1980); (b) observed distribution (Crutcher and Meserve, 1970; Taljaad et al., 1969).

brief description of climate as simulated by the Manabe–Stouffer model described in Section 4.1.

The model climate is obtained as a statistically stationary state emerging from a 15-yr integration of the model. The initial condition for this time integration is an isothermal and dry atmosphere at rest with an isothermal mixed-layer ocean. Accordingly, it is clear that the success or failure of the climate simulation does not depend critically upon the specific choice of the initial condition.

In Fig. 4, the geographical distribution of monthly mean surface air temperature in February obtained from the standard $(1 \times CO_2)$[1] experiment is compared with the corresponding observed distribution from the data compiled by Crutcher and Meserve (1970) and Taljaad et al. (1969). On the basis of this comparison, one can identify unrealistic features in the model distribution. For example, the computed surface air tempera-

[1] $n \times CO_2$ denotes n times the normal CO_2 concentration.

ture is too low over the North Atlantic Ocean due to the absence of the effect of poleward heat transport by ocean currents in the model ocean. On the other hand, it is too high in the Southern Hemisphere, particularly along the periphery of Antarctica, probably because the prescribed cloud amount in the Southern Hemisphere is too small as compared with the observed amount. Nevertheless, it is clear that the model successfully reproduces the large-scale characteristics in the global distribution of the observed surface air temperature.

Figure 5 illustrates the geographical distribution of the difference in surface air temperature between August and February. It is expected that the difference yields information on the approximate magnitude of the seasonal variation except in the equatorial region where semiannual variation predominates. This figure indicates that the model fails to simulate the belt of relatively large temperature variation around 30° S. It does,

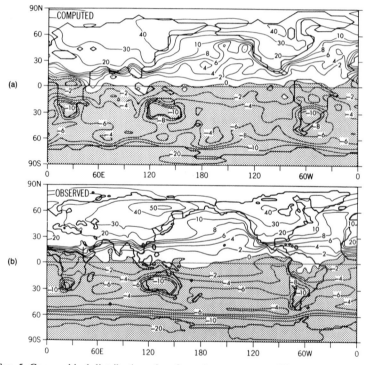

FIG. 5. Geographical distribution of surface air temperature difference (kelvins) between August and February: (a) computed distribution from the 1 × CO₂ experiment (Manabe and Stouffer, 1980); (b) observed distribution [from the data compiled by Crutcher and Meserve (1970) and Taljaad et al. (1969)]. Note that the contour interval is 2 K when the absolute value of the difference is less than 10 K and is 10 K when it is more than 10 K.

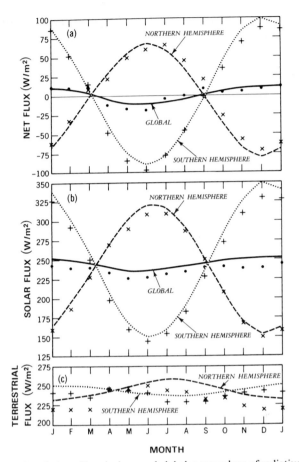

FIG. 6. Seasonal variation of hemisphere and global mean values of radiative fluxes at the top of the atmosphere (Manabe and Stouffer, 1980): (a) net radiative flux (i.e., net downward solar flux minus upward terrestrial flux); (b) net downward solar flux (i.e., incoming solar flux minus reflected flux); (c) upward terrestrial flux. Observed fluxes are deduced from satellite observation by Ellis and Vonder Haar (1976) in the Northern Hemisphere (×), Southern Hemisphere (+), and the entire globe (●).

however, reproduce the region of large annual temperature range over the northeastern part of the Eurasian continent and the northern part of North America. In general, the simulated distribution resembles the observed distribution quite well.

To evaluate the performance of the model in reproducing the overall radiation budget of the atmosphere–ocean system, Fig. 6 is constructed. This figure illustrates the seasonal variation of hemispheric and global mean values of solar, terrestrial, and net radiative fluxes at the top of the model atmosphere. For comparison, the values of the radiative fluxes deduced from satellite observation by Ellis and Vonder Haar (1976) are added to the figure. This comparison reveals that the agreement between the computed and observed fluxes is good, although both the solar and terrestrial fluxes of the model are slightly larger than the observed fluxes. The agreement is particularly noteworthy if one recalls that the prescribed distribution of cloud cover in the model atmosphere has no seasonal variation. These results suggest that the variation of cloud cover has a relatively small role in determining the seasonal variation of hemispheric mean values of radiation fluxes and surface air temperature of the atmosphere. This encourages one to speculate that the cloud cover may not have a dominant effect in determining the sensitivity of global (or hemispheric) mean temperature.

4.3. Design of Sensitivity Experiment

The climatic effect of an increase in atmospheric CO_2 is investigated based upon the comparison between two climates with normal and four times the normal CO_2 concentration. One could have investigated the consequence of a smaller increase of the CO_2 concentration, which is more likely to occur in the future. Instead, the atmospheric CO_2 concentration is increased by a factor of four for the ease of discriminating the CO_2-induced change from the natural fluctuation of the model climate. Figure 7 illustrates how the global mean sea surface temperature of the model evolves with time during the courses of two time integrations of the model with normal ($1 \times CO_2$) and four times the normal ($4 \times CO_2$) concentration of CO_2. This figure clearly indicates that, in both experiments, the global mean sea surface temperature stops changing about 15 yr after the beginning of each time integration. In Section 4.4, the climatic effect of the increase in CO_2 concentration is discussed based upon analysis of the differences between the two statistically stationary states which emerge from the $1 \times CO_2$ and $4 \times CO_2$ experiments.

4.4. Thermal Response

4.4.1. Annual Mean Response. The latitude–height distribution of the CO_2-induced change of zonal mean, annually averaged temperature is illustrated in Fig. 8. In qualitative agreement with the results from the study of radiative, convective equilibrium described earlier, the temperature of the model troposphere increases whereas that of the model strato-

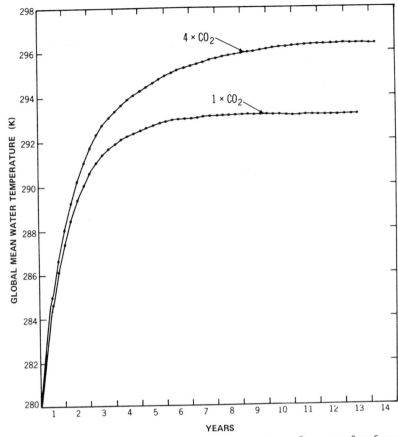

FIG. 7. Time variation of the global mean water temperature of ocean surface from the $1 \times CO_2$ and $4 \times CO_2$ experiments. (From Manabe and Stouffer, 1980.)

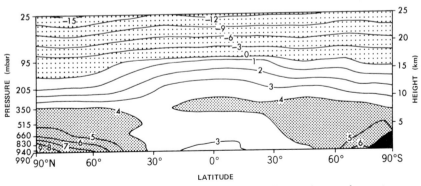

FIG. 8. Latitude–height distribution of the zonal mean difference in annual mean temperature (K) between the $4 \times CO_2$ and $1 \times CO_2$ model atmospheres. (From Manabe and Stouffer, 1980.)

sphere decreases in response to the increase of the CO_2 concentration in air. The tropospheric warming is particularly pronounced in the surface layer at high latitudes where the poleward retreat of snow cover and sea ice with high albedo enhances the warming. At low latitudes, the CO_2-induced heating spreads over the entire model troposphere via the effect of moist convection. Furthermore, the warming in the upper model troposphere is larger than the warming near the earth's surface because of the moist convective control of the vertical temperature distribution in low latitudes. (Note that the moist adiabatic lapse rate decreases with increasing air temperature.) Accordingly, the increase of surface air temperature at low latitudes is much less than the corresponding increase at high latitudes where the heating enhanced by the reduction of surface albedo is limited to the lower model troposphere due to stable stratification. The distribution of the annual mean response described above is qualitatively similar to the earlier results discussed by Manabe and Wetherald (1975, 1980).

One of the important factors which determines the sensitivity of surface temperature to the change of the atmospheric CO_2 concentration is the parameter B, defined by

$$B = \partial F / \partial T_s \qquad (4.1)$$

where F is the outgoing terrestrial radiation at the top of the atmosphere and T_s is surface temperature. Because of the latitudinal variation of the vertical profile of the CO_2-induced warming in the model troposphere described in the preceding paragraph, the values of B at high latitudes are much smaller than the corresponding values at low latitudes. Therefore, in order to compensate for the decrease of F due to the CO_2-induced increase in the atmospheric opacity, the surface temperature at high latitudes must increase much more than the surface temperature at low latitudes. In short, the latitudinal variation of B discussed here partly accounts for the difference in the sensitivity of surface temperature between high and low latitudes, as pointed out by Held (1978).

Figure 8 also reveals that at high latitudes of the model, the lower tropospheric warming in the Northern Hemisphere is larger than the corresponding warming in the Southern Hemisphere. This interhemispheric difference in warming partly results from the absence of the snow albedo feedback mechanism over Antarctica covered by continental ice sheets. It should be kept in mind, however, that the simulated sea surface temperature is too high and the area coverage of sea ice is too narrow in the Southern Hemisphere of the model. Thus, this model tends to underestimate the effect of sea ice albedo feedback and, accordingly, the CO_2-

induced warming in the Southern Hemisphere. A better simulation of the sea ice distribution is required before one can discuss the interhemispheric asymmetry of the CO_2-induced warming.

Owing to the interhemispheric difference in the CO_2-induced warming discussed above, the area mean change of surface air temperature in the Northern Hemisphere is +4.5°C and is larger than the area mean warming of +3.6°C in the Southern Hemisphere. Thus, the global mean increase of surface air temperature is 4.1°C. This result suggests that the warming due to doubling of the CO_2 concentration would be about 2°C provided that the warming is proportional to the increase in the logarithm of the atmospheric CO_2 concentration as explained earlier. This is similar to the corresponding warming of 1.9°C obtained from the radiative, convective model of Manabe (1971) (see Fig. 1). In the global model atmosphere, the surface warming at high latitudes is enhanced by the albedo feedback mechanism and stable stratification, whereas the surface warming at low latitudes is reduced by moist convection as discussed earlier. Because of these two opposing influences, the area mean surface warming of the global model turns out to be similar to the surface warming of the one-dimensional model.

Table II contains the estimates of area mean increases of surface air temperatures (in response to the doubling of atmospheric CO_2 concentration) which were obtained by various authors from their general circulation models of climate. This table indicates that the magnitude of the warming obtained from the seasonal model of Hansen (1979) is 3.9°C and is significantly more than the warming of 2.0°C from the model of Manabe and Stouffer. Since the details of Hansen's study have not been published, it is difficult to determine the basic causes for the difference. According to the report of the Climate Research Board (1979), Hansen obtained larger warming partly because the positive feedback effect of snow cover is exaggerated in his model. In his standard (1 × CO_2) experiment, the simulated surface air temperature is colder than the observed temperature and the area of snow cover is too extensive, resulting in the larger sensitivity mentioned above. In addition, it is speculated in the report that the difference in warming may also result from the difference in the treatment of cloud cover in the two models. Cloud cover is a prognostic variable in the Hansen model, whereas it is prescribed in the model of Manabe and Stouffer. Further comparative assessment of the two studies appears to be necessary before one can pinpoint the real cause of the difference between the results from these two studies.

It is significant that the estimates of the CO_2-induced warming obtained by Gates et al. (1981) and Mitchel (1979) are much smaller than the esti-

TABLE II. ESTIMATES FROM NUMERICAL MODEL EXPERIMENTS OF THE WARMING OF AREA-MEAN SURFACE AIR TEMPERATURE $\overline{\Delta T}^A$ (°C) RESULTING FROM DOUBLING OF CO_2 CONCENTRATION IN THE ATMOSPHERE

Reference	Geography	Sea surface temperature	Insolation	Cloud	$\overline{\Delta T}^A$ (°C) doubling
Interactive ocean					
Manabe and Wetherald (1975)	Idealized	Predicted	Annual	Prescribed	2.9
Manabe and Wetherald (1980)	Idealized	Predicted	Annual	Predicted	3.0
Wetherald and Manabe (1981)	Idealized	Predicted	Annual	Prescribed	3.0[a]
Wetherald and Manabe (1981)	Idealized	Predicted	Seasonal	Prescribed	2.4[a]
Manabe and Stouffer (1979, 1980)	Realistic	Predicted	Seasonal	Prescribed	2.0[a]
Hansen et al. (1979a)	Realistic	Predicted	Annual	Predicted	3.9
Hansen et al. (1979b)	Realistic	Predicted	Seasonal	Predicted	3.5
Noninteractive ocean					
Gates et al. (1981)	Realistic	Prescribed	Seasonal	Predicted	0.3
Mitchell (1979)	Realistic	Prescribed	Seasonal	Prescribed	0.2

[a] Temperature changes inferred from the results of numerical experiments in which the atmospheric CO_2 concentration is increased by a factor of four.

mates obtained from other studies. This is because they assumed that sea surface temperature is unaltered by the increase of the atmospheric CO_2 concentration, imposing a strong constraint upon the changes of the temperature of their model atmospheres.

Table II also reveals that a model with the seasonal variation of insolation is less sensitive to the increase of the atmospheric CO_2 concentration than another version of the model with an annual mean insolation. [Note the difference between the results from the annual and seasonal models of Hansen and the corresponding difference in the results of Wetherald and Manabe (1981).] This is partly because the contribution of the albedo feedback mechanism is smaller in a seasonal model in which snow cover is absent over the Northern Hemisphere continents during most of summer. For further discussion of this topic, refer to Wetherald and Manabe (1981).

Based upon the results from both one-dimensional models and general circulation models of climate, the report of the Climate Research Board (1979) concluded that the global mean increase of surface air temperature resulting from the doubling of the atmospheric CO_2 concentration is near 3°C with a probable error of 1.5°C. In view of the difficulty in developing a reliable scheme of cloud prediction and a realistic model of the joint ocean–atmosphere system, it will take some time to reduce the uncertainty in the estimate of the CO_2-induced change in the atmospheric temperature.

4.4.2. Seasonal Response. Figure 9 shows the seasonal variation of the difference between the zonal mean surface air temperatures of the 4 × CO_2 and the 1 × CO_2 atmospheres. At low latitudes, the warming due to the quadrupling of the atmospheric CO_2 concentration is relatively small and depends little on season, whereas at high latitudes it is generally larger and varies markedly with season, particularly in the Northern Hemisphere. Over the Arctic Ocean and its vicinity, the warming is at a maximum in early winter and is small in summer. This implies that the range of seasonal variation of surface air temperature in these regions decreases significantly in response to the CO_2 increase.

An analysis of heat fluxes over the Arctic Ocean of the model indicates that the CO_2-induced reduction of sea ice thickness is mainly responsible for the large early winter warming of surface air mentioned above. In early winter, the upward conductive heat flux through sea ice in the 4 × CO_2 experiment is larger than the corresponding flux in the 1 × CO_2

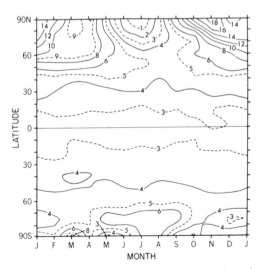

FIG. 9. Latitude–time distribution of the zonal mean difference in surface air temperature (K) between the 4 × CO_2 and 1 × CO_2 model atmospheres. (From Manabe and Stouffer, 1980.) Zonal averaging is made over both oceans and continents.

experiment because of the difference in sea ice thickness. Accordingly, the flux of sensible heat from the ice surface to the atmosphere in the former experiment is also larger than the corresponding flux in the latter experiment. This difference in the upward flux of sensible heat accounts for the large CO_2-induced warming of surface air in early winter. The winter warming is further enhanced by a stable stratification which limits the heating to the lowest layer of the model troposphere at high latitudes.

As Fig. 9 indicates, the magnitude of the warming in summer is much less than the corresponding warming in winter. Because of the reduction of sea ice, the surface albedo decreases significantly from the $1 \times CO_2$ to the $4 \times CO_2$ experiment. However, the additional energy of solar and terrestrial radiation absorbed by ocean surface is used either for melting the sea ice or warming the ice-free mixed-layer ocean which has a large heat capacity. Thus, the summer warming of surface air turns out to be relatively small. It should be noted, however, that the additional energy absorbed during summer in the $4 \times CO_2$ case, delays the appearance of sea ice or reduces its thickness, thereby increasing the conductive heat flux in early winter when the air–water temperature difference becomes very large.

The seasonal variation of the CO_2-induced warming of surface air temperature over the model continents is significantly different from the variation over the model oceans. According to Fig. 10, which shows the latitude–time distribution of the increase in zonal mean surface air temperature over continents, the CO_2-induced warming of high latitudes is at a maximum in early winter, being influenced by the large warming over the Arctic Ocean discussed above. However, Fig. 10 also indicates a secondary center of relatively large warming around 65° N in April. This results from a large reduction in surface albedo during spring when the insolation acquires a near-maximum intensity. The CO_2-induced reduction of snow cover area is responsible for this albedo change.

The discussion in the preceding paragraphs indicates that, as compared with snow cover, sea ice has a quite different influence on the seasonal variation of the CO_2-induced increase of surface air temperature at high latitudes. This is because sea ice not only reflects a large fraction of solar radiation, but also reduces the heat exchange between the atmosphere and underlying seawater. Thus, the CO_2-induced warming of surface air over the Arctic Ocean is at a maximum during early winter as discussed earlier. On the other hand, the warming over continents is also large in spring, when the heating due to the snow albedo feedback mechanism is largest.

In assessing the results of Manabe and Stouffer described here, it is desirable to keep in mind that the mixed layer of their model does not

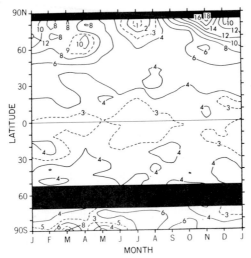

FIG. 10. Same as Fig. 9 except that the zonal averaging of surface air temperature is computed only over continents. (From Manabe and Stouffer, 1980.)

exchange heat with the deeper layer of ocean. Nevertheless, the qualitative aspect of their results is probably valid because a stably stratified halocline, which exists beneath the mixed layer of the Arctic Ocean, markedly reduces the heat exchange between the mixed layer and the deeper layer of the ocean.

4.4.3. Geographical Response. The geographical distribution of the difference between the annual mean surface air temperatures of the 4 × CO_2 and 1 × CO_2 atmospheres is shown in Fig. 11a. This figure indicates that the distribution of the CO_2-induced annual mean warming is almost zonal and reveals the characteristics which are identified in Section 4.4.2 on zonal mean response (i.e., large polar warming, relatively small warming in the tropics, and the interhemispheric asymmetry in the warming between the two polar regions).

The difference in the surface air temperature for the December–February (DJF) period is shown in Fig. 11b. According to this figure, the CO_2-induced warming is larger than its zonal average along the east coast of both Eurasian and North American continents. The large Arctic warming discussed in Section 4.4.2 reduces the cooling effect of the southward advection of cold air along the periphery of the Aleutian and Icelandic lows. This accounts for the relatively large CO_2-induced warming along the east coasts of continents mentioned above.

The distribution of the CO_2-induced change of surface air temperature

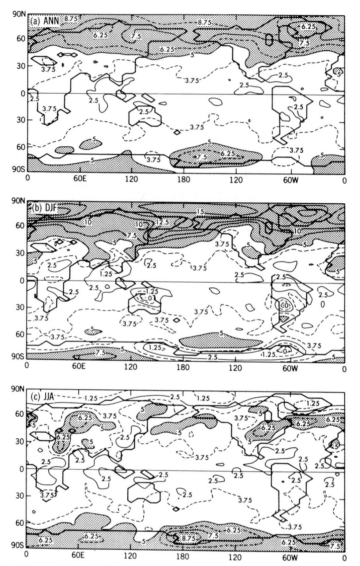

FIG. 11. Geographical distribution of the difference in surface air temperature (K) between the $4 \times CO_2$ and $1 \times CO_2$ model atmospheres: (a) annual mean difference; (b) December–January–February difference; (c) June–July–August difference. Shaded areas identify the regions where the difference exceeds 5°C. (From Manabe and Stouffer, 1980.)

for the June–August (JJA) period is shown in Fig. 11c. In this figure, one can identify the regions of relatively large warming over the continents at middle and high latitudes of the Northern Hemisphere. These regions approximately coincide with the regions where soil moisture decreases in response to the increase of the atmospheric CO_2. The reduction of soil moisture results in the reduction of evaporative ventilation and, accordingly, the warming of the continental surface, particularly during summer when the insolation is at or near maximum intensity. As will be discussed in Section 4.6.3, the geographical distribution of soil moisture change as determined by this numerical experiment is influenced by the large natural fluctuation of the model hydrology. Therefore, the details of the summer distribution of surface air temperature difference should be regarded with caution.

4.5. Hydrologic Response

4.5.1. Annual Mean Response. The increase in CO_2 concentration affects not only the thermal structure of the model atmosphere but also the hydrologic behavior of the model. One of the basic hydrologic changes is the overall intensification of the hydrologic cycle. In response to the quadrupling of the atmospheric CO_2 content, the global mean rates of both evaporation and precipitation increase by as much as 7%.

One of the important factors responsible for the intensification of the hydrologic cycle is the change in the surface radiation budget. For example, an increase in atmospheric CO_2 enhances the downward flux of terrestrial radiation reaching the earth's surface. In addition, the CO_2-induced warming of the troposphere results in the increase of absolute humidity as discussed in Section 3 and also contributes to the increase of downward flux of terrestrial radiation. Thus, a larger amount of radiative energy is received by the earth's surface, to be removed as turbulent fluxes of sensible and latent heat. This accounts for the increase in the global mean rate of evaporation mentioned above.

One can identify another important factor which is responsible for the enhancement of evaporation. According to the Clapeyron–Clausius relationship between the saturation vapor pressure and temperature, the saturation vapor pressure increases almost exponentially with a linear increase of surface temperature. Thus, when surface temperature is high, evaporation becomes a more effective means of ventilating the earth's surface as compared with the upward flux of sensible heat. Accordingly, a larger fraction of radiative energy received by the earth's surface is removed as latent heat rather than sensible heat. This implies the CO_2-induced increase in the evaporation rate.

To satisfy the balance requirement of water vapor in the model atmosphere, the increase in the global mean rate of evaporation discussed above should be matched by a similar increase in the global mean rate of precipitation. This explains why the global mean rates of both precipitation and evaporation increase in response to an increase of atmospheric CO_2 content. For further discussion of the physical mechanisms which are responsible for the CO_2-induced increase of the intensity of the hydrologic cycle, see the articles by Manabe and Wetherald (1975) and by Wetherald and Manabe (1975).

The global intensification of the hydrologic cycle is evident in Fig. 12, which illustrates the latitudinal distributions of annually averaged, zonal mean rates of precipitation and evaporation from both the $1 \times CO_2$ and the $4 \times CO_2$ experiments. This figure indicates that at high latitudes, the increase of the precipitation rate is much larger than that of the evaporation rate, whereas the former tends to be smaller than the latter at low latitudes of the model. This result implies that, as compared with the $1 \times CO_2$ atmosphere, the $4 \times CO_2$ atmosphere receives more moisture from the earth's surface at low latitudes and returns it to high latitudes in the form of increased precipitation. The poleward transport of moisture in the model atmosphere increases significantly in response to the increase of atmospheric CO_2 and compensates for the difference in moisture exchange between the earth's surface and the atmosphere of the model as described above. The increase in the moisture content of air resulting from the CO_2-induced warming of the model troposphere accounts for the increase of poleward moisture transport as discussed by Manabe and Wetherald (1980). The increase of the precipitation rate at high latitudes is responsible for the enhanced wetness of soil over the Arctic and subarctic region during most of the year as discussed below.

4.5.2. *Seasonal Response.*   The CO_2-induced change in the hydrology of the model has a significant seasonal dependence. Figure 13 illustrates the latitude–time distribution of the difference in zonal mean soil moisture between the $4 \times CO_2$ and $1 \times CO_2$ experiment. According to this figure, the difference in zonal mean soil moisture in high latitudes of the model has a large positive value throughout most of the year with the exception of the summer season. As discussed above, this CO_2-induced increase in soil moisture results from the penetration of warm, moisture-rich air into high latitudes of the model. Figure 13 also indicates two zones of reduced soil wetness at middle and high latitudes during the summer season. Qualitatively similar results were obtained by Wetherald and Manabe (1981) from a model with an idealized geography. The characteristics of the CO_2-induced change of soil moisture identified above have relatively high statistical significance (Manabe et al., 1981; Hayashi, 1982).

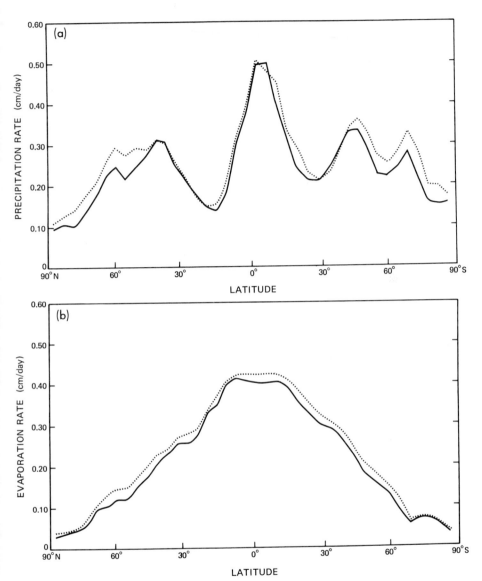

FIG. 12. Latitudinal distribution of (a) zonal mean precipitation rate and (b) zonal mean evaporation rate (in units of centimeters per day) from the $4 \times CO_2$ (·····) and $1 \times CO_2$ (——) experiments. (From Manabe and Stouffer, 1980.)

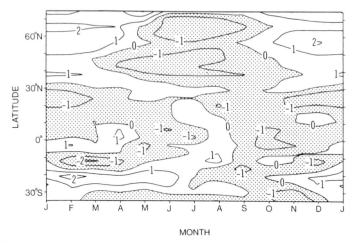

FIG. 13. Latitude–time distribution of zonal mean difference in soil moisture (cm) over continents between the $4 \times CO_2$ and $1 \times CO_2$ experiments. Note that the field capacity of soil moisture is assumed to be 15 cm everywhere. The zonal mean difference is not shown in high-latitude regions of both hemispheres where the longitudinal span of continents is either zero or very small. (From Manabe and Stouffer, 1980.)

To appreciate the relative magnitude of the CO_2-induced soil moisture change, the percentage change of soil moisture is computed by averaging the results from two sets of CO_2 sensitivity experiments and is shown in Fig. 14. This figure indicates that during summer the percentage reduction of zonal mean soil moisture exceeds 20% at 60° N and 44° N. The percentage increase of zonal mean soil moisture, which occurs at high latitudes except during summer, exceeds 10%. In short, the changes in zonal mean soil moisture described in the preceding paragraph, constitute a substantial fraction of the amount of soil moisture itself. Another feature of interest in Fig. 14 is a large fractional reduction of zonal mean soil moisture at about 25° N in winter. Although one notes some fractional changes of zonal mean soil moisture in tropics of the model, the changes have relatively small statistical significance as compared with the changes at middle and high latitudes as discussed by Manabe et al. (1981) and Hayashi (1982).

To determine the mechanisms responsible for the CO_2-induced summer dryness at middle and high latitudes described above, Manabe et al. (1981) made an extensive analysis of the seasonal variation of the soil moisture budget from the model. Their analysis reveals that at high latitudes, the snowmelt season in the $4 \times CO_2$ experiment ends earlier than the corresponding season in the $1 \times CO_2$ experiment. Thus, the warm season of rapid soil moisture depletion begins earlier due to enhanced evaporation, resulting in less soil moisture during summer in the $4 \times CO_2$

experiment. The mechanism described above is mainly responsible for the CO_2-induced dryness in summer at high latitudes of the model.

At middle latitudes, the earlier timing of the spring maximum in snow-melt also contributes to the summer dryness as at high latitudes. In addition, the spring-to-summer reduction in the precipitation rate occurs earlier, contributing to the earlier beginning of the drying season and CO_2-induced summer dryness. In the $4 \times CO_2$ experiment, the period of weak baroclinicity during summer begins earlier than the standard experiment, resulting in the earlier occurrence of the spring-to-summer reduction in the precipitation rate mentioned above.

One can identify another factor which is responsible for the earlier occurrences of the spring maximum and of the spring-to-summer reduction of the precipitation rate. In a CO_2-rich atmosphere, the middle-latitude rainbelt is located poleward of the corresponding rainbelt in the normal CO_2 atmosphere because of the penetration of the moisture-rich, warm air into high-latitude regions. Since the middle-latitude rainbelt moves poleward from winter to summer, this poleward shift of the rainbelt implies an earlier arrival of the rainbelt in spring. It also accounts for the CO_2-induced winter dryness in the model subtropics mentioned above.

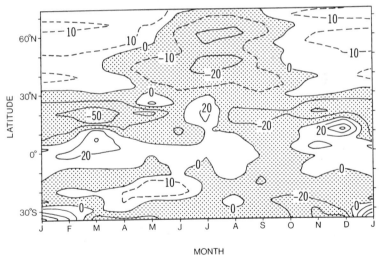

MONTH

Fig. 14. Latitude–time distribution of the percentage change in zonal mean soil moisture over continents in response to the quadrupling of the atmospheric CO_2 concentration. Here, the percentage of difference is computed by averaging the results from two versions of the global model with two different computational resolutions. The zonal mean difference is not shown in high-latitude regions of both hemispheres where the longitudinal span of continents is either zero or very small. See Manabe et al. (1981) for further details.

4.5.3. Geographical Distribution. So far, the CO_2-induced change of zonal mean soil moisture has been discussed. However, one of the ultimate goals of the CO_2–climate sensitivity study is the determination of the geographical distribution of the climate change. Manabe *et al.* (1981) made a preliminary attempt to evaluate the geographical distribution of the CO_2-induced change of soil moisture. Unfortunately, the local hydrology of a general circulation model has much larger temporal variability than the zonal mean hydrology. In order to distinguish the CO_2-induced local hydrologic change from the natural fluctuation of the model hydrology, it is necessary to perform a very long-term integration of the model. The periods of the numerical time integrations conducted by Manabe *et al.* are too short for this purpose. Furthermore, the geographical distribution of hydrologic variables (i.e., precipitation rate) as simulated by current climate models contains many unrealistic features. Therefore, the geographical details of the CO_2-induced hydrologic change is a topic for future studies.

4.6. Signal-to-Noise Ratio[2]

In Sections 4.4 and 4.5, the distribution of the CO_2-induced change in a global model climate is discussed. To appreciate the practical implications of such a climate change and assess its detectability, it is desirable to compare the CO_2-induced change with the magnitude of natural fluctuation of climate. For this purpose, one can estimate the signal-to-noise ratio S/N for the CO_2-induced change of a climatic variable from the following equation:

$$S/N = \Delta_c q / \tilde{\sigma}_q \qquad (4.2)$$

where $\Delta_c q$ indicates the CO_2-induced change of a variable q and $\tilde{\sigma}_q$ denotes a weighted mean standard deviation of q defined by

$$\tilde{\sigma}_q = \frac{(N^{1c} - 1)\sigma_q^{1c} + (N^{4c} - 1)\sigma_q^{4c}}{(N^{1c} - 1) + (N^{4c} - 1)} \qquad (4.3)$$

Here, σ_q^{1c} and σ_q^{4c} are the standard deviations of q in the $1 \times CO_2$ and $4 \times CO_2$ experiments, respectively. N^{1c} and N^{4c} are the number of samples (i.e., number of years in the analysis period) from the two experiments (10 and 3, respectively).

When the signal-to-noise ratio as defined above is much larger than 1,

---

[2] The results described in this section have not been previously published. They are obtained through a recent collaboration between S. Manabe and R. J. Stouffer.

the CO_2-induced change in q is much larger than the magnitude of its natural fluctuation and can easily be detected. On the other hand, if the signal-to-noise ratio is much less than unity, it is practically impossible to detect the CO_2-induced change because it is obscured by the natural variation. In this situation, the CO_2-induced change has little practical implication.

The signal-to-noise ratio is computed for the monthly averaged, zonal mean values of surface air temperature and soil moisture which are discussed in the preceding sections. For this purpose, the periods of the $1 \times CO_2$ and $4 \times CO_2$ time integrations are extended to year 20 and year 16, respectively (see Fig. 7). The data from the last 10-year and 3-year periods of the two time integrations are used for this analysis.

Figure 15a illustrates the latitudinal and seasonal variation of the signal-to-noise ratio for the CO_2-induced change of zonal mean surface air temperature over continents (see Fig. 10 for the distribution of the signal). According to this figure, the signal-to-noise ratio of this quantity in middle latitudes of the Northern Hemisphere is at a maximum during summer when the natural variability of surface air temperature is relatively small. Although the CO_2-induced change of surface air temperature is particularly large at high latitudes with the exception of summer season, the distribution of the signal-to-noise ratio does not share the similar characteristics because the temporal fluctuation of zonal mean surface air temperature is very large near the poles. Earlier, Wigley and Jones (1981) obtained a qualitatively similar result from the CO_2-induced signal shown in Fig. 10 and the observed variability of surface air temperature in the real atmosphere. (See Fig. 1 of their paper. One can approximately deduce the signal-to-noise ratio defined here by dividing their version of signal-to-noise ratio by $\sqrt{5}$.) The similarity between the two results implies that the noise level of the model is not very different from the noise level in the actual atmosphere as determined by Wigley and Jones. In the Southern Hemisphere, where the latitudinal span of continents is narrow, the signal-to-noise ratio over continents is small because of high noise level. Although it is not shown here, the signal-to-noise ratio of zonally averaged surface air temperature over oceans is much larger than the corresponding ratio over continents due to the small variability of surface air temperature over oceans.

The signal-to-noise ratio for zonal mean surface air temperature over both oceans and continents is illustrated in Fig. 15b. In the Northern Hemisphere, the signal-to-noise ratio is large in middle latitudes during the summer season as it is over continents. In the Southern Hemisphere, it is relatively large in middle latitudes throughout most of the year. The signal-to-noise ratio is relatively small in the polar regions where the

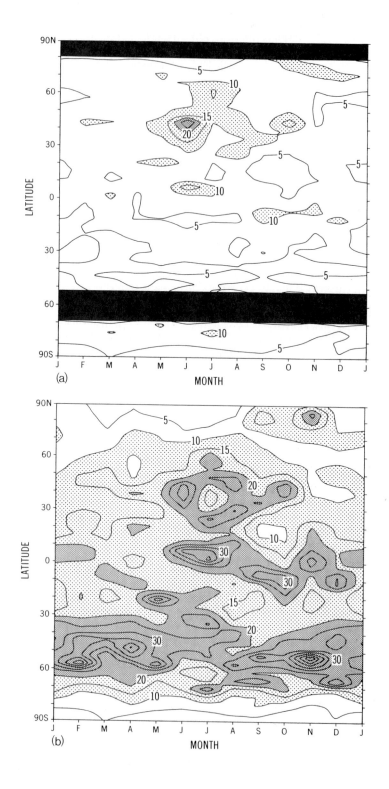

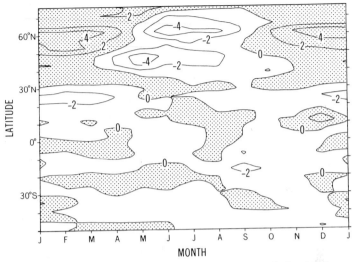

FIG. 16. The latitude–time distribution of the signal-to-noise ratio for the change of zonally averaged, monthly mean soil moisture over continents in response to the quadrupling of the atmospheric CO_2 concentration.

variability of zonally averaged surface air temperature is particularly large, as discussed by North *et al.* (1982).

Figure 16 illustrates the latitudinal and seasonal variation of the signal-to-noise ratio for the CO_2-induced change in zonal mean soil moisture over continents. (See Fig. 13 for the distribution of the signal.) According to this figure, some of the features with relatively large signal-to-noise ratio include the enhanced summer dryness in middle and high latitudes of the Northern Hemisphere, the enhanced winter dryness around 25° N, and the enhanced wetness around 60° N during the fall–winter–spring period. (In the Southern Hemisphere, the signal-to-noise ratio is small because the longitudinal span of continents is small.) Figure 16 also indicates that the changes of soil moisture, which are caused by the quadrupling (or doubling) of atmospheric CO_2 may have a magnitude comparable to its natural variability. Comparison of Figs. 15a and 16 reveals that the signal-to-noise ratio of this variable is much less than the corresponding ratio for the surface air temperature. In summary, the results from the present study suggest that it is harder to detect the CO_2-induced change in soil moisture than the corresponding change in surface air temperature.

FIG. 15. The latitude–time distribution of the signal-to-noise ratio for the change of zonally averaged, monthly mean surface air temperature in response to the quadrupling of the atmospheric CO_2 concentration. The zonal averaging is made over (a) continents and (b) both oceans and continents.

[The statistical significance of the CO_2-induced change may be evaluated by use of the signal-to-noise ratio obtained here. If the signal-to-noise ratio in Figs. 15 and 16 is equal to 1.45, the change is statistically significant at 95% confidence level. For a discussion of the Student's t test, see the book by Panofsky and Brier (1968).]

5. Transient Response

In Section 4, the discussions focused upon the difference between the two equilibrium climates of a model with normal and above-normal atmospheric CO_2 concentrations. However, this equilibrium response of climate will not be realized immediately mainly because of the large thermal inertia of oceans. Therefore, it is important to determine the features of the transient response of climate resulting from the future increase of atmospheric CO_2.

Preliminary studies of CO_2-induced transient climate change have been conducted by the use of simple energy balance models of the ocean–atmosphere system (Schneider and Thompson, 1981; Hoffert *et al.*, 1980). Based upon the results from their study, Schneider and Thompson suggested that the latitudinal distribution of the transient change of climate may be quite different from the equilibrium response described in the preceding sections. For a further investigation of this problem, it is desirable to develop a comprehensive model of the ocean–atmosphere system.

Since the atmosphere–mixed-layer ocean model described in Section 4 lacks the deeper layers of an ocean which have large heat capacity, it is not suitable for the transient response study mentioned above. In order to investigate this problem, it is necessary to construct a full ocean–atmosphere model in which general circulation models of the atmosphere and a full ocean model are coupled with each other. Such a model has been developed at the Geophysical Fluid Dynamics Laboratory during the last 15 yr (Manabe and Bryan, 1969; Manabe *et al.*, 1979). Recently, Bryan *et al.* (1982) used a simplified version of the model for a study of the CO_2 transient response problem. Because of the many assumptions and simplifications adopted in the construction of the model, the qualitative aspect of the result should be received with caution. Nevertheless, the results from this study yield some preliminary insight into the nature of this problem. A brief description of this study follows.

5.1. Coupled Ocean–Atmosphere Model

As pointed out already, major simplification and idealization are incorporated into the coupled ocean–atmosphere model used in this prelimi-

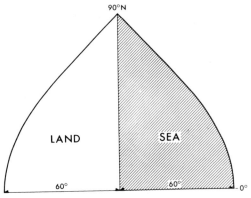

FIG. 17. Computational domain of the ocean–atmosphere model (Bryan *et al.*, 1982).

nary study. For example, seasonal variation of solar insolation is not included, and the geometry of land and sea is highly idealized as shown in Fig. 17. To minimize the amount of calculation, the model atmosphere and ocean are constrained to be triply periodic in the zonal direction with mirror symmetry across the equator. The basic structure of the model is illustrated in Fig. 18 as a box diagram.

The atmospheric component of the model is identical to that of the atmosphere–mixed-layer ocean model used for the study described in Section 4.2 (with the exception of simplifications identified in the preceding paragraph). The oceanic component of the model does not have a sufficient horizontal resolution to simulate mesoscale eddies, which are the dynamic counterparts of cyclones and anticyclones in the atmo-

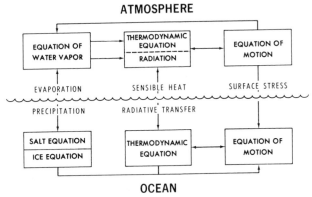

FIG. 18. Box diagram illustrating the structure of the ocean–atmosphere model (Manabe and Bryan, 1969).

sphere. The resolution is sufficient, however, to include the main features of the observed wind-driven and thermohaline circulation. A simple model of sea ice and land snow cover is incorporated as described in Section 4.2.

5.2. Design of Switch-On Experiment

Since the main topic of this study is the response of the coupled model as it makes the transition from one equilibrium state to another, the first requirement was to calculate equilibrium climates corresponding to the normal and above-normal atmospheric CO_2 concentrations. In order to obtain a climatic equilibrium of the model, it was necessary to use an economical method of numerical time integration developed by Manabe and Bryan (1969), Bryan and Lewis (1979), and Manabe et al. (1979) because it takes an extremely long time (several thousand years) for the coupled system to reach a statistically stationary equilibrium climate. This method, which is called the "nonsynchronous method," may be thought of as a relaxation procedure to hasten the convergence to equilibrium. In this study, 1 yr in the atmosphere model was taken to correspond to 110 yr in the upper ocean. One year in the upper ocean, in turn, was taken to be equivalent to 25 yr in the deepest level of the ocean.

The determination of two climatic equilibrium states provides the basis for a switch-on experiment in which the normal CO_2 equilibrium climate is perturbed by a sudden quadrupling of the atmospheric CO_2 concentration. The atmospheric and oceanic components of the model are time-integrated over a period of 25 yr in a synchronous mode as opposed to the nonsynchronous, economical method used to obtain a climate equilibrium.

5.3. Temporal Variation

Figure 19 illustrates the zonally averaged response at the 10th year of the time integration. The atmosphere exhibits the largest response at low levels near the pole, whereas the temperature of the underlying ice-covered seawater hardly changes (i.e., remaining at the freezing point). This result indicates that sea ice insulates the surface air from the underlying seawater. Around the sea ice margin, i.e., at about 75° N, the warming of the sea surface temperature is limited to the surface layer of the water because of a stably stratified halocline which insulates the shallow surface layer from the deeper layers of the ocean. The deepest penetration of the

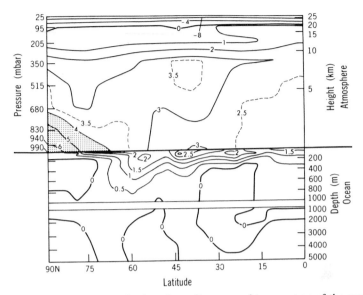

FIG. 19. Latitude–height distribution of zonally averaged temperature of the ocean–atmosphere model at the 10th year minus initial temperature (K), showing the response to a step function increase in atmospheric CO_2 (Bryan *et al.*, 1982).

heat anomaly occurs at about 60° N, where the stratification of the ocean is relatively weak and convective overturning occurs frequently. The greater stratification of the tropical ocean accounts for the smaller penetration of the heat anomaly at low latitudes.

To evaluate the response of the model atmosphere in comparison with its equilibrium response, the change of surface air temperature T may be normalized as follows:

$$R = (T - T_0)/(T_\infty - T_0) \qquad (5.1)$$

where R is the normalized response; T_0 and T_∞ are the initial and final values of the surface air temperature, respectively. T_0 and T_∞ are equal to the equilibrium surface air temperatures of the model atmospheres with the normal and above-normal concentrations of CO_2, respectively. The zonal mean values of R are computed separately over oceans and continents, and are illustrated as a function of latitude and time in Fig. 20a and b. According to Fig. 20a, the rise in surface air temperature over the oceans is initially more rapid at low latitudes than at 60° latitude, where the downward penetration of the heat anomaly is deeper. However, the normalized response becomes nearly uniform at all latitudes and is about 65% at about the 20th year of the time integration. This result suggests

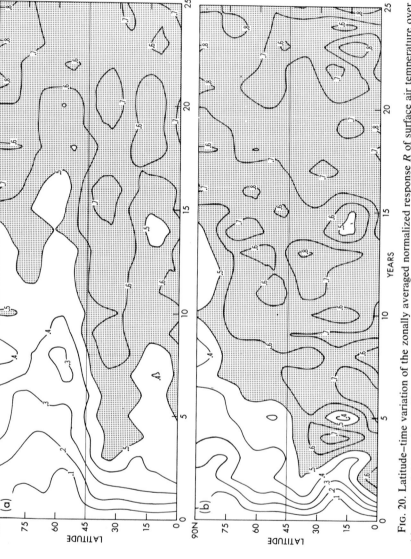

FIG. 20. Latitude–time variation of the zonally averaged normalized response R of surface air temperature over (a) oceans and (b) continents (Bryan et al. 1982). The normalized zonal mean response is $([T] - [T_0])/([T_\infty] - [T_0])$, where square brackets denote zonal mean operator.

that beyond the 20th year, the penetration of the heat anomaly into the deeper ocean is very slow and hardly influences the latitudinal profile of the normalized response of surface air temperature. Over the model continents which have zero thermal inertia, the normalized response of zonal mean surface air temperature is even faster and becomes nearly uniform as early as the 10th year (see Fig. 20b).

This result suggests that the latitudinal dependence of the transient response of surface air temperature is very similar to that of the equilibrium response provided that the time scale for the increase of the atmospheric CO_2 concentration is longer than ~ 10–20 years.

In view of many simplifications and shortcomings of the coupled model used for this study, the results from this computation must be regarded as tentative. Nevertheless, the study of Bryan et al. seems to indicate that, barring an unseen drastic acceleration of the rate of increase of atmospheric CO_2, sensitivity studies of climate equilibrium may be used as an approximate guide for predicting the latitudinal pattern of surface air temperature change in response to a gradual increase of atmospheric CO_2.

6. Concluding Remarks

In this presentation, I have reviewed some of the CO_2–climate sensitivity studies which have been conducted by various authors, in particular, by the staff members at the Geophysical Fluid Dynamics Laboratory of the NOAA. Based upon the results from mathematical models of climate with various degrees of complexities, it has been possible to identify the changes of climate resulting from an increase of atmospheric CO_2 content and to discuss the physical mechanisms responsible for the changes. Some of these CO_2-induced changes of climate are listed below.

(1) The temperature of the troposphere increases whereas that of the stratosphere decreases.
(2) The annual mean warming of the surface air at high latitudes is two to three times as large as the corresponding warming at low latitudes.
(3) Over the Arctic Ocean and the surrounding regions, the CO_2-induced warming has a large seasonal dependence. It is at a maximum in winter and at a minimum in summer. The warming has little seasonal dependence at low latitudes.
(4) The global mean rates of both precipitation and evaporation increase.
(5) The coverage and thickness of sea ice in the polar regions decrease.

(6) The snowmelt season arrives earlier.
(7) The annual mean rate of runoff increases at high latitudes.
(8) During summer, the zonal mean value of soil moisture in the Northern Hemisphere reduces in two belts of middle and high latitude, respectively.

The CO_2-induced changes of climate listed above have been identified by comparing two equilibrium climates of a mathematical model with normal and above-normal atmospheric CO_2 concentrations, respectively. Therefore, these changes may be qualitatively different from the transient response of climate to an increase of the atmospheric CO_2 concentration. In this presentation, I have also reviewed the results from a study in which the transient response of climate to a sudden increase of atmospheric CO_2 was investigated by the use of an ocean–atmosphere model with an idealized geography. The results from this preliminary study suggest that the latitudinal distribution of climate change from an equilibrium response study may be qualitatively similar to the transient response of climate, unless the CO_2 increase accelerates much faster than expected.

In order to make reliable assessments of both the equilibrium and the transient responses of climate to the future increase of atmospheric CO_2, it is necessary to further improve various components of the ocean–atmosphere model, in particular the oceanic component. For example, the thermocline as simulated by a current ocean model is too deep at low latitudes. Therefore, it is probable that a sensitivity study with a current model may yield a distorted assessment of the transient response of climate. Recent measurements of transient tracers such as tritium in the oceans provide data which are very suitable for validation of the transient behavior of ocean models. Preliminary studies of this kind have already been started by Sarmiento (1982).

Although the development of the atmospheric component of the model is at a more advanced stage than the oceanic component, it also has many shortcomings. For example, the geographical distributions of hydrologic variables (i.e., precipitation rate) as simulated by a current general circulation model of the atmosphere contain many unrealistic features, particularly in the vicinity of major mountain ranges (i.e., the Rockies, Himalayas, and Andes). This suggests that the dynamical computation of the flow over and around these mountain ranges requires further improvements.

In the global model described here, the distribution of cloud cover is prescribed. Accordingly, use of this global model implies the assumption that cloud cover is unaltered by change in atmospheric CO_2. Although the success of the global model in simulating the seasonal variation of hemispheric mean values of radiative fluxes and surface air temperature encourages the author to believe that cloud cover may not have a dominant influence on the sensitivity of the global mean climate, it is far from obvious that this is the case (Smagorinsky, 1977). This is the subject of active debate in the current literature (Cess, 1976; Ohring and Clapp, 1980). The development of a reliable scheme of cloud cover prediction to be incorporated into a climate model is urgently needed for the study of this critical problem.

The CO_2-induced climate change described in this study has not been identified by observation mainly because such a change is obscured by the change of climate caused by other factors. These factors include the nonlinear interaction between the atmosphere and oceans, the long-term variation of solar luminosity, and the changes in the concentrations of aerosol and other minor atmospheric constituents (i.e., nitrous oxide, chrolofluorocarbon, etc.). Recently, Hansen *et al.* (1981) claimed to have detected the CO_2-induced climate change by separating the portion of the change attributable to some of these factors from the observed change of climate. However, a more careful determination of the contributions of these factors is required before one is convinced of the validity of their conclusion. For this purpose, it is necessary to carefully monitor the solar constant, the atmospheric loading of aerosols, and the spectral distributions of solar and terrestrial radiation fluxes at the top and bottom of the atmosphere. It is clear that the successful assessment of the CO_2-induced climate change requires an evenhanded emphasis upon both modeling research and monitoring of the climate-controlling parameters.

ACKNOWLEDGMENTS

It is a pleasure to acknowledge the major contribution of Mr. R. T. Wetherald, who has collaborated with the author on the CO_2–climate sensitivity study during the past 20 years. Dr. J. Smagorinsky, the Director of the Geophysical Fluid Dynamics Laboratory, instigated and has wholeheartedly supported this long-term project. His foresight and leadership have been indispensable for the success of the project. The author is very grateful to Mr. R. J. Stouffer, Dr. K. Bryan, and Mr. M. J. Spelman with whom he conducted the very fruitful and enjoyable studies described in this review. Prof. J. Peixóto kindly persuaded the reluctant author to write this review. Drs. H. Levy II and J. Sarmiento reviewed the earlier version of the manuscript and made many valuable suggestions for its improvement. Finally, the assistance of Ms. J. Kennedy and Messrs. P. Tunison and J. Conner has been essential for the preparation of the manuscript.

References

Arrhenius, S. (1896). On the influence of carbonic acid in the air upon the temperature of the ground. *Philos. Mag.* [5] **41**, 237–276.

Augustsson, T., and Ramanathan, V. (1977). A radiative, convective model study of the CO_2 climate problem. *J. Atmos. Sci.* **34**, 448–451.

Bryan, K., and Lewis, L. J. (1979). A water mass model of the world ocean. *JGR, J. Geophys. Res.* **84**(C3), 2503–2517.

Bryan, K., Komro, F. G., Manabe, S., and Spelman, M. J. (1982). Transient climate response to increasing atmospheric CO_2. *Science* **215**, 56–58.

Callendar, G. S. (1938). The artificial production of carbon dioxide and its influence on temperature, *Q. J. R. Meteorol. Soc.* **64**, 223–240.

Cess, R. D. (1976). Climate change: An appraisal of atmospheric feedback mechanisms employing zonal climatology. *J. Atmos. Sci.* **33**, 1831–1843.

Climate Research Board (1979). "Carbon Dioxide and Climate: A Scientific Assessment." Nat. Acad. Sci., Washington, D.C.

Crutcher, H. L., and Meserve, J. M. (1970). "Selected Level Heights, Temperature, and Dew Points for the Northern Hemisphere," NAVAIR 50-IC-52. U.S. Nav. Weather Ser., Washington, D.C.

Ellis, J. S., and Vonder Haar, T. H. (1976). "Zonal Average Earth Radiation Budget Measurement from Satellite for Climate Studies," Atmos. Sci. Pap. 240. Colorado State University, Fort Collins.

Gates, W. L., Cook, K. H., and Schlesinger, M. (1981). Preliminary analysis of experiments on the climatic effects of increasing CO_2 with the OSU atmospheric general circulation model. *JGR, J. Geophys. Res.* **86**(C7), 6385–6393.

Geophysics Study Committee (1977). "Energy and Climate." Nat. Acad. Sci., Washington, D.C.

Hansen, J. E. (1979). "Proposal for Research in Global Carbon Dioxide Source/Sink Budget and Climate Effects." Goddard Institute for Space Studies, New York.

Hansen, J. E., Johnson, D., Lacis, A., Lebedeff, S., Lee, P., Rind, D., and Russel, G. (1981). Climatic impact of increasing atmospheric carbon dioxide. *Science* **213**, 957–966.

Hayashi, Y. (1982). Confidence interval of a climate signal. *J. Atmos. Sci.* **39**, 1895–1905.

Held, I. M. (1978). The tropospheric lapse rate and climate sensitivity experiments with a two-level atmospheric model. *J. Atmos. Sci.* **35**, 2083–2098.

Hoffert, M. I., Callegari, A. J., and Hsieh, C. T. (1980). The role of deep sea heat storage in the secular response to climatic forcing. *JGR, J. Geophys. Res.* **85**, 6667–6679.

Kaplan, L. D. (1960). The influence of carbon dioxide variation on the atmospheric heat balance. *Tellus* **12**, 204–208.

Kondratiev, K. Y., and Niilisk, H. I. (1960). On the question of carbon dioxide heat radiation in the atmosphere. *Geofis. Pura Appl.* **46**, 216–230.

Manabe, S. (1971). Estimates of the future changes of climate due to increase of carbon dioxide concentration in the air. *In* "Man's Impact on the Climate" (W. H. Matthews, W. W. Kellogg, and G. D. V. Robinson, eds.), pp. 249–264. MIT Press, Cambridge, Massachusetts.

Manabe, S., and Bryan, K. (1969). Climate calculation with a combined ocean-atmospheric model. *J. Atmos. Sci.* **26**, 786–789.

Manabe, S., and Stouffer, R. J. (1979). A CO_2-climate sensitivity study with a mathematical model of the global climate. *Nature (London)* **282**, 491–493.

Manabe, S., and Stouffer, R. J. (1980). Sensitivity of a global model to an increase of CO_2 concentration in the atmosphere. *JGR, J. Geophys. Res.* **85**(C10), 5529–5554.

Manabe, S., and Wetherald, R. T. (1967). Thermal equilibrium of the atmosphere with a given distribution of relative humidity. *J. Atmos. Sci.* **24**, 241–259.

Manabe, S., and Wetherald, R. T. (1975). The effects of doubling the CO_2-concentration on the climate of a general circulation model. *J. Atmos. Sci.* **32**, 3–15.

Manabe, S., and Wetherald, R. T. (1980). On the distribution of climate change resulting from an increase in CO_2-content of the atmosphere. *J. Atmos. Sci.* **37**, 99–118.

Manabe, S., Bryan, K., and Spelman, M. (1979). A global ocean-atmosphere climate model with seasonal variation for future studies of climate sensitivity. *Dyn. Atmos. Oceans* **3**, 393–426.

Manabe, S., Wetherald, R. T., and Stouffer, R. J. (1981). Summer dryness due to an increase of atmospheric CO_2-concentration. *Clim. Change* **3**, 347–386.

Mitchel, J. F. B. (1979). "Preliminary Report on the Numerical Study of the Effect on Climate of Increasing Atmospheric Carbon Dioxide," Met. O. 20, Tech. Note No. II/137. Meteorological Office, Bracknell, Berkshire, U.K.

Möller, F. (1963). On the influence of changes in the CO_2 concentration in the air on the radiation balance of the earth's surface and on climate. *JGR, J. Geophys. Res.* **68**, 3877–3886.

Newell, R. E., and Dopplick, T. G. (1979). Questions concerning the possible influence of anthropogenic CO_2 on atmospheric temperature. *J. Appl. Meteorol.* **18**, 822–824.

North, G. R., Moeng, F. J., Bell, T. L., and Cahalan, R. F. (1982). The latitude dependence of the variance of zonally averaged quantities. *Mon. Weather Rev.* **110**, 319–326.

Ohring, G., and Clapp, P. (1980). The effect of changes in cloud amount on the net radiation at the top of the atmosphere. *J. Atmos. Sci.* **84**(C4), 3135–3147.

Panofsky, H. A., and Brier, G. W. (1968). "Some Application of Statistics to Meteorology." Penn. State Univ. Press, University Park, Pennsylvania.

Phillips, N. A. (1956). The general circulation of the atmosphere: A numerical experiment. *Q. J. R. Meteorol. Soc.* **82**, 132–164.

Plass, G. N. (1956). The influence of the 15 carbon dioxide band on the atmospheric infra-red cooling rate. *Q. J. R. Meteorol. Soc.* **82**, 310–324.

Ramanathan, V. (1976). Radiative transfer within the earth's troposphere and stratosphere: A simplified radiative-convective model. *J. Atmos. Sci.* **33**, 1330–1346.

Rasool, S. I., and Schneider, S. H. (1971). Atmospheric carbon dioxide and aerosols: Effects of large increases on global climate. *Science* **173**, 138–141.

Sarmiento, J. (1982). A simulation of bomb tritium entry into the Atlantic Ocean. *JGR, J. Geophys. Res.* (in press).

Schneider, S. H., and Thompson, S. L. (1981). Atmospheric CO_2 and climate: Importance of the transient response. *JGR, J. Geophys. Res.* **86**(C4), 3135–3147.

Smagorinsky, J. (1963). General circulation experiments with the primitive equations. I. The basic experiment. *Mon. Weather Rev.* **91**, 99–164.

Smagorinsky, J. (1977). Modeling and predictability. "Energy and Climate." Geophys. Res. Comm., Nat. Acad. Sci., Washington, D.C.

Taljaad, J. J., Van Loon, H., Crutcher, H. L., and Jenne, R. L. (1969). "Climate of the Upper Air. 1. Southern Hemisphere," NAVAIR 50-IC-55. U.S. Nav. Weather Ser., Washington, D.C.

Tyndall, J. (1863). On radiation through the Earth's atmosphere. *Philos. Mag.* [4] **22**, 160–194, 273–285.

Wang, W. C., Yung, Y. L., Lacis, A. A., No, T., and Hansen, J. E. (1976). Greenhouse effect due to man-made perturbation of trace gases. *Science* **194**, 685–690.

Wetherald, R. T., and Manabe, S. (1975). The effect of changing the solar constant on the climate of a general circulation model. *J. Atmos. Sci.* **32**, 2044–2059.

Wetherald, R. T., and Manabe, S. (1981). Influence of seasonal variation upon the sensitivity of a model climate. *JGR, J. Geophys. Res.* **86**, 1194–1204.

Wigley, T. M. L., and Jones, P. D. (1981). Detecting CO_2-induced climate change. *Nature (London)* **292**, 205–208.

Part II

STATISTICAL-DYNAMICAL MODELS

ALMOST EMPIRICAL APPROACHES TO THE PROBLEM OF CLIMATE, ITS VARIATIONS AND FLUCTUATIONS

G. S. Golitsyn

Institute of Atmospheric Physics
Academy of Sciences of the USSR
Moscow, USSR

1. Introduction

The problem of climate and its variations is becoming one of the leading problems in world science, because the variations affect a great variety of human activities of economic and social character. Dozens of conferences, symposia, and workshops have been organized, the Geneva World Climate Conference of 1979 being the most representative in scope and in attending scientists and officials. The World Meteorological Organization (WMO) and the International Council of Scientific Unions (ICSU) adopted the World Climate Program, effective since January 1, 1980. Many countries have national climate programs and a few governments approved corresponding laws. The Preliminary Plan for the World Climate Research Program (WCRP) was issued in January 1981 and adopted by the Joint Scientific Committee of ICSU–WMO at its March 1981 meeting.

The problem of climate and its changes may be the most complicated that science has faced up to now. Its solution requires the efforts of meteorologists and oceanographers, geographers and biologists, chemists and paleoclimatologists, space technologists and applied mathematicians, among others. Observations of the state of the atmosphere, ocean, land surface, and biosphere must be global and require periods of many years

85

and decades before the signal representing the climate change due to anthropogenic sources can be rectified from the noise or natural variability of the climate system. Such observations can be performed only by satellite measurements and require the whole strength of the modern engineering technique and technology. It is evident that the problem can be solved only by the mutual efforts of scientists from many countries.

The earth's climatic system consists of atmosphere, land, ocean, cryosphere, and biosphere. Therefore, a sufficiently complete description should include the consideration of various processes within each of the subsystems as well as multiple processes of interaction between the subsystems. The degree of detail in the description of the processes depends on the time scales in which we are interested. The World Climate Program has made achievement of an understanding of the climate changes during the next few decades a first goal. During this period of several decades the main factor leading to the climate change will be the increase of CO_2 content in the atmosphere due to an increase in fossil fuel burning.

Present (summer 1981) estimates of the mean global surface temperature change, $\overline{\Delta T_s}$, due to doubling of atmospheric CO_2 content are within the limits $3 \pm 1.5°C$ [National Academy of Sciences (NAS), 1979] and the most reliable estimates are now bounded by 2 and 3°C (USSR–USA Workshop, 1981). Although these changes are small from a thermodynamic point of view compared to the mean global temperature $\overline{T_s} = 288$ K, they may lead to large social and economic consequences (e.g., World Climate Conference, 1979; ICSU/WMO/UNEP 1981; Kellogg and Schware, 1981). Therefore, any specification of the limits for $\overline{\Delta T_s}$ is very important.

The envisaged climate changes will take place simultaneously with natural variations of climatic parameters. Therefore, to separate the real changes from the natural variability, one should know the level of "natural noise" of the climatic system. These two questions, the CO_2-induced climate change and the "noise level" of the system, are the main topics of this article. However a considerable part of this article consists of presenting empirical results based on various sets of data. These data sets are usually averaged over the (Northern) hemisphere in order to exclude or damp, to some extent, influences of various transport processes within the system, and are considered in their annual course. There is a hope that by doing this one may get estimates of climatically important relationships. We will present the results of relating the incoming solar and outgoing thermal radiation as measured by satellites to the mean surface temperature, humidity of the atmosphere, and cloudiness. Such relationships, which are important for the so-called energy balance climate models (Saltzman, 1978; North et al., 1981), can also be calculated from more complicated models (GARP, 1975). The comparison of the radiation values with empirically derived ones will serve as an additional check of the model efficiency.

The author believes that the usefulness of this paper is methodological, i.e., in illustrating the use of the simplest balance considerations supplemented by relations determined from empirical data. In the future such data will undoubtedly be more extensive in time and of a better quality, and hence statistically more significant. Nevertheless, even now such an approach can produce results which probably will not change qualitatively.

2. Determination of Mean Climate Characteristics

The simplest result in climate theory is the estimate of the globally averaged temperature corresponding to the outgoing thermal radiation of a planet:

$$T_e = [Q(1 - A)/\sigma]^{1/4} = (Q_A/\sigma)^{1/4} \qquad (2.1)$$

Here $4Q$ is the solar constant for the planet, A the planetary integral albedo, σ the constant in the Stefan–Boltzmann radiation law. For the earth at the mean distance from the sun, $4Q = 1376$ W m^{-2} (Hickey et al., 1980) and $A = 0.3$ (Ellis and Vonder Haar, 1976). Averaging the solar radiation over the planet surface, one gets $Q(1 - A) = Q_A = 241$ W m^{-2}. According to Eq. (2.1) this gives $T_e = 255$ K.

The uncertainty in our knowledge of the value of Q is still of order of 1%, and the uncertainty of the value of A is of the order of several percent because one may still find in the literature a value of albedo in the range from 0.29 (Raschke, 1975) to 0.32 (Freeman and Liu, 1979). These uncertainties lead to a dispersion in the value of Q_A within the limits 230–244 W m^{-2}. This dispersion is significant because in the problem of determining the earth's surface temperature change due to doubling of the atmospheric CO_2 content, we shall deal (Section 6) with the changes in the energy inflow to the troposphere and surface of only about 4 W m^{-2} (NAS, 1979; Ramanathan et al., 1979; Ackerman, 1979). The situation is saved to some extent by the fact that we shall be interested only in the *changes* of the value of the energy inflow and the basic state.

An important problem still remains of determining the value of T_s itself. It is now solved by semiempirical tuning of various parameterizations when calculating the "greenhouse" effect, the measure of which is the quantity

$$\alpha = T_e/\bar{T}_s \qquad (2.2)$$

88   G. S. GOLITSYN

For $T_e = 255$ K and $\bar{T}_s = 288$ K the value of α is 0.885. The effective temperature of outgoing radiation T_e can serve as a measure of the mean global surface temperature T_s. However, the very meaning of this ancient Greek word "climate" assumes a dependence of the insolation on "latitude." A measure of the real climate is the temperature difference δT between equator and pole. Evidently, the value of δT is determined not only by the amount of solar radiation reaching the polar regions but also by the thermal inertia of the atmosphere and underlying surface and by heat transport within the atmosphere and ocean.

An attempt at a general approach to the problem was undertaken by the author in the late 1960s (Golitsyn, 1968, 1970, 1973). These studies were inspired mainly by a desire to understand the thermal and dynamic regimes of the atmospheres of Venus and Mars. However, due to the generality of the approach, important results were obtained for the earth also. The bases of the approach were similarity and dimensional theory methods supplemented by some thermodynamic arguments. The quantitative calibration of the theoretical results, i.e. the determination of some nondimensional constants that always arise in such an approach, was done using empirical data for the earth's atmospheric circulation (Oort, 1964).

The analysis of the equations describing the atmospheric circulation (Golitsyn, 1973) shows that, properly nondimensionalized, these equations have three main similarity criteria constructed only from the external parameters:

$$\Pi_\omega = \omega r/c_e, \qquad c_e = (\kappa R T_e/\mu)^{1/2}, \qquad \kappa = c_p/c_v \qquad (2.3)$$

$$\Pi_g = H/r, \qquad M = R T_e/\mu g \qquad (2.4)$$

$$\Pi_M = \tau_e/\tau_0, \qquad \tau_e = r/c_e, \qquad \tau_0 = c_p T_e M/Q_A \qquad (2.5)$$

Here ω is the angular velocity of the planet rotation, r its radius, c_e the sound velocity, H the atmospheric scale height, M the mass of an atmospheric unit column, τ_e the time for reaching local thermodynamic equilibrium on the global scale, τ_0 the characteristic time for the thermal inertia of an atmosphere. For all planets Π_g and Π_M are much less than unity. Therefore, patterns of atmospheric circulation are mainly determined by the value of the rotational similarity criterion, Π_ω, the so-called rotational Mach number. Its value is small for Venus, of order unity for Earth and Mars, and large for the giant planets.

Let us assume self-similarity with respect to all three similarity criteria if they are small (this is true only for Venus). This means that their exact values are not essential (at least for some circulation characteristics), or that the exact values of ω, g, and M are not important. Then dimensional arguments give the following formula for the total kinetic energy of circulation

$$E = 2\pi B_1 \sigma^{1/8} c_p^{-1/2} Q_A^{7/8} r^3 \sim 4\pi r^2 Q_A T_e \qquad (2.6)$$

where B_1 is a constant. For the mean wind velocity one obtains

$$U = B_1^{1/2}(\kappa - 1)^{1/2} \Pi_M^{1/2} c_e \qquad (2.7)$$

From the energy equation one may estimate

$$\delta T \sim \Pi_M^{1/2} T_e \qquad (2.8)$$

For Venus these formulas imply winds of about 1 m sec^{-1} and for the equator–pole temperature difference $\delta T \sim 1$ K. These numbers have been confirmed in the 1970s by direct measurements within the main (by mass) lower bulk of the atmosphere.

For Earth and Mars one should take rotation into account. This means that Eq. (2.6) should be multiplied by a function $f(\Pi_\omega)$. An analysis of data for the general circulation of the earth's atmosphere and of numerical experiments for the Martian atmosphere showed that $f(\Pi_\omega) = 1 + \Pi_\omega^2$ for $\Pi_\omega \sim 1$ (Golitsyn, 1973). Then Eq. (2.7) gives $U = 17$ m sec^{-1} (at $B_1 = 0.3$[1]) which agrees with empirical data from Oort (1964). Similarly from (2.8) one can get $\delta T \approx 45$ K.

Equations (2.6)–(2.8) give estimates of wind velocity and the temperature difference as functions of external parameters of a planet and its atmosphere. The specification of various coefficients such as B_1 in Eq. (2.6) or α in Eq. (2.2) is the end result of a vast number of internal processes that require all the physical and computing power of modern science to represent. It is, evidently, impossible to succeed on purely theoretical grounds without invoking various parameterizations validated by observational data. Nevertheless, the approach using the similarity and dimensional arguments supplies an orientation and some understanding of the diversity of planetary climates without which one cannot understand principal features of our own climate. The importance of the last circumstance was stated by the GARP Climate Study conference in 1974 (GARP, 1975).

3. Sensitivity of Climate to Variations of External Factors

In the problem of the climate change an important question is the determination of the sensitivity of climate, i.e., of the rate and intensity of the

---

[1] This value is the only constant of the theory obtained from empirical estimates of the atmospheric kinetic energy dissipation rate (Oort, 1964).

response of the climatic system to the change of any external parameter. Such a response of a model should be calibrated using empirical data or results of more complicated models. A popular test in the 1970s for various climate models was the calculation of the change of global mean temperature $\bar{T}_s$ in response to a change of the solar constant Q, say, by 1%. The scale of the change may be estimated using Eq. (2.1), i.e., $\Delta T = T_e\, \Delta Q/4Q$, or, if the greenhouse effect and albedo are not changing, then Eq. (2.2) gives

$$\Delta \bar{T}_s = \alpha^{-1} T_e\, \Delta Q/4Q = \bar{T}_s\, \Delta Q/4Q \tag{3.1}$$

At $\bar{T}_s = 288$ K and $\Delta Q/Q = 10^{-2}$ one gets $\Delta \bar{T}_s = 0.72$ K. This estimate is obtained without taking into account the response of the atmosphere and ocean (e.g., we assumed a fixed albedo). Some idea of such a response, at least a lower-bound estimate of its magnitude and time scale, may be obtained by considering the annual course of the mean hemispheric surface temperature $\bar{T}_s$. During this annual course (see Appendix) the insolation of the Northern Hemisphere changes by about 29%, and for the Southern Hemisphere the change in insolation reaches 42%. The variation of $\bar{T}_{sNH}$, from January to July reaches 14° (Crutcher and Meserve, 1970). We obtain from this that changing the solar flux F by 1% leads to a change in $\bar{T}_{sNH}$ of 0.2 K. This value is only about one-fourth of the one calculated from Eq. (3.1). Much smaller values are obtained for the Southern Hemisphere, where the temperature $\bar{T}_{sSH}$ changes only by 6° (Taljaard et al., 1969).

The main difference between the two hemispheres is that the area of ocean is considerably larger in the Southern Hemisphere than in the Northern Hemisphere. This circumstance shows the importance of the ocean in the earth's climate formation and in its changes. The Southern Hemisphere requires separate and thorough investigation.

All the differences show that the annual course of the surface temperature $\bar{T}_s$ alone cannot serve as an analog for estimates of climatic changes, as is still sometimes proposed. The cause of this is evident: it is the thermal inertia of the ocean. The calculation just made demonstrates that the time constant for the ocean thermal inertia must be larger than half a year.[2] For the atmosphere the time scale of the thermal inertia is the time τ_0 defined in Eqs. (2.5). For $c_p = 10^3$ J kg^{-1} K^{-1}, $M = 10^4$ kg m^{-2}, $T_e = 255$ K, and $F\downarrow = Q_A = 240$ W m^{-2} one gets $\tau_0 = 10^7$ sec $= 4$ months. This is a characteristic time of the total cooling of the atmosphere; i.e., to change the temperature by, say, 15 K the time should be multiplied by 15/

[2] The cause for changing T_s is not so much the course of the insolation itself as the annual course of the intensity of the outgoing thermal radiation, for which the amplitude is less than half that for solar flux. But this is an observational fact which required years of satellite measurements for its establishment.

255, which gives 6×10^5 sec = 1 week. However, this estimate does not take into account the interaction of the atmosphere with the ocean, which is a powerful heat reservoir.

Let us consider the time scales for the ocean. The part with the least inertia is the upper mixed layer (UML) directly interacting with the atmosphere. Its depth has been determined to be about 70 m on the average (see Manabe and Stouffer, 1980). Let us estimate the time required to change its temperature by 15 K while changing the insolation by 30%, i.e., $\Delta F \downarrow$ = 72 W m^{-2}. This time is $\tau = c_p M \, \Delta T / \Delta F \downarrow$ = 6×10^7 sec ≈ 2 yr. It follows from this that amplitudes of the seasonal changes of temperature for the UML averaged over the hemisphere should be on the order of a few degrees, which agrees well with the observational data (see Monin, 1975).

The UML is connected with the ocean main thermocline, a layer of about 700-m depth, by the mechanism of the Ekman pumping. Under the influence of the Coriolis force, large subtropical gyres are formed where the water acquires a vertical velocity component of order 10–20 cm day^{-1} (NAS, 1979), i.e., of order 50 m yr^{-1}. Therefore, the time for the UML to reach an equilibrium, thermal or chemical, with the main thermocline will be about 15 yr. This will lead to a phase shift between an increased level of atmospheric CO_2 and sea surface temperature (SST). For instance, in the Report of the USSR–USA Workshop (1981) on CO_2 and climate it is stated, on the basis of several model studies, that over 15 yr the value of $\bar{T}_s$ (understood to be the SST) may reach about a half of its equilibrium value.

Because the ocean is heated from above, it is almost everywhere stably stratified. Therefore, the turbulence within it is of small scale and intensity and of great intermittance. As a result the coefficient of the vertical turbulent mixing K is only of order 1 cm^2 sec^{-1} (Östlund et al., 1974). Therefore, the time scale for mixing the whole ocean is $\tau_{oc} \sim l^2 / K = 5000$ yr at the mean ocean depth $l = 4$ km. If one takes into account the downwelling of cold waters in high latitudes and upwelling in lower latitudes, then this time scale could be decreased by several times (see ICSU/ WMO, 1981), i.e., it is of order 1000 yr.

The last characteristic time scale in the earth climate system which can be estimated by simple means is that of the cryosphere. The ice sheets of Greenland and Antarctica have a height of about 2 km. The North American and Scandinavian ice sheets during the last ice age had similar heights (Budyko, 1980). The heat of melting for ice is $L = 3.3 \times 10^5$ J kg^{-1}, and the net radiation balance in summer in high latitudes is positive and reaches 60 W m^{-2}. With this value of the energy inflow, to melt 2 km of ice would require 300 yr. This is evidently only a far lower limit because 60 W m^{-2} acts only in summer but in winter the sheets grow again. As a result the sheets during ice ages grow and melt during periods of order 10^4 yr although times of melt are always several times less than the times of their formation.

The closeness of characteristic times for ice sheets and for deep ocean suggests a thought: is not an interaction of these two subsystems an important factor in the origin of ice ages? An attempt to construct a quantitative model for this interaction was undertaken recently by Saltzman (1977).

Let us return, however, to estimates of the climate sensitivity and, in particular, to the estimate of the change of the mean surface air temperature, $\Delta \bar{T}_s$, in response to a change of the radiation $\Delta F\downarrow /F\downarrow$ by 1%. A detailed review of the problem may be found in the book by Budyko (1980). The first realistic estimate of the value $\Delta \bar{T}_{s1} = (d\bar{T}_s/dF\downarrow)\Delta F\downarrow$ for $\Delta F\downarrow /F\downarrow = 1\%$ was obtained by Manabe and Wetherald (1967). They took into account the increase of absolute humidity with the growth of temperature and its influence on the outgoing thermal radiation and obtained $\Delta \bar{T}_{s1} = 1.2$ K. This relationship between absolute humidity and near-surface temperature is one of the basic processes in the climate problem. The values of $\Delta \bar{T}_{s1}$ obtained by Budyko in 1968–1979 (see his book of 1980) using his energy balance model and data on climatic fluctuations (by correlation of atmospheric turbidity with the changes of $\bar{T}_s$) are all within the limits 1.1–1.5 K. Cess (1976) obtained the somewhat higher value 1.7 K after estimating the intensity of the feedback between albedo and temperature. Two years later Cess (1978), using paleoclimatic reconstructions, estimated changes of albedo for the underlying surface caused by changes in vegetation and found the value of $\Delta \bar{T}_s$ to be closer to 3 K. We shall come back to this effect in Section 4, because the Cess (1978) value appears to be somewhat overestimated. Wetherald and Manabe (1975) found the value to be 1.5 K in their general circulation model (GCM) with the ocean of small heat capacity (so-called "swamp").

The most essential concern in the simple energy balance approach to the problem of climate and its change is the parameterization of intensity of the outgoing thermal radiation $F\uparrow$ as a function of the temperature of the near-surface air. Budyko (1968, 1969) was the first to suggest the linear parameterization of the form

$$F\uparrow = A_f + BT_s \qquad (3.2)$$

Satellite data (Ellis and Vonder Haar, 1976) and temperature data (Crutcher and Meserve, 1970) allow one to find empirical values of the coefficients A_f and B.

To check the empirical equation (3.2), we used two versions of the same observational data. First, satellite data of $F\uparrow$ were averaged for the whole Northern Hemisphere for every month and compared with the temperature data averaged in similar fashion. The linear regression of $F\uparrow$ on $\bar{T}_s$ produced $A_f = 204.2$ W m^{-2} and $B = 1.944$ W m^{-2} K^{-1} with the correlation coefficient $r = 0.830$.

Then the same regression was taken between values of $F\uparrow$ averaged throughout the year over 9 zones of 10° width each and similarly averaged values of T_s. This produced $A = 203.9$ W m^{-2} and $B = 1.986$ W m^{-2} K^{-1} with $r = 0.987$. The regression between all the points of the two sets gave $A_f = 203.9$ W m^{-2} and $B = 1.987$ W m^{-2} K^{-1} with $r = 0.969$.

Therefore, with satisfactory correlation and accuracy we shall adopt for further use

$$F\uparrow = 204 + 2T_s \quad \text{W m}^{-2} \tag{3.3}$$

The use of empirical relationship (3.2) or (3.3) instead of the Stefan–Boltzmann law $F\uparrow = \sigma T^4$ can be explained by the fact that the formation of the thermal radiation leaving the real atmosphere is influenced mainly by the moisture. Because there is less moisture in higher latitudes than in lower latitudes, the effective level of radiation is closer to the surface in higher latitudes (Budyko, 1980; North *et al.*, 1981). This decreases the meridional gradient of the local temperature for outgoing radiation. As a result Eq. (3.2) may be considered simply as the first two terms of a Taylor expansion of the exact radiation law, or as a simple empirical regression of values of $F\uparrow$ on T_s. Note that for the Southern Hemisphere Eq. (3.3) will have different values of A_f and B (Cess, 1976; North *et al.*, 1981).

Now we may return to the question of the change of the mean global (Northern Hemisphere) temperature $\bar{T}_s$ with the change of the solar constant. Because the value of $F\uparrow$ averaged throughout the year is equal to incoming solar radiation $F\uparrow = Q_A$, from Eq. (3.2) it follows that

$$\Delta \bar{T}_s = \Delta Q (dF\uparrow/dT_s)^{-1} = \Delta Q(1 - A)/B \tag{3.4}$$

For $\Delta Q/Q = 10^{-2}$, $\Delta F\downarrow = 2.4$ W m^{-2}, and $B = 2$ W m^{-2} K^{-1} we obtain at once $\Delta \bar{T}_{s1} = 1.2$ K, in accord with Budyko (1980) and Manabe and Wetherald (1967). In this calculation we neglected the change of albedo, cloudiness, and other factors, which we shall consider in Sections 4 and 5.[3]

4. EMPIRICAL ESTIMATES OF FEEDBACK BETWEEN ALBEDO AND TEMPERATURE OF UNDERLYING SURFACE

Many feedbacks act in the climatic system. Some of them are positive, others are negative, i.e., they increase or attenuate the response of the

[3] The same procedure using the more complete satellite data set by Stephens *et al.* (1981) produced $A = 204$ W m^{-2} and $B = 1.8$ W m^{-1} K^{-1}, which makes the climate somewhat more sensitive.

system. A review and qualitative description of many of the feedbacks may be found in Schneider and Dickinson (1974). Even the most complicated GCMs now used for climate studies do not take into account all the diversity of these relationships. Therefore, the most reliable way to estimate them now is from empirical data, though here one can also encounter "underwater stones" connected, for example, with differences of temporal and spatial scales of causes and responses of the system. In order to decrease the amplitudes of such effects it is safer to study strongly averaged values, for instance, monthly means of a parameter averaged over a hemisphere, in trying to establish its dependence, say, on $\bar{T}_s$, as was done in determining Eq. (3.3).

Of all feedbacks acting in the climatic system the most well known is the relationship between albedo A of the system and temperature of the underlying surface T_s. The greater the albedo A, the more solar radiation is reflected back to space, the lower is T_s, and the greater is the area of the snow–ice cover. Clearly it is a positive feedback. However, recent numerical experiments by Manabe and Stouffer (1980) and Wetherald and Manabe (1981) showed the action of this mechanism is not as simple as may appear at first sight. The seasonal behavior is an important factor here and the matter requires further thorough investigation.

Let us look at what observational data show. In Table I we present the values obtained by Mokhov (1981) for the albedo A averaged over the Northern Hemisphere, based on data obtained by Ellis and Vonder Haar (1976), and corresponding temperatures $\bar{T}_s$ for every month of the year obtained by Crutcher and Meserve (1970). The values of $\bar{A}$ were calculated from the expression

$$\bar{A}(t) = \int_0^1 A(x, t)S(x, t)\,dx \tag{4.1}$$

where $A(x, t)$ is the albedo of a zone with a width of 10° latitude, $x = \sin\theta$, θ is latitude, t is time, and

$$S(x, t) = 1.24 - 0.72 \sin^2\theta + 0.80 \sin\theta \cos 2\pi t \tag{4.2}$$

Function (4.2) describes the distribution of the insolation with latitude and time which is counted in fraction of a year from the moment of the (North-

TABLE I. NORTHERN HEMISPHERE AVERAGES OF ALBEDO AND SURFACE AIR TEMPERATURES T_s FOR DIFFERENT MONTHS[a]

Month	$\bar{A}$	$\bar{T}_s$ (°C)	Month	$\bar{A}$	$\bar{T}_s$ (°C)
1	0.2954	7.92	7	0.3105	21.55
2	0.3048	8.69	8	0.2992	21.30
3	0.3055	10.77	9	0.2872	19.48
4	0.3116	14.12	10	0.2807	16.16
5	0.3289	17.46	11	0.2924	12.22
6	0.3271	20.12	12	0.2990	9.37

[a] Calculated after Ellis and Vonder Haar (1976) and Crutcher and Meserve (1970).

ern) summer solstice (North *et al.*, 1981). The calculations by I. I. Mokhov and those by K. Ya. Vinnikov (private communication, June 1981) give values of $\bar{A}$ which seldom differ from unity in the fourth decimal place.

The linear regression

$$\bar{A} = A_0 - A_1 \bar{T}_s, \qquad A_1 = -d\bar{A}/d\bar{T}_s \qquad (4.3)$$

has the maximum correlation coefficient at a lag of $\bar{A}$ relative to $\bar{T}_s$ of 3 or 4 months. For a lag of 3 months we obtain $A_0 = 0.339$, $A_1 = 0.00240$ K^{-1} with the correlation coefficient $r = 0.84$, and for 4 months $A_0 = 0.340$, $A_1 = 0.00246$ K^{-1} with $r = 0.86$. For time lags $\Delta t = 2$ or 5 months, $r = 0.60$ and 0.65, respectively. The lag effect is due to the thermal inertia of the snow, mainly of the Arctic ice cap. In any case, in the annual course it probably causes a weakening of the relationship between albedo and temperature (see Mokhov, 1981).

Now we estimate the influence of the temperature dependence of albedo (4.3) on the incoming solar radiation $F\downarrow = Q[1 - \bar{A}(\bar{T}_s)]$, from which we have (disregarding the time lag)

$$\frac{dF\downarrow}{d\bar{T}_s} = -Q \frac{d\bar{A}}{d\bar{T}_s} = QA_1 \qquad (4.4)$$

With $A_1 = 0.0024$ K^{-1} and $Q = 340$ W m^{-2} we obtain $dF\downarrow/d\bar{T}_s = 0.8$ W m^{-2} K^{-1}. This value can be compared with data of paleoclimatic reconstructions by Efimova (1980). She estimated values of the annual mean planetary albedo for a warm epoch, the early Pliocene (about 6 m.y. B.P.), and for the maximum of the last glaciation (18,000–20,000 yr B.P.). These reconstructions took into account changes in the extent and duration of the snow–ice cover and changes in the character of the vegetation cover (the albedo of vegetation also changes with time of a year). Cess (1978) was the first to consider this effect but Efimova argues that he overestimated it somewhat.

According to Efimova, during the warm period, when mean global temperature was 4.6–4.8 K higher than at present, the planetary albedo was 0.0104 lower than now. From these numbers we obtain $d\bar{A}/d\bar{T}_s = -A_1 \approx \Delta A/\Delta \bar{T}_s = -0.0022$ K^{-1}. For the cold period the value of A was estimated to be 0.0215 but an appreciable part of the change (0.0090) was due to growth of sea ice in the Southern Hemisphere. We shall consider only the Northern Hemisphere, for which we obtain $A = 0.0125$. A number of estimates have been given for the mean hemispheric surface temperature of the Northern Hemisphere (see, e.g., Budyko, 1980) during the maximum of the last glaciation. It appears that it was some 5 K less than now. Then we get $d\bar{A}/d\bar{A}_s \approx 0.0125/(-5 \text{ K}) = -0.0025$ K^{-1}.

Such a close agreement of paleoclimatic estimates of $d\bar{A}/d\bar{T}_s$ with the value found from the annual course may be fortuitous because of a number of natural uncertainties in paleoreconstructions. Nevertheless, the order of magnitude of the value $d\bar{A}/d\bar{T}_s$ hopefully is determined correctly. It may be compared to the value $dA/d\bar{T}_s = -0.0016$ K^{-1} obtained by Cess (1976) by treating the results of numerical experiments by Wetherald and Manabe (1975), who studied the changes of their model climate due to the solar constant changes.

Unfortunately, as G. Ohring noted, our estimates of the value $dA/d\bar{T}_s$ using the annual course of A and $\bar{T}_s$ include the dependence of A on the mean solar zenith angle the cosine of which varies for the hemisphere from 0.55 to 0.41. The exclusion of such a dependence requires the construction of a model for albedo of the atmosphere–underlying surface system. Such an exclusion should appreciably decrease the value determined from the annual course. Before this is done it is difficult to judge how the value of $dA/d\bar{T}_s$ determined from the annual course will be related to the value resulting from real climate change which we are seeking. Possibly our value, 0.0024 K^{-1}, may be considered as an upper limit, and one should have this in mind in further use of it. Note, however, that the time lag between T_s and A in the annual course is a real phenomenon not related to the zenith angle.

5. EMPIRICAL ESTIMATES OF THE ROLE OF CLOUDS IN RADIATION BALANCE

Practically all estimates of changes in the temperature regime obtained up to now (except Petukhov et al., 1975; Wetherald and Manabe, 1980) have been made with fixed cloudiness. The influence of clouds on the radiation balance is an important and contradictory problem. In the Preliminary Plan for WCRP (ICSU–WMO, 1981) the problem was listed first among several others. Various mechanisms by which the cloud influences the energy balance (e.g., water content, height of the upper boundary, cloud amount, and distribution with altitude; see Schneider and Dickinson, 1974; Paltridge, 1980) give results of different sign and amplitude. Therefore, the answer should be sought first of all in analyzing observational data.

However, due to data incompleteness, insufficient length of cloud time series, and difficulties in interpreting the satellite observations, for examples, even when working with empirical material one must use some additional hypotheses. A number of such empirical studies have been performed but their results differ, suggesting the need for additional analysis.

Cess (1976), using the data of Ellis and Vonder Haar (1976), found that the changes in cloud amount averaged over the (Northern) hemisphere do not influence the net radiation balance at the top of the atmosphere. This confirms the conclusions obtained by Budyko (1980) at the end of the 1960s. Later a similar conclusion was reached by Manabe and Wetherald

(1980) and by Wetherald and Manabe (1980), based on a GCM with fully interacting clouds subject to changed CO_2 content in the atmosphere and changed values of the solar constant. The conclusion of these two papers is that although the influence of clouds on the radiation balance is high locally, it is small when averaged globally. At the same time Ohring and Clapp (1980) and Hartmann and Short (1980), using the NOAA satellite data for 1974–1978, concluded that the increase of the cloud amount will cool the atmosphere.[4]

The procedures used in all these papers can be debated. For instance, Cess (1976) performed his correlation analysis not temporally for the annual course, but spatially using the mean annual data for various latitudinal zones. But the interlatitudinal changes of variables do not necessarily correspond to climatic changes because there are changes with latitude of geography, circulation patterns, and the cloud distribution. However, as we showed in our analysis in Section 3 of the dependence of the outgoing thermal radiation on the near-surface temperature, the effect of cloudiness (at least on the outgoing radiation) differs little in estimates obtained from the annual course of hemispheric characteristics and from annual averages of various latitudinal belts. Ohring and Clapp (1980) did not use direct data on clouds, but rather some indirect characteristics depending on cloudiness. Hartmann and Short (1980) do the same, also using an assumption that is not evident for climatic studies, i.e., that the variations of the radiation balance are determined mainly by variations in the cloud amount.

Here we shall describe results of an analysis of the role of cloudiness in the radiation balance, performed by Mokhov with the author's participation. Monthly mean empirical data will be used averaged over the Northern Hemisphere.

Schneider (1972) proposed the following partial derivative as a measure of the influence of cloudiness on the net radiation balance at the top of the atmosphere, $\Delta F = F\downarrow - F\uparrow$:

$$\delta = \partial(F\downarrow - F\uparrow)/\partial n \qquad (5.1)$$

The other variables, temperature and humidity, for example, are fixed. Components of the radiation balance at the top of the atmosphere are determined by Ellis and Vonder Haar (1976). From results of numerical experiments by Manabe and Wetherald (1980) and by Wetherald and Manabe (1980) it follows that variations of the thermal radiation $\delta F\uparrow$ are mainly determined by variations of temperature $T_s = T$, humidity q, and cloudiness n. Therefore, in the linear approximation

$$\delta F\uparrow = a_1 \delta T + a_2 \delta n + a_3 \delta q \qquad (5.2)$$

---

[4] A discussion of the different satellite data sets on radiation balance and effects of clouds may be found in Cess et al. (1982).

where $\delta X = X - \bar{X}$, $\bar{X}$ is the annual mean of X, and

$$a_1 = \frac{\partial F\uparrow}{\partial T}\bigg|_{n,q}, \qquad a_2 = \frac{\partial F\uparrow}{\partial n}\bigg|_{T,q}, \qquad a_3 = \frac{\partial F\uparrow}{\partial q}\bigg|_{T,n} \qquad (5.3)$$

the subscripts indicating parameters that are held constant.

From the above numerical experiments the variations of the solar radiation $F\downarrow$ absorbed by the system are found to depend mainly on cloud amount, albedo of underlying surface A_s, and atmospheric humidity. Instead of A_s we take the temperature of the surface $T_s = T$ as a basic independent variable. All these variables determine the total planetary albedo A of the earth. Because $F\downarrow = Q(1 - A)$,

$$\delta F\downarrow = - Q\, \delta A, \qquad \delta A = b_1\, \delta T + b_2\, \delta_n + b_3\, \delta q \qquad (5.4)$$

The quantities b_i are determined similarly to a_i in Eqs. (5.3):

$$b_1 = \frac{\partial A}{\partial T}\bigg|_{n,q}, \qquad b_2 = \frac{\partial A}{\partial n}\bigg|_{T,q}, \qquad b_3 = \frac{\partial A}{\partial q}\bigg|_{T,n} \qquad (5.5)$$

When analyzing the empirical data one should have in mind the possibility of a time lag between the variations of the value of A and the variations of its arguments in Eqs. (5.4), as was found for the empirical dependence $\bar{A} = \bar{A}(\bar{T}_s)$ in the previous section.

Before coming to a determination of a_i and b_i, we estimate the quality and mutual consistency of data on the cloud amount and its variation, from different sources. We shall use the data compilations by (1) Berlyand and Strokina (1980), (2) Curran et al. (1978), (3) Hoyt (1976), and (4) Thompson (1979). They all present monthly means for various latitudinal belts and from this one can calculate the mean hemispheric cloud amount $\bar{n}$, which varies from 0.6 for (1) to 0.4 for (3). However, the variations δn in their annual course are well correlated with each other. For instance the correlation coefficient of δn from (1) and (4) is equal to $r_{1,4} = 0.98$. Similarly $r_{1,3} = 0.84$ and $r_{3,4} = 0.96$. In the results of calculations described below we adopt the cloudiness values from (1), the temperature from Crutcher and Meserve (1970), and the near-surface specific humidity from Oort and Rasmusson (1971).

Using multiple linear regression (5.2), we found

$$a_1 = \frac{\partial F\uparrow}{\partial T}\bigg|_{n,q} = 6.12(\pm 0.13) \quad \text{W m}^{-2}\,\text{K}^{-1}$$

$$a_2 = \frac{\partial F\uparrow}{\partial n}\bigg|_{T,q} = 270(\pm 16) \quad \text{W m}^{-2} \qquad (5.6)$$

$$a_3 = \frac{\partial F\uparrow}{\partial q}\bigg|_{T,n} = -15.6(\pm 0.4) \quad \text{W m}^{-2}\,(\text{g/kg})^{-1}$$

The standard errors of the estimates are shown in parentheses. The correlation coefficient of the multiple linear regression is $r = 0.93$.

For the linear regression (5.4) the multiple regression correlation coefficient is maximal when δA is shifted by 3 months relative to its arguments, and at that lag the standard deviation of b_2 is minimal. Thus, the situation reported in Section 4 concerning the dependence of the albedo on T_s is repeated. One may say that such an effect can be attributed to inertia in the annual course of the high-albedo cryosphere and to the seasonal changes in the albedo of vegetation. With such a time lag

$$b_1 = \left.\frac{\partial F\uparrow}{\partial T}\right|_{n,q} = 0.859(\pm 0.035) \quad \text{W m}^{-2}\,\text{K}^{-1}$$

$$b_2 = \left.\frac{\partial F\uparrow}{\partial n}\right|_{T,q} = 278(\pm 1) \quad \text{W m}^{-2} \tag{5.7}$$

$$b_3 = \left.\frac{\partial F\uparrow}{\partial q}\right|_{n,T} = 1.87(\pm 0.10) \quad \text{W m}^{-2}\,(\text{g/kg})^{-1}$$

The values of partial derivatives in Eqs. (5.6) and (5.7) agree with the values of full derivatives $dF\uparrow/d\bar{T}_s = 2.0$ W m^{-2} K^{-1} and $dF\downarrow/d\bar{T}_s = 0.8$ W m^{-2} K^{-1} obtained in Sections 3 and 4. In fact,

$$\frac{dF\uparrow}{dT} = \left.\frac{\partial F\uparrow}{\partial T}\right|_{n,q} + \left.\frac{\partial F\uparrow}{\partial n}\right|_{T,q}\frac{dn}{dT} + \left.\frac{\partial F\uparrow}{\partial q}\right|_{T,n}\frac{dq}{dT}$$
$$= a_1 + a_2\frac{dn}{dT} + a_3\frac{dq}{dT} \tag{5.8}$$

$$-Q\frac{dA}{dT} = \frac{dF\downarrow}{dT} = \left.\frac{\partial F\downarrow}{\partial T}\right|_{n,q} + \left.\frac{\partial F\downarrow}{\partial n}\right|_{q,T}\frac{dn}{dT} + \left.\frac{\partial F\downarrow}{\partial q}\right|_{T,n}\frac{dq}{dT}$$
$$= b_1 + b_2\frac{dn}{dT} + b_3\frac{dq}{dT} \tag{5.9}$$

The value of dn/dT determined by Mokhov (1981) is 0.004 K^{-1}, the value of dq/dT can be calculated using the data of Oort and Rasmusson (1971) and of Crutcher and Meserve (1970) and is 0.333 (g/kg) K^{-1}. Substituting all these values into Eqs. (5.8) and (5.9), we obtain $dF\uparrow/dT = 1.96$ W m^{-2} and $dF\downarrow/dT = 0.83$ W m^{-2}, which agree very well with the values $1.98 \approx 2$ and 0.8 W m^{-2} found directly for the full derivatives. This analysis quantitatively demonstrates, perhaps for the first time, the role of moisture which is found to be important.

We see that in spite of the large sensitivity of both components of the radiation balance to the changes of the cloud amount, their contribution to the value $\delta = b_2 - a_2 = 278 - 270 = 8$ W m^{-2} is compensated with the 3% accuracy, which is below the 5% accuracy of the satellite measurements.

This conclusion supports the results of Cess (1976) and of Wetherald and Manabe (1980), although one should consider the annual mean balance only in which there is a cancellation of the two effects of clouds on the solar and thermal radiation.

It is evident that for a more reliable determination of the value of δ a longer series of better measurements is necessary. But now it seems plausible that on a global annually averaged scale the influence of clouds on the net radiation balance is not large. However, for lesser spatial and time scales such a conclusion cannot be extended. For local climate and its change the influence of clouds might be quite important.

6. Sensitivity of Climate to Changes of Atmospheric CO_2 Content

The history of the CO_2 problem, the climate change related to its variations, is about 120 yr old already and connected with the names of many great scientists. The first to note that changes in CO_2 concentration in air can cause variations of climate was Tyndall (1861). Afterward, near the turn of the century Arrhenius (1896, 1903) and Chamberlin (1897, 1898, 1899) propounded a hypothesis that the cause of the quaternary glaciations could be the change of the CO_2 content in the atmosphere.

The first modern statement of the problem was given by Callendar (1938). He proceeded from the fact that the increase in burning of the fossil fuels would increase the emission of CO_2 into the atmosphere. He noted that the ocean could not absorb the excess CO_2 from the atmosphere. By his estimates, about one-quarter of the emitted CO_2 could be left in the atmosphere. The present estimates increase this atmospheric part to about one-half [more precisely to $\frac{22}{39} = 56.5\%$, see USSR–USA Workshop Report (1981)], and in the future, due to saturation of the warming ocean, this part can only increase further. Nevertheless, in retrospect Callendar's estimate seems more than satisfactory. He was the first to account for the fact that the growth of the near-surface temperature due to the increase in CO_2 content should enhance the evaporation, which would increase the greenhouse effect. As we now know, this is the leading feedback in the whole problem. Using a very simple radiation model, Callendar estimated that on doubling the CO_2 concentration in the atmosphere from 300 to 600 ppm the mean surface temperature would increase by 1.6 K. One should consider this result as a global mean because he neglected (consciously, and with regret) dynamical factors and cloudiness. Now we know that this estimate is rather close to the results of exceedingly more complicated numerical models that take into account modern data on the absorption band structure for CO_2 and H_2O, the large-scale atmospheric dynamics, and air–sea interaction.

Callendar was in error only on one point. He believed that the rate of CO_2 emission into the atmosphere would not grow markedly and his Table VI gave 330 ppm for the 21st century (a value achieved in 1973) and 360 ppm for the 22nd century (a value expected around 1990). Evidently, he believed in the progress of technology leading to increasing efficiency of power machines: he was a specialist on steam engines.

One can find a survey of the present state of the CO_2–climate problem in many publications (see, e.g., USSR–USA Workshop, 1981; ICSU/WMO/UNEP, 1981; NAS, 1979). The most detailed model used to tackle the problem seems to be the one by Manabe and Stouffer (1980). This model has realistic topography, an upper mixed layer in the ocean, and the annual course of insolation. It yields the result that by quadrupling the CO_2 content in the atmosphere the global mean temperature T_s increases by 4.1 K (by doubling CO_2 the surface temperature increases by 2 K). It is interesting to compare these results with the ones obtained by Manabe and Wetherald (1980), who for a model geography found that the doubling of CO_2 gave $\Delta \bar{T}_s = 3.0$ K and the quadrupling of CO_2 gave $\Delta \bar{T}_s = 5.9$ K, for the annual mean insolation. Finally, Wetherald and Manabe (1981) obtained $\Delta \bar{T}_s = 6$ K for the quadrupling of CO_2 content for the annual mean insolation and 4.8 K for the seasonal changes of the insolation. The difference is explained by the finding of R. D. Cess (private communication, 1980) that in the model of Manabe and Stouffer (1980) the feedback between albedo and T_s is practically absent. Qualitatively, the absence is explained by the fact that in winter, when the area covered by snow and ice is large in high latitudes, the insolation is small (or absent) and in summer this area decreases. Some role may also be played by an inadequate parameterization of the sea ice (S. Manabe, private communication, 1981).

The addition of CO_2 to the atmosphere lifts the effective level of thermal outgoing radiation, decreasing the radiation leaving the troposphere (Ramanathan *et al.*, 1979; Ackerman, 1979; NAS, 1979). This means the heating of the troposphere and underlying surface. This decrease, $\Delta F\uparrow$, averaged globally, is about 4 W m^{-2}. Using our parameterization (3.3), one gets

$$\Delta \bar{T}_s = \Delta F\uparrow / B \qquad (6.1)$$

if one disregards the dependence $A(T_s)$. With $\Delta F\uparrow = 4$ W m^{-2} and $B = 2$ W m^{-2} K^{-1} we immediately obtain $\Delta \bar{T}_s = 2$ K. Accounting for albedo dependence on T_s gives

$$\Delta \bar{T}_s = \Delta F{\uparrow}/(B - dF{\downarrow}/d\bar{T}_s) \tag{6.2}$$

With $dF{\downarrow}/d\bar{T}_s = 0.8$ W m^{-2} K^{-1} (see the end of Section 4) Eq. (6.2) yields $\Delta T_s = 3.3$ K. Both estimates (obtained by Mokhov, 1981) are quite close to the results obtained from large numerical GCMs. The closeness can be explained by the fact that the empirical parameterization (3.3) for $F{\uparrow} = F{\uparrow}(T_s)$ captures the main feedback in the atmosphere between the outgoing thermal radiation, the temperature of the underlying surface, and the humidity content in the atmosphere.

More complicated is the question of the feedback $A = A(T_s)$. We have already noted that the intensity of this feedback is about twice that found by Cess (1976) without considering changes of the albedo due to the biosphere, and less than Cess (1978) estimate when considering this biosphere effect. It appears that with a lag of about 3 months, the role of this feedback can be decreased (see Mokhov, 1981). However, with a systematic melting of the ice during many years, the resulting decrease of the area and duration of the snow cover should activate this feedback. This poses a question concerning the time scale of this mechanism. Evidently the question requires additional study.

Some remarks concerning the similarity of the responses of the climatic system, at least, a model one, to the different types of thermal drive of the system are in order. Wetherald and Manabe (1975) already noted that the response to the change of the solar constant is analogous in many details to the response to changing the CO_2 content in the atmosphere. In our simplest mean hemispheric balance consideration these two forcings are not distinguished. Neglecting the influence of T_s upon albedo ($A = 0.3$), the value 4 W m^{-2} (for doubling of the CO_2 content) corresponds to change of the solar constant by $[4/340(1 - 0.3)] \times 100\% = 1.7\%$. Accounting for the feedback $A(\bar{T}_s)$ does not change this value. In fact,

$$\Delta \bar{T}_s = \Delta Q(1 - A)/(B - Q\, dA/d\bar{T}_s), \tag{6.3}$$

from which, for $B = 2$, $Q\, dA/d\bar{T}_s = 0.8$ W m^{-2} K^{-1}, and $\Delta Q(1 - A) = 2.4$ W m^{-2}, we get $\Delta \bar{T}_s = 2$ K. This value is somewhat higher than the values mentioned in Section 3, but lower than the value of Cess (1978). From Eq. (6.2) we obtain $\Delta \bar{T}_s = 3.3$ K for the doubling of the CO_2 content, and $3.3/2 \approx 1.7$ because this number is simply the ratio $4/2.4$.

We conclude this section with a few words on the character of the temperature changes due to the increase of the CO_2 concentration $\bar{c}$ in the atmosphere. A number of authors starting with Schneider (1972) noted that the change in T_s is approximately proportional to the logarithm of the relative change in $\bar{c}$. The explanation of the dependence was given by

Augustsson and Ramanathan (1977). It is connected with the saturation of the main CO_2 absorption band 15 mcm. As a result the change of the heat balance of the troposphere–underlying surface system, i.e., $\Delta F\uparrow$, is proportional to $\ln(c/c_0)$; where c_0 is a reference concentration of CO_2. The proportionality coefficient is (4/ln 2) W m^{-2} and we may write

$$\Delta F\uparrow \approx 5.8 \ln(c/c_0) \quad \text{W m}^{-2} \tag{6.4}$$

This relationship can change with a considerable lowering of c because the 15-mcm band would cease to be saturated. With a large increase of c the hot bands at 12 and 18 mcm would become important (Cess et al., 1980) and that would change the relationship somewhat.

With the burning of fossil fuels other optically active gases besides CO_2 are emitted into the atmosphere, such as carbon monoxide, methane, and NO_x. The last two have absorption bands in the thermal infrared themselves and all of them react with ozone in a long chain of reactions. Ozone has an appreciable absorption at its band 9.6 mcm. From the computations by Hameed et al. (1980) the total value of $\Delta F\uparrow$ can increase by several tens of percent compared with the change in $\Delta F\uparrow$ from CO_2 alone (see also USSR–USA Workshop, 1981). One can estimate the climatic significance of these additional changes by the method described in this section.

7. STATISTICAL PROPERTIES OF CLIMATIC SYSTEM

Like any large dynamical system, the climatic system has its own noise. The value of its various characteristics fluctuate. For instance, the annual mean surface temperature of the Northern Hemisphere, as can be calculated using data from 1881 to 1981 (see Table II), fluctuates with the standard deviation $\sigma_T = 0.23$ K. The total range of the fluctuations during the period reached 1 K. The temperature was minimal in the middle of the 1880s and maximal in the second half of the 1930s. From the middle of the 1960s a small warming trend is observed: during the 1970s the value of T_s was almost 0.1 K above the norm for all of the period considered.

Investigations of statistical properties of climatic elements, such as temperature and precipitation, have acquired increasing importance in considering the problem of anthropogenic forcing of the climate. It is necessary to extract the "signal," the climate change, from the natural variability of the climate. For this, one must understand and be able to describe the variability, i.e., the "noise" of the system. Studies of the possible changes of the climate are performed now with models of different complexity. A good model must reproduce not only the mean distributions of climatic elements but also their higher statistical moments. This

TABLE II. DEVIATIONS OF ANNUAL MEAN SURFACE AIR TEMPERATURE T_s (K) FROM NORM FOR THE LATITUDINAL BELT 17.5–87.5° N FOR 1881–1981[a]

Year	0	1	2	3	4	5	6	7	8	9
1880	—	−0.14	−0.28	−0.21	−0.50	−0.50	−0.42	−0.42	−0.30	−0.08
1890	−0.15	−0.27	−0.28	−0.31	−0.29	−0.18	−0.16	−0.05	−0.18	−0.03
1900	0.03	0.03	−0.30	−0.28	−0.28	−0.32	−0.08	−0.32	−0.16	−0.19
1910	−0.27	−0.06	−0.26	−0.21	0.00	0.04	−0.12	−0.32	−0.23	−0.10
1920	−0.01	0.17	0.05	0.12	0.09	0.16	0.24	0.16	0.23	0.09
1930	0.27	0.29	0.26	−0.11	0.30	0.15	0.17	0.39	0.53	0.33
1940	0.32	0.17	0.26	0.35	0.33	0.07	0.18	0.29	0.20	0.16
1950	0.05	0.22	0.21	0.41	0.12	0.09	−0.12	0.11	0.22	0.24
1960	0.22	0.15	0.16	0.13	−0.19	−0.11	−0.07	0.15	−0.05	−0.19
1970	0.15	−0.01	−0.29	0.17	0.03	0.14	−0.12	0.17	0.08	0.22
1980	0.15	0.50								

[a] After Vinnikov *et al.* (1980) and USSR–USA Workshop (1981). The 1981 value was supplied by G. V. Gruza.

section will describe our attempts to estimate, at least, second moments of the hemispherically averaged distributions of temperature $\bar{T}_s$. Some of the results have been obtained by the author's graduate student, P. F. Demchenko, and some of them were obtained jointly.

The physical basis for such an approach was formulated by Hasselmann (1976). In the earth's climatic system its various parts are characterized by different time scales. The most variable is the atmosphere: the synoptic weather perturbations have periods of several days. Speaking more precisely, the characteristic times of decay for the time correlation functions of meteorological elements measured at a point have values of about 3 days (see, e.g., Gandin *et al.*, 1976). At the same time the ocean has a much greater time inertia because its enthalpy is some three orders of magnitude larger than the enthalpy of the atmosphere. Representing the climatic system as the interacting inertial ocean and fastly fluctuating atmosphere, Hasselmann (1976) suggested considering the climatic fluctuations as the motion of a heavy Brownian particle under the action of molecular collisions. The sources of fluctuations of climatic charcteristics (the mean surface temperature of the ocean covering 71% of the earth's surface) will be the weather fluctuations integrated by the ocean, the inertial part of the system.

Because of the large difference in time scales of the ocean and atmosphere, we can consider the weather fluctuations as stochastic forcing δ-correlated in time. Mathematical techniques for the description of such processes are now well developed (see, e.g., Klyatskin, 1975, 1980). In the development and testing of such an approach it is expedient to use the simplest models that allow a clear interpretation of the results and at the same time reproduce some main characteristic features of modern climate and its change.

For such a model we shall use the nonstationary radiation balance equation averaged hemispherically. Fluctuations of the solar radiation are connected mainly with the albedo fluctuations caused on short time scales mainly by clouds. Fluctuations of the outgoing thermal radiation are caused by clouds and by temperature variations of the atmosphere and underlying surface. For small deviations of temperature T_s from its mean climatic value $\bar{T}_s$, $\Delta T = T_s - \bar{T}_s$, we may write an evolution equation

$$C \frac{d \Delta T}{dt} = -B \, \Delta T(t) + R'(t) \qquad (7.1)$$

Here $B = 2$ W m^{-2} K^{-1} is the climate sensitivity parameter [see Eq. (3.2)]. The value

$$R'(t) = -QA'(t) - F{\uparrow}' \qquad (7.2)$$

describes synoptic fluctuations of the hemispherically averaged radiation balance, where $Q = 340$ W m^{-2} is one-quarter of the solar constant, A' the fluctuations of planetary albedo, and $F{\uparrow}'$ the fluctuations of mean hemispheric outgoing flux. We shall consider the thermal inertia C to be identified only with the inertia of the upper mixed layer (UML) of the ocean through which the atmosphere interacts with the ocean. This is the usual procedure in energy balance models (e.g., Thompson, 1979). For the depth of the UML we take $l = 50$ m. Manabe and Stouffer (1980) determined that for the whole World Ocean $l = 68$ m on the average. But taking into account that about 30% of the globe is land, the value 50 m seems to be reasonable. Then $C = \rho c_p l = 2 \times 10^8$ J m^{-2} K^{-1}. Note that Eq. (7.1) is the Langevin equation for the case considered (see, e.g., Kittel, 1958).

Now we shall consider the fluctuations of the radiation balance R' to be δ-correlated in time. Then their correlation function is

$$B_R(t_2 - t_1) = \langle R'(t_1)R'(t_2) \rangle = 2D \, \delta(t_2 - t_1) = 2\sigma_R^2 \tau_c \, \delta(t_2 - t_1) \qquad (7.3)$$

Here, angular brackets denote time averaging, σ_R^2 is the variance of fluctuations of the radiation balance R, and τ_c the characteristic value of the correlation time of R.

Let us now multiply Eq. (7.1) at time t_1 by $\Delta T(t_2)$ for $t_1 - t_2 = \tau > 0$, and average with time. Assuming stationarity and accounting for Eq. (7.3), we obtain the equation for the time correlation function B_T of fluctuations of mean hemispheric temperature $\bar{T}_s$:

$$C \frac{dB_T}{d\tau} = -BB_T(\tau) \qquad (7.4)$$

For integration of this equation one must know the initial condition $B_T(\tau = 0) = \sigma_T^2$, i.e., the variance of temperature fluctuations. For the Langevin-type equation (7.1) with random forcing R, having a "white noise" spectrum, this is a standard problem of the theory of stochastic processes (see, e.g., Klyatskin, 1975, 1980; Kittel, 1958). The solution is

$$\sigma_T^2 = \sigma_R^2 \tau_c / BC \tag{7.5}$$

where we made use of Eq. (7.3), and the time correlation function is

$$B_T(\tau) = \sigma_T^2 e^{-\tau/\tau_0}, \qquad \tau_0 = C/B \tag{7.6}$$

The power spectrum corresponding to a stochastic process with such a correlation function is

$$F(\omega) = \frac{1}{\pi} \int_0^\infty B(\tau) \cos \omega\tau \, d\tau = \frac{2\sigma_T^2 \tau_0/\pi}{1 + \omega^2 \tau_0^2} \tag{7.7}$$

This is a typical spectrum of a "red noise" where the spectral density decreases for high frequencies and saturates for small ω. The characteristic time τ_0 for the values C and B assumed above is 10^8 sec, which is about 3 yr.

Now, by Eq. (7.5) we expressed the variance of the fluctuations of the hemispherically averaged surface temperature $\bar{T}_s$ in terms of the variance of fluctuations of the hemispherically averaged radiation budget σ_R^2, the parameter of the ocean thermal inertia C, and the parameter of the climate sensitivity parameter B. Unfortunately, we do not know the empirical value of σ_R^2. The only paper known to us dealing with this problem is the one by Hartmann and Short (1980). They presented maps for albedo, outgoing thermal radiation, and the net radiation balance ($F\uparrow - F\downarrow$) for summer and winter seasons obtained from satellite data. Using these maps, we estimated, for a number of latitudes, the zonally averaged values of the variance of albedo caused by synoptic fluctuations (see Table III) and of the variance of outgoing thermal radiation $F\uparrow$ (Table IV). Afterward, Dr. Hartmann kindly supplied us with maps of the variance of these values, which revealed that the simple procedure used by us to

TABLE III. ZONALLY AVERAGED VALUES OF STANDARD DEVIATIONS OF ALBEDO WEATHER FLUCTUATIONS σ_A AT A POINT (IN W m^{-2})

	Latitude (°N)				
	0	20	40	60	75
Winter	0.14	0.11	0.14	0.12	—
Summer	0.12	0.11	0.13	0.16	0.11

TABLE IV. ZONALLY AVERAGED VALUES OF STANDARD DEVIATIONS OF OUTGOING
THERMAL RADIATION $\sigma_{F\uparrow}$ AT A POINT (IN W m^{-2})

	Latitude (°N)				
	0	20	40	60	75
Winter	23	21	22	16	13
Summer	29	25	24	23	11

estimate the variances systematically underestimates them by about 30%. Nevertheless, all the maps (and tables) show that the intensity of the albedo synoptic variability does not reveal any considerable latitudinal and seasonal behavior, whereas the variability of the outgoing thermal radiation decreases from the equator toward the poles. We shall try to use these data to estimate the variance of the mean hemispheric net radiation budget, i.e., the value of σ_R^2.

For this we have to know the spatial correlation structure of the fluctuations of the net radiation budget. To calculate the standard deviation of the net radiation budget at a point σ_{R0}, we must make an assumption concerning the correlation coefficient between fluctuations of albedo and of the outgoing thermal radiation as follows from Eq. (7.2):

$$\sigma_{R0}^2 = \sigma_{F\uparrow}^2 + 2QS(\theta, t)\sigma_A\sigma_{F\uparrow}r_{AF\uparrow} + Q^2S^2(\theta, t)\sigma_A^2 \qquad (7.8)$$

The insolation function $S(\theta, t)$ is described by Eq. (4.2). If one considers that the fluctuations both of albedo and of the outgoing thermal radiation are caused only by the fluctuations of the cloud amount, then $r_{AF\uparrow} = -1$, which is clearly a lower bound. Ohring and Clapp (1980) present data on interannual variability of monthly mean values of A and $F\uparrow$ (for a point, of course). Using their data, one can estimate that $r_{AF\uparrow} = -0.8$, although for such time scales many climatic feedbacks may already be important.

In Table V we present annual and seasonal mean values of the standard deviations of the net radiation budget calculated from Eq. (7.8) with the

TABLE V. ZONAL AND ANNUAL AVERAGES OF STANDARD DEVIATION OF NET
RADIATION BALANCE SYNOPTIC FLUCTUATIONS AT A POINT (IN W m^{-2}) FOR TWO
VALUES OF THE CORRELATION COEFFICIENT $r_{AF\uparrow}$

	Latitude (°N)				
$r_{AF\uparrow}$	0	20	40	60	75
−1	27	21	21	35	26
−0.8	37	29	31	28	27

two values of the correlation coefficient $r_{AF\uparrow} = -1$ and -0.8. Any pronounced latitudinal behavior in σ_{R0} is not apparent and therefore, for preliminary rough estimates we shall later consider the value σ_{R0} to be constant and equal to the mean hemispheric value $\bar{\sigma}_{R0}$. From the data of Table V we find that for $r_{AF\uparrow} = -1$ we have $\bar{\sigma}_{R0} = 25$ W m^{-2}, and for $r_{AF\uparrow} = -0.8$ we have $\bar{\sigma}_{R0} = 33$ W m^{-2}. Accounting for the likelihood of underestimating $\bar{\sigma}_{R0}$ by our procedure, we shall take $\bar{\sigma}_{R0} = 40$ W m^{-2} as a representative value for the globe.

For simplicity and as a first approximation we assume that the two-dimensional field of fluctuations of the net radiation budget is statistically homogeneous and isotropic. This assumption may find some justification in results of the treatment of cloud observations by Sonechkin and Khandurova (1969). These authors present the spatial correlation function for the cloud amount computed using satellite observations over the European part of the USSR. This function has a characteristic scale of decay of about 1000 km and is practically isotropic up to this distance. Then for the sought value of σ_R^2 we may write

$$\sigma_R^2 = \left\langle \left[\frac{1}{2\pi} \int_0^{2\pi} d\lambda \int_0^{\pi/2} R'(\theta, \lambda) \cos\theta \, d\theta \right]^2 \right\rangle \approx \frac{2\sigma_{R0}^2}{\pi} \frac{r_c^2}{a^2} \qquad (7.9)$$

where r_c is the radius of the spatial correlation of the fluctuations (about 1000 km) and a the radius of the earth. Substituting Eq. (7.9) into Eq. (7.5), we obtain the final expression for the estimate of the variance of the mean hemispheric temperature

$$\sigma_T^2 = \frac{2\sigma_{R0}^2\tau_c}{\pi BC} \frac{r_c^2}{a^2} \qquad (7.10)$$

For the time scale τ_c we take 3 days, which is a characteristic time of the cloud field variability (Sonechkin and Khandurova, 1969). With $r_c = 1000$ km, $a = 6370$ km, $\bar{\sigma}_{R0}^2 = 40$ W m^{-2}, and the aforementioned values of B and C, we obtain $\sigma_T = 0.13$ K. This value is almost half the value $\sigma_T = 0.23$ for 1881–1980. This may be considered reasonable because we do not take into account the destabilizing albedo–temperature feedback and many other effects. For instance, the atmospheric turbidity fluctuations for 1883–1977 (Pivovarova, 1977; Bryson and Goodman, 1980) are sufficiently well correlated with the temperature fluctuations for these same years. These turbidity fluctuations can explain about half of the temperature variability (Vinnikov et al., 1980). Anyway, it seems that Eq. (7.5) or (7.10) can give a useful estimate of a random component of climatic variations caused by assimilation of the net radiation budget weather fluctuations by the oceanic UML which has a much larger inertia.

An analogous approach was developed by the author together with P. F. Demchenko for the description of joint random fluctuations of the snow–ice boundary latitude and the outgoing thermal radiation averaged over the hemisphere (see Golitsyn and Demchenko, 1980). One of the results of this study was a determination of a similarity parameter for the fluctuations of the latitude of the snow–ice boundary (or its area) and the variability of the mean hemispheric temperature:

$$\Pi_{sT} = \frac{\sigma_\theta Q}{\sigma_T B} = \frac{\sigma_S Q}{2\pi a^2 \sigma_T B} \qquad (7.11)$$

Here σ_θ^2 is the variance of the sine of latitude of snow–ice boundary, $\sigma_S^2 = (2\pi a^2 \sigma_\theta)^2$ the variance of the snow–ice-covered area, and the other notations are as above. We do not have a sufficiently long time series for the position of the boundary latitude, or the area confined by it, that is comparable in duration to the hemispheric mean temperature time series. Nevertheless, Vinnikov et al. (1980) present a series for mean ice-cover area in the Arctic basin and adjacent seas for 1947–1980. The data are presented for selected months as well as for an annual average over the 34 yr. The data allow one to calculate that the annual average ice area is 7.8 $\times$ 10^6 km^2, with the standard deviation $\sigma_S = 0.2 \times 10^6$ km^2. This number can be appreciably increased due to variations of the snow-covered area. In fact, data by Kukla (1978) show that the number for 1968–1975 can be doubled, as a minimum, if one disregards the large change in the area around 1971–1972. If we take $\sigma_S = 0.4 \times 10^6$ km^2 and $\sigma_T = 0.15$ K, as can be calculated from Vinnikov et al. (1980) for 1947–1980, then $\Pi_{sT} = 1.8$. This empirical estimate of the similarity parameter (7.11) should be considered a very rough one.

Data by Kukla (1978) describe too short a series of the snow–ice area. He used a 12-month running mean and the data treated this way do not show much variability except for a large change in the winter of 1971–1972 in the Northern Hemisphere when there was much persistent snow present in Asia. It should be noted that before this time the mean snow area was about 33 $\times$ 10^6 km^2, reaching about 38 $\times$ 10^6 km^2 in that winter and decreasing to only about 36 $\times$ 10^6 km^2 in subsequent winters. If this is not an effect of changes in instrumentation or interpretation it must represent a marked climatic fluctuation. As a result the formal value of the standard deviation σ_S is 1.9 $\times$ 10^6 km^2 and with $\sigma_T = 0.23$ K (for the complete 100-yr series) we get $\Pi_{sT} = 5$.

It would be useful while experimenting with various numerical models of climate to compute the value Π_{sT} from the model generated data. The comparison of this value with the empirical value can be a valuable characteristic of the noise level of the model. As satellite data on snow–ice cover accumulate, one should reestimate the value of Π_{sT}. The present data allow one to conclude that the value might be within the limits of 2 to 5.

8. Conclusion

The author has tried to demonstrate that simplified approaches to the problem of climate and its change based on a use of observational data for the present climate can produce reasonable estimates of a number of important integral characteristics of the climatic system. For instance, the use of similarity and dimensional arguments allowed us to estimate the temperature difference between equator and pole and mean winds in the atmosphere. A parameterization of the outgoing thermal radiation in terms of its dependence on the surface temperature, an analysis of albedo–temperature dependence, and the use of a mean hemispheric radiation budget, allowed us to obtain estimates of the sensitivity of the climate system to an increase of atmospheric CO_2 content. The net radiation budget fluctuation estimates also give realistic values for the mean hemispheric climatic temperature fluctuations when using the "diffusional" concept of climate.

The value of simplified approaches is, of course, methodological, first of all (and could also be considered pedagogical—"educational toys" is the term used by one climatologist for energy balance models). However, such approaches or models can appreciably clarify the physics and the relative importance of various processes acting in the system. Sometimes they are able to give quite new results showing directions for further studies, as has happened with the "white earth" or "deep freeze" concept (Budyko, 1968, 1969, 1980). Various relationships emerged from such considerations and if checked empirically can be useful in calibrating more complicated models.

Why can the simplified approaches give reasonable results for such a multifaceted problem as the one considered here? The exact answer to the question is as difficult to obtain as to solve the problem itself. Evidently the basic relationships, such as the global radiation balance, are valid when averaged over a sufficiently long time. For such global averaging, all energy transfers within the system are not very essential. All changes of the balance are brought about by the system with an intensity which can be estimated from the annual course of some parameters of the system if only some feedbacks with much longer time constants do not interfere. Perhaps when climatic changes are relatively small such approaches can be useful and justified.

What information is in fact necessary not only for "simple" approaches but for more sophisticated studies of the problem of climate and its change? This information includes sufficiently long observations for all components of the radiation budget, albedo, cloudiness, snow and ice cover, humidity, temperature of atmosphere and underlying surface including ocean, and the knowledge of the ocean response, among others. In this review we tried to demonstrate only the simplest ways of using the climatic information.

APPENDIX

The insolation coming into a hemisphere varies during the year due to changes in the sun's declination, as well as due to changes of the sun–earth distance arising because the orbit is an ellipse with the eccentricity $\epsilon = 0.0167$. The earth passes its perihelion at the beginning of January and the distance is the shortest at this time.

If the orbit of the earth's revolution around the sun were circular, then the total amount of the solar radiation received by the planet would not vary. But its distribution between the hemispheres is changing. Let us consider the time of the solstice, e.g., the winter one for the Northern Hemisphere. Consider the plane tangent to the earth's sphere at the subsolar point. For this plane an elementary calculation shows that the area of the Northern Hemisphere illuminated by the sun is equal to

$$S_r = \tfrac{1}{2}\pi a^2(1 - \sin \alpha) \qquad (A.1)$$

where $\alpha = 23°27'$ is the angle between the axis of the earth's rotation and the perpendicular to its orbit plane. The value of $\sin \alpha$ is 0.3979. Due to the present proximity of the earth to the sun at the perihelion, the flux of the solar radiation is larger by a factor of $1 + 2\epsilon = 1.0334$. Therefore, the value in the parentheses in Eq. (A.1) changes from 1.365 in summer to 0.635 in winter.

For the Southern Hemisphere the effects of the axis declination and of the orbital eccentricity are the reverse (i.e., additive) and the amplitude of the insolation variation is equal to 43%. Therefore, from the summer solstice to the winter solstice the total insolation coming to the Southern Hemisphere changes by a factor of about 1.43/0.57 = 2.5.

ACKNOWLEDGMENTS

I would like to thank my younger colleagues, especially I. I. Mokhov and P. F. Demchenko. We began working on the problem together, at first in awe at its complexity and then trying to find ways to do something meaningful. Most constructive and useful for us were numerous discussions with Profs. M. I. Budyko and R. D. Cess. Their contributions showed us that in the problem of climate and its change there are still possibilities to obtain important results without large expenditures of material and human effort, using only scientific common sense and already accumulated observational data in a consistent and persistent manner. For this we are extremely grateful to both. We are also very thankful to K. Ya. Vinnikov for many discussions related to empirical data on climate, and to Dr. G. Ohring, who made some useful comments that led to improvements of the first version of the text.

References

Ackerman, T. R. (1979). On the effect of CO_2 on atmospheric heating rates. *Tellus* **30**, 115–123.

Arrhenius, S. (1896). On the influence of carbonic acid in the air upon the temperature of the ground. *Philos. Mag.* **41** [5], 237–275.

Arrhenius, S. (1903). "Lehrbuch der kosmischen Physik." Hirzel, Leipzig.

Augustsson, T., and Ramanathan, V. (1977). A radiative–convective model study of the CO_2 climate problem. *J. Atmos. Sci.* **34**, 448–451.

Berlyand, T. G., and Strokina, L. A. (1980). "Global Distribution of the Total Cloud Amount." Hydrometeorol. Publ. House, Leningrad.

Bryson, R. A., and Goodman, B. M. (1980). Volcanic activity and climatic changes. *Science* **207**, 1041–1044.

Budyko, M. I. (1968). On the origin of ice ages. *Meteorol. Gidrol.* No. 2, pp. 3–8.

Budyko, M. I. (1969). The effect of solar radiation variations on the climate of the Earth. *Tellus* **21**, 611–619.

Budyko, M. I. (1980). "Climate in the Past and Future." Hydrometeorol. Publ. House, Leningrad.

Callendar, G. S. (1938). The artificial production of carbon dioxide and its influence on temperature. *Q.J.R. Meteorol. Soc.* **64**, 223–240.

Chamberlin, T. C. (1897). A group of hypotheses bearing on climatic changes. *J. Geol.* **5**, 653–683.

Chamberlin, T. C. (1898). The influence of great epochs of limestone upon the constitution of the atmosphere. *J. Geol.* **6**, 609–621.

Chamberlin, T. C. (1899). An attempt to frame a working hypotheses of the cause of glacial periods on an atmospheric basis. *J. Geol.* **7**, 545–584.

Cess, R. D. (1976). Climate change: An appraisal of atmospheric feedback mechanisms employing zonal climatology. *J. Atmos. Sci.* **33**, 1831–1843.

Cess, R. D. (1978). Biosphere-albedo feedback and climate modeling. *J. Atmos. Sci.* **35**, 1765–1768.

Cess, R. D., Ramanathan, V., and Owen, T. (1980). The Martian paleoclimate and enhanced atmospheric carbon dioxide. *Icarus* **41**, 159–165.

Cess, R. D., Briegleb, B. P., and Lian, M. S. (1982). *J. Atmos. Sci.* **39**, No. 1.

Crutcher, H. L., and Meserve, J. M. (1970). "Selected Level Heights, Temperatures and Dew Point Temperatures for the Northern Hemisphere," NAVAIR 50-1C-52. NOAA Environ. Data Serv., Washington, D.C.

Curran, R. J., Wexler, R., and Nack, M. L. (1978). Albedo climatology analysis and the determination of fractional cloud cover. *NASA Tech. Memo.* **NASA TM-X-79576**, 1–45.

Efimova, N. A. (1980). Influence of the surface albedo changes on the Earth's thermal regime. *Meteorol. Gidrol.* No. 7, pp. 50–56.

Ellis, J., and Vonder Haar, T. H. (1976). "Zonal Average Earth Radiation Budget Measurements from Satellites for Climatic Studies," Atmos. Sci. Pap. No. 240. Colorado State University, Fort Collins.

Freeman E., and Liu, Y.-C. (1979). Climatic effects of cirrus clouds. *Adv. Geophys.* **21,** 311–359.

Gandin, L. S., Zakhariev, V. I., and Kagan, R. L. (1976). "Statistical Structure of Meteorological Fields." Budapest.

GARP (1975). The physical basis of climate and climate modelling. *GARP Publ Ser.* **16,** 1–265.

Golitsyn, G. S. (1968). Estimates of some characteristics of general circulation in atmospheres of terrestrial planets. *Izv., Acad. Sci., USSR, Atmos. Oceanic Phys. (Engl. Transl.)* **4,** 1131–1138.

Golitsyn, G. S. (1970). A similarity approach to the general circulation of planetary atmospheres. *Icarus* **13,** 1–24.

Golitsyn, G. S. (1973). "Introduction into the Dynamics of Planetary Atmospheres." Hydrometeorol. Publ. House, Leningrad [Engl. transl.: *NASA Tech. Transl.* **NASA TT F-15,** 1–172, (1974)].

Golitsyn, G. S., and Demchenko, P. F. (1980). Statistical properties of a simple energy balance climate model. *Izv., Acad. Sci., USSR, Atmos. Oceanic Phys. (Engl. Transl.)* **16,** 1235–1242.

Hameed, S., Cess, R. D., and Hogan, J. S. (1980). Response of the global climate to change in atmospheric chemical composition due to fossil fuel burning. *JGR, J. Geophys. Res.* **85,** 7537–7545.

Hartmann, D. L., and Short, D. A. (1980). On the use of Earth radiation budget for studies of clouds and climate. *J. Atmos. Sci.* **37,** 1233–1250.

Hasselmann, K. (1976). Stochastic climate models. Part I. Theory. *Tellus* **28,** 473–485.

Hickey, J. R., Stove, L. L., Jacobovitz, H., Pellegrino, P., Maschhoff, R. H., House, F., and Vonder Haar, T. H. (1980). Initial solar irradiance determination from Nimbus 7 cavity radiometer measurements. *Science* **208,** 281–283.

Hoyt, D. V. (1976). The radiation and energy budgets of the Earth using both ground based and satellite-derived values of total cloud cover. *NOAA Tech. Rep.* **ERL 362-ARL 4,** 1–124.

ICSU–WMO (1981). "Preliminary Plan for the World Climate Research Programme." World Meteorol. Organ., Geneva.

ICSU–WMO–UNEP (1981). "On the Assessment of the Role of CO_2 on Climate Variations and Their Impact," Joint Meeting of Experts. World Meteorol. Organ., Geneva.

Kellogg, W., and Schware, R. (1981). "Climate Change and Society." Westview Press, Boulder, Colorado.

Kittel, C. (1958). "Elementary Statistical Physics." Wiley, New York.

Klyatskin, V. I. (1975). "Statistical Description of Dynamical Systems with Fluctuating Parameters." Fizmatgiz, Moscow.

Klyatskin, V. I. (1980). "Stochastic Equations and Waves in Randomly Inhomogeneous Media." Fizmatgiz, Moscow.

Kukla, G. J. (1978). Recent changes in snow and ice. *In* "Climatic Change" (J. Gribbin, ed.), pp. 115–129. Cambridge Univ. Press, London and New York.

Manabe, S., and Stouffer, R. J. (1980). Sensitivity of a global climate model to an increase of CO_2 concentration in the atmosphere. *JGR, J. Geophys. Res.* **85,** 5529–5554.

Manabe, S., and Wetherald, R. T. (1967). Thermal equilibrium of the atmosphere with a given distribution of relative humidity. *J. Atmos. Sci.* **24,** 241–259.

Manabe, S., and Wetherald, R. T. (1975). The effect of doubling the CO_2 concentration on the climate of a general circulation model. *J. Atmos. Sci.* **32,** 3–15.

Manabe, S., and Wetherald, R. T. (1980). On the distribution of climate change resulting from an increase in CO_2 content of the atmosphere. *J. Atmos. Sci.* **37,** 99–118.

Mokhov, I. I. (1981). On the CO_2 influence on the thermal regime of the Earth's climatic system. *Meteorol. Gidrol.* No. 7, pp. 50–56.

Monin, A. S. (1975). The role of the oceans in climate models. *GARP Publ. No. 16, ICSU-WMO* pp. 201–205.

National Academy of Sciences (NAS) (1979). "Carbon Dioxide and Climate: A Scientific Assessment." Natl. Acad. Sci., Washington, D.C.

North, G. R., Cahalan, R. F., and Coakley, J. A., Jr. (1981). Energy-balance climate models. *Rev. Geophys. Space Phys.* **19,** 91–122.

Ohring, G., and Clapp, P. (1980). The effect of changes in cloud amount on the net radiation at the top of the atmosphere. *J. Atmos. Sci.* **37,** 447–454.

Oort, A. H. (1964). On the estimates of the atmospheric energy cycle. *Mon. Weather Rev.* **92,** 483–493.

Oort, A. H., and Rasmusson, E. M. (1971). Atmospheric circulation statistics. *NOAA Prof. Pap.* No. 5, pp. 1–323.

Östlund, H. G., Dorsey, H. G., and Rooth, C. C. (1974). GEOSECS North Atlantic radiocarbon and tritium results. *Earth Planet. Sci. Lett.* **23,** 69–86.

Paltridge, G. W. (1980). Cloud-radiation feed-back to climate. *Q.J.R. Meteorol. Soc.* **106,** 895–899.

Petukhov, V. K., Feigelson, E. M., and Manuilova, N. I. (1975). Regulating role of cloudiness in thermal effects of anthropogenic aerosol and carbon dioxide. *Izv., Acad. Sci., USSR, Atmos. Oceanic Phys. (Engl. Transl.)* **11,** 802–808.

Pivovarova, Z. I. (1977). "Radiation Characteristics of the Climate of USSR." Hydrometeorol. Publ. House, Leningrad.

Ramanathan, V., Lian, M. S., and Cess, R. D. (1979). Increased atmospheric CO_2: Zonal and seasonal estimates of the effect on the radiation energy balance and surface temperature. *JGR, J. Geophys. Res.* **84,** 4949–4958.

Raschke, E. (1975). Satellite capability in monitoring climate parameters. *GARP Publ. No. 16, ICSU-WMO* pp. 252–257.

Saltzman, B. (1977). Global mass and energy requirements for glacial oscillations and their implications for mean ocean temperature oscillations. *Tellus* **24,** 205–212.

Saltzman, B. (1978). A survey of statistical–dynamical models of the terrestrial climate. *Adv. Geophys.* **20,** 183–304.

Schneider, S. H. (1972). Cloudiness as a global climate feedback mechanism: The effects on the radiation balance and surface temperature of variations of cloudiness. *J. Atmos. Sci.* **29,** 1413–1422.

Schneider, S. H. (1975). On the carbon dioxide–climate confusion. *J. Atmos. Sci.* **32,** 2060–2066.

Schneider, S. H., and Dickinson, R. (1974). Climate modeling. *Rev. Geophys. Space Phys.* **12,** 447–493.

Sonechkin, D. M., and Khandurova, N. S. (1969). Results of a study of spatial–time variability of cloudiness over European part of the USSR. *Tr. Hydrometeorol. Cent.* No. 50, pp. 37–46.

Stephens, G. L., Campbell, G. G., and Vonder Haar, T. M. (1981). *JGR, J. Geophys. Res.* **86,** C10, 9739–9760.

Taljaard, J. J., van Loon, H., Crutcher, H. L., and Jenne, R. L. (1969). "Climate of the Upper Air: Southern Hemisphere. V.1. Temperatures, Dew Points, and Heights at Selected Pressure Levels," NAVAIR 50-1C-55. NOAA Environ Data Serv., Washington, D.C.

Thompson, S. L. (1979). Development of a seasonally-verified planetary albedo parametrization for zonal energy balance climate models. *GARP Publ. Ser.* **22,** 1002–1023.

Tyndall, I. (1861). On the absorption and radiation of heat by gases and vapour and on the physical connection of radiation absorption and conduction. *Philos. Mag.* [4] **22,** 167–194, 273–285.

USSR–USA Workshop (1981). "Climatic Effects of Increased Atmospheric Carbon Dioxide," Report of the Workshop held in Leningrad, June 1981.

Vinnikov, K. Ya., Gruza, G. V., Zakharov, V. F., Kirillov, A. A., Kovyneva, N. P., and Rankova, E. Ya. (1980). Modern changes of climate in the Northern Hemisphere. *Meteorol. Gidrol.* No. 6, pp. 5–17.

Wetherald, R. T., and Manabe, S. (1975). The effect of changing the solar constant on the climate of the general circulation model. *J. Atmos. Sci.* **32,** 2044–2055.

Wetherald, R. T., and Manabe, S. (1980). Cloud cover and climate sensitivity. *J. Atmos. Sci.* **37,** 1485–1510.

Wetherald, R. T., and Manabe, S. (1981). Influence of seasonal variation on the sensitivity of a model climate. *JGR, J. Geophys. Res.* **86,** 1194–1204.

World Climate Conference (1979). "Extended Summaries of Papers Presented at the Conference." World Meteorol. Organ., Geneva.

PARAMETERIZATION OF TRAVELING WEATHER SYSTEMS IN A SIMPLE MODEL OF LARGE-SCALE ATMOSPHERIC FLOW

G. J. Shutts[1]

Atmospheric Physics Group
Department of Physics
Imperial College of Science and Technology
London, United Kingdom

1. INTRODUCTION

The science of weather forecasting has at its foundation the notion that the evolution of the state of the atmosphere is governed by deterministic, Newtonian physics and hence is, in principle, predictable. Representation of the real world by a model with a finite number of degrees of freedom is considered to be the major stumbling block to successful long-term weather forecasting. Model air-parcel trajectories locally (where resolution is insufficient) soon depart from real trajectories, and this error rapidly spreads throughout the model. The problem, however, is even deeper than this, due to the presence of resonances and instabilities in dynamical systems. In a suitable phase-space representation of the system state, the trajectory describing its evolution may run into "pathological" regions from which it emerges on a course that cannot be predicted with absolute

---

[1] Present address: Meteorological Office, Bracknell, Berkshire RG12 2SZ, United Kingdom.

precision (Prigogine, 1980, p. 40). The estimation of the maximum time period over which weather forecasts can feasibly remain useful has been the subject of much interest (Lorenz, 1969). This ultimate limit to predictability is thought to be of the order of 2 weeks, yet at present forecasts typically remain useful for about 1 week (Bengtsson, 1980). A time limit on weather forecasting skill, however, does not imply that climate prediction is similarly restricted, since weather might be regarded as noise on a predictable climatic signal.

The primary goal for dynamical climatologists is to be able to account for the observed averaged state of the atmospheric motion, temperature, and moisture fields over a time scale shorter than the seasonal time scale yet longer than the characteristic lifetime of a depression ($\sim$5–7 days). This understanding must necessarily come from an appreciation of the physical structure and transport properties of idealized models of real atmospheric motion systems and the role they play in determining the climate. A secondary, though perhaps more important, long-term goal would be to understand why climate, defined on this time scale, varies from year to year, leading to good or bad winters and summers, and how ultimately these could be predicted.

A common approach adopted by climatologists is to use numerical weather prediction models in rather the same way that a biologist performs experiments. One "control" model integration, over a period of months, is made under fixed "external" conditions (insolation, sea surface temperatures, snow cover, etc.) and is compared to an integration for which one aspect of the model has been changed (e.g., removal of mountains). The relative importance of different processes in the model can, in principle, be assessed by this approach, though interpretation is often difficult. A fundamental problem is that two control integrations starting from different initial states will not give exactly the same climates, and so experimentation is meaningless unless the "signal" can be distinguished from the climatic "noise" of the control integration. From an economic point of view, these computer simulations are costly and do not obviously lead to improved understanding because of their complexity.

The philosophy adopted here is that understandable models are preferable, however simplistic. Great economies can be made by parameterization of the cyclone scale of motion, by which we mean the traveling depressions and anticyclones of middle latitudes, so that low-resolution statistical-dynamical models can be built with long time steps. Low-resolution models can be used as a test-bed for parameterization theories and, together with observations of transient eddy fluxes, illuminate the nature of interaction between forced stationary long waves and traveling weather systems.

In Section 2, salient features of the January and July monthly mean climatology are discussed in the context of existing theories of the forcing of large-scale motion. In addition to the classical theories of orographic and thermal forcing, the mechanism of warm anticyclone maintenance put forward by Green (1977) is reviewed. This is particularly relevant to the more general question of the mechanism of interaction between forced planetary-scale waves and traveling weather systems.

Section 3 is a critique of the philosophy behind the parameterization of weather systems, together with a brief history of mixing and transfer theories.

Section 4 outlines the transfer theory of heat and vorticity in its zonal-average and local time-average three-dimensional forms, together with approximations for ease of implementation. Its relation to other wave/mean-flow interaction theories, such as the Eliassen–Palm fluxes, is discussed, and physical interpretation of relevant terms is sought.

In Section 5 the details of a three-level quasi-geostrophic climate model are described, where all but the largest baroclinically stable scales of motion are parameterized. Three experiments are described in which the steady-state response to a given Newtonian heating distribution is required. In particular, the role of eddy vorticity transfer in determining the mean surface pressure pattern is investigated.

2. Dynamics of Monthly Mean Climatology

The nonuniform heating of a planet by a distant source of heat trivially explains, through geometrical considerations, why equatorial regions are hotter than polar regions, but the dynamics of the ensuing convective circulation is far more complex. A theory of climate is required to explain the observed average distribution of winds—the trades, polar easterlies and temperate westerlies—together with their longitudinal variations and pressure anomalies. Apart from the time-average conditions, the positions of the major storm tracks (where the variance of the pressure field about its time average is greatest) must also be accounted for since these represent the regions of weather changeability and high rainfall.

Two aspects of the climate problem are usefully considered in isolation: the zonal-mean and the longitudinally varying distributions of motion and temperature. In its simplest form, the problem of understanding the zonal-mean climate reduces to the determination of the zonal-mean temperature distribution and surface winds, since the atmosphere closely satisfies thermal wind equilibrium (even after taking a zonal average).

2.1. The Zonal-Mean Climate

The latitudinal imbalance of diabatic heating between poles and equator is offset by the bodily transport of heat in large-scale convective motion systems. This transport can be split into stationary and transient eddy contributions, both of which are thought to play an important part in the northern winter hemispheric heat balance, though transient eddies dominate in summer and at all times in the Southern Hemisphere. The transient contribution arises predominantly from amplifying weather systems in which warm, tropical air moves poleward and upward ahead of a trough in the height contours and cold, polar air moves equatorward and sinks in the rear. The stationary contribution comes mainly from long planetary waves forced by topographic inhomogeneities and is particularly evident in winter. The maintenance of time-averaged zones of surface winds with a westerly component in middle latitudes and an easterly component elsewhere demands transport of westerly momentum into middle latitudes to balance the frictional transfer of angular momentum to the earth. Again, this is primarily achieved by the large-scale eddy motion—as depressions and anticyclones on synoptic charts—though for reasons that are rather subtle compared to those given for heat transport. Transportation of heat and momentum by the zonally averaged circulation is dominant only in the tropics.

Special significance is attached to the surface winds as a climatic parameter, since they indicate through frictional convergence the positions of wet and arid zones. Furthermore, the zonal-mean surface winds are determined by the height-integrated momentum flux convergence, thereby elevating the importance of momentum transport. The height variation of the zonal-mean eddy momentum flux convergence, together with surface friction, drives the three-cell mean meridional circulation (i.e., Hadley, Ferrel, and polar cells) through the necessity of maintaining geostrophic and hydrostatic balance.

For these reasons, dynamical climatologists have devoted a great deal of attention to those aspects of the structure of large-scale motion systems that determine their ability to transport momentum. It has been found that there are many ways of rationalizing the momentum-concentrating properties of traveling weather systems: kinematic arguments based on the "bowing" of trough lines (Kuo, 1952), Rossby-wave radiation and Eliassen–Palm fluxes (Edmon et al., 1980), two-dimensional turbulence as an inverse energy cascade (Kraichnan, 1967; Rhines, 1979), or potential vorticity transfer (Green, 1970). All seem equally plausible ways of looking at the momentum- or energy-transporting properties of large-scale eddies, though none provides a complete, quantitative picture that is entirely satisfying. The problem at hand is to model the zonal-mean climate by representing the average transport properties of the traveling eddies in terms of the time- and zonally averaged fields. It might turn out that this is not possible if we insist that the model must have some predictive capability, but at the very least it would be of didactic value.

2.2. Longitudinal Variations of Climate: Quasi-Geostrophic Theories

The zonal-mean climate, important as it is, conceals the great longitudinal variations in the Northern Hemisphere associated with the existence of oceans and continents. Interannual differences in climate at any place in temperate latitudes are more attributable to anomalies of the longitudinal rather than zonal distribution of climate. Because of the relative warmth and large thermal capacity of the oceans, in winter, air moving off the cold continents rapidly warms through the upward transmission of sensible and latent heat, so that the surface air temperature approaches that of the sea. Conversely, as air moves on to the continental land masses, a more gradual cooling due to radiation takes place without significant sensible energy transfer from the surface. The horizontal temperature gradients set up imply an accompanying motion system through thermal wind balance.

Another aspect of continents important to the forcing of longitudinal asymmetries in climate is the variation of surface elevation, which causes horizontal motion by the forced compression of vortex tubes through ascent and descent.

In addition to these mechanisms for producing longitudinal variations in climate is the preferential growth of weather systems in relation to an existing pattern of long waves or topography itself. This takes the form of two major storm tracks extending northeastward across the Atlantic and Pacific oceans from the east coasts of the major continents. It is along these tracks that the majority of eddy heat and vorticity transport takes place.

2.2.1. Thermal and Orographic Forcing. The existence of intense winter anticyclones over the Asian continent has long been recognized to be due, at least in part, to the high density of cold air formed by radiational cooling. One of the earliest three-dimensional models of "thermal forcing" based on the linearized motion and thermodynamic equations was given by Smagorinsky (1953). A cold surface anticyclone was shown to be maintained by vortex compression due to large-scale buoyant descent downstream of cooling regions. The amplitude of the associated wave field decreases with height (to zero in frictionless solutions) up to the midtroposphere, thereafter increasing with height. Phase change with height is rapid near the level of minimum amplitude (discontinuous in the frictionless case) with a 180° difference between the surface and tropopause (i.e., cold anticyclone overlain by cold low).

Numerous studies have been made since which extend and complement these results with more realistic model parameters (Döös, 1962; Saltzman, 1965, 1968; Sankar-Rao and Saltzman, 1969; Sankar-Rao, 1965b; Bates, 1977; Shutts, 1976, 1978). The two last-named authors show the importance of the stratospheric sink of wave energy propagated away from the source region of the lower troposphere for stationary waves of wave numbers 1 and 2. Phase lines of these lowest wave numbers tilt westward with height up to at least 30 km in winter and result in an important poleward transport of heat in the heating and cooling regions below. In these simple models the Icelandic and Aleutian lows of the surface-pressure maps tilt rapidly to the west with height and appear at 500 mbar as the major troughs over the east coasts of North America and Asia. At higher levels the troughs amplify with height and appear in the lower stratosphere as intense troughs over the two continents. Similarly, the Siberian anticyclone can be traced upward as a ridge tilting westward into the lower stratosphere.

Simple analytical solutions to the linearized quasi-geostrophic potential vorticity (QGPV) equation contain most of the relevant physical processes necessary to describe thermal and orographic forcing. Linearizing about a constant zonal wind $\bar{U}$, the QGPV and thermodynamic equations are

$$\left(\frac{\partial}{\partial t} + \bar{U}\frac{\partial}{\partial x}\right)\left\{\frac{\partial^2\psi'}{\partial x^2} + \frac{\partial^2\psi'}{\partial y^2} + f_0 + \beta_0 y + \frac{f_0^2}{\rho_0 g}\frac{\partial}{\partial z}\left(\frac{\rho_0}{B}\frac{\partial\psi'}{\partial z}\right)\right\}$$
$$= \frac{f_0}{\rho_0}\frac{\partial}{\partial z}\left(\frac{\rho_0 S}{B}\right) \tag{2.1}$$

$$\frac{f_0}{g}\left(\frac{\partial}{\partial t} + \bar{U}\frac{\partial}{\partial x}\right)\frac{\partial\psi'}{\partial z} + wB = S \tag{2.2}$$

where t is time; x, y, and z are Cartesian coordinates representing the eastward, northward, and vertically upward directions; ψ' is the nondivergent eddy stream function; w is the vertical velocity; f_0 and β_0 are midlatitude values of the Coriolis and beta parameters; $\rho_0(z)$ is the basic-state density field; B is the basic-state static stability; S is the rate of change of log potential temperature due to thermal sources; and g is the acceleration due to gravity.

For a source function S given by

$$S = S_0 \exp(-bz) \sin \lambda x \cos \mu y$$

where S_0 is a measure of the intensity, b^{-1} is an exponential depth scale of heating, and λ and μ are components of a horizontal wave vector in the x and y directions, respectively, a stationary solution is

$$\psi' = \text{Re}\left\{ \frac{(gS_0/f_0\bar{U}\lambda) \cos \mu y e^{i\lambda x}}{b^2 + (b/H_0) + v_0^2 + (4H_0^2)^{-1}} \right.$$
$$\left. \times \left[\left(b + \frac{1}{H_0} \right) \exp(-bz) + \left(iv_0 - \frac{1}{2H_0} \right) \exp\left(iv_0 + \frac{1}{2H_0} \right) z \right] \right\} \quad (2.3)$$

where

$$v_0^2 = \frac{gB}{f_0^2} \left(\frac{\beta_0}{\bar{U}} - \lambda^2 - \mu^2 \right) - \frac{1}{4H_0^2}$$

and it has been assumed that B is constant and that density ρ_0 decreases exponentially with height with a depth scale H_0. It can readily be shown that solutions with real and positive v_0 propagate wave energy upward and have westward-tilting phase lines. Solutions with $v_0^2 < 0$ are evanescent with specific kinetic energy $\frac{1}{2}\rho_0|\nabla_H\psi'|^2$ decreasing with height (subscript H denotes horizontal components only), though this need not demand that the amplitude decrease with height. The rapid increase in static stability as wave motion transmitted upward penetrates the stratosphere, has little effect on planetary-scale motion, but is strongly reflective for cyclone-scale motion.

One of the first models of orographic forcing was that of Charney and Eliassen (1949), who calculated the response of an equivalent-barotropic atmosphere to a realistic mountain profile on a given latitude circle. With a strongly dissipative Ekman boundary layer (represented parametrically), they found a realistic longitudinal profile of geopotential height which agreed well with observed winter 500 mbar contour height patterns. Successive improvement of model parameters and formulation have failed, however, to provide a more realistic three-dimensional picture of observed stationary-wave structure (Sankar-Rao, 1965a), and it is now recognized that the contribution of orographic forcing is probably important only in the upper troposphere and stratosphere (Manabe and Terpstra, 1974).

The prototype model of orographic forcing is obtained by setting $S \equiv 0$ in Eq. (2.1) and using the linearized lower boundary condition $w =$

$\bar{u}(\partial h_0/\partial x)$ at $z = 0$, where $h_0(x, y)$ is the height of the surface. The appropriate solution can easily be shown to be

$$\psi' = \mathrm{Re}\left\{-\frac{agB}{f_0[iv_0 + (2H_0)^{-1}]}\exp\left[\left(iv_0 + \frac{1}{2H_0}\right)z + i\lambda x\right]\right\}\cos\mu y \quad (2.4)$$

where $h_0 = a\cos\lambda x\cos\mu y$ and a is the amplitude of height variation. Again, if v_0 is real, then a positive value corresponds to a westward-tilting wave radiating energy upward. High surface pressure occurs on the windward side of the orographic slope, consistent with a net wave drag on the surface in the direction of the wind. If $v_0^2 < 0$ (strong zonal wind) and $v_0 = i|v_0|$, then

$$\psi' = \frac{agB}{f_0[|v_0| - (2H_0)^{-1}]}\exp\left[-\left(|v_0| - \frac{1}{2H_0}\right)z\right]\cos\lambda x\cos\mu y \quad (2.5)$$

and two cases are of interest: $|v_0|$ greater or less than $(2H_0)^{-1}$. In the strongly evanescent case, $|v_0| > (2H_0)^{-1}$, the wave amplitude dies out with height, and high pressure is in phase with the orographic peaks. Thermal wind balance demands that they be cold anticyclones. In the weakly evanescent case, $|v_0| < (2H_0)^{-1}$ and the wave amplitude increases with height in spite of the absence of energy propagation. High pressure now coincides with the orographic troughs, and since their amplitude increases with height they must be warm anticyclones.

For an atmosphere of height-varying $\bar{U}$, mixed behavior can occur, so that the effective value of v_0^2 may change from positive near the surface to negative higher up, and standing waves can exist in the lower layer. This type of mode can be identified by the appearance of nodes in the solution where the phase changes discontinuously by 180°, and the amplitude passes through zero.

The possibility of resonance exists for waves satisfying the criterion $|v_0| = (2H_0)^{-1}$ in Eq. (2.5), which simplifies to $\lambda^2 + \mu^2 = \beta_0/\bar{U}$. The importance of the propagating solution [Eq. (2.4)] in the troposphere is doubtful since most of the variance of $h_0(x, y)$ is contained in sufficiently high wave numbers for v_0^2 to be negative (Sankar-Rao, 1965b). Large-scale wave motion can also be induced by tall mountain blocks, such as the Himalayas, through the forced lateral displacement of airstreams and their subsequent oscillation and dispersion as Rossby waves.

2.2.2. Vorticity Forcing by Transient Eddy Motion. Edmon (1980) has shown that the contour-height anomalies of several severe winters have vertical phase lines and exist as cold low or warm anticyclone modifications to the normal January mean climate. Furthermore, it is well known that blocking episodes are characterized by warm anticyclones which extend throughout the depth of the troposphere. A natural question to ask, therefore, is What forcing mechanism can induce this type of anomaly? Thermal forcing is an unlikely candidate in view of 180° phase shift in the vertical structure. One circumstance in which thermally forced motion could be consistent with Edmon's conclusion is if heating or cooling anomalies existed in the lower stratosphere. Forced adiabatic descent beneath a lower stratospheric cooling region could then lead to a warm anticyclone, though cooling rates would have to be prohibitively large to account for the observed intensity of blocking (Green, 1977).

Weakly evanescent orographic forcing (as previously defined) is another possible mechanism for generating warm anticyclones, and in the presence of a basic-state westerly sheared flow we might expect conditions to be even more favorable for their generation. In particular, it is possible for equivalent-barotropic anomalies to exist as a remote response to orographic or thermal forcing, as shown by Hoskins and Karoly (1981).

Green (1977) has proposed that warm anticyclone anomalies can be maintained by the action of vorticity transport associated with traveling weather systems (see also Austin, 1978, 1980). Consider the time-averaged quasi-geostrophic vorticity equations

$$\mathbf{\nabla}_H \cdot (\bar{\zeta}\bar{\mathbf{V}}_H) + \mathbf{\nabla}_H \cdot (\overline{\zeta'\mathbf{V}_H'}) + \beta_0\bar{v}$$
$$= (f_0/\rho_0)[\partial(\rho_0\bar{w})/\partial z] + (\mathbf{k}/\rho_0) \cdot \mathrm{curl}(\partial\bar{\mathbf{\tau}}_H/\partial z) \qquad (2.6)$$

where all fields have been split into local time-average (denoted by overbars) and eddy contributions (denoted by primes), with ζ the vertical component of vorticity, $\mathbf{V}_H$ the nondivergent, horizontal velocity field, and $\bar{\mathbf{\tau}}_H$ the horizontal stress vector due to subsynoptic-scale motions. If the area of blocking at any level in the vertical is defined by the region enclosing the zero isopleth of $\bar{\zeta}$, then the horizontal integral over this region (denoted by $\wedge$) of Eq. (2.6) gives

$$\oint \overline{\zeta'\mathbf{V}_H'} \cdot \mathbf{n}\, dl + \beta_0\widehat{\bar{v}}$$
$$= (f_0/\rho_0)\widehat{[\partial(\rho_0\bar{w})/\partial z]} + (\mathbf{k}/\rho_0) \cdot \widehat{\mathrm{curl}(\partial\bar{\mathbf{\tau}}_H/\partial z)}$$

where dl is the length of a line segment in the zero $\bar{\zeta}$ isopleth and $\mathbf{n}$ is the unit normal vector.

Above the boundary layer, $\bar{\mathbf{\tau}}_H$ is small and $\beta_0\widehat{\bar{v}}$ is also likely to be small due to the symmetry of the blocking anticyclone, so that

$$\oint \overline{\zeta'\mathbf{V}_H'} \cdot \mathbf{n}\, dl \doteq (f_0/\rho_0)\widehat{[\partial(\rho_0\bar{w})/\partial z]}$$

Hence it can be seen that the outflow of cyclonic vorticity from the blocking region demands vertical redistribution through the stretching of planetary vorticity. Given that $\bar{w}$ is zero near the tropopause, then large-scale descent must be expected within the block, and the consequent adiabatic warming is balanced by divergence of the total heat flux and radiational cooling. Our interest in vorticity forcing by transient motion is more general, however, and we wish to know to what extent the time-mean patterns of pressure and temperature are influenced by it.

Holopainen (1978) calculated the large-scale Reynolds stresses in the time-mean momentum and vorticity equations for two 5-yr periods and found their contribution to be important compared with the time-mean advection of mean momentum and vorticity. Lau and Wallace (1979), on the other hand, find that an 11-yr average of wintertime transient eddy vorticity flux convergences at 300 mbar are smaller than the corresponding time-mean convergence by a factor of two or three. They attribute this difference to the inclusion of seasonal variation in Holopainen's definition of *transient*. However, they note a strong correlation between the surface-pressure distribution and the pattern of eddy vorticity flux divergence at 300 mbar, suggesting perhaps that the height-average eddy and mean vorticity flux divergences might be of comparable size.

More recently, Holopainen and Oort (1981) have computed the vorticity budget in the Icelandic and Aleutian lows and found the mean advection of mean vorticity to be highly cancellatory in the vertical integral, so that the surface torque is largely sustained by transient eddy vorticity flux convergence. Even if the transient eddy vorticity forcing is smaller than the mean vorticity advection term, this does not necessarily imply that transient eddy motion is unimportant in determining the mean circulation. For instance, transient eddy forcing might excite a resonant response under certain critical conditions for which $\bar{\mathbf{V}} \cdot \nabla\bar{\zeta}$ is strongly balanced by $\beta\bar{V}$ and much greater than $\nabla_H \cdot \overline{\mathbf{V}_H'\zeta'}$.

2.2.3. Proposed Mechanism of Interaction between Thermally Forced Long Waves and Transient Eddies. Consider an idealized atmosphere where isotherms are parallel to latitude circles except for an isolated cold anomaly resulting from a localized upstream cooling region, illustrated schematically in Fig. 1. In the absence of eddy instabilities and their transfer properties, thermal forcing theory requires that the lower tropospheric flow be anticyclonic in the cold region and cyclonic above. How might this structure be modified by cyclone-wave heat and vorticity transport? As indicated in Fig. 1, vorticity is transported into, and perhaps displaced downstream of, the cold anomaly and associated upper contour height trough. Also, observations and theoretical studies (Simmons and

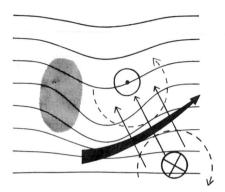

FIG. 1. Schematic diagram illustrating the relation of the storm track (prominent black arrow) to the trough in the isotherms associated with a region of cooling (shaded area). Bold arrows across the storm track indicate the direction of heat and vorticity transport, and curved dashed arrows denote the sense of vorticity forcing due to the traveling weather systems. The circle containing a dot (cross) denotes regions of eddy-induced upward (sinking) motion.

Hoskins, 1978) find that this transport is highly peaked at jet-stream level. By choosing a suitable area over which to average, as in the blocking anticyclone previously described, one can argue that this upper level convergence of eddy vorticity flux must be balanced by mean upward motion in the cold region. If, as before, this mean upward motion vanishes near the tropopause level, then the implied compression aloft and stretching of vortex tubes below lead to a lowering of the height at which the cold anticyclone turns into a cold low.

Furthermore, the warming due to eddy heat-flux convergence in the same region partly balances the adiabatic cooling associated with this induced vertical motion, but also has the effect of adjusting the amplitude and phase of the diabatic sources and sinks "seen" by the long wave. The net effect of this is to reduce the amplitude of the wave and shift the phase of the temperature field westward (White and Green, 1982a).

The self-consistency of this argument can be tested by numerical simulation using a parameterization scheme for representing the cyclone-scale transports. The net effect of eddy interactions with the forced component of the time-mean flow pattern is therefore to render the flow more barotropic by "bringing down" the upper level winds; and the extreme case, where the changeover height from cold anticyclone to cold low is brought down to the surface, corresponds to Green's (1977) mechanism for making deep warm anticyclones (or cold lows).

The cold anomaly of Fig. 1 can be identified with the cold eastern sides of the major continents in winter. Traveling depressions and anticyclones grow in the region of strong temperature gradient to the south of the anomaly and move northeastward defining the major storm track. The fact that mean eddy transports may be downstream of the main baroclinic regions of the atmosphere is consistent with linear instability analyses (Frederiksen, 1979) and the notion that vorticity and momentum transport occur in the latter stages of the life cycle of a depression (Simmons and Hoskins, 1978).

2.3. Interpretation of Observations

Having outlined dynamical mechanisms believed to be important in controlling the monthly mean climate, we can now construct a picture to explain the observed climatological features appearing on synoptic charts. This scenario need not necessarily be the "whole truth," but it will give a framework in which to discuss the observations and will eventually be testable by numerical experimentation.

First, the most obvious feature of the winter surface-pressure maps is the intense anticyclone of central Asia known as the Siberian High, in which temperatures are very low (-20 to $-40°C$), particularly in the northeast of the continent. Simple thermal forcing theories seem to explain quite adequately its magnitude and vertical structure, if diabatic cooling is about 1–2 K/day over the continent. At 500 mbar an intense trough exists over the east coast of Asia in association with the rapid falloff of pressure with height in cold air and is in better agreement with the energy-radiating, thermally forced wave ($v_0^2 > 0$) than the evanescent solution for which the upper trough lies directly above the anticyclone (Shutts, 1978). Also, observations of wave number 1 (Muench, 1965; Van Loon and Jenne, 1973) show no sign of nodal behavior at the latitude of the Siberian High, though admittedly this might be difficult to detect without data of high vertical resolution.

It should be noted that general circulation models of the weather prediction type often fail to reproduce the Siberian High (Gilchrist *et al.,* 1973), particularly if the Himalayan mountain block is absent (Mintz, 1964; Manabe and Terpstra, 1974). Mintz attributed this to the exaggerated baroclinic wave activity between Siberia and the Indian Ocean (permitted if mountains are not included), which allowed radiational cooling on the continent to be balanced by eddy heat transport. A further possibility is that orographic forcing itself contributes substantially to the maintenance of the anticyclones, though its absence in summer would therefore be difficult to explain.

Of less clear-cut explanation are the Icelandic and Aleutian lows of the winter time-mean charts. A synoptic meteorologist might be inclined to say that they represent the graveyard of depressions at the ends of the two major storm tracks across the Atlantic and Pacific oceans, whereas a planetary-wave theoretician might view them as "heat lows" formed by the warming of air moving across relatively warm oceans. Certainly in accord with the latter hypothesis is the fact that these low-pressure areas are absent at 500 mbar (or at least do not occur above the same spots) and that in summer months anticyclones are pronounced over the oceans with little evidence of an Aleutian low and only a suggestion of an Icelandic low. Also, in summer low pressure dominates over southern Asia (the monsoon) in conjunction with high mean surface temperatures (30–35°C) and is unmistakable as a thermal low.

In spite of these facts, Holopainen's calculations do indicate strong eddy flux convergence of cyclonic vorticity to the north of the storm tracks of a magnitude at least comparable with the mean flux convergence when height-averaged. This would be consistent with the appearance of cyclonic vorticity downstream of the convergence region in the time mean together with an anticyclone to the south of the storm track. The remnants of the summer Azores anticyclone on the winter surface-pressure maps may be thought of as such an eddy-driven feature.

Observations of the latitude–height distributions of the phases of wave numbers 1 and 2 (Van Loon and Jenne, 1973) show that their westward tilt with height is less rapid in middle latitudes than would be indicated by thermal forcing alone. This might be due in part to orographically forced wave contributions with vertical phase lines or, alternatively, might result from the effects of transient eddy interactions which tend to make long waves more barotropic, particularly at these latitudes (near 50° N), where transient eddy activity is greatest.

Stationary long-wave structure in the Southern Hemisphere is quite different from that in the Northern Hemisphere (Van Loon and Jenne, 1973). The July mean contour-height fields are dominated by wave numbers 1 and 3, which have vertical phase lines and amplitude increasing with height up to the tropopause at least, and so resemble Edmon's anomalies and Green's vorticity-forced long waves. In particular, wave number 1 has an interesting meridional structure with an amplitude minimum at about 40° S where the phase changes rapidly such as to give a difference of about 150° between low and high latitudes. This, again, strongly suggests Green's vorticity-forcing mechanism whereby cyclonic vorticity is locally pumped by eddies out of low latitudes and into high latitudes so as to give a dipole structure to the resulting vorticity-forced component of the pressure field. Wave number 3, although having vertical phase lines, apparently does not have the same meridional structure.

Although Green's mechanism is expressed in rather abstruse terminology, it merely expresses the tendency of developing depressions to move poleward with respect to their traveling anticyclone companions, so that in the time mean, cyclonic (anticyclonic) vorticity appears poleward (equatorward) of the storm track.

A matter for our further concern is the tendency noted by several authors (Wiin-Nielsen and Brown, 1962; Brown, 1964; Holopainen, 1970) for the diabatic heating to destroy available potential energy (APE) in the wintertime long-wave pattern. The simple thermal forcing models considered so far do not exhibit this behavior. Nonradiating, frictionless solutions have zero APE-generation rate, whereas energy-radiating solutions with friction at the ground usually have positive generation rates. From a synoptic point of view this destruction of APE occurs when very cold continental air moves out across the oceans and is rapidly warmed, leading to a strong correlation between low temperature and diabatic heating when zonally averaged. Holopainen (1970) shows that this destruction of APE is balanced by that drawn from the zonal-mean flow in the manner of a baroclinic wave disturbance.

In defense of the thermal forcing theories, it should be noted that wave solutions which transmit energy into the stratosphere also draw APE from the zonal-mean temperature field in the sense that the associated long-wave transport of heat is poleward, down the normal zonal-mean temperature gradient.

3. Parameterization: How and Why

It is essential when modeling a physical system with infinite degrees of freedom to represent a physically distinct part of that system (an infinite subset) by inaccurate or semiempirical mathematical expressions that express the irreversible effect of omitted processes. Because climate modeling involves the interaction of motion scales ranging from molecular to planetary dimensions, parameterization of many physically distinct processes is required. In principle, the atmosphere could be treated as a collection of atoms and molecules whose states and interactions are defined within the bounds of quantum theory and where radiative and diffusive processes are described at the atomic level. Climate would be defined in an ensemble-average sense and calculated from a wave function representing the entire state of the atmosphere. This approach could be considered as the general-circulation-model philosophy taken to the limit of absurdity.

Not only is it unthinkable to perform such calculations—it is also undesirable from the point of view of scientific methodology. For example, the diffusion of heat along a metal bar can be succinctly represented by a diffusion law for which the rate of heat transfer is proportional to the local temperature gradient. This well-known law not only is a practical convenience, but also embodies our understanding of heat conduction in contrast to a calculation on the molecular level. In this way we regard parameterization not only as expedient, but as a valuable expression of scientific knowledge. Unfortunately, the reality of the situation is that turbulent fluid motions do not readily succumb to mathematical analysis, and even if they did, their most useful statistical properties might not even be parametrically representable. The point is that parameterization of some physical processes need only be as good as the climate model requires. For instance, the adoption of an isentropic boundary layer conveniently avoids the subtlety of cumulus dynamics and associated heat transport properties—energy budget considerations suffice. Momentum transport, however, is not accounted for by this scheme, though its omission might not be important. Many species of atmospheric motion (e.g., tides and acoustic waves) have negligible impact on the global transfer of energy and momentum within the troposphere and can safely be neglected.

All numerical climate models parameterize the interaction of boundary-layer turbulence with the synoptic and planetary scales of motion by crude stress laws based on notions of momentum mixing. Even though in reality upgradient transfer of momentum might occur locally, the overall sense and magnitude of energy dissipation and momentum transfer can be arranged to agree with observed estimates.

Radiative transfer of heat and its interaction with clouds is another aspect of climatology requiring parameterization, and for which the determination of the required accuracy of representation is of paramount importance.

For our purposes, *climate* will be taken as the average over at least 1 month of various aspects of the motion and temperature fields. The question now arises as to what motion scales can and should be parameterized and by what criterion they should be selected. The dominance of wave numbers 1, 2, and 3 in the monthly mean 500-mbar contours suggests that all smaller scales of motion should be parameterized. This choice would be consistent with the differing physical mechanisms associated with the long, stationary waves forced by orography and topographic thermal contrasts and the traveling weather systems dependent on baroclinic instability in the horizontal density field. In terms of spatial scale, however, the distinction is not nearly so clear, with most of the kinetic energy occurring in the lowest wave numbers and only a suggestion of a double-peaked energy spectrum corresponding to the two types of motion system (Cheng and Wiin-Nielsen, 1978, Burrows, 1976).

The distinction between mean and wave motion is clear for zonal-mean climate models where all nonzonal motion is parameterized—including the stationary, forced planetary-wave motion. In this case the concept of *eddy* is unambiguous in contrast to the difficulty inherent in defining the eddy part of the flow at any specific time and place. For instance, "eddy" may be defined as the deviation from some chosen flow field which could be the climatological mean flow, the time-mean flow based on some arbitrary time interval, or even some flow pattern which has a special property such as satisfying the conservative equations of motion. Consider, for instance, an averaging period of 1 year such that all deviations from the annual mean are called transient. It is clear that seasonal variation in the amplitude of forced, quasi-stationary long waves will be classified as transient motion along with traveling weather systems. Since it is intended to parameterize only the latter, it is necessary to choose a time period long compared to the lifetime of a depression but short compared to the seasonal time scale so that transient motion is mostly due to the baroclinic instability mechanism. Even so, it is possible for the forced long waves themselves to contribute to the measured transient motion if oscillation between different equilibria of the system exists (e.g., Charney and Straus, 1980).

In spite of these difficulties, the aim of our proposed large-scale parameterization of cyclone-scale motion is to represent adequately the magnitude and spatial distribution of their associated transport properties (second-moment statistics) in terms of monthly mean flow fields. Our central assumption is that the average fluxes of dynamically important quantities are uniquely related to these fields. Compare this assumption, for instance, with the idea advanced by Stone (1978) that the temperature gradient in middle and high latitudes is close to the critical value for baroclinic instability (Pedlosky, 1964) and that any necessity for an increased poleward heat transport is easily met (without altering the temperature gradient substantially) by virtue of the "superefficiency" of baroclinic wave transfer. The implication is that the poleward rate of transfer of heat is governed by the magnitude of the sources and sinks of heat energy rather than the zonal-mean temperature gradient in an analogous way to the vertical transfer of heat in an isentropic boundary layer.

Herein lies a difference in philosophy that may prove crucial in assessing the feasibility of parameterization. Does large-scale atmospheric motion respond freely to local thermal imbalance in a passive manner, or is it sufficiently constrained to dictate the magnitude of heat transfer and hence the heat sources and sinks? The prototype of the latter behavior is purely conductive transfer (i.e., no convection or radiation), for which the inflexible thermal conductivity of air would force a temperature distribution close to radiative equilibrium. This highly inefficient mode of heat transfer should be contrasted with Paltridge's maximum-entropy-production state (Paltridge, 1978) for which (in essence) the conversion rate of potential to kinetic energy is maximized. At least locally the atmosphere probably operates between these two extremes.

If Stone's "baroclinic adjustment" hypothesis proves correct, then parameterization of heat flux in terms of mean-flow parameters must be wrong.

From a practical point of view, the main advantages of parameterization in climate modeling are the substantial economy in computer effort that can be achieved with low spatial resolution and that the results are usually more amenable to physical interpretation than those of general circulation experiments.

4. Transfer Theory of Geostrophic Turbulence

4.1. A Brief Review

The classical theory of turbulence developed by Prandtl likened the action of eddies to that of molecular collisions in the kinetic theory of gases. The concepts of mean free path and molecular velocity were replaced by those of average eddy scale and typical fluid velocity. Moreover, those properties that molecules were known to preserve between collisions were supposed to be conserved by parcels of fluid. Thus the mixing length theory for the transfer of heat, particulates, and momentum evolved. The theory was successful mainly in describing boundary-layer flow, or rather suitable rationalizations could be found to account for the diffusion coefficients observed. Among the general disquiet were questions of (1) the validity of assuming conservation of momentum for a parcel of fluid when the pressure force between it and its surroundings was clearly large and correlated with the motion; (2) the origin of the "mixing" processes which finally erased the contrast between parcel and environment; and (3) the fact that the fluxes were expressed in terms of local gradients, whereas the motion pictured was of larger scale.

Taylor (1915) apparently circumvented the problem of nonconservation of momentum by introducing the notion of vorticity transfer by two-dimensional mixing, but as recognized in several later studies (e.g., Eady, 1954; Stewart and Thompson, 1977) a new problem appears in the form of

G. J. SHUTTS

nonconservation of total momentum. In the context of the general circulation of the atmosphere, absolute vorticity mixing was once a popular concept invoked to explain the polar-front jet stream and surface polar easterlies (Rossby, 1947). The approach was limited, however, by the general nonconservation of absolute vorticity in developing wave systems. The Prandtl theory of momentum transfer had to be abandoned when applied to the large-scale atmospheric transports, since the implied eddy mixing coefficient was frequently negative (Starr, 1968).

Transport of heat from equator to pole has been considered less of a problem since the mixing length theory gives the correct sense of transfer, down the temperature gradient, particularly in the zonal mean.

It has emerged, however, that large-scale exchange of cold and warm air masses in the atmosphere is not a mixing process at all, and, as discussed by Ludlam (1966), weather systems are organized to take warm tropical air (from near the surface) to polar regions (at upper levels), where it remains while giving up its thermal energy before coming back to the tropics as cold polar air in another weather system (Fig. 2). Consideration of the relative streamlines on an isentropic surface is instructive. Warm boundary-layer air in low latitudes moves westward relative to a growing depression, subsequently turning northward where it rises rapidly up the cold front. Convective available potential energy is converted into kinetic energy of horizontal motion as the air passes upward into the jet stream and moves ahead of the storm system. In a case study, Ludlam traces the movement of potentially warm boundary-layer air over the Caribbean through a weather system, arriving later as a northwesterly jet

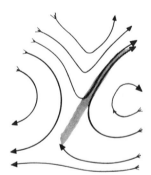

FIG. 2. Schematic representation of relative flow in a major trough of large-scale slope convection over an ocean in a surface of constant potential temperature (about 30°C). The shaded zone denotes the cold frontal zone at the confluence between two principal airstreams. Since the isentropic surface slopes upward toward the pole, northward flowing air generally rises and equatorward flow is sinking (adapted from a diagram by F. H. Ludlam).

stream over Western Europe, and calculates the release of kinetic energy along the trajectory by an extension of the parcel theory method.

Green (1970) describes such a large-scale horizontal exchange process in terms of a "transfer" rather than mixing theory, though formally the representations are identical. Potential vorticity is believed to be well conserved in traveling weather systems, and a particularly useful form of this quantity, the quasi-geostrophic potential vorticity, was used by Green to generalize the vorticity mixing theories. The essence of transfer theory is to relate the zonal- or time-mean eddy fluxes of conserved quantities to their associated mean gradients by a matrix of "transfer" coefficients whose spatial form can be inferred from stability analyses of relevant laminar flows. Hence the rate of transfer of potential temperature is expressed as an anisotropic diffusion law, involving the mean horizontal and vertical potential-temperature gradients. The rate of transfer of quasi-geostrophic potential vorticity can be expressed in terms of the horizontal potential-vorticity gradient only since conservation occurs following the horizontal wind vector. This, together with the heat transfer rate, can be related to the horizontal eddy flux of momentum, providing a rational scheme for deducing the large-scale Reynolds stresses and ultimately the surface winds.

4.2. Parameterization of Zonal-Mean Eddy Heat and Vorticity Fluxes

Mathematically, the conservation of quasi-geostrophic potential vorticity q is expressed in rectangular, Cartesian geometry (with x, y, and z the eastward, poleward, and height coordinates) by $(\partial q/\partial t) + (\mathbf{V}_H \cdot \mathbf{\nabla}) q = 0$, where

$$q = \zeta + f_0 + \beta_0 y + \frac{f_0}{\rho_0} \frac{\partial}{\partial z}\left(\frac{\rho_0 \delta \phi}{B}\right)$$

where $\phi = \phi_0(z) + \delta\phi(x, y, z, t)$ and $B = (d/dz) \ln \phi_0$. Here ϕ is the log potential temperature and the other terms are defined as in Section 2. Note that to the accuracy of the quasi-geostrophic scale analysis (i.e., for which the product of Richardson and Rossby numbers is very much greater than unity), vertical advection of q can be neglected.

The zonal-mean poleward flux of potential vorticity as represented by the Eulerian mean $\overline{q'v'}$ (where v' is the meridional component of the nondivergent wind vector) is given by

$$\overline{q'v'} = \overline{\zeta'v'} + \frac{f_0}{\rho_0} \overline{v' \frac{\partial}{\partial z}\left(\frac{\rho_0 \delta \phi'}{B}\right)}$$

where any quantity X can be expressed as the sum of a zonal-mean contribution $\bar{X}$ and an eddy contribution X'.

Use of the thermal wind relation (of the Boussinesq approximate equations), $\partial v'/\partial z = (g/f_0)(\partial\delta\phi'/\partial x)$, and the condition that all fields are cyclic in x gives

$$\overline{q'v'} = \overline{\zeta'v'} + \frac{f_0}{\rho_0}\frac{\partial}{\partial z}\left(\rho_0\frac{\overline{v'\delta\phi'}}{B}\right) \tag{4.1}$$

which forms the basis of Green's transfer theory.

It is easily shown that the vorticity flux is related to the poleward flux of westerly momentum by the expression $\rho_0\overline{\zeta'v'} = -\partial(\rho_0\overline{u'v'})/\partial y$ where u is the zonal component of the wind vector, so that poleward vorticity transfer implies an acceleration of the westerlies. The second term on the right-hand side of Eq. (4.1) can also be interpreted as a direct forcing of the zonal flow.

Consider a basic-state ϕ distribution $\phi_* = \phi_*(y, z)$ and a perturbation field $\delta\phi'$ resulting from an arbitrary infinitesimal vector displacement $\xi(x, y, z)$ then $\delta\phi' = -\xi\cdot\nabla\phi_*$ (Taylor expansion to first order) so that $\overline{\delta\phi'v'} = -\overline{v'\xi}\cdot\nabla\phi_*$ or, using the geostrophic wind relation,

$$\rho_0\overline{\delta\phi'v'} = \overline{\frac{\delta p'}{f_0}\frac{\partial\xi}{\partial x}}\cdot\nabla\phi_*$$

where $\delta p'$ is the pressure perturbation associated with the wave. Now $\overline{\delta p'(\partial\xi/\partial x)}$ represents the zonal-average wave stress acting on the corrugated ϕ surface, and so we may write

$$f_0\rho_0\overline{\delta\phi'v'}\left(\frac{\partial\phi_*}{\partial n}\right)^{-1} = \overline{\delta p'\frac{\partial}{\partial x}\xi_n}$$

where ξ_n is the displacement normal to the ϕ surface and $\partial\phi_*/\partial n$ is the magnitude of the ϕ gradient.

Hence the heat flux term in Eq. (4.1) represents a force due to the vertical variation of wave stress. This interpretation is more commonly invoked in oceanography to explain why horizontal heat transfer due to "mesoscale" eddies drives deep ocean currents, the wave stress acting on the thermocline.

In the context of small amplitude wave disturbances to a zonal current, Eliassen and Palm (1961) identify a vector

$$\mathbf{F} = \left(0, -\rho_0\overline{u'v'}, \frac{f_0\rho_0}{B}\overline{v'\delta\phi'}\right)$$

(in the quasi-geostrophic approximation) whose divergence is zero in the absence of transience and forcing dissipation.

The use of this "Eliassen–Palm flux" as a diagnostic superior to those based on energy conversions has been recommended in a paper by Edmon *et al.* (1980).

The role of $\mathbf{F}$ as a forcing stress is apparent in the zonal-mean zonal momentum equation $\partial \bar{U}/\partial t - f\bar{v}_* = \nabla \cdot \mathbf{F}$, where

$$\bar{v}_* = \bar{v} - \frac{1}{\rho_0} \frac{\partial}{\partial z} \left(\rho_0 \frac{\overline{\delta\phi'v'}}{B} \right)$$

is called the *residual* mean circulation and is zero for steady waves in the absence of irreversible eddy physics (Andrews and McIntyre, 1976).

In the zonal-mean version of the transfer theory, transient heat and potential vorticity fluxes are parameterized according to the following equations (overbars denote zonal *and* time averages):

$$\overline{v'\delta\phi'} = -K_{vy} \frac{\partial \bar{\phi}}{\partial y} - K_{vz} \frac{\partial \bar{\phi}}{\partial z}$$

$$\overline{w'\delta\phi'} = -K_{wy} \frac{\partial \bar{\phi}}{\partial y} - K_{wz} \frac{\partial \bar{\phi}}{\partial z} \qquad (4.2)$$

$$\overline{v'q'} = -K_{vy} \frac{\partial \bar{q}}{\partial y}$$

where K_{vy}, K_{vz}, K_{wy}, and K_{wz} are transfer coefficients proportional to the local temperature gradient. (Note again that only K_{vy} is relevant to potential vorticity transfer since q is conserved following the horizontal motion.) These relations are correct to second order for small amplitude motion, with $K_{vy} = \overline{v'\xi_y}$ and $K_{vz} = \overline{v'\xi_z}$ where ξ_y and ξ_z are the y and z components of the displacement vector $\boldsymbol{\xi}$. The notion that K_{vy} might depend linearly on the pole–equator temperature difference ΔT follows from considering the available potential energy of an idealized system and its subsequent release as kinetic energy. A typical eddy velocity is then proportional to ΔT and it follows that for similar meridional particle displacements $K_{vy} \propto \Delta T$.

The averaging time scale for which Eqs. (4.2) are valid has been the subject of work by Lorenz (1979), who introduced the idea of forced and free variations in heat-transfer fluctuations. Forced variations in heat transfer occur when increased sources and sinks of heat lead to increased temperature gradients and a concomitant increase in heat transport. All other reasons for a changed temperature gradient are called free variations and are typically associated with dynamical instabilities. Lorenz analyzed data and showed that diffusive laws like Eqs. (4.2) are valid only on seasonal or longer time scales and at large space scales corresponding to forced variations of the heat flux. A study by Stone and Miller (1980) appears to verify the correctness of Green's deduction that the transfer coefficient is linearly dependent on the latitudinal temperature gradient in middle latitudes at the seasonal time scale, especially in the correlation of the total (eddy and mean meridional circulation) height-integrated heat transport with the height-averaged (or surface) temperature gradient. They also found that the sum of stationary and transient heat fluxes led to a better correlation with temperature gradient than either component alone.

The transfer equations (4.2) may be substituted into Eq. (4.1), to provide a relationship between the zonal-mean eddy vorticity flux (and hence the momentum flux) and the mean fields of potential temperature and vorticity. If it is assumed that the static stability B is independent of height then

$$\rho_0 \overline{v'\zeta'} = -\frac{\partial}{\partial y}(\rho_0 \overline{u'v'}) = \frac{f_0}{B}\frac{\partial}{\partial z}\left\{\rho_0 K_{vy}\frac{\partial \bar{\phi}}{\partial y} + \rho_0 K_{vz}\frac{\partial \bar{\phi}}{\partial z}\right\} - \rho_0 K_{vy}\frac{\partial \bar{q}}{\partial y}$$

which can be simplified using the definition of $\bar{q}$ to give

$$\rho_0 \overline{v'\zeta'} = -\rho_0 K_{vy}\left(\beta_0 + \frac{\partial \bar{\zeta}}{\partial y}\right) + \frac{\rho_0 f_0}{B}\frac{\partial \bar{\phi}}{\partial y}\frac{\partial K_{vy}}{\partial z} + f_0\frac{\partial}{\partial z}(\rho_0 K_{vz}) \qquad (4.3)$$

In principle, all that would now be required to determine the vorticity flux distribution is the spatial form and magnitude of K_{vz} and K_{vy} based perhaps on baroclinic stability theory and energy arguments. The evaluation of these quantities is complicated by the existence of a powerful dynamical constraint on the vorticity flux. For reentrant channel flow geometry, the poleward vorticity must satisfy the condition $\int_0^L \rho_0 \overline{v'\zeta'}\,dy = [-\rho_0\overline{u'v'}]_0^L = 0$ at all heights when, as is reasonable, the horizontal Reynolds stress $\overline{u'v'}$ vanishes on northern and southern walls located at $y = 0$ and L.

Without simplifying assumptions, the problem of determining the form of the transfer coefficients under this constraint becomes as difficult as performing the appropriate zonal stability analysis, thereby defeating the purpose of parameterization. Clearly, it is going to be necessary to make approximations and assumptions before a useful form of the parameterization can be found.

In a channel bounded by rigid, plane upper and lower boundaries, at $z = 0$ and H, the height integral of Eq. (4.3) gives

$$\int_0^H \rho_0\overline{v'\zeta'}\,dz = -\int_0^H \rho_0 K_{vy}\left(\beta_0 + \frac{\partial \bar{\zeta}}{\partial y}\right)dz + \frac{f_0}{B}\int_0^H \rho_0\frac{\partial \bar{\phi}}{\partial y}\frac{\partial K_{vy}}{\partial y}\,dz \qquad (4.4)$$

using the fact that $K_{vz} = \overline{v'\zeta_z'} = 0$ at $z = 0, H$, since ζ_z is zero there.

Air parcels "caught up" in developing baroclinic wave systems suffer large horizontal and vertical displacements (Ludlam, 1966) such that the average transfer coefficient depends on the mean flow properties integrated over the distribution of trajectories. For this reason it is convenient to suppose that the K's averaged over the life cycles of many real baroclinic wave systems are the same as those in a similar system for which the baroclinity is constant. More precisely, this related system will be one in which the horizontal and vertical gradients of potential temperature are constant and equal to the average appropriate to the real system. Furthermore, the horizontal gradient of absolute vorticity in the related system is chosen constant and equal to the average in the real system.

Now it is a well-known result of quasi-geostrophic stability analyses that the vorticity transport due to perturbations of the above type of flow field is identically zero, which suggests from Eq. (4.4) that

$$\int_0^H \rho_0 K_{vy} \, dz = -\frac{H}{\gamma} \int_0^H \rho_0 \frac{\partial K_{vy}}{\partial z} \, dz$$

where γ is a parameter characteristic of the related flow state.

A further simplification is made by taking the terms $\beta_0 + \partial \bar{\zeta}/\partial y$ and $\partial \bar{\phi}/\partial y$ outside the integrals on the right-hand side of Eq. (4.4) and replacing them with a constant height-averaged value. This is permissible since neither term varies greatly with respect to height in its respective time- and zonal-mean average, and so Eq. (4.4) reduces to

$$\int_0^H \rho_0 \overline{v'\zeta'} \, dz = -\left[\beta_0 + \frac{\partial \bar{\bar{\zeta}}}{\partial y} + \frac{f_0 \gamma}{B} \frac{\partial \bar{\bar{\phi}}}{\partial y} \right] \int_0^H \rho_0 K_{vy} \, dz \qquad (4.5)$$

where the double overbar denotes the height and zonal-mean value. The γ parameter is now conveniently determined by the vorticity flux constraint, which states that $\int_0^L \overline{\rho_0 v'\zeta'} \, dy = 0$.

Equation (4.5) tells us that poleward eddy vorticity transfer will occur in those regions of largest temperature gradient and elsewhere equatorward transfer predominates—the extent of the regions being determined by the vorticity flux constraint.

Since in quasi-geostrophic theory the height-integrated vorticity flux determines the sense of the zonal-average surface winds, it can readily be seen that this parameterization will lead to westerlies in midlatitudes and easterlies elsewhere, in accord with the observed distribution, though other regimes are possible, particularly when the baroclinic zones are highly concentrated (White, 1977).

K_{vy} exhibits a strong maximum in the vicinity of the steering level of a baroclinic wave, where air parcels are traveling at approximately the same speed as the wave and thereby remain in a certain phase of motion for long periods. For waves growing in a zonal jet, the steering level will tend to slope upwards away from the jet maximum (Kane, 1973), which implies that the γ parameter must vary with respect to latitude. This effect is offset to some extent by the integrative property discussed earlier, whereby the form of the transfer coefficients appropriate to real finite-amplitude waves depends on the properties of the flow integrated over trajectories. Some allowance for γ variation will be made in parameterization schemes used later for climate simulation.

4.3. Parameterization of Local Time-Mean Eddy Heat and Vorticity Fluxes

Extension of the original, zonal-mean transfer theory to include longitudinal variations has been the subject of work by White and Green (1982a). A relationship between the local time-mean vector fluxes and their corresponding time-mean vector gradients is sought in the form $\overline{\mathbf{V}_H' \phi'} = -\mathbf{K} \cdot \nabla \bar{\phi}$ or

$$\overline{u'\phi'} = -K_{ux}\frac{\partial \bar{\phi}}{\partial x} - K_{uy}\frac{\partial \bar{\phi}}{\partial y} - K_{uz}\frac{\partial \bar{\phi}}{\partial z}$$

$$\overline{v'\phi'} = -K_{vx}\frac{\partial \bar{\phi}}{\partial x} - K_{vy}\frac{\partial \bar{\phi}}{\partial y} - K_{vz}\frac{\partial \bar{\phi}}{\partial z}$$

(4.6)

where $\phi = \bar{\phi}(x, y, z) + \phi'(x, y, z, t)$; $\mathbf{K}$ represents the tensor array of coefficients defined in Eq. (4.6); and the overbars now denote the time-mean, and primes the deviation therefrom.

One might justifiably ask, Is there any observational evidence for or against a diffusive type of transport law in the local time-mean fluxes? Superficially the answer is no, but I shall argue that this is primarily due to an incorrect interpretation of the relevant observations.

Several investigators have sought relationships between local eddy heat transfer and the local temperature gradient (Clapp, 1970; Tucker, 1977; Savijarvi, 1977; Lau and Wallace, 1979). In general, these investigations have failed to reveal any unique relationship and show only that the horizontal vector heat flux tends on average to be down the local temperature gradient, particularly in the lower troposphere (Lau and Wallace, 1979). One important fact that has emerged from the Lau and Wallace paper is that a substantial proportion of the vector heat flux field is non-divergent (or rotational) and as such is of no consequence to the parameterization problem. At 200 mbar, Lau and Wallace found that the mean nondivergent component of the transient eddy heat flux is two or three times larger than the divergent component. The irrotational vector heat flux is apparently directed down the temperature gradient to a much greater extent than the total heat flux vector. Consistent with this is the finding by Lau (1979) that the divergence of the transient eddy heat flux is closely in phase with the time-mean temperature field. The moral to be drawn from these results is that it is more sensible to parameterize the dynamically "active" divergent part of the transient eddy fluxes than the measured total fluxes.

A similar dominance of rotational fluxes in the vector fields of transient vorticity and potential vorticity fluxes was found. The origin of large nondivergent components in the vector fields was explained quite simply by the authors in the following manner.

In pressure (p) coordinates, the geostrophic wind $\mathbf{V}_g$ and hydrostatic equations are given by $\mathbf{V}_g = (\mathbf{k}/f) \times \nabla_p \Phi$ and $\partial \Phi / \partial p = RT/p$ where $\mathbf{k}$ is the upward pointing unit vector, Φ is the geopotential, R is the gas constant, and f is the Coriolis parameter.

If it is assumed that the eddy geopotential Φ' is a separable function of the horizontal coordinates (x, y) and pressure p, then it can easily be verified that the eddy heat flux by transient motion is

$$\overline{T'\mathbf{V}_g'} = -\frac{p}{4Rf} \mathbf{k} \times \nabla_p \left(\frac{\partial \overline{\Phi'^2}}{\partial p} \right) \tag{4.7}$$

The separability approximation applies to equivalent-barotropic disturbances whose phase lines are vertical, as in mature and decaying cyclones. A similar nondivergent eddy vorticity flux exists by virtue of the strong correlation between relative vorticity ζ_g' and geopotential Φ' since $\zeta_g' = f^{-1} \nabla_p^2 \Phi' \simeq -\mu_*^2 \Phi'$, where μ_*^{-1} is a characteristic eddy length scale, and so

$$\overline{\zeta_g' \mathbf{V}_g'} \simeq -\frac{\mu_*^2}{2f} \mathbf{k} \times \nabla_p \overline{\Phi'^2} \tag{4.8}$$

The maxima of $\overline{\Phi'^2}$ correspond to the major storm tracks, and so Eqs. (4.7) and (4.8) demand that the nondivergent vorticity flux vectors circulate cyclonically around these regions while the eddy heat flux vectors circulate anticyclonically (cyclonically) at levels below (above) the height of maximum $\overline{\Phi'^2}$. The appearance of countergradient heat fluxes in upper levels over Eastern Europe might be accounted for by the dominance of the (southward) nondivergent heat flux vector. Numerical models of the wind-driven ocean circulation show that widespread regions of countergradient heat flux occur in the ocean interior (Holland and Rhines, 1980). They conclude that these "anomalous" fluxes are associated with the local, reversible decay of baroclinic eddies which restore the undisturbed temperature field existing before instability sets in. If, however, they are associated with the rotational flux component, this interpretation is unlikely to be correct.

My overall conclusion, therefore, is that diffusive parameterization as represented by Eqs. (4.6) is not obviously inconsistent with observations provided that $\overline{u'\phi'}$ and $\overline{v'\phi'}$ are interpreted as components of a divergent heat flux vector.

To derive a parameterization of the horizontal flux of relative vorticity, the vector equivalent of Eq. (4.1) is required. This can easily be shown to be

$$\overline{q'\mathbf{V_H'}} = \overline{\zeta'\mathbf{V_H'}} + \frac{f_0}{\rho_0}\frac{\partial}{\partial z}\left(\rho_0\,\overline{\frac{\phi'\mathbf{V_H'}}{B}}\right) - \frac{f_0}{B}\overline{\frac{\phi'\partial\mathbf{V_H}}{\partial z}}$$

or using the thermal wind relation $\partial\mathbf{V_H}/\partial z = g\mathbf{k}/f_0 \times \nabla\phi'$ becomes

$$\overline{q'\mathbf{V_H'}} = \overline{\zeta'\mathbf{V_H'}} + \frac{f_0}{\rho_0}\frac{\partial}{\partial z}\left(\rho_0\,\overline{\frac{\phi'\mathbf{V_H'}}{B}}\right) - \frac{g}{2B}\mathbf{k} \times \overline{\nabla\phi'^2}$$

where $\mathbf{V_H}$ is the horizontal nondivergent wind vector.

Splitting the vector fluxes into divergent and nondivergent parts such that a vector $F = (F)_D + (F)_{ND}$ we may write

$$(\overline{q'\mathbf{V_H'}})_D = (\overline{\zeta'\mathbf{V_H'}})_D + \frac{f_0}{\rho_0}\frac{\partial}{\partial z}\left(\frac{\rho_0}{B}(\overline{\phi'\mathbf{V_H'}})_D\right)$$

$$(\overline{q'\mathbf{V_H'}})_{ND} = (\overline{\zeta'\mathbf{V_H'}})_{ND} + \frac{f_0}{\rho_0}\frac{\partial}{\partial z}\left(\frac{\rho_0}{B}(\overline{\phi'\mathbf{V_H'}})_{ND}\right) - \frac{g}{2B}\mathbf{k} \times \overline{\nabla\phi'^2}$$

$$(4.9)$$

Diffusive parameterization may now be used for $(\overline{q'\mathbf{V_H'}})_D$ and $(\overline{\phi'\mathbf{V_H'}})_D$. Following White and Green (1982a) it is convenient to assume $K_{ux} = K_{vy} = K_h$ and $K_{uy} = K_{vx} = 0$ so that Eqs. (4.6) become

$$(\overline{u'\phi'})_D = \left(-K_h\frac{\partial\bar{\phi}}{\partial x} - K_{uz}\frac{\partial\bar{\phi}}{\partial z}\right)_D$$

$$(\overline{v'\phi'})_D = \left(-K_h\frac{\partial\bar{\phi}}{\partial y} - K_{vz}\frac{\partial\bar{\phi}}{\partial z}\right)_D$$

$$(4.10)$$

$$(\overline{q'\mathbf{V_H'}})_D = -(K_h\,\nabla_H\bar{q})_D.$$

Substitution of the above expressions into Eq. (4.9) leads to

$$(\overline{\zeta'\mathbf{V_H'}})_D = \left[-K_h\,\nabla_H(\bar{\zeta} + \beta_0 y) + \frac{f_0}{B}\frac{\partial K_h}{\partial z}\,\nabla_H\bar{\phi}\right.$$

$$\left. + \frac{f_0}{\rho_0}\frac{\partial}{\partial z}\left\{\frac{\rho_0}{B}\,[\mathbf{i}K_{uz} + \mathbf{j}K_{vz}]\frac{\partial\bar{\phi}}{\partial z}\right\}\right]_D$$

after a little manipulation using

$$\bar{q} = \bar{\zeta} + f_0 + \beta_0 y + \frac{f_0}{\rho_0} \frac{\partial}{\partial z} \left(\frac{\rho_0 \bar{\phi}}{B} \right)$$

The height integral of the density-weighted vorticity flux is therefore given by

$$\int_0^H \rho_0 \overline{(\zeta' \mathbf{V'_H})}_D \, dz = - \int_0^H \rho_0 K_h \, \mathbf{\nabla}_H (\bar{\zeta} + \beta_0 y) \, dz$$

$$+ \int_0^H \frac{\rho_0 f_0}{B} \, \mathbf{\Delta}_H \bar{\phi} \, \frac{\partial K_h}{\partial z} \, dz \tag{4.11}$$

where again we have used $K_{uz} = K_{vz} = 0$ at $z = 0$ and H, consistent with the plane rigid-lid assumption.

The vector vorticity flux at any height is subject to a similar integral constraint to that used in the zonal-mean formulation. White and Green (1982a) show how this leads to two integral equations (one for each component) which must be satisfied by K_h. In the next section we shall describe another intermediate scheme in which only the meridional transfer by transient motion is parameterized.

We have chosen to find parameterizations for the eddy transport of heat and vorticity separately, but it is important to realize that the potential vorticity transport alone determines the evolution of the ψ field and that this involves only the transfer coefficients associated with horizontal gradients. Inaccuracy in the representation of K_{vz}, for instance, will only affect the relative sizes of heat and vorticity transport and the vertical velocity field.

A numerical model with parameterized cyclone waves can be built by integrating the potential vorticity equation directly with horizontal diffusion of quasi-geostrophic potential vorticity. It turns out that in multilevel quasi-geostrophic numerical models (Haltiner, 1971) the sum of the potential vorticity fluxes on all levels at a certain point is equal to the sum of the relative vorticity fluxes. This enables the vorticity flux constraint to be applied directly in terms of the height-integrated potential-vorticity flux. The two-level equations are particularly convenient for this purpose, and Marshall (1981), in a channel model of the circumpolar Antarctic current, uses this fact to determine the relative magnitudes of the transfer coefficients on each level (their horizontal profile is given by the modulus of the horizontal temperature gradient).

Sela and Wiin-Nielsen (1971) built a zonally averaged climate model using a quasi-geostrophic scheme that diffused potential vorticity according to empirically prescribed transfer coefficients. Some difficulty was found in reproducing a realistic seasonal cycle, which they attributed to the inadequacy of their Newtonian heating. Also, their parameterization did not automatically satisfy the constraint on vorticity flux and required an arbitrary adjustment to the global angular momentum field.

In spite of these problems, the model was capable of representing the injection of westerly momentum into midlatitudes by transient eddy motion, and I believe that a generalization of this approach to three-dimensional flow would be fruitful.

5. A HIGHLY PARAMETERIZED CLIMATE MODEL

The diffusive, low-resolution climate models of the type to be discussed here have a pathology quite different from that of conventional general circulation models. Explicitly represented motion scales of the model do not exhibit any form of instability in a parameter regime typical of the terrestrial climate, such as that associated with weather systems. This is a desirable property, since it is these shorter, unstable scales that are to be parameterized. The absence of rapidly amplifying wave motions leads to the dying out of any free climatic oscillation (as defined by Lorenz, 1979) in the course of a time integration, and ultimately an absolutely steady state is achieved, where all local time derivatives are identically zero. General circulation models behave quite differently for fixed external forcing—continuously spawning new weather systems and usually tending only to some statistically steady state where time means based on some suitable interval length are similar.

The stochastic character of real atmospheric flow dynamics, therefore, does not appear in our type of climate model, and it is clear that only the forced component of climatic variation approaching the seasonal time scale can be modeled. This, of course, is consistent with the definition of climatic time scale adopted here, where fluctuations of the free type are to be regarded as weather variability.

Bearing this in mind, it is now possible to decide upon the type of question the model might ultimately be capable of answering. If it were known, for instance, that diabatic heating over the Atlantic Ocean due to subsynoptic scale processes was to be below average because of lower than average sea-surface temperatures, this type of model would, in principle, be capable of determining the associated change in wind and temperature fields. The problem of climate prediction then becomes equivalent to the forecasting of ocean surface temperatures given the mean sea-level wind and temperature. Perhaps a "weather prediction" ocean model, with much longer time steps than its atmospheric counterpart, could be used in conjunction with a parameterized climate model to give seasonal climate predictions.

5.1. Model Description

The model equations are those of standard quasi-geostrophic theory expressed in spherical polar geometry for a dry atmosphere and using $h = \log(p_0/p)$ as a height coordinate, where p is the pressure and p_0 is a constant value of average surface pressure:

vorticity equation,

$$\frac{\partial}{\partial t} \nabla^2_H \psi + \mathbf{V}_H \cdot \{(\nabla^2_H \psi + 2\Omega \sin \theta) \mathbf{V}_\psi\} = \frac{f_0}{p_*} \frac{\partial}{\partial h} (p_* \omega) \qquad (5.1)$$

hydrostatic balance,

$$\partial \Phi'/\partial h = RT' \qquad (5.2)$$

continuity of mass,

$$\mathbf{V}_H \cdot \mathbf{V}_H + \frac{1}{p_*} \frac{\partial}{\partial h} (p_* \omega) = 0 \qquad (5.3)$$

thermodynamic equation,

$$\frac{\partial T'}{\partial t} + \mathbf{V}_H \cdot \{T' \mathbf{V}_\psi\} + \omega B = Q/C_p \qquad (5.4)$$

quasi-geostrophic balance equation,

$$\nabla^2_H \psi = \nabla^2_H (\Phi'/f_0) \qquad (5.5)$$

where

- θ is latitude
- ψ is the stream function of the nondivergent part of the horizontal wind vector $\mathbf{V}_\psi = \mathbf{k} \times \nabla_H \psi$
- $\mathbf{V}_H$ is the horizontal wind vector
- Φ' is the geopotential perturbation from a hemispheric average
- ω is the pseudovertical velocity Dh/Dt
- T' is the temperature perturbation from some basic state $T_0(h)$
- p_* is a normalized pressure p/p_0
- B is a static stability parameter equal to $\kappa T_0 + (dT_0/dh)$, where κ is the gas constant R divided by the specific heat at constant pressure, C_p

Q is the rate of diabatic heating (energy/unit time)
Ω is the angular rotation rate of the earth

Subscripts H denote the use of horizontal components only, so that all divergences and Laplacians are two-dimensional and calculated in h surfaces.

With some manipulation, the vorticity and thermodynamic equations can be written in terms of ψ and ω and nondimensionalized to give

$$\frac{\partial}{\partial t} \nabla_H^2 \psi + \nabla_H \cdot \{(\nabla_H^2 \psi + 2\sin\theta)\, \mathbf{V}_\psi\} = \frac{1}{p_*}\frac{\partial}{\partial h}(p_*\omega) \qquad (5.6)$$

$$\frac{\partial}{\partial t}\left(\frac{\partial\psi}{\partial h}\right) + \nabla_H \cdot \left\{\frac{\partial\psi}{\partial h}\mathbf{V}_\psi\right\} + \omega B_* = Q_* \qquad (5.7)$$

using Ω^{-1} as a time scale; a, the radius of the earth, as a length scale, choosing $f_0 = 2\Omega\sin(30°\text{ latitude}) = \Omega$, and with $Q_* = \kappa Q/\Omega^3 a^2$ and

$$B_* = R\left(\kappa T_0 + \frac{dT_0}{dh}\right)\Big/\Omega^2 a^2$$

Instead of solving two prognostic equations (5.6) and (5.7), it is convenient to obtain ω from a diagnostic equation and use the vorticity equation for integration in time. The resulting nondimensional "omega equation" is

$$B_*\nabla_H^2\omega + \frac{\partial}{\partial h}\left[\frac{1}{p_*}\frac{\partial}{\partial h}(p_*\omega)\right] = \nabla_H^2 Q_* + \frac{\partial}{\partial h}\left\{\nabla_H \cdot \left[(\nabla_H^2\psi + 2\sin\theta)\,\mathbf{V}_\psi\right]\right\}$$

$$- \nabla_H^2\left[\nabla_H \cdot \left(\frac{\partial\psi}{\partial h}\mathbf{V}_\psi\right)\right] \qquad (5.8)$$

which together with Eq. (5.6) forms a closed pair of equations for ψ and ω.

The conventional, multilevel vertical structure for quasi-geostrophic models is adopted (Haltiner, 1971) with, in our case, two ω (or thermodynamic) levels sandwiched midway between three kinematic ψ levels. The ground and model top coincide with extreme pseudo-ω levels, where boundary conditions are expressed directly in terms of ω. The horizontal structure of ψ and ω at any appropriate level is expanded in a truncated set of spherical harmonics, such that the wave harmonics are of odd parity and the zonal harmonics are of even parity (odd-parity harmonics are antisymmetric about the equator). Such a mixed-parity representation turns out to be essential for a hemispheric model in which nondivergent and zonal-mean cross-equatorial flow are to be forbidden. The hemispheric average of pressure and temperature perturbations together with the vertical velocity must be zero for consistency with the definition of perturbation and to prevent cross-equatorial flow. In this mixed parity formulation this is automatically satisfied if no hemispheric-average forcing is present.

If $P_n^m(\sin \theta)$ is the associated Legendre function of rank m and degree n, then the horizontal structure of the model fields is represented by the following spherical harmonics $Y_n^m (= P_n^m(\sin \theta) \exp(im\lambda)$, with $\lambda =$ longitude):

Y_2^0, Y_4^0, Y_6^0	Y_2^1, Y_4^1, Y_6^1	Y_3^2, Y_5^2, Y_7^2	Y_4^3, Y_6^3
Zonal modes	Wave number 1	Wave number 2	Wave number 3

The following properties were used:

$$\nabla_H^2[Y_n^m] = -n(n + 1) \, Y_n^m$$

$$2 \int_0^1 P_n^m P_{n'}^{m'} \, d\mu = \delta_{m'n'}^{mn}$$

if $n - m$ and $n' - m'$ are either both odd or both even, where

$$\delta_{m'n'}^{mn} = \begin{cases} 0 & \text{if } (m, n) \neq (m', n') \\ 1 & \text{if } (m, n) = (m', n') \end{cases}$$

The velocity fields, however, are more naturally expressed in spherical harmonics *after* multiplication by $\cos \theta$. Flow over the pole becomes possible in this representation.

Evaluation of the divergence terms in Eqs. (5.6) and (5.8) is accomplished by the grid-transform technique (Bourke, 1972), whereby the relevant fluxes are calculated by "gridding" the appropriate fields at equal intervals along selected Gaussian latitudes. Exact calculation of the divergences is possible by Fourier expansion of the fluxes at grid points on these latitude circles, followed by integration with respect to sin latitude using Gaussian quadrature.

For instance, consider the evaluation of the vorticity flux divergence given in spherical polar form by

$$\nabla_H \cdot [\zeta \mathbf{V}_\psi] = \frac{\partial}{\partial \mu} (\zeta V) + \frac{1}{1 - \mu^2} \frac{\partial}{\partial \lambda} (\zeta U) = D_z(\mu, \lambda)$$

where $U = u \cos \theta$, $V = v \cos \theta$, $\mu = \sin \theta$, and u and v are the eastward and northward components of the nondivergent wind vector. Expanding the vorticity flux components in a Fourier series up to wave number J gives

$$\zeta U = \tfrac{1}{2}A_0(\mu) + \sum_{m=1}^J (A_m(\mu) \cos m\lambda + B_m(\mu) \sin m\lambda)$$

$$\zeta V = \tfrac{1}{2}C_0(\mu) + \sum_{m=1}^J (C_m(\mu) \cos m\lambda + D_m(\mu) \sin m\lambda)$$

so that

$$D_z = \frac{1}{2}\frac{dC_0}{d\mu} + \sum_{m=1}^{J}\left[\left(\frac{dC_m}{d\mu} + \frac{mB_m}{1-\mu^2}\right)\cos m\lambda + \left(\frac{dD_m}{d\mu} - \frac{mA_m}{1-\mu^2}\right)\sin m\lambda\right]$$

where A_m, C_m and B_m, D_m are the respective cosine and sine coefficients treated as functions of latitude.

If now the divergence D_z is expanded into spherical harmonics according to the relation

$$D_z(\mu, \lambda) = \sum_{\substack{n=2 \\ n\ \text{even}}}^{N} D_{cn}^0 P_n^0(\mu) + \sum_{m=1}^{J}\sum_{\substack{n=m+1 \\ (n-m)\ \text{odd}}}^{N} (D_{cn}^m \cos m\lambda + D_{sn}^m \sin m\lambda)\, P_n^m$$

it follows using the orthonormality condition that

$$D_{cn}^0 = \int_0^1 \frac{dC_0}{d\mu} P_n^0\, d\mu \quad \text{and} \quad D_{cn}^m = 2\int_0^1 \left(\frac{mB_m}{1-\mu^2} + \frac{dC_m}{d\mu}\right) P_n^m\, d\mu$$

$$D_{sn}^m = 2\int_0^1 \left(\frac{dD_m}{d\mu} - \frac{mA_m}{1-\mu^2}\right) P_n^m\, d\mu$$

and with the n summation truncated at N. Finally, integration by parts yields expressions suitable for numerical evaluation:

$$D_{cn}^0 = -\int_0^1 C_0 \frac{dP_n^0}{d\mu}\, d\mu$$

$$D_{cn}^m = 2\int_0^1 \left(\frac{mB_m}{1-\mu^2} P_n^m - C_m \frac{d}{d\mu} P_n^m\right) d\mu$$

$$D_{sn}^m = -2\int_0^1 \left(\frac{mA_m}{1-\mu^2} P_n^m + D_m \frac{d}{d\mu} P_n^m\right) d\mu$$

and it can be shown that their integrands are polynomial in μ and so are amenable to exact integration (using the fact that $C_m = D_m = 0$ at $\mu = 0, 1$).

Substitution of the spectral expansions of ψ and ω into Eq. (5.6) leads to an ordinary differential equation for the time rate of exchange of the spectral coefficients of ψ. Time integration was initiated using a Miyakoda start (Miyakoda, 1960) and continued by the leapfrog method incorporating a Robert filter to suppress "splitting" of the solution after many time steps (Robert, 1966).

In the experiments to be described here, ω was taken to be zero at the model top corresponding to a height of about 12 km. The lowest pseudo-ω level of the model is identified with the top of a surface Ekman layer where ω is set proportional to the vorticity extrapolated to that level.

Extrapolation of the ψ field to this "surface" level is given by the formula $\psi_0 = (15\psi_1 - 10\psi_2 + 3\psi_3)/8$, where ψ_i is the stream function on level i, subscript 0 corresponding to the surface; and the Ekman layer expression for ω_0 is given by $\omega_0 = \alpha \nabla_H^2 \psi_0$, where α can be related to (boundary layer eddy diffusivity/f_0)$^{1/2}$.

The diabatic heating field Q is prescribed by a Newtonian heating law of the form

$$Q = -\gamma_* C_p(T' - T'_*) \tag{5.9}$$

where $T'_*(\lambda, \mu, h)$ is a perturbation equilibrium temperature field to be specified and γ_*^{-1} is a time constant. In the absence of any convective or parameterized heat transfer, the model temperature distribution would equal the equilibrium temperature. It is appropriate to split this formula into zonal-mean and nonzonal components with differing time constants associated with each. Very cold continental air moving out over warm oceans in winter adjusts toward equilibrium at a much faster rate than the hemispheric-scale temperature field adjusts to equilibrium. This results from the differing physical processes establishing equilibrium, namely, radiation and small-scale convection, of which the latter allows more rapid thermal adjustment. Hence, Eq. (5.9) has been written as

$$\begin{aligned} \bar{Q}^\lambda &= -\gamma_z C_p(\bar{T}'^\lambda - \bar{T}'^\lambda_*) \\ Q_E &= -\gamma_E C_p(T'_E - T'_{*E}) \end{aligned} \tag{5.10}$$

where $^{-\lambda}$ denotes the zonal-average and subscript E denotes the remaining eddy component with corresponding zonal-mean and eddy time constants γ_z^{-1} and γ_E^{-1} respectively.

Before considering the parameterization of weather systems, it is worthwhile discussing some of the weaknesses of the quasi-geostrophic formulation. Perhaps the most restrictive assumption of quasi-geostrophic theory is the constancy of the Coriolis parameter in all terms except the one that represents the meridional advection of planetary vorticity (the beta effect). The stretching term on the right-hand side of Eq. (5.1) misrepresents the vertical velocities needed to balance the advection terms on the left-hand side, particularly in the tropics. Thus, a vorticity source in low latitudes requires much weaker vertical stretching velocities than in the real atmosphere, and as a partial consequence of this, the model Hadley circulation is weaker than observed. The opposite of this happens in high latitudes, where the induced mean-meridional circulation is more intense in the model than observed.

Of more direct concern is the neglect of Coriolis variability in the thermal and geostrophic wind equations which demand that the latitudes of maximum zonal wind and its shear coincide with the latitudes of maximum pressure and temperature gradients, respectively. In the real atmosphere, the maxima of the mean zonal wind and its shear occur to the south of the pressure and temperature-gradient maxima.

Another, apparently rather serious assumption is the neglect of the latitudinal variation of static stability. Mean static stabilities in high latitudes are at least twice those in tropical regions, and proper account of the role of moisture only serves to accentuate this difference. That quasi-geostrophic models can and do give sensible results when used to simulate hemispheric climate is a remarkable fact. One possible explanation of their forgiving nature lies in the regular appearance of the constant factor f_0/B. Since the latitudinal variation of the Coriolis and static stability parameters is in the same sense, it is possible that their ratio is a more slowly varying function of latitude and that there is some compensation in the two approximations. In spite of the weaker than observed Hadley circulation in the model, poleward heat transport is still effective, since the (dry) static stability is larger than observed.

5.2. Model Parameterization

Lau and Wallace (1979) showed that the divergent part of the observed time-mean vector heat and vorticity fluxes due to transient motion are predominantly poleward, and in view of the difficulties associated with the satisfaction of the vector vorticity flux constraint described in Section 3, it was decided to parameterize the poleward components only.

In spherical polar geometry, Eqs. (4.10) and (4.11) for the heat and vorticity fluxes, respectively, become

$$\rho_0(\overline{\phi'V'})_D = \left(-\rho_0 K_h \frac{(1-\mu^2)}{a} \frac{\partial \bar{\phi}}{\partial \mu} - \rho_0 K_{V'z} \frac{\partial \bar{\phi}}{\partial z}\right)_D$$

$$\int_0^H \rho_0(\overline{\zeta'V'})_D = \left\{-\frac{(1-\mu^2)}{a}\left[\int_0^H \rho_0 K_h \left(2\Omega + \frac{\partial \bar{\zeta}}{\partial \mu}\right) dz\right.\right.\qquad(5.11)$$
$$\left.\left.-\frac{f_0}{B}\int_0^H \rho_0 \frac{\partial K_h}{\partial z} \frac{\partial \bar{\phi}}{\partial \mu} dz\right]\right\}_D$$

where V' is the poleward eddy component of velocity × cos latitude and using similar approximations to those introduced in the zonal-mean case, the vorticity-flux expression may be simplified to

$$\int_0^H \rho_0(\overline{\zeta'V'})_D \, dz = \left\{-\frac{(1-\mu^2)}{a}\left[2\Omega + \frac{\partial \bar{\bar{\zeta}}}{\partial \mu} + \frac{f_0\gamma}{B}\frac{\partial \bar{\bar{\phi}}}{\partial \mu}\right]\int_0^H \rho_0 K_h \, dz\right\}_D \qquad(5.12)$$

where the double overbar now denotes a time and height average and γ is determined by the hemispheric constraint on the vorticity flux

$$\int_0^{2\pi} \int_0^1 \rho_0 (\overline{\overline{\zeta'V'}})_D \, d\mu \, d\lambda = 0$$

The use of Eq. (5.11) for the transient entropy flux is complicated by the term $\rho_0 K_{V'z}(\partial\bar{\bar{\phi}}/\partial z)$, which introduces a further transfer coefficient into the scheme. There are, however, strong geometrical reasons for believing that $K_{V'z}$ should be simply related to K_h. Eady (1949) showed that the maximum rate of release of potential energy by finite amplitude displacements in a cyclone wave occurs if the trajectory slopes are half the initial slope of the isentropic surfaces, and he suggested that this is a feature of real amplifying baroclinic waves near the steering level.

Consider an initially zonal flow so that if ϵ is the slope of the isentropic surfaces then the optimum vertical displacement of an air parcel ξ_z from its rest level is related to its meridional displacement ξ_θ by

$$\xi_z = \frac{1}{2} \epsilon \xi_\theta = - \frac{1}{2a} \frac{\partial\bar{\phi}^\lambda/\partial\theta}{\partial\bar{\phi}^\lambda/\partial z} \xi_\theta$$

and so

$$K_{V'z} = \overline{V'\xi_z}^\lambda = - \frac{1}{2a} \frac{\partial\bar{\phi}^\lambda/\partial\theta}{\partial\bar{\phi}^\lambda/\partial z} \overline{V'\xi_\theta}^\lambda$$

or

$$K_{V'z} \frac{\partial\bar{\phi}^\lambda}{\partial z} = - \frac{1}{2a} K_h \frac{|\bar{\phi}^\lambda}{\partial\mu} (1 - \mu^2)$$

On average, the trajectory slopes are more like $\frac{1}{3}\epsilon$, since they must approach zero at the ground and tropopause so that

$$K_{V'z} \frac{\partial\bar{\phi}^\lambda}{\partial z} \doteq - \frac{1}{3}(1 - \mu^2) \frac{K_h}{a} \frac{\partial\bar{\phi}^\lambda}{\partial\mu}$$

suggesting the following expression for eddy heat flux:

$$(\overline{\phi'V'})_D \doteq - \frac{2}{3} \frac{(1 - \mu^2)}{a} K_h \frac{\partial\bar{\phi}}{\partial\mu} \tag{5.13}$$

As discussed earlier, there are good theoretical reasons to suppose that the average transfer coefficient should be proportional to the local temperature difference across a baroclinic zone. Inspired by this supposition, the following expression for the height-averaged transfer coefficient was used in the model integrations:

$$\frac{1}{(\rho_S - \rho_H)H_0} \int_0^H \rho_0 K_h \, dz = \frac{4 \times 10^7}{T_0} \left|\frac{\partial \bar{\bar{T}}}{\partial \theta}\right| G(\theta) \quad [\text{m}^2 \, \text{sec}^{-1}] \quad (5.14)$$

where $\rho_S = \rho_0(z = 0)$, $\rho_H = \rho_0(z = H)$; T_0 is a constant mean atmospheric temperature taken to be 250 K. The log pressure coordinate form of this equation can be obtained by replacing ρ_0/ρ_S with p_* and z/H_0 with h; $G(\theta)$ is an arbitrary function equal to 1 if $\theta > 30°$ N and tailing off to zero south of $30°$ N, as shown in the tabulation:

θ (°N)	$G(\theta)$	θ (°N)	$G(\theta)$
>30	1	6.6	0.07
25.6	0.6	2.7	0.03
18.4	0.35	0.5	0.0
11.9	0.15		

Without it the transfer coefficient in low latitudes is predicted to be too large, conspiring with the large beta there and ultimately leading to an overestimation of the momentum and vorticity transport due to transient eddies. This is at least partly the fault of the quasi-geostrophic formulation, which allows unrealistically large horizontal temperature gradients in low latitudes through the largeness of the Coriolis parameter there (see White, 1977).

The height variation of parameterized vorticity transfer remains undetermined in this formulation of the transfer theory, and so it is chosen to be in accord with observations. On the three kinematic model levels, the density-weighted vorticity flux was apportioned in the ratio 1.8 : 0.9 : 0.3 (with the largest on the uppermost level) in such a way that the height average is consistent with Eq. (5.12).

Equation (5.13) was applied on the two model thermodynamic levels with K_h equal to the height integral [Eq. (5.14)] multiplied by an appropriate constant so as to give a 1 : 4 ratio of the heat flux $\rho_0 \overline{V'T'}$ when $\partial \bar{\phi}/\partial \mu$ is independent of height. This low-level bias of the heat flux was chosen to be consistent with the observed bias (Oort and Rasmusson, 1971) and in such a way as to preserve the correct height integral of the heat flux. Note that ϕ' and $\bar{\phi}$ can be replaced by T' and $\bar{T}$ in Eq. (5.13) when differentiation with respect to μ is performed holding h constant. Also, since $\phi' = (T'/T_0) + \kappa h'$, $\overline{\phi' \mathbf{V}_\psi'} = (\overline{T' \mathbf{V}_\psi'}/T_0) + \kappa \overline{h' \mathbf{V}_\psi'}$; but $\overline{h' \mathbf{V}_\psi'} = (gH_0/f_0)\mathbf{k} \times \nabla(\frac{1}{2}\overline{h'^2})$, which is a nondivergent vector, and so $(\overline{\phi' \mathbf{V}_\psi'})_D = (\overline{T' \mathbf{V}_\psi'}/T_0)_D$.

A problem of the spherical polar formulation of the vorticity transfer theory is the absence of equatorward transport in polar regions, so that polar easterlies do not appear. For middle-latitude zonal jets, stability analyses indicate that the steering level rises to high levels on the northern and southern flanks of a baroclinic zone, implying that K_h, which is large near the steering level, should peak higher up, and hence γ should be negative there. By specifying a latitudinal shape for γ of the form $\gamma = \gamma_0[\mu^2(1 - \mu^2) - \frac{1}{8}]$, where γ_0 is now the constant to be determined by the vorticity flux constraint, White (1974) found that polar easterlies did occur.

The introduction of empirically based "shape functions" does detract from the original transfer theory and, one imagines, restricts its use to circulation regimes rather close to the observed terrestrial climate. Even so, their role is based on understandable physical concepts, and there seems no fundamental reason why this empiricism should not eventually be removed and embodied in a more elaborate scheme. Some progress has been made in this direction by White and Green (1982b). There is a real danger, however, that this elaboration might take the form of a disguised linear-stability analysis and be at least as time consuming to perform.

What evidence is there that this longitudinally dependent form of the transfer theory will give sensible results?

First, it should be noted that Eq. (5.12) implies that poleward vorticity transport will occur in the regions of largest poleward temperature gradient, where "largeness" is defined with respect to the mean poleward absolute vorticity gradient. In this sense, our parameterization of vorticity flux is in agreement with the linear stability analyses of nonzonal flows carried out by Frederiksen (1979). He showed that transient momentum flux convergence occurred in and slightly downstream of regions where the vertical wind shear is most "supercritical," as defined by Pedlosky (1964).

Second, there is some observational evidence of a dipole structure to the transient eddy vorticity flux divergence field about the jet-stream maxima at 300 mbar (Lau, 1979). This would seem to indicate that cyclonic vorticity is pumped out of regions to the south of jet maxima (those off the east coast of North America and Asia and that of the North African jet) and into regions immediately to the north.

Thirdly, regarding the heat transfer expression, Eq. (5.13), Stone and Miller (1980) quote observational evidence for a square-law dependence of the height-integrated heat flux on the height-averaged local temperature gradient in middle and high latitudes.

My overall opinion is that in spite of the rather ad hoc adjustments necessary to make the vorticity transfer expression work, it still stands as a quantitatively correct expression consistent with our intuitive picture of large-scale vorticity dynamics.

5.3. Experiments and Results

Our primary interest is in the role played by transient eddies in forcing the time-mean flow given a distribution of diabatic sources of heat that approximates that of the observed Northern Hemisphere winter. In particular, the dynamical forcing associated with the longitudinal variations of eddy transports is required for comparison with observed estimates.

In the following experiments to be described, the distribution of radiative equilibrium temperature T_*^i was fixed with $\gamma_z = 1/(30 \text{ days})$ and $\gamma_E = 1/(10 \text{ days})$—Eq. (5.10). The resulting steady-state heating distribution on level 1 associated with it in the first experiment (E1) is given in Fig. 3. It is to be regarded as suggestive of the real distribution of diabatic heating rather than a best fit to a calculated distribution. Lau (1979) finds typical diabatic heating rates at 700 mbar of about 1 K day^{-1} over large areas of the ocean and cooling rates of about 1–2 K day^{-1} over the continental land masses. Even larger sources and sinks are found at 1000 mbar. Since the diabatic forcing is thought to decrease with height, T_* on level 2 was reduced by a factor of 0.4 over its level 1 magnitude. The friction coefficient α was chosen to be 300 m, consistent with a boundary-layer eddy diffusivity of 6.6 m^2 sec^{-1} and the basic-state static stability $d\phi_0/dz$ was chosen to be 1.8×10^{-5} m^{-1}.

In E1, the model—with parameterization as described in the previous section—is integrated to steady state in about 120 days. The resulting surface-pressure map, Fig. 4, shows a high-pressure belt extending around most longitudes at 30° N with cells downstream of the two cooling

FIG. 3. Distribution of diabatic heating on level 1 in K/day at the steady state in E1.

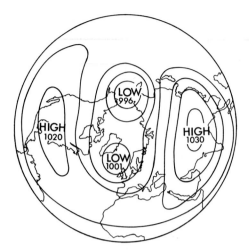

FIG. 4. Steady-state surface pressure field for E1. Contour interval, 5 mbar.

regions over the western side of the continents. The high-pressure cell to the north of India is the model's attempt at making a cold Siberian anticyclone. Low-pressure areas near Iceland and over the eastern extremity of the USSR are rather convincing manifestations of observed climatological lows in these regions.

Figure 5 shows the height contours on the middle level of the model, which can be thought of as the 500 mbar surface. Troughs extend south-

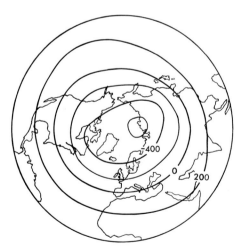

FIG. 5. Geopotential height contours on level 2 (deviation from hemispheric mean) for E1. Contour interval, 200 m.

ward across the eastern side of the two major continents in association with the coldest lower atmosphere air. The Siberian trough is to the west of its observed position merely because heating extends inland in the model, in contrast to the real atmosphere, where heating is confined to the sea. This effect is also noticeable in the lower level temperature field, Fig. 6, where the coldest air extends into central Siberia at the longitude of zero diabatic heating. Air is coldest after it has traversed the whole extent of the cooling region from west to east.

The eddy heat flux divergence on the upper thermodynamic level is shown in Fig. 7. Warming due to eddy convergence of heat exists between 50° and 75° N with a small region of divergence at the pole due to a reversed temperature gradient there. This reversal occurs on the upper thermodynamic level in the model only, near both the pole and equator. Cooling due to eddy heat flux divergence occurs south of 50° N, peaking at 30° N. Longitudinal variations are not great, ranging from −1.4 to −0.6 K day^{-1} at 60° N and considerably smaller than the range of diabatic heating variation. On level 1, however, longitudinal variations in eddy heat flux divergence are about twice as large as on level 2 and make an important contribution to the thermodynamic balance. Convergence of eddy heat flux into the northeastern corner of the USSR is balanced by a divergence maximum to the south at about 30° N, consistent with a strong poleward flux of heat in between (i.e., where the poleward temperature is large).

Eddy vorticity flux divergence on level 3 is shown in Fig. 8. The general picture of cyclonic vorticity forcing for the latitude zone 45°–75° N is in

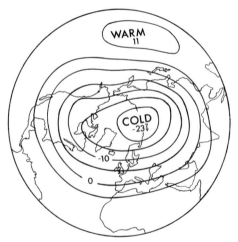

FIG. 6. Lower level perturbation temperature field for E1. Contour interval, 5 K.

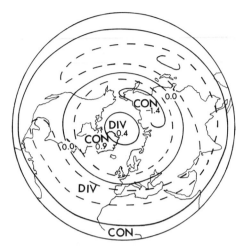

FIG. 7. Parameterized eddy heat flux divergence on the upper thermodynamic level for E1. Contour interval (dashed), 0.5 K/day; DIV, divergence; CON, convergence.

agreement with the findings of Lau (1979) at 300 mbar, though observed longitudinal variations are more pronounced.

The zonal-mean distribution of eddy vorticity flux convergence is, of course, very similar to the zonal-mean surface-pressure distribution, so that shaded areas are on average low pressure. Also, the meridional circulation driven by the vorticity transport implies, on average, rising motion in shaded regions.

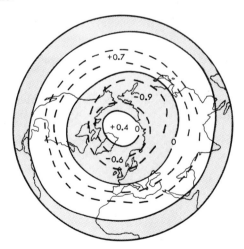

FIG. 8. Parameterized eddy vorticity flux divergence on level 3 for E1. Contour interval (dashed), 0.5×10^{-10} sec^{-2}; shading indicates regions of convergence.

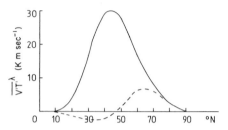

FIG. 9. Latitudinal distribution of zonal-mean eddy heat flux due to parameterized motion (bold curve) and explicit stationary long-wave motion (dashed curve) on a lower thermodynamic level.

In Fig. 9, zonal-mean latitudinal distribution of poleward heat transport, due to transient (parameterized) and stationary (explicit) waves, is given. It is clear that in this model most of the eddy transport of heat poleward is achieved by parameterized motion. In fact, explicit wave transport of heat is actually equatorward south of 45° N, although rather small. One possible cause of this reduced stationary-wave heat transport compared to observations is the inability of model planetary waves to propagate energy into the stratosphere because of the wave-reflecting boundary condition, (Shutts, 1978). It is hoped that the use of a radiation boundary condition for ω at the model top will increase the heat transport by forcing the wave to tilt to the west more rapidly with height.

Zonal-mean poleward flux of vorticity is also dominated by the parameterized component and is likely to be in conflict with observations, since it is known that the stationary-wave transport of momentum poleward is at least as important as the transient eddy transport in winter. One gets the general impression that the explicit waves prefer to arrange themselves in phase relationships that demand the least possible interaction with the zonal flow and each other.

Figures 10 and 11 show the zonal-mean zonal component of the wind field and the vertical velocity field for E1. Apart from the zonal-mean jet

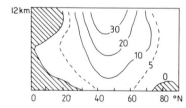

FIG. 10. Latitude–height cross section of the zonally averaged zonal component of the wind field in m sec^{-1}. Shaded areas indicate regions of easterly zonal-mean winds.

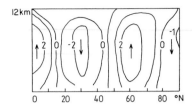

FIG. 11. Latitude–height cross section of the zonally averaged vertical velocity field in mm sec^{-1}.

maximum being a little too far north and the surface westerly maximum of 7 m sec^{-1} being rather large, the overall picture is in agreement with the observed zonal-mean profile given by Oort and Rasmusson (1971). The distribution of zonal-mean vertical velocity is in qualitative agreement with the observed three-cell structure, though the relative magnitudes of the cell circulation rates are incorrect. In reality, upward motion in the Hadley cell is two or three times larger than that in the Ferrel cell at 60° N. This can, as has been mentioned before, be attributed mainly to the neglect of the latitudinal variation of the Coriolis and static stability parameters. The influence of water vapor transport and latent heat release must also play a large part in determining the structure and intensity of the Hadley cell—probably by decreasing the horizontal scale of the upward branch (Vàllis, 1982).

The amplitude and phase of wave numbers 1, 2, and 3 at 60° and 45° N are displayed in Fig. 12a and b. For all waves, amplitudes generally increase with height though the extrapolated surface amplitude is nearly always larger than that on level 1. The existence of a lower tropospheric amplitude minimum (where the cold anticyclone changes to a cold low) is a regular feature of the linearized theory of thermally forced planetary waves, though in this case the height of the minimum is reduced by the transient eddy vorticity forcing (see Section 1). At 60° N, wave number 1 is trapped with vertical phases lines in the model's upper troposphere and rapid westward tilt with height at level 1 in the region of amplitude minimum. The phase difference of about 170° between the top and bottom of the model troposphere is typical of a thermally forced wave with cold high overlain by cold low.

Since the distribution of diabatic heating has a strong wave number 2 component, it is no surprise to find wave number 2 dominating the stream function field. Unlike wave number 1, it does tilt westward with height between levels 2 and 3 at 60° N, and it is due almost entirely to this wave that the explicit stationary waves transport heat in high latitudes. At 45° N, wave numbers 1 and 2 lean slightly eastward with height between

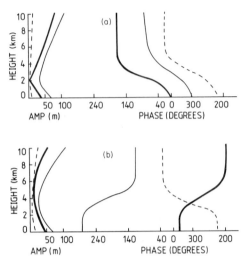

FIG. 12. Amplitude and phase of wave numbers 1 (bold curve), 2 (thin curve), and 3 (dashed curve) at (a) 60° N and (b) 45° N.

levels 2 and 3, consistent with equatorward heat transport and in direct conflict with observations (Van Loon and Jenne, 1973). At both latitudes, wave number 3 is trapped and of relatively small amplitude.

The noninteractive role of the explicit long waves is interesting and has also been found by White and Green (1982a) in similar experiments with beta-plane channel model. Would the long waves behave similarly using model equations that incorporate the real spatial variability of the Coriolis parameter? Parameterized motion is important mainly in determining the zonal-mean state of the model, and the long-wave response is similar to that given by linear theory with little feedback to the zonal flow.

Some further experiments have been carried out to assess the importance of longitudinal variations of parameterized vorticity transport. The procedure adopted was to take the calculated distribution of the poleward component of vorticity on a latitude circle and spread it evenly by replacing the grid-point values by the zonal mean. Repeating E1 with the vorticity flux treated in this led to a steady state of very similar flow pattern with surface-pressure values different by less than a millibar. This was to be expected, since longitudinal variations of the parameterized eddy vorticity flux divergence in E1 were typically about seven times smaller than their explicit motion counterpart at level 3. This ratio of transient eddy to time-mean flow vorticity forcing is a useful quantity in gauging the importance of transient eddy motion. Lau and Wallace (1979) found on average that the time-mean flow vorticity forcing in the Northern Hemisphere winter was two to three times larger than the transient eddy forcing.

The relative smallness of the model's longitudinal variations of eddy vorticity forcing must be in part due to the neglect of east–west transfer and partly due to failure in representing the local intensity of temperature gradient on the east coasts of the continents. Whatever the reason, we should not necessarily conclude that transient eddies are unimportant in forcing longitudinal variations in the flow fields by vorticity transfer.

Because our main interest is in the mechanism of transient eddy vorticity forcing, experiments have been carried out whereby the longitudinal variations in this forcing have been increased by a certain factor. How much do they have to be increased before Green's warm anticyclone mechanism becomes important and to what extreme values of eddy vorticity flux divergence does this correspond?

We imagine the eddy vorticity flux at a point in the atmosphere to be a very noisy quantity in time and that for some periods of time its value will considerably exceed the climatological norm. It is during these periods that Green's mechanism might apply and it is with regard to these conditions that my experiments were conceived. Clearly this is a departure from the philosophy previously adopted whereby the model was used to simulate the long-term "forced" climatic state rather than the "free" climatic anomalies.

In the first of these experiments (to be called E2), the longitudinal deviations of the parameterized vorticity flux from the appropriate zonal-mean are increased by a factor of three. This brings it into line with observed transient eddy vorticity forcing as found by Lau and Wallace (1979). The resulting surface-pressure pattern is shown in Fig. 13. The general pattern is still very similar to the normal case portrayed in Fig. 4, except that the "Aleutian" low is now more intense and about 30° farther west. The anticyclone over the southern United States has a strong ridge extending into Canada. Time-mean flow vorticity forcing is now typically $2\frac{1}{2}$ times larger than eddy vorticity forcing. Figure 14 is the distribution of transient eddy vorticity forcing on level 3 which shows a region of strong convergence above the main surface depression and a region of divergence to the south at 20° N. Comparing this with the "normal" distribution, Fig. 8, it can be seen that cyclonic vorticity forcing over Canada has been replaced with anticyclonic forcing.

The phase difference of wave number 1 between level 3 and the surface at 60° N is now only 30° compared to 170° previously, showing how the changeover height (where cold high turns to cold low) has been brought down to the surface by the vorticity forcing mechanism described in

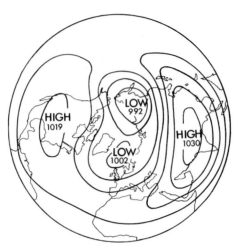

FIG. 13. Surface-pressure distribution for E2. Contour interval, 5 mbar.

Section 2. The zonal-mean state for E2 was not substantially different from the normal E1 case.

In the final experiment, E3, the vorticity flux deviations from its zonal-mean value were multiplied by four. The model integration did not settle down to an absolutely steady state, and so instead Fig. 15 depicts the average-surface-pressure distribution over the second integration month (subsequent monthly averages are not greatly different). The flow pattern

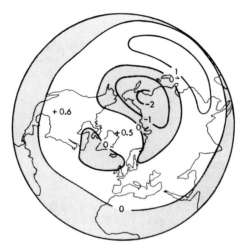

FIG. 14. Distribution of parameterized eddy vorticity flux divergence on level 3 for E2. Contour interval, 10^{-10} sec^{-2}; shading denotes convergence regions.

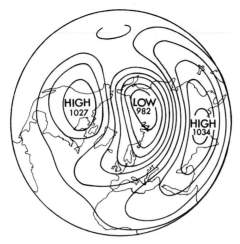

FIG. 15. Mean surface-pressure distribution in second integration month of E3. Contour interval, 5 mbar.

is now quite different, with a deep cold low over northern Siberia and a warm anticyclone over western Canada. Consulting the corresponding eddy vorticity flux divergence distribution (Fig. 16) confirms that they are primarily eddy-forced features. The phase difference of wave number 1 between level 3 and the surface is now only 5°, showing that the motion is virtually equivalent barotropic. Extremes of transient eddy vorticity forc-

FIG. 16. Mean parameterized eddy vorticity flux divergence in second integration month of E3 on level 3. Contour interval 10^{-10} sec^{-2}; shading denotes convergence regions.

ing of about 3×10^{-10} sec^{-2} at level 3 are required to have this effect, and it would be revealing to see if this forcing magnitude is reached in anomalous circulation months. Edmon (1980) has shown that anomalous winter circulation patterns are associated with equivalent-barotropic anomalies, which strongly suggests that they are transient-eddy-forced and as such are, according to Lorenz's definition, free variations. The vertical phase coherence of the pressure field can be assessed by comparing the surface-pressure field with the level 2 height contours, Fig. 17. Contour heights over Canada are between 200 and 300 m higher than the "normal" climatology. The high pressure system over Canada can be thought of as a wave number 1 blocking anticyclone, and although the scale is much larger than observed, I believe that the mechanism involved in its formation is essentially the same as that for real blocks.

In this experiment the transient eddy and mean-flow vorticity forcing are of comparable size. It should be noted, however, that it is the height integral of these quantities that is important in determining the sense of the surface vorticity. In thermally forced quasi-geostrophic planetary-wave systems, there is a strong tendency for cancellation in the height integral of the mean-flow vorticity flux divergence associated with the opposite signs of the pressure perturbation in the upper and lower atmosphere. In fact, in the frictionless case of uniform zonal wind, linear theory predicts zero height-integrated mean-flow divergence. In the absence of beta effect this must always be true, whatever the zonal flow profile, since the height integral of the stretching term $f_0(\partial w/\partial z)$ vanishes if w is zero at sufficient height and at the surface.

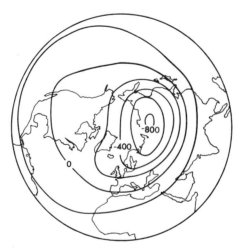

FIG. 17. Mean height contours (deviation from hemisphere-average height) on level 2 for the second integration month of E3. Contour interval, 200 m.

It has been tacitly assumed in the above experiments that transport of heat is governed by the normal transfer theory expression, even when longitudinal variations of vorticity transfer are increased.

5.4. Remarks on Further Experiments

The relative cheapness of model integrations (the cost is several orders of magnitude smaller than the running cost of a general circulation model) has allowed well over 100 integrations to be performed during the course of its existence. Cynics might argue that such a large sample of model experiments gives one more scope for selecting an "accidentally good" calculation which displays as many features of the real atmosphere as possible, whereas expensive models have much less choice. It would be dishonest to suggest that the results described here are free from this type of selection, but it should be emphasized that in the majority of experiments carried out, the same overall flow patterns were found. Climatic "noise" is not a problem since most integrations run to an absolutely steady state. Experiments over a fairly wide range of parameter space have provided us with a general picture of the model's psychology.

By far the most delicate aspect of the model's behavior is the calculation of the zonal-mean vorticity fluxes from the parameterization formula, Eq. (5.12). Without curtailing the magnitude of the transfer coefficient K_h in low latitudes, vorticity transport is overestimated, which in turn leads to strong surface winds and frictionally driven mean meridional circulation. In some experiments, a vigorous Ferrel cell was able to transport heat as effectively as the parameterized eddies, except in the reverse direction, which consequently led to an increase in poleward temperature gradient and, through Eq. (5.12), larger vorticity fluxes. The resulting positive feedback mechanism naturally rendered the integration unstable.

Because of the passive nature of the model's long waves, the zonal-mean state was essentially determined by transfer theory parameters. Explicit wave motion resembled the trapped, thermally forced planetary waves of linearized planetary wave theory except for some modification by parameterized motion.

This type of behavior did not always occur, however, and in some experiments with large-amplitude longitudinal variations in diabatic heating (about double those given here), explicit wave vorticity transport produced a radically different climate. In an experiment with large-amplitude diabatic heating variations which were specified as fixed rather than obtained from a Newtonian heating law, strong southward transport of vorticity in polar regions by the long waves resulted in strong easterlies extending from the polar cap to about 50° N throughout the model troposphere. Very large amplitude wave motion occurred about the region centered on the stagnation line, where the zonal-mean wind is zero.

Strangely enough, no intermediate cases have been found: the flow fields were either of the "normal" type described earlier, or this anomalous circulation state when thermal forcing was large. The details of this phenomenon are still under investigation.

The effects of longitudinal variations in the parameterized eddy heat transfer have also been investigated. It was found that if a longitudinally independent form of the eddy heat transfer law, Eq. (5.11), was used, the integration did not tend to an exactly steady state. Apparently, the highest wave number mode n approaches the critical wave number for baroclinic instability in the model, and longitudinal variations in the eddy transfer of heat are sufficient to suppress this.

The main effect of longitudinal variations in eddy heat flux divergence is to shift the phase of the temperature field upstream and reduce the amplitude of the wave motion. Destruction of long-wave APE by parameterized motions is compensated for by a greater production rate through the stronger phase correlation of diabatic heating with the temperature field.

As remarked earlier, observations show that the long quasi-stationary waves are associated with APE destruction due to a *negative* correlation between heating rate and temperature (Holopainen, 1970; Brown, 1964). That this should be the case for traveling cyclone waves is perfectly understandable, since warm air moving into polar regions must cool and similarly cold air moving toward the equator must warm, but this is not so for stationary thermally forced waves. Holopainen concludes that stationary waves are maintained energetically by the extraction of APE from the zonal-mean flow. It should be borne in mind, however, that the sign of the correlation is rather sensitive to phase errors (Clapp and Winninghoff, 1963) since the heating and temperature phases are approximately 90° out of phase [see the linear solution, Eq. (2.3)]. Yao (1979) studied a highly truncated spectral model of large-scale flow forced by orography in which quasi-stationary wave patterns resulted from the phase locking of unstable baroclinic waves with the orography. From an energetic point of view these quasi-stationary waves were maintained by drawing APE from the zonal-mean flow, as observations suggest for the real atmosphere. It might turn out that this energy generation discrepancy between linear thermal forcing models of stationary long waves and observations is unimportant to their structure, since it might only involve a small phase shift.

Another aspect of transient eddy parameterization that has been investigated with the model is the effect of fluxes being displaced downstream of strongly baroclinic regions. Since momentum and vorticity transports are thought to occur predominantly in the mature and decaying phase of the life cycle of a depression (Simmons and Hoskins, 1978), there is reason to believe that the greatest fluxes of vorticity will occur downstream of the storm's birthplace (typically where the poleward temperature gradient is largest). Experiments have been carried out with all calculated vorticity fluxes due to parameterized motion's being displaced downstream by a certain angular distance. No important effect resulted until the vorticity flux was displaced downstream by 60° longitude in the case of strong thermal forcing when the anomalous circulation pattern previously described returned.

6. CONCLUSION

One of the results of my experiments is that vorticity forcing due to the effects of traveling weather systems averaged over time plays a secondary role compared to the effect of diabatic heating anomalies in determining the longitudinal variations in climate. Forcing of the time-average flow by orography has been entirely neglected, but even so, the model climate is at least qualitatively correct. It is interesting that a common deficiency of general circulation models is their tendency to overestimate the intensity of the Icelandic and Aleutian lows. In synoptic terms, this is apparently related to the formation of intense, occluded depressions which refuse to fill because of continual rejuvenation by successive weather systems. Analysis of the local time-mean vorticity budget would presumably reveal this effect as due to an abnormally large eddy vorticity flux convergence. It is vaguely amusing to entertain the idea that a parametric representation of transient eddy motion might be more satisfactory than an over-energetic explicit description.

If eddy vorticity flux parameterization were completely neglected, the upper tropospheric flow of the model would not look significantly different—only the zonal-mean component of the surface-pressure and vertical velocity fields would be fundamentally changed. Wind and temperature fields are primarily governed by thermodynamic balance, the thermal wind equation, and the fact that surface winds are generally small compared to the thermal wind (see Adem, 1975). Parameterization of eddy heat transport is essential if the model is to maintain a realistic temperature field and provide the necessary export of energy from the tropics.

Considerable scope for improvement of the model remains; for instance, the primitive equations of motion could replace the quasi-geostrophic set used here. As previously argued, one would expect a better description of the mean-meridional circulation and mean zonal wind profile. The form of the transfer theory used in the model requires further elaboration so that the full spatial variability of the parameterized fluxes is determined without appeal to observation. In addition to this, the whole gamut of model "physics" used in general circulation models (radiative heat transfer, surface energy budget, hydrological cycle, etc.) could be incorporated, and with a realistic representation of the earth's topography a more serious comparison with observations might be sought.

Since my parameterized motion systems are defined according to mechanism rather than scale, it would be tempting to increase the model's resolution so that regional climatology could be better described in the model (e.g., Mediterranean climate). The required shorter wave modes would normally be baroclinically unstable, but it seems possible that this instability might be eliminated by the diffusive form of eddy heat flux parameterization. An optimum degree of resolution could be inferred from the amplitude spectrum of the monthly mean height contours.

The model is based very much on Adem's philosophy that the climatic-mean state of the atmosphere is primarily governed by local thermodynamic balance. Even so, dynamical processes do influence the mean thermodynamic state of the atmosphere by simple advection by the mean wind—included in a later version of his model (Adem, 1970)—and vertical motion induced by vorticity transfer. In connection with the latter process it is interesting to note that if a zonal-mean version of the parameterized model is run without parameterized vorticity transfer, the steady-state solution has no mean meridional circulation and zero surface wind speed everywhere.

The model described in this article can be viewed as an extension (to include dynamics) of Adem's model, albeit with a greatly simplified thermodynamic forcing function. Eddy parameterizations similar to those used here have been incorporated into a low-resolution wind-driven ocean circulation model (Marshall, 1981) to represent transfer by the "mesoscale" eddy field, and it is hoped eventually to build a coupled ocean–atmosphere climate model in which air–sea heat and moisture exchange is represented.

The stochastic nature of the real atmosphere is not represented in any way by the model, and one wonders what the repercussions of its omission are. Lorenz (1975) and Leith (1975) envisage future parameterized climate models that artificially mimic this effect by the inclusion of some type of random-forcing function. We could, for instance, allow the local value of the transfer coefficients to vary randomly about a mean and concede a reinterpretation of transfer theory philosophy.

At the expense of some empiricism, particularly associated with the correct representation of the zonal-mean vorticity fluxes, I believe that parameterized models can at least be competitive with general circulation models in simulating the present day climate. The present form of the transfer theory is not reliable enough for it to be valid in conditions far removed from the terrestrial climate. For instance, Williams (1979) has simulated the multiple jet structure of Jupiter's atmosphere using a conventional quasi-geostrophic model with appropriate Jovian planetary parameters (gravitational acceleration, rotation rate, radius, etc.). It seems rather unlikely that a parameterized model such as that described here would be capable of reproducing this jet structure. Perhaps the main difficulty is that eddy scales and meridional particle displacements on Jupiter are much smaller than the planetary scale, unlike the terrestrial atmosphere. Turbulent transfer is therefore much more localized on a scale comparable with the width of Jupiter's bands—which Williams identifies with the Rhines length scale k_β^{-1} [$= (2\hat{u}/\beta)^{1/2}$], $\hat{u}$ being a typical particle velocity. In such cases, the horizontal structure of the transfer coefficients should reflect this narrowness of meridional scale.

Improvement of quasi-geostrophic parameterization schemes must come from further diagnostic studies of observed and numerically modeled eddy transfer of potential vorticity and, to this end, experiments are to be carried out using a high-resolution, explicitly resolving version of the three-level model in conjunction with the parameterized model described here. In this way, the statistically steady states of the two models can be compared so that the transfer theory may be tested and refined.

Acknowledgments

I would like to thank Dr. J. S. A. Green for his comments on the text and Jean Ludlam for typing the manuscript. I also gratefully acknowledge funding by the British Natural Environmental Research Council.

References

Adem, J. (1970). Incorporation of advection of heat by mean winds and by ocean currents in a thermodynamic model for long-range weather prediction. *Mon. Weather Rev.* **98**, 776–786.

Adem, J. (1975). A critical appraisal of simple climate models. *GARP Publ. Ser.* **16**, 163–170.

Andrews, D. G., and McIntyre, M. E. (1976). Planetary waves in horizontal and vertical shear: the generalized Eliassen–Palm relation and the mean zonal acceleration. *J. Atmos. Sci.* **33**, 2031–2048.

Austin, J. F. (1978). The blocking of mid-latitude westerly winds by stationary anticyclones. Ph.D. thesis, University of London.

Austin, J. F. (1980). The blocking of middle latitude westerly winds by planetary waves. *Q. J. R. Meteorol. Soc.* **106,** 327–350.

Bates, J. R. (1977). Dynamics of stationary ultra-long waves in middle latitudes. *Q. J. R. Meteorol. Soc.* **103,** 397–430.

Bengtsson, L. (1980). Numerical prediction of atmospheric blocking—a case study. *Tellus* **33,** 19–42.

Bourke, W. (1972). An efficient one-level primitive equation spectral model. *Mon. Weather Rev.* **100,** 683–689.

Brown, J. A. (1964). A diagnostic study of tropospheric climatic heating and the generation of available potential energy. *Tellus* **16,** 371–388.

Burrows, W. R. (1976). A diagnostic study of atmospheric spectral kinetic energies. *J. Atmos. Sci.* **33,** 2308–2321.

Charney, J. G., and Eliassen, A. (1949). A numerical method for predicting the perturbations of the middle latitude westerlies. *Tellus* **1,** 38–54.

Charney, J. G., and Straus, D. M. (1980). Form-drag instability, multiple equilibria and propagating, planetary waves in baroclinic, orographically forced, planetary wave systems. *J. Atmos. Sci.* **37,** 1157–1176.

Cheng, T.-C., and Wiin-Nielsen, A. (1978). On nonlinear cascades of atmospheric energy and enstrophy in a two-dimensional spectral index. *Tellus* **30,** 313–322.

Clapp, P. F. (1970). Parameterization of macroscale transient heat transport for use in a mean-motion model of the general circulation. *J. Appl. Meteorol.* **9,** 554–563.

Clapp, P. F., and Winninghoff, F. J. (1963). Tropospheric heat sources and sinks at Washington D.C., Summer, 1961, related to the physical features and energy budget of the circulation. *Mon. Weather Rev.* **91,** 494–504.

Döös, B. R. (1962). The influence of exchange of sensible heat with the earth's surface on the planetary flow. *Tellus* **14,** 133–147.

Eady, E. T. (1949). Long waves and cyclone waves. *Tellus* **1,** 33–52.

Eady, E. T. (1954). The maintenance of the mean zonal surface currents. *Proc. Toronto Meteorol. Conf., 1953.*

Edmon, H. J. (1980). A study of the general circulation over the northern hemisphere during the winters of 1976–77 and 1977–78. *Mon. Weather Rev.* **108,** 1538–1553.

Edmon, H. J., Hoskins, B. J., and McIntyre, M. E. (1980). Eliassen–Palm cross-sections for the troposphere. *J. Atmos. Sci.* **37,** 2600–2616.

Eliassen, A., and Palm, E. (1961). On the transfer of energy in stationary mountain waves. *Geophys. Norv.* **22,** 1–23.

Frederiksen, J. S. (1979). The effects of long planetary waves on the regions of cyclogenesis: Linear theory. *J. Atmos. Sci.* **36,** 195–204.

Gilchrist, A., Corby, G. A., and Newson, R. L. (1973). A numerical experiment using a general circulation model of the atmosphere. *Q. J. R. Meteorol. Soc.* **99,** 2–34.

Green, J. S. A. (1970). Transfer properties of the large-scale eddies and the general circulation of the atmosphere. *Q. J. R. Meteorol. Soc.* **96,** 157–185.

Green, J. S. A. (1977). The weather during July 1976: some dynamical considerations of the drought. *Weather* **32,** 120–126.

Haltiner, G. J. (1971). "Numerical Weather Prediction." Wiley, New York.

Holland, W. R., and Rhines, P. B. (1980). An example of eddy-induced ocean circulation. *J. Phys. Oceanogr.* **10,** 1010–1031.

Holopainen, E. O. (1970). An observational study of the energy balance of the stationary disturbances in the atmosphere. *Q. J. R. Meteorol. Soc.* **96,** 626–644.

Holopainen, E. O. (1978). On the dynamic forcing of the long-term mean flow by the large-scale Reynolds' stresses in the atmosphere. *J. Atmos. Sci.* **35,** 1596–1604.

Holopainen, E. O., and Oort, A. H. (1981). On the role of large-scale transient eddies in the maintenance of the vorticity and enstrophy of the time-mean flow. *J. Atmos. Sci.* **38,** 270–280.

Hoskins, B. J., and Karoly, D. J. (1981). The steady linear response of a spherical atmosphere to thermal and orographic forcing. *J. Atmos. Sci.* **38,** 1179–1196.

Kane, I. D. (1973). A numerical investigation into the stability of zonal flows in the atmosphere. Ph.D. thesis, University of London.

Kraichnan, R. H. (1967). Inertial ranges in two-dimensional turbulence. *Phys. Fluids* **10,** 1417–1423.

Kuo, H. (1952). Dynamical aspects of the general circulation and the stability of zonal flows. *Tellus* **3,** 268–284.

Lau, N.-C. (1979). The observed structure of tropospheric stationary waves and the local balances of vorticity and heat. *J. Atmos. Sci.* **36,** 996–1016.

Lau, N.-C., and Wallace, J. (1979). On the distribution of horizontal transports by transient eddies in the northern hemisphere wintertime circulation. *J. Atmos. Sci.* **36,** 1844–1861.

Leith, C. E. (1975). The design of a statistical dynamical climate model and statistical constraints on the predictability of climate. *GARP Publ. Ser.* **16,** 137–141.

Lorenz, E. N. (1969). The predictability of a flow which possesses many scales of motion. *Tellus* **21,** 289–307.

Lorenz, E. N. (1975). Climatic predictability. *GARP Publ. Ser.* **16,** 132–136.

Lorenz, E. N. (1979). Forced and free variations of weather and climate. *J. Atmos. Sci.* **36,** 1367–1376.

Ludlam, F. H. (1966). The cyclone problem: A history of models of the cyclonic storm. Inaugural lecture, 1966, University of London.

Manabe, S., and Terpstra, T. B. (1974). The effects of mountains on the general circulation of the atmosphere as identified by numerical experiments. *J. Atmos. Sci.* **31,** 3–42.

Marshall, J. (1981). On the parameterization of geostrophic eddies in the ocean. *J. Phys. Oceanogr.* **11,** 257–271.

Mintz, Y. (1964). Very long term global integration of the primitive equations of atmospheric motion. *Proc. WMO/IUGG Symp., Boulder, Colorado.*

Miyakoda, K. (1960). The methods of numerical time integration of one dimensional linear equations and their inherited errors. *J. Meteorol. Soc. Jpn.* **38,** 259–287.

Muench, H. S. (1965). On the dynamics of the wintertime stratospheric circulation. *J. Atmos. Sci.* **22,** 349–360.

Oort, A. H., and Rasmusson, E. M. (1971). Atmospheric circulation statistics. NOAA, U.S. Dept. of Commerce.

Paltridge, G. W. (1978). The steady-state format of global climate. *Q. J. R. Meteorol. Soc.* **104,** 927–945.

Pedlosky, J. (1964). The stability of currents in the atmosphere and the ocean: Part II. *J. Atmos. Sci.* **21,** 342–353.

Prigogine, I. (1980). "From Being to Becoming—Time and Complexity in the Physical Sciences." Freeman, San Francisco, California.

Rhines, P. B. (1979). Geostrophic turbulence. *Annu. Rev. Fluid Mech.* **11,** 401–441.

Robert, A. J. (1966). The integration of a low order spectral form of the primitive meteorological equations. *J. Meteorol. Soc. Jpn.* **44,** 237–245.

Rossby, C.-G. (1947). On the distribution of angular velocity in gaseous envelopes under the influence of large-scale horizontal mixing processes. *Bull. Am. Meteorol. Sci.* **28**(2), 53–68.

Saltzman, B. (1965). On the theory of the winter-average perturbations in the troposphere and stratosphere. *Mon. Weather Rev.* **93,** 195–211.

Saltzman, B. (1968). Surface boundary effects on the general circulation and macroclimate:

A review of the theory of the quasi-stationary perturbations in the atmosphere. *Meteorol. Monogr.* **30,** 4–19.

Sankar-Rao, M. (1965a). Continental elevation influence on the stationary harmonics of the atmospheric motion. *Pure Appl. Geophys.* **60,** 141–159.

Sankar-Rao, M. (1965b). On the influence of the vertical distribution of stationary heat sources and sinks in the atmosphere. *Mon. Weather Rev.* **93,** 417–420.

Sankar-Rao, M., and Saltzman, B. (1969). On a steady state theory of global monsoons. *Tellus* **21,** 308–330.

Savijarvi, H. (1976). The interaction of the monthly-mean flow and large-scale transient eddies in two different circulation types. Part II. *Geophysica* **14,** 207–229.

Sela, J., and Wiin-Nielsen, A. (1971). Simulation of the atmospheric energy cycle. *Mon. Weather Rev.* **99,** 460–468.

Shutts, G. J. (1976). The generation and propagation of planetary Rossby waves. Ph.D. thesis, University of London.

Shutts, G. J. (1978). Quasi-geostrophic planetary wave forcing. *Q. J. R. Meteorol. Soc.* **104,** 331–350.

Simmons, A. J., and Hoskins, B. J. (1978). The life cycles of some nonlinear baroclinic waves. *J. Atmos. Sci.* **35,** 414–432.

Smagorinsky, J. (1953). The dynamical influence of large-scale heat sources and sinks on the quasi-stationary mean motions of the atmosphere. *Q. J. R. Meteorol. Soc.* **79,** 342–366.

Starr, V. P. (1968). "Physics of Negative Viscosity Phenomena." McGraw-Hill, New York.

Stewart, R. W., and Thompson, R. E. (1977). Re-examination of vorticity transfer theory. *Proc. R. Soc. London, Ser. A* **354,** 1–8.

Stone, P. H. (1978). Baroclinic adjustment. *J. Atmos. Sci.* **35,** 561–571.

Stone, P. H., and Miller, D. A. (1980). Empirical relations between seasonal changes in meridional temperature gradients and meridional fluxes of heat. *J. Atmos. Sci.* **37,** 1708–1721.

Taylor, G. I. (1915). Eddy motion in the atmosphere. *Philos. Trans. R. Soc. London, Ser. A* **215,** 1–26.

Tucker, G. B. (1977). An observed relation between the macroscale local eddy heat flux and the mean horizontal temperature gradient. *Q. J. R. Meteorol. Soc.* **103,** 157–168.

Vallis, G. K. (1982). A statistical dynamical climate model with a simple hydrology cycle. *Tellus* **34,** 211–227.

Van Loon, H., and Jenne, R. L. (1973). Zonal harmonic standing waves. *JGR, J. Geophys. Res.* **78,** 4463–4471.

White, A. A. (1974). Large scale momentum transfer in a general circulation model of the atmosphere. Ph.D. thesis, University of London.

White, A. A. (1977). The surface flow in a statistical climate model—a test of parameterization of large-scale momentum fluxes. *Q. J. R. Meteorol. Soc.* **103,** 93–119.

White, A. A., and Green, J. S. A. (1982a). A nonlinear atmospheric long wave model incorporated parameterization of transient baroclinic eddies. *Q. J. R. Meteorol. Soc.* **108,** 55–85.

White, A. A., and Green, J. S. A. (1982b). Transfer coefficient eddy flux parameterizations in a simple model of the zonal-average atmospheric circulations. *Q. J. R. Meteorol. Soc.* (to be submitted).

Wiin-Nielsen, A., and Brown, J. A. (1962). On diagnostic computations of atmospheric heat sources and sinks and generation of available potential energy. *Proc. Int. Symp. Numercial Weather Predict., Tokyo, 1960* pp. 593–613.

Williams, G. P. (1979). Planetary circulations: 2. The Jovian quasi-geostrophic regime. *J. Atmos. Sci.* **36,** 932–968.

CLIMATIC SYSTEMS ANALYSIS

Barry Saltzman

Department of Geology and Geophysics
Yale University
New Haven, Connecticut

1. Introduction

Figure 1 is a schematic pictorialization of the complete climatic system, comprising an atmosphere, hydrosphere, cryosphere, and biolithosphere. All lines of historical and geological evidence point to the fact that this system is changing continuously on all time scales and that the statistics of the system defining the "climate" are not stationary. We are reasonably certain, for example, that the amount of ice in this system has pulsated dramatically over the last few million years, and probably intermittently over a much longer period. In Fig. 2 we show some standard examples of the types of variability that have been recorded.

ADVANCES IN GEOPHYSICS, VOLUME 25

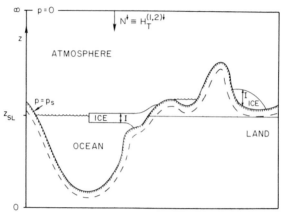

FIG. 1. Schematic representation of the land–ocean–ice–atmosphere climatic system.

There are many possible causes for climatic fluctuations of the kinds shown in Fig. 2, but we can make at least one main subdivision of these possibilities (see Saltzman, 1978; Lorenz, 1979): (1) variable *forcing* due to purely external inputs that affect the climatic state but are themselves unaffected by the climatic state (e.g., astronomical factors such as the solar radiation and earth-orbital changes; solid-earth tectonic factors such as continental drift and volcanic activity) and (2) *free* changes that can occur even if the forcing is steady, due to internal feedbacks and instabilities that can involve all components of the climatic system. It seems likely that the observed climatic variability represents a complex mix of both forced and free effects. We know, for example, that externally imposed diurnal and seasonal variations lead to significant observed variations, and there is some evidence that even the much weaker 19,000–23,000-yr and 41,000-yr earth-orbital periods due to precession and obliquity changes are manifest in the records (Hays *et al.*, 1976). On the other hand, the most significant variation in ice volume revealed in Fig. 2 is one of about 100,000 yr, near which one can find only what appears to be an extremely weak forcing due to earth eccentricity variations. More generally, we find a broad spectrum of variability, only a very small part of which is directly related to known forcing frequencies (see Fig. 3). These observations are indicative of strong *free* variability within the system that can perhaps best be described as "climatic turbulence."

Since any real physical system must be subject to some frictional and diffusive processes, we are led to the realization (see Lorenz, 1963) that we are dealing with an extremely complex, forced, dissipative system that

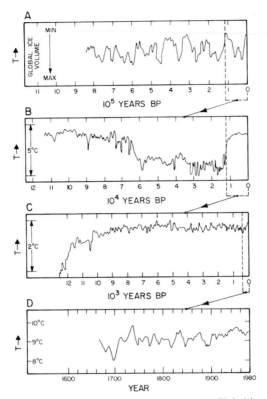

FIG. 2. Selected climatic time series for the past 1 m.y. (A) Global ice volume deduced from oxygen isotope variations of planktonic foraminifera in a deep-sea core (Shackleton and Opdyke, 1973); global mean temperature variations over the last 100,000 yr (B) and over the last 10,000 yr (C), based on oxygen isotope variations in a Greenland ice core (Dansgaard *et al.*, 1971); (D) thermometric measurements in England over the past 300 yr (Mason, 1976).

undoubtedly contains a rich assortment of linear and nonlinear, positive and negative, feedbacks. The theory for the evolution of such a system will inevitably require all the tools of applied mathematics usually included under the broad title of "dynamical systems analysis." In the next section we shall discuss the fundamental physical equations that must provide the deterministic basis for such an analysis of the *climatic* system. In addition to the desire to explain, or account for, the past variations of climate, the main practical objective and ultimate end product of such a climatic system analysis is the *prediction* of future changes, including the consequences of natural, inadvertent, and purposeful changes in various parts of the system.

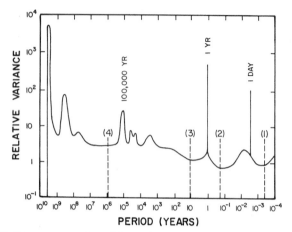

FIG. 3. Qualitative representation of the variance spectrum of climatic change (adapted from Mitchell, 1976). Dashed lines (1)–(4) are located at relative variance minima, representing optimal time periods over which to define average states (see Appendix A and Table I).

2. GENERAL THEORETICAL CONSIDERATIONS AND EQUATIONS: THE BASIS FOR CLIMATE MODELING

The atmosphere, oceans, and ice masses are basically continua describable by field variables, each of these domains having an infinite number of degrees of freedom. The behavior of such continua are generally governed by *partial* differential equations which are the classical fluid dynamical equations expressing conservation of mass (continuity equations), momentum (Navier–Stokes equations of motion), and energy (first law of thermodynamics), along with diagnostic relationships for internal energy and the thermodynamic "state."

Let us assume that such a set of equations governing the nearly "instantaneous" values of all the variables describing the climatic system can be written in an "exact" form. As an example, the following is the set commonly adopted for the main carrier fluids comprising the atmosphere and oceans (see Pedlosky, 1971):

mass:

$$d\rho/dt = -\rho \nabla \cdot \mathbf{V} \tag{2.1}$$

momemtum:

$$d\mathbf{V}/dt = -(1/\rho)\nabla p + \nabla \Phi - 2\mathbf{\Omega} \times \mathbf{V} + \nu[\nabla^2 \mathbf{V} + \tfrac{1}{3}\nabla(\nabla \cdot \mathbf{V})] \tag{2.2}$$

energy:

$$de/dt = -(p/\rho)\nabla \cdot \mathbf{V} + q \tag{2.3}$$

plus the diagnostic relations

$$e = cT \tag{2.4}$$

$$\rho = \begin{cases} p/RT & \text{(atmosphere)} & (2.5a) \\ \rho^*[1 - \mu_T(T - T^*)] & \text{(ocean)} & (2.5b) \end{cases}$$

where ρ is the density; p is the pressure; T is the temperature;

$$\mathbf{V} = u\mathbf{i} + v\mathbf{j} + w\mathbf{k}$$
$$u = a \cos \varphi \, d\lambda/dt \quad \text{(eastward speed)},$$
$$v = a \, d\varphi/dt \quad \text{(northward speed)},$$
$$w = dz/dt \quad \text{(vertically upward speed)},$$

$\mathbf{i}, \mathbf{j}, \mathbf{k}$ are the unit vectors eastward, northward, and upward, respectively; λ is the longitude; φ is the latitude; z is the vertical distance; t is the time; a is the radius of the earth;

$$\nabla = \mathbf{i} \, \partial/a \cos \varphi \, \partial\lambda + \mathbf{j} \, \partial/a \, \partial\varphi + \mathbf{k} \, \partial/\partial z$$

$$\Phi = gz$$

g is the acceleration of gravity; Ω is the angular velocity of the earth; ν is the kinematic viscosity; q is the rate of heat addition per unit mass due to radiation, conduction, phase changes, and viscosity; e is the internal energy; c is the specific heat at constant volume; R is the gas constant for air; x^* is the standard value of x; μ_T is the coefficient of thermal expansion;

$$d/dt = \partial/\partial t + \mathbf{V} \cdot \nabla$$

In practice we cannot measure and are not really interested in "instantaneous" values, but rather in an *average* value of the variables over some time period $\Delta t = \delta$. Typical averaging periods of interest are about an hour (defining *synoptic* values resolved by standard meteorological network measurements) and several (e.g., 10) years (which shall constitute our definition of a *climatic* average). The synoptic-average ($\delta_s = 1$ hr) statistics of the state of the climatic system constitute the "weather," whereas the climatic-average ($\delta_c = 10$ yr) statistics constitute the "climate." The averaging periods of interest are determined by the nature of the variations exhibited by the system, such as are revealed in Figs. 2 and 3, and other considerations to be discussed below. In Appendix A we discuss some systematics of the averaging process for climatic variability based on the discussion in the remainder of this section. A summary is given in Table I.

TABLE I. A RESOLUTION OF CLIMATIC VARIABILITY[a]

Index (n)	Averaging period[b] (δ)	Spectral band	Physical phenomena	Known forcing
(0–1)		(0–1 hr)	Micro- and meso-meteorological eddies: turbulence and convection	
(1)	1 hr		Synoptic average	
(1–2)		(1 hr–3 months)	Diurnal cycle, cyclone waves, blocking, and index variations	Diurnal solar radiation cycle
(2)	3 months		Seasonal average	
(2–3)		(3 months–10 yr)	Annual cycle, year-to-year (interannual variability)	Annual solar radiation cycle
(3)	10 yr		Climatological average	
(3–4)		10 yr–1 m.y.	Historical and paleo-climatological variations, ice ages	Milankovitch earth-orbital radiation cycles
(4)	1 m.y.		Tectonic average	
(4–5)		1 m.y.–4.5 b.y.	Ultra-long-term climate variations influenced by global tectonics and planetary evolution	Continental drift
(5)	4.5 b.y.		Age of the earth	

[a] See Appendix A.
[b] m.y., Million years; b.y., billion years.

In order to obtain dynamical equations governing the evolution of such running time-average variables, the "exact" instantaneous equations must be averaged over the relevant time interval δ. As is well known from turbulence theory (e.g., Monin and Yaglom, 1971), there are at least two major consequences of such an averaging process: (1) eddy stress terms are introduced that must be parameterized to effect closure in terms of the averaged variables, and (2) stochastic forcing terms are introduced due to the effects of nonsystematic random departures of the instantaneous values from the mean values and the impossibility of achieving exact parameterizations.

These consequences can be illustrated by considering a simplified form of the energy equation (3) in which we take $\nabla \cdot \mathbf{V} \approx 0$ (e.g., an ocean):

$$\partial T/\partial t = -\nabla \cdot T\mathbf{V} + q/c \qquad (2.6)$$

The heating function $q = q(\chi_j, t)$ is generally a complex set of terms that are dependent linearly and nonlinearly on many other climatic variables χ_j ($j = 1, 2, 3, \ldots$) and on t itself through external forcing that can be deterministically prescribed (such as periodic earth-orbital radiative changes) or randomly imposed (such as atmospheric turbidity changes due to volcanic eruptions).

We now seek to transform this equation to one governing the 10-yr running average values of the variables, defined by

$$\bar{\chi}(\tau, \delta) \equiv \frac{1}{\delta} \int_{\tau - \delta/2}^{\tau + \delta/2} \chi \, dt \equiv \bar{\chi}^{(\delta)}(\tau)$$
$$\equiv \chi(\tau) - \chi'(\tau, \delta) \tag{2.7}$$

where $\delta = 10$ yr, and the primed quantity is the departure of the instantaneous value from the climatic mean value, which departure includes the effects of all frequencies higher than 10 yr. An important reason for choosing $\delta = 10$ yr is that there appears to be a relative minimum in observed natural variability on this time scale (i.e., as seen in Fig. 3, this period falls within a relative spectral "gap" between more energetic long-term variations such as the ice ages and shorter-term variations such as the annual cycle, mesoscale ocean eddies, and synoptic storms). Formally, we can write

$$P(T') \ll \delta \ll P(\bar{T}) \tag{2.8}$$

where P stands for the dominant period of the fluctuations of its argument. This makes it likely that the 10-yr running mean will vary relatively slowly and quasi-linearly within any averaging interval δ, and that there will be many higher-frequency fluctuations within this averaging interval (cf. Fig. 4). Thus, because of Eq. (2.8) we have as first approximations

$$\bar{\bar{\chi}} \approx \bar{\chi} \tag{2.9}$$

$$\overline{\chi'} \approx 0 \tag{2.10}$$

$$\overline{\partial \chi / \partial t} = \overline{\partial \bar{\chi} / \partial t} + \overline{\partial \chi' / \partial t} \approx \partial \bar{\chi} / \partial t + \overline{\partial \chi' / \partial t} \tag{2.11}$$

$$\overline{\chi_1 \chi_2} = \overline{\bar{\chi}_1 \bar{\chi}_2} + \overline{\chi_1' \chi_2'} + \overline{\chi_1' \bar{\chi}_2} + \overline{\bar{\chi}_1 \chi_2'}$$
$$\approx \bar{\chi}_1 \bar{\chi}_2 + \overline{\chi_1' \chi_2'} \tag{2.12}$$

In Appendix A we generalize and formalize the application of Eqs. (2.7)–(2.12) to the full frequency range exhibited by the climatic system wherein multiple variance minima (i.e., relative "gaps") and bands of variance maxima prevail.

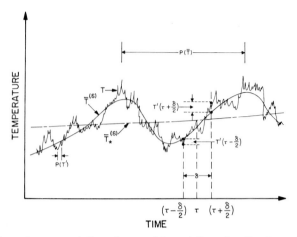

FIG. 4. Schematic representation of a possible variation of a climatic variable (say, T) from which the annual cycle is removed. The running mean of that variable over an averaging interval δ is denoted by the $\bar{T}^{(\delta)}$ curve, the departures from which are denoted by T'. Note the random nature of the quantity $(\overline{\partial T'/\partial t})^{(\delta)} = \delta^{-1}[T'(\tau + \delta/2) - T'(\tau - \delta/2)] = R_1$, where $P(T') \ll \delta \ll P(\bar{T}^{(\delta)})$.

Now if we apply Eqs. (2.7) and (2.9)–(2.12) to Eq. (2.6), we obtain

$$\frac{\partial \bar{T}}{\partial t} = -\boldsymbol{\nabla} \cdot \bar{T}\bar{\mathbf{V}} - \boldsymbol{\nabla} \cdot \overline{T'\mathbf{V}'} + \frac{1}{c}\overline{q(\chi_j, t)} - \left(\overline{\frac{\partial T'}{\partial t}}\right) \qquad (2.13)$$

At any time τ,

$$\left(\overline{\frac{\partial T'}{\partial t}}\right) = \frac{T'(\tau + \delta/2) - T'(\tau - \delta/2)}{\delta} \equiv \mathcal{R}_1 \qquad (2.14)$$

which, as can be seen from Fig. 4, will be essentially *random* if T' contains an aperiodic component along with the seasonal cycle. At any time step the amplitude of this random term can be of the same order as $\partial \bar{T}/\partial t$, but will be reduced as δ is increased, thereby increasing the "signal-to-noise" ratio. As we have indicated in Eq. (2.14), we denote this random component by $\mathcal{R}_1$.

Another source of random "forcing" must be introduced when we parameterize the stress term $\overline{\mathbf{V}'T'}$ and the stress terms that are included in $q(\chi_j, t)$. In general, we have

$$\overline{(\chi_1'\chi_2')} = f(\bar{\chi}_j) + \mathcal{E} \qquad (2.15)$$

where $\bar{\chi}_j$ can be any set of mean climatic variables, f is a deterministic formula usually obtained by some mixture of theory and empiricism (e.g., the "diffusion" approximation, $\overline{\mathbf{V}'T'} = -K \boldsymbol{\nabla}\bar{T}$, where K is a constant eddy diffusivity), and $\mathcal{E} = \phi + \mathcal{R}_2$ is an error which contains a systematic part ϕ and a random part $\mathcal{R}_2$. For the diffusion approximation the random component will be present if for no other reason than the variability of K when estimated over different time periods (cf. Robock, 1978).

Additional sources of random forcing enter the problem, even at the level of the instantaneous equations. These are due, for example, to irregular external inputs (e.g., volcanism) that constitute an unknown aperiodic forcing, which we can denote by $\mathcal{R}_3$, and to the uncertainty of initial conditions. Thus, after parameterization the partial differential equation (2.6) can be written in schematic form:

$$\partial \bar{T}/\partial t = F(\bar{T}, \bar{\chi}_j; t) + \mathcal{R} \qquad (2.16)$$

where F is a collection of terms involving *partial derivatives,* and $\mathcal{R} = \mathcal{R}_1 + \mathcal{R}_2 + \mathcal{R}_3$ is a representation of the random forcing that must be included in equations governing variables averaged over any time period. For some averaging periods, however, the amplitude of the random component may be small enough to be neglected [as is usually done in the general circulation models (GCMs) applied to synoptic average variables]. On the other hand, it seems likely that for the statistical-dynamical models (SDMs) formulated for longer term mean values (e.g., 10 yr) such random forcing cannot be neglected. A fairly complete presentation of the deterministic parts of the fundamental fluid dynamical equations of the type illustrated by (2.13), governing synoptic and climatic average variables in all domains of the climatic system, is given by Saltzman (1978).

As noted by Lorenz (1963), in order to solve these partial differential equations they must be reduced either to a finite system of *ordinary* differential equations governing the amplitudes of a truncated orthogonal spatial expansion of the variables, or to a system of difference equations governing values at a discrete space–time grid. Thus, in practice we are forced to deal with a more approximate, reduced, system that can always be written in the form

$$d\bar{\chi}_j/dt = f_j(\bar{\chi}_j, t) + \mathcal{R}_j \qquad (2.17)$$

where $f_j(\bar{\chi}_j, t)$ is the deterministic part that, in addition to linear and nonlinear terms in $\bar{\chi}_j$, can contain time-dependent nonautonomous forcing components, and $\mathcal{R}_j$ is the stochastic part. Depending on whether this system is written for grid point values or orthogonal (e.g., Fourier) components, the variables $\bar{\chi}_j$ will be functions of (λ, ϕ, z) or wave numbers (k, l, m) in the case of a Fourier expansion.

Such a set of time-dependent ordinary differential equations governing averaged quantities and containing stochastic forcing terms constitutes a "statistical-dynamical system" characterized by a very large but discrete number of degrees of freedom. As we have already indicated, if the sys-

tem governs the synoptic average variables with the synoptic spatial resolution (thereby requiring parameterization only of subsynoptic frequencies and spatial scales), we speak of the system as a GCM. In this case the
solution can be iterated forward in time to generate a full set of statistics
(e.g., means, variances, frequencies of rarer events) that constitute the
climate. This, of course, is a demanding procedure in time and resources.

If one wishes to study long-term climatic change, i.e., evolution over
hundreds of years and more, it is natural to consider equations governing
a longer-term average such as 10 yr. In this "SDM" case we must parameterize not only subsynoptic phenomena but all phenomena up to and
including interannual variations (see Appendix A)—a most difficult and
challenging task. As we have noted, whereas the stochastic amplitude
may be relatively small in the GCM, it will tend to be large in the SDM,
though hopefully not quite of the same order as the deterministic terms.
Moreover, additional equations, including the parameterization formulas,
will be necessary to deduce the higher-order statistics (such as the spatial
and temporal variances and amplitude of the seasonal cycle) that are
needed for a full description of climate. This constitutes a form of "inverse problem of climatology" (see Kim *et al.,* 1981), wherein one seeks
to infer the statistics of higher frequency, higher wave number behavior
from a knowledge of the low-frequency, low-wave-number distribution.

It seems likely that to describe the macrobehavior of the climatic system with the detail in which we are interested (or more properly, with the
accuracy with which observations can reveal it) will require only some
much smaller number of variables (and their governing equations) than
are represented by the full set (2.17) applied to all the variables, in all the
domains, at a synoptic spatial grid. Thus, a major challenge in developing
the theory of long-term climatic fluctuations will be to learn how to trun-
cate the system in ways that capture the main variabilities while at the
same time permitting closure of the system by physically valid parameterizations. Related to this is the need to identify groupings of variables that
are "coherent" or diagnostically related so that they can be represented
by much reduced sets of variables or even a single variable. A formal
procedure for accomplishing such reductions by successive spatial integrations is described in Saltzman (1978). One strategy is to start with the
very simplest one-variable model obtained by integrating over the entire
climatic system (e.g., Budyko, 1969; Sellers, 1969; Imbrie and Imbrie,
1980) and systematically expanding the model with added variables and
equations. A partial test of the success of this process is the degree to
which the unexplained variability is of a random white-noise variety on
which no further determinism can be brought to bear.

In any event, systems of the form (2.17) relevant for the study of climate can be developed with the number degrees of freedom (i.e., levels of complexity) ranging from one to the huge number involved in GCMs. At this point in the development of a theory of climate there is ample justification for studies involving all levels of complexity. It would appear that we are now at the rather primitive stage of simple "getting a feel" for the role of all the competitive feedbacks and forcings that are possible in the system depending on the choice of time scale of variability.

Whatever the level of complexity, a complete analysis of the system treated must involve certain elements, including determination of the following:

(1) The steady states (i.e., the *equilibrium positions* or fixed points) of the system.

(2) The *stability* of the equilibria for small displacements. (Note: For very complex systems, such as GCMs, it may be impossible to determine the existence of *un*stable equilibria, or to distinguish nearby multiple equilibria, in the face of all the numerically and physically generated fluctuations.)

(3) The *sensitivity of the equilibria* to changes in parameters, with respect to (a) their *position* in phase space and (b) the *stability* of the equilibria, including the bifurcations (i.e., structural stability).

(4) The *time-dependent evolution* of the system from specific initial conditions, for various choices of parameters, portrayed along a time axis, and as trajectories in phase space where this can be instructive.

(5) The effects of the random fluctuations (stochastic noise) on the above deterministic time-dependent solutions (i.e., structural stochastic stability). This includes a probabilistic description of the system solution, which in fact is the only rigorous description that can be given about the time evolution in view of the inherent presence of stochastic noise due to internal aperiodic high-frequency behavior, uncertain initial conditions, and uncertain forcing.

Based on the considerations discussed in Section 3 to follow, we shall in Section 4 formulate a simple two-component climatic feedback system containing some speculative elements that can serve as a prototype illustrating the application of the above sequence of steps in a dynamical systems analysis.

3. TIME CONSTANTS AND INTEGRAL CONSTRAINTS

In this section we shall discuss some physical considerations that are relevant in forming climate models in general, focusing on the consequences of the disparate "time constants" of the various domains of the climate system, and the integral mass and energy constraints on the system. The discussion will conclude with a presentation of some suggestive results from studies of atmospheric equilibration to its lower boundary state, leading up to the presentation in Section 4 of a special two-component system for longer-term variability that will serve as our prototype for the application of the systems analysis approach outlined above.

3.1. Equilibration Times for the Different Climatic Domains

The climatic system is extremely heterogeneous, containing domains (or subsystems) that are all interactive to some extent but have vastly different properties and modes of behavior if considered alone. One important property distinguishing each of the various domains is its "time scale," i.e., the time it would take for dissipative processes acting alone in the absence of continued forcing to remove departures from equilibrium. A short time scale indicates that the system (or subsystem) has the inherent capability of responding quickly to any perturbations from its equilibria. Thus, the time scale is directly related to the so-called "equilibration time" (also called the "response," "relaxation," or "adjustment" time), which is a measure of the time it takes for the system (or subsystem) to reequilibrate after a small change in its boundary conditions or forcing. A domain with a short equilibration time can be considered quasi-statically equilibrated to a neighboring domain that has a much longer equilibrium time. In this sense, the component with the longer equilibration time can be considered to drag along or "carry" the component with the more rapid equilibration time. In Appendix B we put these statements on a more quantitative basis, the main conclusions of which are summarized as conditions I and II below based on the following definitions:

Let $\bar{\chi}^{(\delta)}$ be the average of χ over a time interval δ satisfying Eq. (2.8) (where χ can be temperature T or mass of a domain M, for example), $P(\bar{\chi}^{(\delta)})$ be the characteristic period of variation of $\bar{\chi}^{(\delta)}$, and ϵ_χ be the *equilibration time* (i.e., the time it would take for a displacement of $\bar{\chi}^{(\delta)}$ from its long-term quasi-equilibrium value $\bar{\chi}_*^{(\delta)}$ to be reduced to e^{-1} of its displacement, after the forcing is removed). [The *thermal* equilibration time is shown to be approximately of the form

$$\epsilon_T \approx \left[b + \left(\frac{\kappa}{D^2}\right) + \left(\frac{K_v}{D^2}\right)_\delta + \left(\frac{K_h}{L^2}\right) \right]^{-1} = \epsilon_T(\delta)$$

where b is a cooling coefficient for long-wave radiation; κ is the molecular thermal diffusivity; $K_v(\delta)$ and $K_h(\delta)$ are the vertical and horizontal thermal diffusivities, respectively, due to *eddies* of all periods shorter than δ; D and L are characteristic vertical and horizontal spatial scales of $\bar{T}^{(\delta)}$, respectively.]

With these definitions we have the following conditions determining the nature of the equations governing $\bar{\chi}^{(\delta)}$:

I. $\epsilon_\chi(\delta) < \delta$ is likely to be sufficient to ensure that the equation governing $\bar{\chi}^{(\delta)}$ is *diagnostic,* i.e., that we can assume a quasi-steady-state $\partial\bar{\chi}^{(\delta)}/\partial t \approx 0$.

II. $\epsilon_\chi(\delta) \gg \delta$ is a necessary condition that the equation governing $\bar{\chi}^{(\delta)}$ be *prognostic* (i.e., $\partial\bar{\chi}^{(\delta)}/\partial t$ be of comparable magnitude to the other terms), the additional condition for sufficiency being that $\epsilon_\chi \gg P(\bar{\chi}^{(\delta)})$. If $\epsilon_\chi \ll P(\bar{\chi}^{(\delta)})$, then $\partial\bar{\chi}^{(\delta)}/\partial t$ will be negligibly small compared to the other terms (i.e., the system is "overdamped") and $\bar{\chi}^{(\delta)}$ can be determined diagnostically.

In Table II we list the main subsystems and their characteristic present dimensions and masses, thermal constants, estimates of $\epsilon_T(\delta)$ [in most cases for both $\delta = \delta_s \equiv 1$ hr (the *synoptic* average) and $\delta = \delta_c \equiv 10$ yr (the *climatic* average)], and rough estimates of ϵ_M for the ice domains. In the atmosphere, for example, it can be seen that, whereas for $\delta = \delta_s$ we have $\epsilon_T(\delta_s) \gg \delta_s$, for $\delta = \delta_c$ we have $\epsilon_T(\delta_c) \ll \delta_c$, implying that for the climatological mean the atmospheric thermal response is so fast that its mean can be considered as an equilibrium state. A schematic portrait showing the linkages and equilibration times for δ_c of all the domains comprising the complete climatic system is given in Fig. 5. In general, as we proceed from top (i.e., the atmosphere) to bottom (deep ocean and ice sheets) in this figure we encounter increasingly longer equilibration times (i.e., slower "response" times).

In forming Table II we have considered each climatic domain separately. Actually, all adjacent domains are linked by complex physical processes involving cross-boundary fluxes of mass, momentum, and energy that constitute "forcing" and "feedback" in the system. These linkages are indicated by the connecting lines in Fig. 5. As we said at the beginning of this subsection, domains with large ϵ (slow response time) will tend to "carry along" the domains with smaller ϵ, which because of their fast response times tend to adjust quasi-statically to the changing boundary conditions imposed by the slow-response domain. This follows from the fact that if $\epsilon_y \gg \epsilon_x$, for two domains x and y, one can always

TABLE II. PROPERTIES

Climatic domain	A (10^{12} m²) δ_c	D (m)	ρ (kg/m³)	M (10^{18} kg)	c (10^3 J/kg K)	(Mc) (10^{21} J/K)	κ (m²/sec)[b]
Atmosphere							
Free	510	10^4	0–1	5	1	5	−5
Boundary layer	510	10^3	1	0.5	1	0.5	−5
Ocean							
Mixed layer	334	10^2	10^3	34	4	10^2	−7
Deep	362	$4 \cdot 10^3$	10^3	1400	4	$5 \cdot 10^3$	−7
Sea ice (pack-shelf)	30	$(1–10^2)$	10^3	0.5	2	$(10^{-1}–1)$	−6
Continents							
Lakes and rivers	2	10^2	10^3	0.2	4	1	−7
Lithobiosphere	131	2	$3 \cdot 10^3$	1	0.8	1	−6
Snow and surface ice layer	80	1	$5 \cdot 10^2$	10^{-1}	2	10^{-1}	−7
Mountain glaciers	1	10^2	10^3	0.1	2	10^{-1}	−6
Ice sheets	14	10^3	10^3	10	2	10	−6

[a] Characteristic values of the following quantities are given (some of which are derived from a table given by Hoffert et al., 1980): present climatic mean horizontal area A, present climatic mean depth D, density ρ, present climatic mean mass $M = \rho DA$, specific heat c_p, heat capacity (Mc_p), molecular thermal diffusivity κ, vertical eddy thermal diffusivity $K_v(\delta)$, horizontal eddy thermal diffusivity $K_h(\delta)$, horizontal eddy length scale $L(\delta)$, vertical eddy diffusive time scale (D^2/K_v), horizontal eddy diffusive time scale (L^2/K_h), longwave radiative time scale b^{-1} for the atmosphere and surface layers of the land and ocean, thermal equilibration time $\epsilon_T(\delta) = [b + (\kappa/D^2) + (K_v/D^2)_\delta + (K_h/L^2)_\delta]^{-1}$, mass equilibration

adopt an averaging period $\delta = \epsilon_x$ making the equation governing x diagnostic but leaving the domain for which $\epsilon_y \gg \delta$ essentially prognostic. In this sense we speak of the deep ocean and ice sheets, for example, as the "carriers" of long-term climatic variability, the prognostic equations for these domains governing the nonequilibrium evolution of the system. The nature and path of this evolution, however, can be markedly influenced by all the higher frequency effects of the fast-response domain, the deterministic parts of which must be parameterized and the random parts of which must be included as stochastic forcing (cf. Section 2 and Appendix A). By the same token, the low-frequency, long-response-time phenomena can be considered as fixed conditions in deducing the separate behavior of the domain with higher frequencies and shorter response times. As

OF CLIMATIC DOMAINS[a]

K_v (m²/sec)[b]		K_h (m²/sec)[b]		L (m)[b]		D^2/κ (sec)[b]	D^2/K_v (sec)[b]		L^2/K_h (sec)[b]		b^{-1} (sec)[b]	ϵ_T (sec)[b]		ϵ_M (sec)[b]	$P_{T,M}$ (sec)[b]
δ_s	δ_c	δ_s	δ_c	δ_s	δ_c		δ_s	δ_c	δ_s	δ_c		δ_s	δ_c	δ_c	
1	2	1	6	6	7	13	7	6	11	8	6	7	6	—	5, 7, 9–12
1	1	1	6	6	6	11	5	5	11	6	6	5	5	—	5, 7, 9–12
−3	−2	−1	3	5	6	11	7	6	9	9	8	7	6	—	6, 7, 9–12
−4	−3	−2	2	(6–7)	6	13	11	10	14	10	—	11	10	—	7–12
—	—	—	—	—	—	(6–10)	—	—	—	—	—	(6–10)		(6–10)	7, 9–12
−3	−3	—	−2	—	2	11	—	6	—	6	6	6	6	—	7–12
—	—	—	—	—	—	6	—	—	—	—	6	6	6	—	5, 7, 9–12
—	—	—	—	—	—	5	—	—	—	—	5	5	5	5	5, 7, 9–12
—	—	—	—	—	—	10	—	—	—	—	—	10	10	9	9–12
—	—	—	—	—	—	12	—	—	—	—	—	12	12	11	10–12

time ϵ_M for the ice domains, and observed or inferred major periods of variation P. The averaging periods considered are $\delta = \delta_s \approx 1$ hr $\approx 10^3$ sec (the synoptic average) and $\delta = \delta_c \approx 10$ yr $\approx 10^8$ sec (the climatic average). Sources other than Hoffert *et al.* (1980) used in constructing this table include Nace (1969) for water inventory, Flint (1971) for ice inventory, Priestley (1959) for κ and K, and Birchfield (1977) and Sergin (1979) for estimates of ϵ_M.

[b] Values given are exponents of 10 for these variables.

a general rule, if two domains have similar response times, the behavior in both domains must be solved for simultaneously, i.e., two adjacent domains independently in equilibrium for a given δ ($\epsilon \leq \delta$) must be in equilibrium with each other.

One consequence of the above remarks, which we have already mentioned, is that the climatic mean ($\delta = \delta_c$) state of the *atmosphere* should be deducible as a "steady-state" problem subject to fixed boundary conditions imposed by a neighboring domain for which $\epsilon \gg \delta$. From the values of ϵ given in Table II we see that the response of the surface of the earth (i.e., the biolithosphere temperature, surface snow and ice masses, and to a slightly lesser extent the mixed-layer ocean temperature) is fast enough to be of an equilibrium nature comparable to the atmosphere.

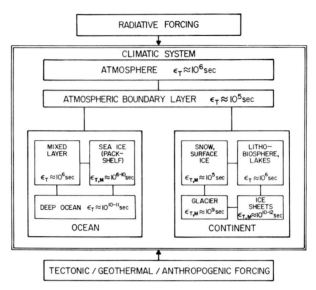

FIG. 5. Schematic representation of the domains of the climatic system showing the estimated equilibration times.

Therefore these aspects of the surface state should be deduced simultaneously with the atmospheric state, especially insofar as there are significant feedbacks between the atmosphere and these domains (e.g., the ice-albedo feedback). If we arbitrarily prescribe the surface snow–ice distribution, for example, one can indeed compute the atmospheric state that is required for equilibrium with it, but this equilibrium may not be consistent with the prescribed snow–ice field (e.g., the equilibrium may require surface temperatures that are well below freezing in regions of ample moisture where no ice is prescribed). Moreover, it is important to realize that even though the atmosphere and its lower-boundary surface layer may be in equilibrium, judging from the geologic record of long-term climatic change this equilibrium most likely entails a small but significant *dis*equilibrium of the total ice mass field and the deep ocean temperature. In the next subsection we set down an integral equation governing this disequilibrium. We also show that the tendency for rapid equilibration implies that any net addition of heat to the atmosphere (e.g., by release of latent heat of fusion due to ice formation or by heat flux from the ocean) must be compensated quasi-statistically by a net radiative imbalance at the top of the atmosphere.

3.2. Some Integral Constraints on Mass and Energy

Rigorous discussions of the globally integrated requirements for mass and energy that must be satisfied by any dynamical system purporting to represent climatic change are given in two previous articles (Saltzman, 1977, 1978). Good approximations to the exact equations for total energy and mass in the entire climatic system derived in these articles are the following:

$$(M_a c_a) \frac{d\hat{T}_a}{dt} + (M_w c_w) \frac{d\hat{T}_w}{dt} + (M_l c_l) \frac{d\hat{T}_l}{dt} + (M_i c_i) \frac{d\hat{T}_i}{dt}$$

$$\approx [L_f + (c_w \hat{\mathcal{T}}_w - c_i \hat{\mathcal{T}}_i)] \frac{dM_i}{dt}$$

$$- L_v \frac{dM_v}{dt} + \sigma(\tilde{N}^{\downarrow} + \tilde{G}^{\uparrow}) \tag{3.1}$$

and

$$dM_a/dt \approx 0 \tag{3.2}$$

$$dM_l/dt \approx 0 \tag{3.3}$$

$$d(M_w + M_i)/dt \approx 0 \tag{3.4}$$

In these equations we have defined the subscripts a, w, l, i, and v to denote the atmospheric, hydrospheric, lithobiospheric, ice, and water vapor components of the climatic system, respectively; M_x is the total mass of the component denoted by x; c is the specific heat at constant pressure; L_v and L_f are the latent heats of vaporization and fusion, respectively; $\mathcal{T} = T - 273$ K; $\sigma = 4\pi a^2$ is the surface area of the earth assumed to be nearly the same at the top of the atmosphere and the base of the climatic system; $\tilde{N}^{\downarrow}$ is the net globally averaged difference between incoming solar radiation $\tilde{H}_T^{(1)\downarrow}$ and outgoing terrestrial radiation $\tilde{H}_T^{(2)\uparrow}$ measured per unit area at the top of the atmosphere (i.e., $\tilde{N}^{\downarrow} = \tilde{H}_T^{(1)\downarrow} - \tilde{H}_T^{(2)\uparrow}$); $\tilde{G}^{\uparrow}$ is the net globally averaged upward geothermal flux of heat; and $(\hat{\,})_x$ denotes a mass-weighted average of $(\,)_x$ over the entire component denoted by x. The values of M, c, and (Mc) for the various components are listed in Table I.

Assuming that a typical variation of temperature of the atmosphere, land surface, and ice that accompanied the ice ages of the Quaternary is of the order of 10 K (probably a high value), that the variation of "precipitable water" M_v is of the order of 10^{15} kg (cf. Saltzman and Vernekar,

1975), and that the ice mass decreased by 50×10^{18} kg since the Wisconsin maximum at 18,000 B.P. (Flint, 1971), the following order of magnitude relations based on *average* rates of change would appear to hold:

$$\frac{1}{\sigma}\left[(M_a c_a)\frac{d\hat{T}_a}{dt},\ (M_l c_l)\frac{d\hat{T}_l}{dt},\ (M_i c_i)\frac{d\hat{T}_i}{dt},\ L_v\frac{dM_v}{dt} \right]$$

$$\ll \frac{L_f}{\sigma}\frac{dM_i}{dt} \sim 10^{-1}\ W\ m^{-2} \tag{3.5}$$

$$0 < (c_w\hat{\mathcal{T}}_w - c_i\hat{\mathcal{T}}_i) \ll L_f \tag{3.6}$$

Whether or not these inequalities are universally valid, it is plausible to assume that because of the comparably short equilibration times of the atmosphere and lithobiosphere, and the strong relation between M_v and $\hat{T}_a$, the quantities $\hat{T}_a$, $\hat{T}_l$, and M_v all vary synchronously 180° out of phase with M_i, so that as a first approximation

$$\left[(M_a c_a)\frac{d\hat{T}_a}{dt} + (M_l c_l)\frac{d\hat{T}_l}{dt} + (M_i c_i)\frac{d\hat{T}_i}{dt} + L_v\frac{dM_v}{dt} \right] \approx -k\frac{dM_i}{dt} \tag{3.7}$$

where k is a positive constant.

Thus, we obtain the following equations relating the bulk ocean temperature $\hat{T}_w$, total ice mass M_i, net radiation $(\sigma\tilde{N}^\downarrow)$, and net geothermal flux $(\sigma\tilde{G}^\uparrow)$:

$$(c_w M_w)\frac{d\hat{T}_w}{dt} - \Pi\frac{dM_i}{dt} - (\sigma\tilde{N}^\downarrow) - \sigma\tilde{G}^\uparrow \approx 0 \tag{3.8}$$

where

$$\Pi = L_f + (c_w\hat{\mathcal{T}}_w - c_i\hat{\mathcal{T}}_i) + k$$

and if we further invoke (3.5) and (3.6),

$$\Pi \approx L_f$$

If we consider the atmospheric and oceanic domains separately, we have the following similar constraints:

First, for the *atmosphere,*

$$(c_a M_a)\frac{d\hat{T}_a}{dt} \approx \sigma(\tilde{N}^\downarrow + \tilde{H}_S^\uparrow) + L_f\frac{dM_i}{dt} - L_v\frac{dM_v}{dt} \tag{3.9}$$

where $\tilde{H}_S^\uparrow\ (=\tilde{H}_S^{(1)\uparrow} + \tilde{H}_S^{(2)\uparrow} + \tilde{H}_S^{(3)\uparrow} + \tilde{H}_S^{(4)\uparrow})$ is the net upward flux of heat at the earth's surface due to shortwave radiation $H_S^{(1)\uparrow}$ (generally negative), outgoing longwave radiation $H_S^{(2)\uparrow}$, conductive–convective upward flux of sensible heat $H_S^{(3)\uparrow}$, and water phase transformations $H_S^{(4)\uparrow} = H_S^{(4E)\uparrow} + H_S^{(4F)\uparrow}$, where $H_S^{(4E)\uparrow} = L_v E$, $H_S^{(4F)\uparrow} = -L_f F$, and E and F are the rates of surface evaporation/ablation and freezing/sublimation per unit area, respectively. At each point on the surface, the net upward heat flux to the atmosphere must be nearly balanced by the net upward flux from below, which we denote by $H_S^{(5)\uparrow}$, i.e.,

$$H_S^\uparrow = H_S^{(5)\uparrow} \tag{3.10}$$

Thus, consistent with (3.5) we have for an "equilibrium atmosphere" (i.e., $d\hat{T}_a/dt = dM_v/dt = 0$), the relation

$$\sigma \tilde{N}^\downarrow + \sigma \tilde{H}_S^\uparrow + L_f \frac{dM_i}{dt} \approx 0 \tag{3.11}$$

Second, for the *ocean* as a whole, we have

$$c_w M_w \frac{d\hat{T}_w}{dt} \approx \sigma_w \tilde{H}_{Sw}^\downarrow - \sigma_{wi} \tilde{H}_{Swi}^\uparrow + \sigma_w \tilde{G}_w^\uparrow \tag{3.12}$$

where σ_w and σ_{wi} are the surface areas of seawater and sea ice, respectively; $\sigma_w = \sigma_w + \sigma_{wi}$; $(\tilde{\ })_w$ and $(\tilde{\ })_{wi}$ denote averages over the seawater and sea ice areas, respectively.

3.3. Climatic Equilibrium of the Atmosphere and Surface of the Earth

In Section 3.1 we demonstrated that the atmosphere and upper layer of the earth's surface can respond so quickly to perturbations that we can consider the climatic-average (i.e., $\delta = 10$ yr) properties of these domains to be in a quasi-equilibrium state with the much more slowly varying deep oceans and ice sheets. Numerous models have been formulated and solved to deduce such equilibrium states, with relatively high degrees of fidelity to the presently observed climate. As already noted in Section 2, there are two basic modeling approaches to achieving these solutions: GCMs and SDMs.

3.3.1. The General Circulation Models (GCMs). These models (sometimes also called *explicit-dynamical models*) are based on equations governing the synoptic-average variables. As seen from Table II, these equations must be *prognostic* equations. The time-dependent solutions of these equations represent a sequence of weather maps that are allowed to evolve numerically until the statistically steady state is achieved. The output is then averaged to yield the model "climate" in much the same manner as real weather records are processed to determine climatic norms. It is assumed that although the individual details of the evolving numerically generated maps cannot be accurate by the standards of "weather prediction," their ensemble statistics can be fairly accurate. Because these models treat a good deal of phenomena included in the high-energy frequency band between 1 hr and 1 week (see Fig. 3) by *explicit* deterministic physics, this approach represents the most rigorous and complete theory of the equilibration of the atmosphere and surface layer to the quasi-steady boundary conditions imposed by the deep ocean and ice sheets. There are now many examples of such solutions, a good review and discussion of which is given by Smagorinsky (1974).

Difficulties. As noted in this latter reference, the computer time required for even a single experimental run to equilibrium is enormous— e.g., a model containing grid points at about 10 atmospheric levels spaced every 200 km in the horizontal, run for about 500 days would take about 50 days (i.e., only 10 times faster than real time)! Given the present costs of computer time and the human resources needed, it is clearly prohibitive to do very much experimentation and long-term integration with these models. Moreover, to simulate long-term climatic change, for which the carrier time-dependent equations are those governing the deep ocean and ice masses, it would be necessary to calculate such equilibria at some regular interval, say every few hundred years. Even this would seem to be prohibitive for an analysis of the Quaternary glaciations.

Also, with reference to this problem of long-term climatic change, it was noted at the end of Section 2 that along with the atmospheric climatic portrait deduced at any time, the most significant result of such an atmospheric equilibrium calculation is the determination of (1) the net flux of heat into or out of the oceans (that must be balanced by release or consumption of the latent heat of fusion and/or by a net radiative imbalance at the top of the atmosphere), and (2) the net rate of accumulation or melting and ablation of ice mass. At present GCMs are generally constrained by the requirements that the net globally averaged fluxes of heat across the top of the atmosphere and the lower boundary are identically zero. Although this would seem to be a reasonable constraint since the departures to be expected are small and perhaps not within the resolving power of the methods of computation now used, it is probably fair to say that unless some effort is made to relax these constraints and make such estimates (albeit crudely and subject to large errors or noise), the GCM will never be capable of being used to simulate long-term climate change in which the bulk of the ocean participates thermodynamically.

Finally, from the viewpoint of dynamical systems analysis, we must recognize that the GCM typically contains a huge number of degrees of freedom as represented, for example, by the thousands of difference equations for each dependent variable at each grid point in the three-dimensional spatial lattice. With such a system, subject to a finite computational noise level, it would be most difficult to distinguish nearby multiple stable equilibrium states if they existed, and impossible to isolate unstable equilibria. It seems reasonable to expect, however, that the macroturbulent behavior represented by these many equations is not altogether "free," but possesses some coherent organization to effect the bulk transports of mass, momentum, and energy required by the conservation principle applied to the climatic-average state.

All these considerations point to the desirability of advancing the development of SDMs in which the task of generating the synoptic eddy statistics by continuous hour-by-hour integration of a huge dynamical system is supplanted by the development of physically based, deterministic representations of the parameterized effects of these eddies on the climatic-mean state. These SDMs will be the focus of our attention in the remainder of this article.

3.3.2. Statistical-Dynamical Models (SDMs). These models are based on equations governing the climatic-mean atmospheric and surficial variables, which from Table II are essentially diagnostic. As we have noted above, SDMs are characterized by the need to parameterize not only the effects of fluctuations of frequencies of less than about an hour as in the GCMs, but also the effects of all frequencies up to the climatic averaging period chosen (e.g., 10 yr). This can include the diurnal and seasonal cycles, mesoscale phenomena, synoptic weather waves, and even some interannual variations. In view of the complexity and high energy of all this subclimatic variability, it is clear that the development of physically sound parameterizations must constitute a major challenge of dynamical meteorology. It is also likely that the stochastic terms in the governing SDM equations will be significant, so that one could not expect to observe that the climatic system resides at any deterministic equilibrium even for fairly good parameterizations. Thorough reviews of the general foundations of SDMs are given by Schneider and Dickinson (1974) and by Saltzman (1978); at this point we note, simply, that in spite of the parameterization difficulties, SDMs based on rather crude approximations can account for a good deal of the large-scale spatial variability of the climatic-mean state. For example, Saltzman and Vernekar (1971a) demonstrated that the present zonal-average climatic statistics for the atmosphere and surface can be deduced from an SDM to roughly the same accuracy as from a GCM. In the next subsection we describe a few significant findings from the Saltzman–Vernekar model that are of relevance for long-term climatic change.

3.3.3. Some Results from Application of the Saltzman–Vernekar Model. A fairly complete SDM of the zonal-average equilibrium response of the atmosphere to a prescribed surface distribution of ocean, ice, and land and a prescribed oceanic temperature distribution at the base of the seasonal thermocline has been developed over nearly a 20-yr period by A. D. Vernekar and the writer (see Saltzman, 1964, 1968; Saltzman and Vernekar, 1968, 1971a,b, 1972, 1975). This model includes not only the thermodynamic energy equation which is the basis of the so-called energy balance SDM, or EBM (see reviews by Schneider and Dickinson 1974; Saltzman, 1978; North *et al.,* 1981), but also the quasi-geostrophic momentum equations and the water vapor continuity equation from which mean poloidal motions and the hydrologic cycle can be deduced. By omitting these latter equations, the model is reducible to an EBM. In accordance with the remarks in the last paragraph of Section 3.1, we note that the surface temperature distribution obtained from this model is generally consistent with the prescribed ice coverage to the same level of accuracy that the commonly used parameterization of the ice extent proposed by Budyko (1969) and Sellers (1969) can represent this ice edge. To study the sensitivity of the Saltzman–Vernekar model to changes in the solar constant, we have recently incorporated the Budyko–Sellers parameterization linking surface temperature and ice edge in an explicit manner (Saltzman and Vernekar, 1983).

In brief review, the deduced dependent variables of the model include the climatic means and variances of the temperature and wind, the humidity, and all components of the surface and atmospheric heat, momentum, and water balances (e.g., precipitation and evaporation, horizontal and vertical fluxes of sensible and latent heat and momentum due to eddies and mean poloidal motions). The vertical fluxes are determined within the atmosphere and at the interface between the atmosphere and subsurface media (ocean, land, ice), by means of a full set of parameterizations of short- and longwave radiation, convection of sensible heat, and latent heat processes. To effect closure of the system, baroclinic and barotropic wave theory is used in a crude manner to parameterize the horizontal transports of heat, momentum, and water vapor in the atmosphere.

More detailed and comprehensive reviews of the model specifications and results have been given recently by the writer (Saltzman, 1978, 1979). Here we cite only the following result, first reported in Saltzman (1977), that is of significance in considerations of the long-term climatic change.

When present lower-boundary ice coverage is specified, it is found that the deduced equilibrium atmospheric climatic state is in reasonably good agreement with present observations, but this solution entails a net imbalance of the climatic system as a whole, i.e., a net upward radiative flux at the top of the atmosphere, $\bar{N}^{\uparrow} \approx 6$ W m^{-2}, that goes along with a net loss of heat from the oceans, $\bar{H}_S^{\downarrow}$, of the same magnitude [as required by Eq. (3.11) since $dM_i/dt = 0$ in the model]. This result is not completely without observational support [see satellite radiation measurements reported by Gruber (1977) and by Smith *et al.* (1977)], but in view of the many possible errors in such net radiation measurements and the many inadequacies of our model, it is probable that this agreement is fortuitous. Moreover, the magnitude of this imbalance seems inordinately large since the implied change in mean ocean temperature is at the rate of 0.01 K yr^{-1}. However, it is perhaps of greater significance that this same model requires a *reverse* set of fluxes, $\bar{N}^{\downarrow} = \bar{H}_S^{\downarrow} \approx 4$ W m^{-2}, for atmospheric equilibrium to be achieved when the ice age conditions that prevailed at 18,000 B.P. are specified (Saltzman and Vernekar, 1975). The latter result is largely due to the fact that large regions of the high-latitude ocean, which under present ice-free conditions are losing heat rapidly, are "insulated" by sea ice in an ice age (cf. Newell, 1974). Thus, even if only the *direction* of the theoretically computed changes in $\bar{N}^{\uparrow} (= \bar{H}_S^{\downarrow})$ is valid, it would appear that the insulating effect of the sea ice may be of importance. The possibility exists that in a relatively ice-free state the ocean as a whole tends to lose heat, whereas in a heavily glaciated state the ocean as a whole tends to gain heat, suggesting that there can be free oscillations between mean ocean temperature and sea-ice extent if mean ocean temperature can affect the ice extent. These ideas were the starting point for a series of models of longer-term climatic variations developed by the writer and his colleagues over the last few years, which were designed to serve as illustrative prototypes of the systems analysis approach. We shall review the results in the following two sections. A detailed review of the model itself is given in Appendix C, including a more general formulation that allows the possibility for massive ice-sheet development on land (a factor that was excluded from the Saltzman–Vernekar model and the aforementioned sea ice-insulator models).

4. A Prototype Deterministic System

In Appendix C we describe the physical basis of a general SDM governing long-term climatic change, including variations of continental ice sheets as well as marine ice forms. We are presently undertaking a study to explore the properties of the continental ice-sheet model governed by the prognostic equations (C.15) and (C.19) given therein, a schematic feedback flowchart of which is given in Fig. 19. For our immediate purpose of illustrating the application of the systems analysis approach, we shall now review previous results for the *all-ocean* model described in Appendix C in which only sea-ice formation is admitted, but which can also be identified to some extent with more massive ice-shelf formation (cf. Fig. 18B). These results were developed in a series of articles (Saltzman, 1978, 1979, 1982; Moritz, 1979; Saltzman and Moritz, 1980; Saltzman *et al.*, 1981). In spite of obvious inadequacies, this model may still represent enough of the climatically relevant feedbacks to be of physical interest. In any event, the model is meant here to serve as a simple prototype of a more rigorous and complete climatic feedback system that in principle should include many additional dependent variables and corresponding equations. Such a more complex system would undoubtedly contain a more complex attractor set than we shall consider here.

4.1. The "Reference" Deterministic Solution

The system to be considered is of the form (C.27) and (C.28) of Appendix C, i.e.,

$$d\eta/dt = \Phi(\eta, \theta)$$
$$d\theta/dt = \Psi(\eta, \theta)$$

where η is the sine of the latitude of marine ice extent; θ is the mean temperature of the entire ocean; Φ and Ψ are autonomous, nonlinear, functions of η and θ, various versions of which are given in the aforementioned series of articles.

In Fig. 6 we show the deterministic solution for the system, given a set of "reference" values of all the coefficients and fixed parameters that are believed to be reasonable estimates based on present observations (see Saltzman, 1980, Tables 1 and 2, and Saltzman, 1982). This solution, obtained by standard methods for autonomous systems (e.g., Boyce and DiPrima, 1977), is in the form of a $(\eta - \theta)$ phase-plane representation. Two physically admissable equilibria (within the bounds $0 < \eta < 1$) are found: a stable "spiral" equilibrium ($\eta_*^{(1)} = 0.8904$, $\theta_*^{(1)} = 277.505$ K) and an unstable "saddle-point" equilibrium ($\eta_*^{(2)} = 0.6469$, $\theta_*^{(2)} = 277.674$ K). The time-dependent evolution of the solution for any arbitrary initial condition is indicated by the trajectories shown. In Fig. 7 (top) we show one sample trajectory in the attractor region of the stable equilibrium in units of the *departures* from the equilibrium. The initial departure is $\eta' = -0.2$ and $\theta' = 0$. The "residence density" portrait in this figure gives some

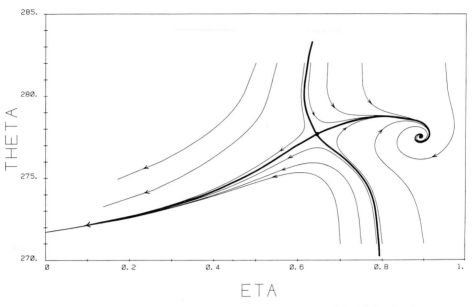

FIG. 6. Deterministic phase-plane solution of the system (C.27) and (C.28) for the all-ocean model, given the "reference" values of parameters. From Saltzman (1982).

measure of the time spent by the system in a particular point in phase space, the time interval between each dot being 60 yr.

In Fig. 8 we show the net heat flux across the ocean–sea ice surface, $H_S^\uparrow(\xi)$, corresponding to the two equilibria.

4.2. Sensitivity of Equilibria to Changes in Parameters: Prediction of the Second Kind

The equilibrium positions and their stability characteristics can change as a consequence of changes in any of the many parameters specified in the problem. If P_i denotes the value of any particular parameter, the change in the equilibrium position (η_*, θ_*) due to a small (say 10%) departure of P_i from its reference value $\hat{P}_i$ is given by

$$\mathscr{S}(\eta_*, \theta_*)_i = 0.1\hat{P}_i\left[\frac{\partial(\eta_*, \theta_*)}{\partial P_i} + \sum_j \frac{\partial(\eta_*, \theta_*)}{\partial P_j}\frac{\delta P_j}{\delta P_i}\right] \tag{4.1}$$

This represents the *total sensitivity* of (η_*, θ_*) to P_i. If we hold fixed all parameters P_j other than the particular parameter P_i (that is, neglect the

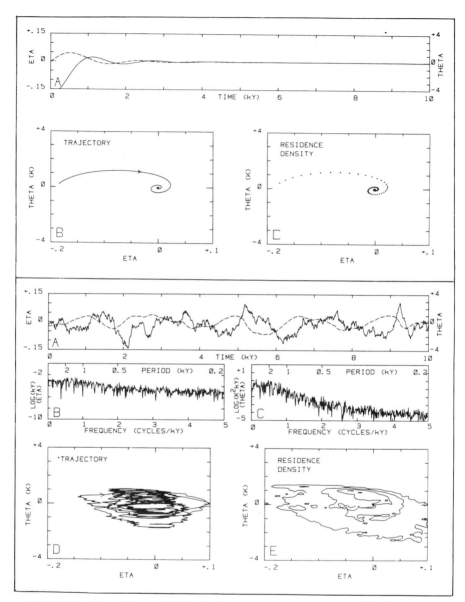

FIG. 7. (Top) Sample deterministic solution near the stable equilibrium ($\eta_*^{(1)}$, $\theta_*^{(1)}$) of Fig. 6, in units of departure from the equilibrium (η', θ'): (A) time evolution of η' (full curve) and θ' (dashed curve); (B) trajectory; (C) trajectory speed (dots are spaced at every 60 yr). (Bottom) Sample stochastic solution for the above deterministic system, obtained by perturbing η with random forcing of an amplitude corresponding to observed interannual variations: (A) sample evolution; (B) the variance spectra for η; (C) the variance spectra θ; (D) the sample phase plane trajectory; (E) the "residence density" of the solution. From Saltzman (1982).

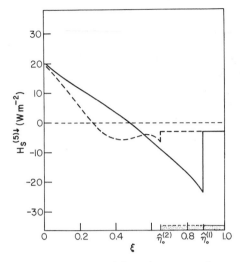

FIG. 8. Distribution of the net downward flux of energy at the ocean surface, $H_S^{(5)\downarrow}(\xi)$, corresponding to the two reference equilibria $\hat{\eta}_\cdot^{(1)}$ (solid curve) and $\hat{\eta}_\cdot^{(2)}$ (dashed curve). From Saltzman and Moritz (1980).

dependence of P_j on P_i), we define a *partial sensitivity* (cf. Saltzman and Pollack, 1977),

$$s(\eta_*, \theta_*) = 0.1\hat{P}_i \frac{\partial(\eta_*, \theta_*)}{\partial P_i} \tag{4.2}$$

As noted in Saltzman and Moritz (1980), a sensitivity analysis for all parameters of a model, based on Eqs. (4.1) and (4.2), is important for two main reasons: (1) it provides some measure of the reliability of deduced equilibria in the face of known levels of uncertainty of the parameters, and (2) by revealing the parameters to which the equilibria are most sensitive, it points out the areas where future research efforts are most needed to establish more precise estimates of the parameters or the physical processes they represent.

In discussions of the possibilities for climatic change, we are often most interested in the shift of the equilibria in response to variations of particular parameters that can be regarded as external forcing (e.g., the solar constant S and CO_2 changes due to anthopogenic sources). For example, a commonly used measure of the overall sensitivity of climatic models is the partial sensitivity,

$$\beta \equiv s(\tilde{T}_{S*})_S = 0.1\hat{S} \frac{\partial \tilde{T}_{S*}}{\partial S} \tag{4.3}$$

where $\hat{S}$ denotes the present (i.e., reference) value of the solar constant and $\bar{T}_S$ is the global-average surface temperature. However, this latter measure is, at best, valid only very close to one of several possible equilibria that could be present. A fuller description of the sensitivity of the equilibria to variations of a particular parameter is provided by a plot of the equilibrium points as a function of the parameter, examples of which are the "equilibrium portraits" shown in Fig. 9 for our present model (Saltzman and Moritz, 1980). This figure shows the variations of the ice-edge equilibrium positions as a function of the solar constant (a) and reference CO_2 concentration (b).

The shift of an equilibrium due to the variation of a single external parameter is an important piece of information concerning the possibilities for climatic change, constituting what has been termed "predictability of the *second* kind" by Lorenz (1975). However, such a prediction is not equivalent to a prediction of the actual trajectory that the climatic system will executue (prediction of the *first* kind). In the first place, it is likely that other external variables are also changing at the same time; in the second place, even if there were no other external changes taking place, it is highly unlikely that the complete climatic system ever actually resides at the deterministic equilibrium point in phase space. Rather, judging from the continuous spectrum of past climatic change, it is probable that climate always represents a transient, nonequilibrium state, the trajectory of which may be close to some stable equilibrium point or stable limit cycle but never locks firmly into one. Such departures from equilibrium must always be present, if only as a consequence of the instabilities and nonlinearities on time scales shorter than the climatic averaging period, which generate ever-present stochastic noise (e.g., "weather" fluctuations and interannual variability). As one possible scenario, we can imagine that the evolution of the climatic state consists of a trajectory, under the constant influence of stochastic forcing, seeking to approach asymptotically a stable equilibrium that is drifting in response to changing external conditions. In Fig. 10 we show a schematic two-dimensional portrayal of such a trajectory, say for η and θ, corresponding to an arbitrary displacement of the equilibrium due perhaps to a changed amount of CO_2 in the atmosphere as depicted in Fig. 9(b). As shown in this diagram, it is entirely possible that at any time the displacement of the trajectory from the equilibrium point is greater than the displacement of the equilibrium from its initial position. In Section 4.3 we shall describe some other possible scenarios, in the context of our prototype model with constant external forcing.

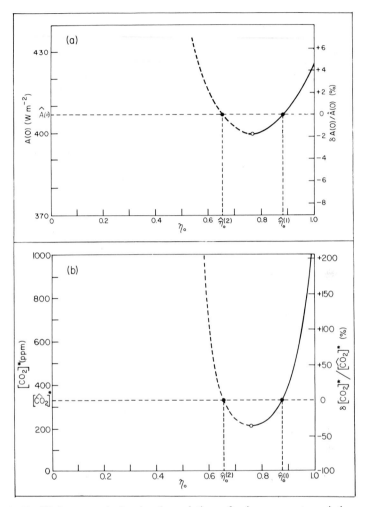

FIG. 9. Equilibrium portrait showing the variations of η_* in response to variations only of the incoming radiation $A(0)$ (a) and to variations only of reference CO_2 content of the atmosphere (b). From Saltzman and Moritz (1980).

4.3. Structural Stability

In addition to the changes in *position* of the equilibrium in phase space described above, the possibility also exists that the *stability* of the equilibrium is changed when parameters of a dynamical system are changed. The

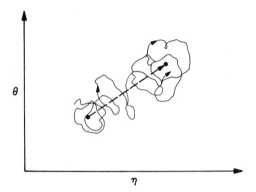

FIG. 10. Hypothetical trajectory of the climatic system corresponding to a displacement of a stable equilibrium point of the system due to external forcing.

most dramatic form of such a stability change is a transition between stability and instability (e.g., a Hopf "bifurcation"; see Marsden and McCracken, 1976). An example of this is illustrated in Fig. 11 for our model. By increasing b_{10} [the coefficient controlling the magnitude of the CO_2-positive feedback—cf. Eq. (C.21) and Fig. 19] from its reference

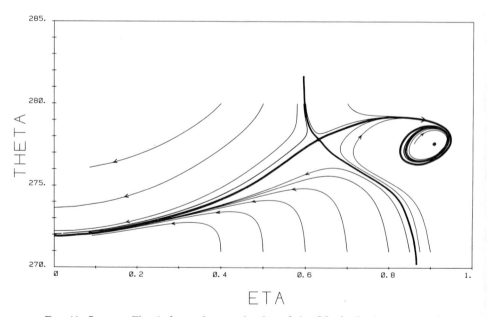

FIG. 11. Same as Fig. 6, for an increased value of the CO_2 feedback parameter b_{10}, showing a bifurcation to a stable limit cycle around $(\eta_*^{(1)}, \theta_*^{(1)})$. From Saltzman (1982).

value of 10^{-3} to a value of 7×10^{-3}, a bifurcation occurs in which the stable spiral equilibrium $(\eta_*^{(1)}, \theta_*^{(1)})$ becomes an unstable spiral equilibrium while retaining stability at large departures from $(\eta_*^{(1)}, \theta_*^{(1)})$. This implies the existence of the stable limit cycle shown in Fig. 11.

Along with the possibility that climatic variability occurs near a stable equilibrium point (as illustrated in Figs. 6 and 7), the possibility also exists that climatic variability occurs near an unstable equilibrium point around which a limit cycle shapes the variations. These possibilities represent two major scenarios for the basic determinism of free climatic variability. Other possibilities, not illustrated here, are the existence of multiple unstable equilibria that form a strange attractor (cf. Lorenz, 1963), or multiple stable equilibria between which the system can flip under the influence of stochastic noise (cf. Sutera, 1980).

In Fig. 12 (top) we depict the behavior of the deterministic limit cycle flow surrounding $(\eta_*^{(1)}, \theta_*^{(1)})$ in more detail, showing in parts B and C the variance spectra for η and θ, respectively. For our particular reference values a period of close to 2000 yr is indicated.

In Fig. 13 we depict the changing physical state of the model climatic system as it executes the limit cycle. Here τ_0 and η_0 represent the equilibrium values of the equatorial surface temperature and ice edge, respectively, and H denotes the net flux of heat into or out of the ocean.

With even more drastic changes in the parameters, the model can be reduced to an extremely simple form described in Saltzman (1978) in which only a single stable equilibrium is admissible (i.e., there is no nearby unstable saddle point equilibrium). The behavior of a sample trajectory for this system is shown in Fig. 14 (top). Similarly, as described in Saltzman *et al.* (1981), it is possible to construct a model in which only a single stable limit cycle is present, again with no nearby unstable saddle point equilibrium [see Fig. 15 (top)]. The remarkable feature of this latter, van der Pol-type, relaxation oscillation is the presence of a biomodality in the trajectory speed shown in part E of Fig. 14 (i.e., the "residence density" portrait).

5. CLIMATE AS A STOCHASTIC-DYNAMICAL SYSTEM: EFFECTS OF RANDOM FORCING

As we have said, the deterministic system just discussed is, at best, a highly speculative and simplified model of some of the feedbacks that are likely to be operating in the real climatic system. Clearly, there is much room for improvement of the parameterizations and for enrichment of the total physical content and fidelity of the model. For the reasons discussed

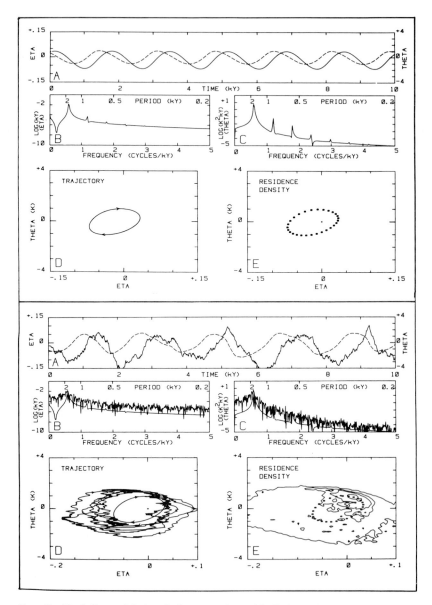

Fig. 12. (Top) Deterministic solution near the stable limit cycle portrayed in Fig. 11, showing the variance spectra for η (B) and θ (C) in addition to the sample evolution (A), phase plane trajectory (D), and trajectory speed shown by dots spaced at every 60 yr (E). From Saltzman (1982). (Bottom) Stochastic solution near the stable limit cycle, using the same format as Fig. 7 bottom. From Saltzman (1982).

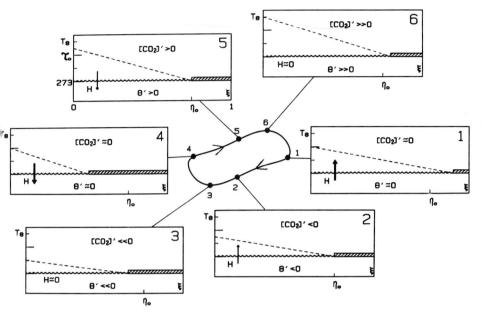

FIG. 13. Schematic sequence of climatic states corresponding to the limit cycle solution. From Saltzman *et al.* (1981).

in Section 2, however, it is unlikely that any amount of added physical rigor would enable us to represent the true behavior of the climatically averaged variables η and θ as a purely *deterministic* system (see also, Hasselmann, 1976; Saltzman, 1982). The most we could hope for is that with such added physical rigor, the residual effects due to inadequate representation of phenomena of higher frequency than the climatic averaging period can be considered "random," i.e., representable by a white-noise (Wiener) process of suitable amplitude. This constitutes the *stochastic* component of any climatic dynamical system.

In essence, this stochastic component represents the effects of determinism on scales that we cannot, or choose not to, consider, the sources of which are threefold as noted in Section 2:

(1) The impossibility of fulfilling the so-called Reynolds conditions for the climatic average because of aperiodic fluctuations on shorter time scales.
(2) The difficulty of adequately parameterizing the nonlinear effects of the higher-frequency phenomena [e.g., each of the parameterizations (C.13) and (C.20)–(C.26) should have an error term added to it, at least a part of which is "random"].

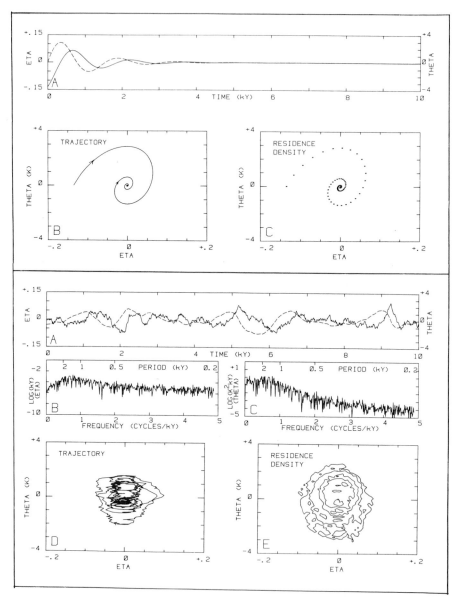

FIG. 14. Same as Fig. 7, but for the simpler model described in Saltzman (1978) containing only a single stable equilibrium. From Saltzman (1982).

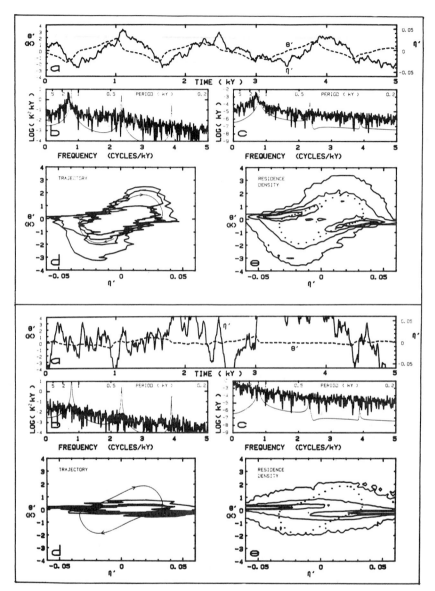

FIG. 15. (Top) Sample stochastic solution for the idealized, van der Pol-type model described in Saltzman *et al.* (1981), using the reference noise amplitude. Format is the same as in Fig. 12. The deterministic solution is shown by light lines. (Bottom) Sample stochastic solution, using a noise amplitude three times larger than the previous reference value. From Saltzman *et al.* (1981).

(3) The impossibility of specifying initial conditions and boundary conditions (i.e., external forcing) without some level of uncertainty and error.

We must therefore conclude that the variations of climate derived from any model must be imprecise to some extent and hence be describable only in some probabilistic form. A schematic flowchart showing the components of a theoretical model of climatic variability, including the stochastic components, is shown in Fig. 16.

We now add to the completeness of the deterministic feedback system discussed in the last section by including the effects of a white-noise stochastic forcing process. The most appropriate manner of writing the resulting "stochastic-dynamical" system is in the so-called Ito form (e.g., Schuss, 1980),

$$\delta\eta = \Phi(\eta, \theta)\,\delta t + \epsilon_\eta^{1/2}\,\delta W \tag{5.1}$$

$$\delta\theta = \Psi(\eta, \theta)\,\delta t + \epsilon_\theta^{1/2}\,\delta W \tag{5.2}$$

where $\delta\eta$ and $\delta\theta$ are increments of η and θ over a finite time step δt; Φ and Ψ are the deterministic functions for which solutions were obtained in the previous section; $\epsilon^{1/2}/\delta t$ is the standard deviation (or typical amplitude) of the random fluctuations; δW is the incremental white-noise Wiener function that can be identified with the output of a random number generator of variance equal to unity.

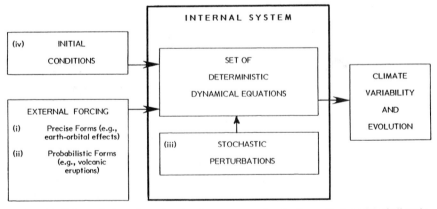

FIG. 16. Schematic flowchart showing the components of a theoretical model of climatic variability.

5.1. The Stochastic Amplitude

The first, and perhaps most critical, requirement for an analysis of the stochastic properties of the model is to estimate a reasonable reference amplitude of the noise as measured by $\epsilon^{1/2}$.

As noted above and in Section 2, there are several contributing factors to this noise amplitude. The simplest approximation is to assume that the past record of aperiodic variability about the running average climatic variations gives an acceptable measure of the combined effects. For $\delta = 10$ yr the interannual variability of the annual mean ice edge should give a good first approximation of $\epsilon_{\eta}^{1/2}$.

In Saltzman and Moritz (1980) and Saltzman (1982) we neglected any direct stochastic forcing of θ (i.e., $\epsilon_{\theta}^{1/2} = 0$), and determined the interannual noise level of η to be of the order of $\epsilon_{\eta}^{1/2} = 10^{-3}$. We shall take values equal or larger than this value to illustrate the possible effects.

5.2. Structural Stochastic Stability

A fundamental question is the degree to which the deterministic properties of a model can survive the presence of a reasonable level of stochastic noise, and to what extent the stochastic noise introduces new features when interacting with the deterministic solutions that would never appear in its absence (e.g., "tunneling" or quasi-periodic "exit" from one stable equilibrium to another). As a general rule, stochastic noise is a source of instability in a system tending to prevent it from ever settling down to a fixed stable point or limit cycle.

In the bottom halves of Figs. 7, 12, 14, and 15 we show the influence of various levels of white noise on the deterministic sample trajectory solutions described in the upper halves. These illustrate a few of the possibilities, the "residence densities" in particular giving evidence of the probabilistic nature of the response to be expected in any model puporting to realistically represent climatic solutions. For our present two-parameter stochastic system the "residence density" portrait is a mapping of the number of 1-yr time steps per 10^4 yr that the solution resides in an elementary rectangle formed by a 60×60 grid of the $\eta{-}\theta$ phase plane shown. More generally, when the dynamical system contains more than two variables, the residence density portrait should be in the form of a multidimensional "cloud" in phase space.

Figure 7 (bottom) illustrates how the presence of stochastic noise can lead to sustained quasi-periodic behavior near a stable spiral equilibrium. Note also the asymmetry of the residence density with respect to the equilibrium point at $(\eta', \theta') = (0, 0)$. This asymmetry results from the influence of the second, unstable, equilibrium shown in Fig. 6, which tends to drive the stochastic trajectories in the manner shown. We can see that *in systems with multiple equilibria under the influence of stochastic perturbations, the solution not only does not permanently coincide with the stable equilibrium point but may not even have a maximum probability of being at that equilibrium point.* In the case where the system contains only a single stable spiral equilibrium shown in Fig. 14 (top), the addition of stochastic perturbations again leads to sustained periodic variations but with a symmetric residence density portrait centered on the equilibrium point (Fig. 14, bottom).

In Fig. 12 (bottom) we show the effects of stochastic perturbations on the limit cycle deterministic flow, obtained by increasing the CO_2 feedback (see Section 4.3) and thereby passing from a stable to an unstable spiral equilibrium (i.e., a Hopf bifurcation). It is of interest to note the similarities of the residence density portraits for the deterministic flows on either side of this bifurcation subjected to similar stochastic perturbations. The possibility for this type of response was pointed out by Sutera (1980) in a discussion of the Lorenz strange attractor. Thus, similar residence density signatures obtained from observations may not uniquely reveal the nature of the nearby equilibria and the underlying deterministic flow.

The addition of stochastic perturbations to a relaxation oscillator of the general form shown in Fig. 12 (top), but containing no other equilibria outside the limit cycle, leads to the interesting bimodal residence density portrait shown in Fig. 15 (top). This represents a good example of an *almost-intransitive* system (cf. Lorenz, 1968) in which, because of the double maximum in the limit cycle trajectory speed, the system tends to reside for relatively long time periods in two distinct regions of phase space.

Finally, in Fig. 15 (bottom) we show the effects on this van der Pol-type system of increasing the amplitude of the stochastic perturbations above what we believe is a "reasonable" level. This exercise illustrates how a high level of stochastic forcing in one part of a dynamical system (η in this case) can damp the response in another part of the system (θ). Thus, the possibility exists that in spite of a basic limit cycle determinism with significant θ variability, no signature of such variability may be observed due to the presence of stochastic noise.

6. Concluding Remarks

The above examples illustrate how the parameterized and random effects of higher frequency phenomena in the climatic system can be incorporated in statistical-dynamical models governing long-period variability. This represents one extreme in the mode of representation of these effects; the other is the use of general circulation type models to generate the relevant higher frequency behavior and, by repeated equilibrations at selected time intervals, to couple this behavior to the lower frequency evolution. In principle, the latter approach seems more desirable and, in fact, would be necessary to determine the geographic spatial details of climatic change. It must be recognized, however, that it is unlikely that any amount of physical rigor regarding the high-frequency behavior of the system will be adequate to generate with sufficient accuracy the extremely low average rates of flux of mass and energy involved in long-term climatic change [10^{-1} W m^{-2}, cf. relation (3.5)]. Moreover, the past spatial variations of climate become increasingly difficult to resolve observationally beyond a relatively short time prior the present (cf. Lamb, 1977).

For these reasons it is probably a matter of necessity to treat somewhat hypothetical and speculative physical scenarios expressed as simple dynamical systems of the type we have illustrated. Many other applications of this approach have been made, most of them quite recently (e.g., Eriksson, 1968; Fraedrich, 1979; Källén et al., 1979; Sergin and Sergin, 1976; Sergin, 1979; Sutera, 1981). In their pioneering work, Sergin and Sergin noted the necessity for treating *two* variables of similarly long time constant if one is to admit the possibility that long-term oscillatory behavior of the type revealed by the paleoclimatic record is of a "free" type. As they noted, and as our equilibrium analysis makes clear, the deep ocean temperature (θ) and ice-sheet and ice-shelf mass are the most likely candidates. Thus, in our previous prototype models we can identify η with combined sea-ice shelf mass and sea-ice pack extent, the relevant "inertia" factor being proportional to the ratio (I/Λ). In the newer model proposed in Appendix C, an explicit separation is made between ice sheet/ice shelf determined by the prognostic quantity ζ, and sea-ice pack of shorter time constant determined by an essentially diagnostic quantity η. The latter variable should, however, be the main contributor to stochastic forcing.

At this time the main observational signal for long-term climatic change is the fluctuation of the *single* variable, global ice volume, determined from the $^{18}O/^{16}O$ analysis of the deep sea cores. As shown in Fig. 2A, this curve exhibits marked fluctuations in a period band centered near 100,000 yr. A fundamental question, related to our above discussion, is posed: Since direct forcing on the scale of 100,000 yr seems at first glance to be negligibly small, it would appear to be necessary that some other factor or factors *internal to the system* be significantly different for the same volume of ice at the time when the ice volume is *growing* compared with the time when it is *declining*. What are these factors? Three possibilities seem plausible. The first is that, whereas the volume of ice may be the same in the growth and decay stage, the *spatial distribution* may be much different. A second possibility is the one referred to above, namely that the bulk ocean temperature is significantly different. A third possibility, perhaps related to the second, is that the CO_2 content of the atmosphere is different in periods of growth and decay of the ice sheets. These latter two possibilities are included in the models discussed in Section 4.

It is clear that to validate models of the kind described in this review, it will be necessary to obtain synchronous long-term records of at least one significant variable in addition to ice volume, the most important of which is probably the bulk ocean temperature θ. Some limited and conflicting inferences regarding the evolution of θ have been obtained based on oxygen isotope records of benthonic foraminifera (e.g., Shackleton and Opdyke, 1973; Shackleton and Kennett, 1975; Duplessy *et al.*, 1975; Kellogg *et al.*, 1978) and based on benthic foraminiferal abundances and assemblages (e.g., Streeter, 1973; Schnitker, 1974; Lohmann, 1978). Another highly promising possibility for estimating the evolution of θ has been suggested by the author's colleague D. C. Rhoads: to examine deep-sea stratigraphic cores for evidence of the O_2 content of the oceans—high (aerobic) values signifying generally cool conditions and low (anoxic) values signifying warm conditions. The sedimentary signatures of these states are discussed by Rhoads and Morse (1971). For example, major anoxic periods can be clearly identified by black shale laminae. To obtain the resolution required to identify variations over the past million years, however, it would be necessary to consider cores in regions of relatively rapid sediment deposition.

Recently there have been some exciting reports concerning the possible variations of atmospheric CO_2 in conjunction with the ice ages, as determined by chemical analyses of glacial ice cores (Berner *et al.*, 1980; Delmas *et al.*, 1980). The results seem to indicate that substantially reduced amounts (of the order of 50%) of CO_2 were present in the atmosphere at the time of the rapid growth in ice volume. These findings are not inconsistent with the dynamical climatic system discussed in Section 4.

As a final point, we note that the prototype dynamical system we have dealt with in this review is unrepresentative of the broader possibilities in at least one respect: the system is autonomous, containing no long-term, time-dependent variations in external forcing. For example, much recent discussion has centered on the role of the variations of the earth's orbital parameters in forcing climatic change (e.g., Hays *et al.*, 1976; Suarez and Held, 1976). More complete climatic systems should include representations of these periodic components that can enrich the response possibilities in most interesting ways, especially when coupled with stochastic forcing (see Benzi *et al.*, 1982; Nicolis, 1982).

Appendix A. A Resolution of Climatic Variability

The variance spectrum shown in Fig. 3 suggests that the total variability of the climatic system can be resolved crudely into several physically distinct frequency bands separated by "gaps" of relatively low variance. In accordance with (2.8) and the general discussion of Section 2, the periods of these gaps represent the optimal time intervals over which to define "average states" governed by dynamical prognostic equations. In particular, there is some justification for partioning the system into bands with limits at about 1 hr (defining the "synoptic" average), 3 months (defining the "seasonal" average), 10 yr (defining the "climatological" average), 1 m.y. (defining the "tectonic" average, over which period the continent–ocean distribution can be taken as essentially constant), and, finally, 4.5 b.y. (the "age of the earth").

In Table I we list these gap or averaging periods (δ), the dominant types of variability included between these gaps, and an indexing scheme for labeling these averaging periods [$\delta = (n) = (1), (2), (3), (4), (5)$, respectively] and the time periods of the fluctuations contained between these gaps. It is possible, perhaps, to make a finer resolution than this, particularly in the band between 10 yr and 1 m.y. However, the data is necessarily poorly sampled over this period, and the indications from the estimates that have been made are that, except for a fairly distinct 100,000-yr oscillation of ice volume over the last 1 m.y., the variability is quite broadly distributed.

Given the resolution shown in Table I, we can expand the instantaneous value of any climatic variable χ in the following form

$$\chi = \bar{\chi}^{(5)} + \chi'_{(4-5)} + \chi'_{(3-4)} + \chi'_{(2-3)} + \chi'_{(1-2)} + \chi'_{(0-1)} \qquad (A.1)$$

where $\bar{\chi}^{(n)}$ is the average of χ over the period indexed by (n) in Table I (e.g., $\bar{\chi}^{(5)}$ is the mean value of χ over the age of the earth), $\chi'_{[n-(n+1)]}$ is the departure of $\bar{\chi}^{(n)}$ from $\bar{\chi}^{(n+1)}$, i.e.,

$$\chi'_{(4-5)} = (\bar{\chi}^{(4)} - \bar{\chi}^{(5)})$$
$$\chi'_{(3-4)} = (\bar{\chi}^{(3)} - \bar{\chi}^{(4)})$$
$$\chi'_{(2-3)} = (\bar{\chi}^{(2)} - \bar{\chi}^{(3)})$$
$$\chi'_{(1-2)} = (\bar{\chi}^{(1)} - \bar{\chi}^{(2)})$$
$$\chi'_{(0-1)} = (\chi - \bar{\chi}^{(1)})$$

In general,

$$\bar{\chi}^{(n)} = \bar{\chi}^{(n+1)} + \chi'_{[n-(n+1)]} \tag{A.2}$$

and

$$\chi'_{[(n-1)-n]} + \chi'_{[n-(n+1)]} = \chi'_{[(n-1)-(n+1)]} \tag{A.3}$$

so that we can always express Eq. (A.1) in the form

$$\chi = \bar{\chi}^{(n)} + \chi'_{[0-n]} \tag{A.4}$$

Assuming $\bar{\chi}^{(n)}$ satisfies Eq. (2.8), the following relationships are valid to a first approximation (χ and y are any climatic variables; n and N are any index numbers):

$$\overline{\chi'_{[(N-1)-N]}}^{(n \geq N)} \approx 0 \tag{A.5}$$

$$\overline{\bar{\chi}^{(N)}}^{(n \geq N)} \approx \bar{\chi}^{(n)} \tag{A.6}$$

$$\overline{\bar{\chi}^{(n)}\bar{y}^{(n)}}^{(n)} \approx \bar{\chi}^{(n)}\bar{y}^{(n)} \tag{A.7}$$

$$\overline{\bar{\chi}^{(N)}y'_{[(N-1)-N]}}^{(n \geq N)} \approx 0 \tag{A.8}$$

$$\overline{\chi'_{[(N-1)-N]}y'_{[N-(N+1)]}}^{[n \geq (N+1)]} \approx 0 \tag{A.9}$$

$$\overline{\bar{\chi}^{(n)}\bar{y}^{(n)}}^{(n+1)} \approx \bar{\chi}^{(n+1)}\bar{y}^{(n+1)} + \overline{\chi'_{[n-(n+1)]}y'_{[n-(n+1)]}}^{(n+1)} \tag{A.10}$$

Equations (A.8) and (A.9) are statements of the *orthogonality* of fluctuations in one frequency band with those in another band.

With these relations any prognostic equation for χ of the following prototypical form, containing quadratic, linear, and constant terms,

$$\partial\chi/\partial t = a\chi y + b\chi + c \tag{A.11}$$

(a, b, c are constants) can be transformed into a sequence of equations governing an average over any period δ indexed by (n):

$$\frac{\partial\bar{\chi}^{(n)}}{\partial t} = a[\bar{\chi}^{(n)}\bar{y}^{(n)} + \overline{\chi'_{(0-n)}y'_{(0-n)}}^{(n)}] + b\bar{\chi}^{(n)} + c \tag{A.12}$$

where

$$\overline{\chi'_{(0-n)}y'_{(0-n)}}^{(n)} = \overline{\chi'_{(0-1)}y'_{(0-1)}}^{(n)} + \overline{\chi'_{(1-2)}y'_{(1-2)}}^{(n)} + \cdots$$
$$+ \overline{\chi'_{[(n-1)-n]}y'_{[(n-1)-n]}}^{(n)} \tag{A.13}$$

contains the stresses due to all the subaverage periods of fluctuation, arranged according to physically distinct bands.

APPENDIX B. TIME CONSTANTS AND CONDITIONS FOR EQUILIBRATION

In order to clarify the qualitative statements made in Section 3.1 concerning the "equilibration times" for the different climatic domains, let us consider again the energy equation (2.6) in which we can assume T stands for *potential temperature* so that the equation is applicable to the atmosphere as well as the hydrosphere, cryosphere, and biolithosphere. When Eq. (2.6) is transformed to govern the evolution of an average value of T over arbitrary period δ satisfying Eq. (2.8), as pictorialized in Fig. 4, the equation takes the form of Eq. (2.16) (see Appendix A):

$$\frac{\partial \bar{T}^{(\delta)}}{\partial t} = -\nabla \cdot (\bar{\mathbf{V}}^{(\delta)} \bar{T}^{(\delta)} + \overline{\mathbf{V}'T'}^{(\delta)})$$

$$+ \frac{1}{c} [\bar{q}_{SW}^{(\delta)} + \bar{q}_{LW}^{(\delta)} + \bar{q}_{COND}^{(\delta)} + \bar{q}_{LAT}^{(\delta)}] + \mathcal{R} \qquad (B.1)$$

where $\mathcal{R}$ is stochastic forcing appropriate to the averaging interval δ. The various modes of heating included in the brackets, due to shortwave radiation, longwave radiation, conduction, and latent heat release, respectively, are rather complex functions of other variables (such as water vapor and cloudiness), but several crude parameterizations are commonly postulated. For example, the mean longwave radiation is often approximated by

$$\frac{1}{c} \bar{q}_{LW}^{(\delta)} \approx a_1 - b\bar{T}^{(\delta)} \qquad (B.2)$$

where $b \approx 4\pi a_1^2 B(Mc)^{-1}$; a_1 is the radius of the earth; $B \approx 2.2$ W m^{-2} K^{-1}; M is the mass of the climatic subsystem; c is the heat capacity of the climatic subsystem (see Hoffert *et al.*, 1980).

Similarly, the mean conductive heating can be parameterized in the form

$$\frac{1}{c} \bar{q}_{COND}^{(\delta)} = \kappa\nabla^2\bar{T}^{(\delta)} \approx -\frac{\kappa}{D^2} (\bar{T}^{(\delta)} - \Theta) \qquad (B.3)$$

where Θ is a spatial mean value of $\bar{T}^{(\delta)}$, κ is the thermal diffusivity, and D is the characteristic vertical spatial scale of $\bar{T}^{(\delta)}$.

For this discussion let us neglect heating due to phase transformations, q_{LAT}, which generally serves to transfer heat from the hydrosphere to the atmosphere. The release of latent heat within the atmosphere is quite difficult to parameterize. There is some tendency for this process to warm those portions that are cooler climatologically (i.e., high latitudes), but this is not true universally, as is evident from the existence of the inter-tropical convergence rainband near the equator and "warm-sector" precipitation in cyclone waves.

To complete our parameterizations, we adopt the diffusion approximation for $\overline{\mathbf{V}'T'}^{(\delta)}$, i.e.,

$$\overline{\mathbf{V}'T'}^{(\delta)} = -K_\delta \, \nabla \bar{T}^{(\delta)} \tag{B.4}$$

so that

$$\nabla \cdot \overline{\mathbf{V}'T'}^{(\delta)} = -K_\delta \, \nabla^2 \bar{T}^{(\delta)}$$

$$\approx \left(\frac{K_v(\delta)}{D_\delta^2} + \frac{K_h(\delta)}{L_\delta^2} \right) (\bar{T}^{(\delta)} - \Theta) \tag{B.5}$$

where L_δ is the characteristic horizontal spatial scale of $\bar{T}^{(\delta)}$ over which the *eddy* transports (i.e., mixing) occur, K_v is the vertical eddy diffusivity, and K_h is the horizontal eddy diffusivity. As we have indicated, both the eddy diffusivity K and the horizontal length L are generally highly dependent on the averaging interval δ which determines the nature of the departures defining the "eddies" (see Appendix A).

Thus, Eq. (B.1) takes the form

$$\frac{\partial \bar{T}^{(\delta)}}{\partial t} \approx -\nabla \cdot \bar{\mathbf{V}}^{(\delta)} \bar{T}^{(\delta)} + \frac{\bar{q}_{SW}^{(\delta)}}{c} + a$$

$$+ \left[\left(\frac{\kappa + K_{v\delta}}{D^2} \right) + \left(\frac{K_h}{L^2} \right)_\delta \right] \Theta - \frac{1}{\epsilon_T} \bar{T}^{(\delta)} + \mathfrak{R} \tag{B.6}$$

where

$$\epsilon_T = \left[b + \left(\frac{\kappa}{D^2} \right) + \left(\frac{K_v}{D^2} \right)_\delta + \left(\frac{K_h}{L^2} \right)_\delta \right]^{-1} \tag{B.7}$$

As indicated in Eq. (B.7), ϵ_T is a function of b, (κ/D^2), $(K_v/D^2)_\delta$, and $(K_h/L^2)_\delta$, the reciprocals of which are the "long-wave radiative time scale," the "molecular diffusive time scale," the "vertical eddy diffusive time scale," and the "horizontal eddy advective time scale," respectively. Because the eddy time scales are functions of the averaging interval δ, it follows that $\epsilon_T = \epsilon_T(\delta)$.

We now let

$$\bar{\chi}^{(\delta)} = \bar{\chi}_*^{(\delta)} + \bar{\chi}^{(\delta)'} \qquad (\chi = T, \mathbf{V}, \ldots) \tag{B.8}$$

where $\bar{\chi}_*^{(\delta)}$ denotes quasi-equilibrium values (assumed here to be unique) corresponding to a very long-term average of $\bar{\chi}^{(\delta)}$, and $\bar{\chi}^{(\delta)'}$ is the departure of $\bar{\chi}^{(\delta)}$ from $\bar{\chi}_*^{(\delta)}$, so that $|\partial\bar{\chi}^{(\delta)}/\partial t| \sim |\partial\bar{\chi}^{(\delta)'}/\partial t| \gg |\partial\bar{\chi}_*^{(\delta)}/\partial t|$ (see Fig. 4).

(In the notation of Appendix A [see Eq. (A.2)] $\bar{\chi}^{(\delta)}_* = \bar{\chi}^{(n+1)}$ and $\bar{\chi}^{(\delta)'} = \chi'_{[n-(n+1)]}$, where n is the *index* for δ.) The quasi-equilibrium, long-term mean values satisfy the relation

$$\frac{\partial \bar{T}^{(\delta)}_*}{\partial t} = -\nabla \cdot [\bar{\mathbf{V}}^{(\delta)}_* \bar{T}^{(\delta)}_* + (\bar{\mathbf{V}}^{(\delta)'} \bar{T}^{(\delta)'})_*]$$

$$+ \frac{\bar{q}^{(\delta)}_{SW*}}{c} + a_* + \left[\left(\frac{\kappa}{D^2}\right) + \left(\frac{K_v}{D^2}\right)_\delta + \left(\frac{K_h}{L^2}\right)_\delta\right]\Theta$$

$$- \frac{1}{\epsilon_T} \bar{T}^{(\delta)}_* + \mathcal{R}_* \tag{B.9}$$

where $\mathcal{R}_*$ is stochastic forcing appropriate to the long-term average of $\partial \bar{T}^{(\delta)}/\partial t$. If we subtract Eq. (B.9) from Eq. (B.6), we obtain the following equation for the rate of change of $\bar{T}^{(\delta)'}$:

$$\frac{\partial \bar{T}^{(\delta)'}}{\partial t} \approx \frac{\partial \bar{T}^{(\delta)}}{\partial t} = \mathcal{F}_T + \mathcal{R}' - \frac{1}{\epsilon_T} \bar{T}^{(\delta)'} \tag{B.10}$$

where

$$\mathcal{F}_T = \left[-\nabla \cdot (\bar{\mathbf{V}}^{(\delta)'} \bar{T}^{(\delta)'} + \bar{\mathbf{V}}^{(\delta)}_* \bar{T}^{(\delta)'} + \bar{\mathbf{V}}^{(\delta)'} \bar{T}^{(\delta)}_*) + \frac{\bar{q}^{(\delta)'}_{SW}}{c} + a'\right]$$

is the deterministic forcing tending to drive $\bar{T}^{(\delta)}$ from its long-term mean value and $\mathcal{R}' = \mathcal{R} - \mathcal{R}_*$ is a stochastic forcing component.

If the dissipative processes are acting alone, in the absence of all forcing $\mathcal{F}_T$ and $\mathcal{R}'$, we have

$$\frac{\partial \bar{T}^{(\delta)'}}{\partial t} \approx -\frac{1}{\epsilon_T} \bar{T}^{(\delta)'} \tag{B.11}$$

the solution of which is

$$\bar{T}^{(\delta)'} = (\bar{T}^{(\delta)})'_{t=0} \exp(-t/\epsilon_T) \tag{B.12}$$

Thus, ϵ_T is the time it would take for an initial departure of $\bar{T}^{(\delta)}$ from its longer-term mean value to be reduced by e^{-1} (or 0.37), representing the order of magnitude of the thermal "equilibration time" or "time scale."

It follows from the condition (2.8), $\delta \ll P(\bar{T}^{(\delta)})$, that

$$\frac{\partial \bar{T}^{(\delta)}}{\partial t} \sim \frac{\partial \bar{T}^{(\delta)'}}{\partial t} \sim \frac{|\bar{T}^{(\delta)'}|}{P(\bar{T}^{(\delta)})/4} < \frac{|\bar{T}^{(\delta)'}|}{\delta} \tag{B.13}$$

Thus, if $\epsilon_T < \delta$, the last term of Eq. (B.10) (i.e., $\epsilon_T^{-1} \bar{T}^{(\delta)'}$) tends to be much greater than $\partial \bar{T}^{(\delta)}/\partial t$ and must be balanced deterministically by $\mathcal{F}_T$. The condition

$$\epsilon_T < \delta \qquad\qquad\qquad (B.14)$$

is therefore likely to be sufficient to ensure that Eq. (B.9) or (B.10) will be a diagnostic equation in which $\partial \bar{T}^{(\delta)}/\partial t$ can be regarded as a very small difference between large forcing and dissipation. Stated differently, the minimum averaging period required to ensure that $\partial \bar{T}^{(\delta)}/\partial t \approx 0$ is $\delta \approx \epsilon_T$.

If, under the condition (B.14), the forcing were to change gradually rather than be removed entirely as in Eq. (B.11), it is clear that $\bar{T}^{(\delta)}$ would adjust quasi-statically to the changing forcing with no time lag between the forcing and response.

From the same considerations it follows that $\epsilon_T \gg \delta$ is a necessary condition that $\partial \bar{T}^{(\delta)}/\partial t$ be of comparable magnitude to the right-hand terms [i.e., that (B.10) be a *prognostic* equation for $\bar{T}^{(\delta)}$]. This is not a sufficient condition, however, since if $\epsilon \ll P(\bar{T}^{(\delta)})$, there must be a near balance (i.e., equilibrium) between $\mathcal{F}_T$ and $\epsilon_T^{-1}(\bar{T}^{(\delta)})'$. More generally, if $\epsilon \gg \delta$, we then have the following further implications denoted by the arrows:

$$\epsilon_T \gg P(\bar{T}^{(\delta)}) \rightarrow \frac{\partial (\bar{T}^{(\delta)})'}{\partial t} \approx \mathcal{F}_T$$

$$\epsilon_T \approx P(\bar{T}^{(\delta)}) \rightarrow \frac{\partial (\bar{T}^{(\delta)})'}{\partial t} \approx \mathcal{F}_T - \frac{1}{\epsilon_T}(\bar{T}^{(\delta)})'$$

$$\epsilon_T \ll P(\bar{T}^{(\delta)}) \rightarrow 0 \approx \mathcal{F}_T - \frac{1}{\epsilon_T}(\bar{T}^{(\delta)})'$$

Arguments similar to these can be applied to the other equations expressing conservation of momentum and mass. Thus, for example, we can replace $\bar{T}^{(\delta)}$, $\mathcal{F}_T$, and ϵ_T in (B.10) by $\bar{M}^{(\delta)}$ (the mass of a domain of the climatic system), $\mathcal{F}_M$ (the mass "forcing"), and ϵ_M (the equilibration time for $\bar{M}^{(\delta)}$), respectively. Since *temperature* and *ice mass* are perhaps the most important climatic variables defining the state of the climatic system, in Section 3.1 we pursue this discussion further by analyzing the implications of the above condition on ϵ_T and ϵ_M for various domains of the climatic system.

APPENDIX C. A GENERALIZED SDM GOVERNING LONG-TERM CHANGES OF THE COMPLETE CLIMATIC SYSTEM

C.1. Preliminary Considerations and Modeling Approximations

Let us consider an idealized earth, the surface of which can consist of four mutually exclusive states: unglaciated land, glaciated land (i.e., continental snow and ice sheet), open ocean, and sea ice. We shall denote variables representing the properties of these four components by the subscripts l, li, w, and wi, respectively, and let the subscripts L denote ($l + li$) and W denote ($w + wi$). A schematic hemispheric map of this system is given in Fig. 17, in which $\xi = \sin \varphi$ (φ is latitude); $\Lambda = \lambda/2\pi$ (λ is longitude); $\alpha(\xi)$ is the fraction of a given latitude occupied by continent; $\xi = \chi$ and γ are the equatorward and poleward limits of the continent, respectively; $\xi = \zeta$ at the zonal-mean ice sheet/snow margin; $\xi = \eta$ at the zonal-mean sea ice margin; and $\xi = (\zeta + \beta)$ at the zonal-mean peak of the ice sheet. In Fig. 18 we show two schematic meridional cross sections: (A) over the continent (e.g., at $\Lambda = A$ in Fig. 17), and (B) over the ocean (e.g., at $\Lambda = B$ in Fig. 17). Defining $z = 0$ at an arbitrary reference level below the deepest point in the ocean, we denote sea level by z_{SL}, the height of the lithosphere by z_l, the height of the surface of the earth above sea level by h, and the "top" of the atmosphere by z_T. The thickness of the ice is denoted by I.

It is to be understood that all variables introduced in the development of this SDM are climatic-mean variables, i.e., variables averaged over about a 10-yr period, so that the annual cycle and subannual periods, including synoptic and diurnal fluctuations, are to be included in any

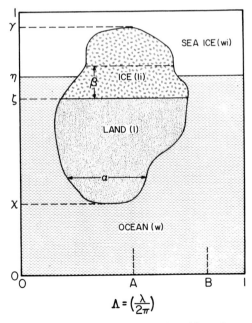

FIG. 17. Schematic hemispheric map of an idealized earth's surface consisting of a single continent (land and ice/snow) and an ocean (open water and sea ice).

parameterized fluxes. These climatic mean variables will be specified at several levels: the surface of the earth (denoted by the subscript S), the top of the atmospheric planetary boundary layer corresponding roughly to a height of 2 km (denoted by the subscript Z), and the top of the atmosphere (denoted by the subscript T).

We also define the following averaging processes for the four domains under consideration:

$$\bar{\psi} \equiv \int_0^1 \int_0^1 \psi \, d\xi \, d\Lambda \qquad \text{(hemispheric area average)}$$

$$= \left(\frac{\sigma_l}{\sigma}\right)\bar{\psi}_l + \left(\frac{\sigma_{li}}{\sigma}\right)\bar{\psi}_{li} + \left(\frac{\sigma_w}{\sigma}\right)\bar{\psi}_w + \left(\frac{\sigma_{wi}}{\sigma}\right)\bar{\psi}_{wi} \qquad \text{(C.1)}$$

where $\sigma = 2\pi a^2$ is the area of the hemisphere, σ_x is the area occupied by the component denoted by x, and $\bar{\psi}_x$ is the average of ψ_x over σ_x:

$$\hat{\psi}_x \equiv \frac{1}{M_x} \int \psi_x \, dM \qquad \text{(mass-weighted average of } \psi_x \text{ over } M_x) \quad \text{(C.2)}$$

where M_x is the hemispheric mass of the component denoted by x, and $dM = \sigma \, d\xi \, d\Lambda \, \rho_x \, dz$ (see Section 3.2).

Our main concern here will be with formulating a simple SDM governing the climatic mean variables, $\bar{\psi}^{(\delta c)}$, that can account for the growth and decay of ice in the system, $M_i = (M_{li} + M_{wi})$. This cryospheric component provides the main signature for longer-term climatic change, having a large amplitude at the longest time scales exhibited by the system as a whole (see Table I). Because of the nearly similar thermal time scale of the deep ocean, it is natural to expect that this oceanic component should also play a prognostic role in a model of long-term climatic change, thereby introducing the possibility for harmonic oscillation. By the same token, because the response time of the sea-ice "pack" is shorter than that of the ice sheets and "shelf" sea ice, we might expect that the sea-ice pack edge η tends to be carried along quasi-statically by the long-term ice-sheet fluctuations, in much the same manner that the climatic mean atmosphere and surface boundary layer quasi-statically follow the ice-sheet behavior. Broadly speaking, the high-inertia ice sheets and ice shelves tend to force an ice flow to the surrounding oceans and to create cold environments over the adjacent open oceans, favoring the formation of the sea-ice pack. As a first approximation, it may be possible to bypass the physics of this complex process by assuming simply that

$$\eta = \eta(\zeta) \qquad \text{(C.3)}$$

the most simple approximation being that the sea-ice edge coincides with the land-ice edge on the idealized hemisphere shown in Fig. 17, i.e., $\eta =$

ζ. It is probable, however, that sea ice plays a prognostic role, also, for shorter-term climatic change and possibly in the earliest stages of longer-term ice-sheet development (see Denton and Hughes, 1981).

For the special case of an all-ocean planet ($\alpha = 0$), the sea-ice extent η becomes a free variable. An equation for the variations of η under these circumstances is given by Saltzman (1978) and is reviewed in Section C.2.

Let us now specify a functional form for the variation of the climatic-mean surface temperature between the equator and this ice-sheet and/or sea-ice boundary which we take to be at a fixed zonal-mean temperature $T_S(\Xi) \equiv \tau_\Xi$, where $\Xi \equiv \zeta$ or η. As depicted in Fig. 17, we shall assume that $\alpha = 0$ at the equator (which is approximately true for the present-day earth), and that the equatorial oceanic surface temperature is dependent on the mean ocean temperature ($\hat{T}_w \equiv \theta$) because of the shallowness of the mixed layer near the equator and the sensitivity of this equatorial temperature to upwelling from below (e.g., Knauss, 1978; Wyrtki, 1981). We adopt the form specified in Saltzman (1982):

$$T_s(0, \vartheta) \equiv \tau_0(\vartheta) = 300 + k\vartheta^3 \ (K) \tag{C.4}$$

where k is an assigned constant ($0.1 \ \mathrm{K}^{-2}$) and ϑ is the departure of θ from the present value of 277 K (see Dietrich, 1963), i.e., $\vartheta = \theta - 277$ K, so that τ_0 equals the presently observed value of 300 K for the presently observed value of θ. Assuming a quadratic dependence of T_s on ξ, we have [see Saltzman, 1982, Eq. (13)]

$$T_s(\xi; \Xi, \vartheta) = \tau_0(\vartheta) - a_s(\vartheta) \ \xi^2/\Xi^2 \tag{C.5}$$

where $a_s(\vartheta) = \tau_0(\vartheta) - \tau_\Xi$ and $\Xi = \delta$ or η. Poleward of the ice edge we can either assume that Eq. (C.5) continues to apply or compute $T_s(\xi)$ from the surface heat balance condition (3.10). For both cases we can require that over an elevated ice sheet the temperature at the surface be reduced in accordance with a fixed lapse rate Γ_i. For the present zonal-mean ice edge $\Xi \approx 0.9$, and taking $\tau_\Xi = 263$ K (Budyko, 1969), the representation (C.5) implies a polar temperature of 259 K, which is close to the observed value.

The atmospheric temperature at the top of an extended planetary boundary layer at $z = Z = 2$ km, $T_Z(\xi)$, can also be represented by the form described by Saltzman [1982, Eq. (14)]:

$$T_Z(\xi; \Xi, \theta) = \tau_{Z0}(\vartheta) - a_Z(\vartheta)\xi^2/\Xi^2 = T_s(\xi; \Xi, \vartheta) - \Gamma Z \tag{C.6}$$

where $\tau_{Z0}(\vartheta) = \tau_0(\vartheta) - \Gamma Z$, $a_Z(\vartheta) = a_s(\vartheta) = \tau_0(\vartheta) - \tau_\Xi$, and $\Gamma = 5.75$ $\mathrm{K \ m}^{-1}$.

Another representation that will be needed is one for fractional cloud coverage, n. For example, we can adopt the form proposed in Saltzman

(1982), which is assumed to be valid only for $0 < \xi < \Xi$:

$$n(\xi; \Xi) = N_0 + N_1\Xi + N_2\Xi^2 + N_3(\xi/\Xi)^3 \qquad (C.7)$$

where $N_0 = 0.1$, $N_1 = 0.8$, $N_2 = -0.4$, and $N_3 = 0.4$. For $\Xi < \xi < 1$, we can assume that n decreases linearly to the pole, i.e.,

$$n(\xi; \Xi) = n(\Xi) - N_4(\xi - \Xi) \qquad (\Xi < \xi < 1) \qquad (C.8)$$

C.2. Ice-Mass Equation

We now set down the fundamental equations for the model, starting with the equation for dM_i/dt which we can write in the following independent forms:

$$\frac{dM_i}{dt} = -\frac{\sigma}{L_f}(\tilde{N}^{\downarrow} + \tilde{H}_S^{\uparrow}) \qquad (C.9a)$$

$$= \sigma(\tilde{P}_i + \tilde{F}) \qquad (C.9b)$$

where $N^{\downarrow}$ and $H_S^{\uparrow}$ are defined in Section 3.2, P_i is the rate of snowfall at the surface, and F is the rate of freezing of surface waters (a negative value of which signifies the rate of melting). The first form (C.9a) is the approximate statement of conservation of energy for the atmosphere that was given previously as Eq. (3.11), and the second form (C.9b) is the continuity equation expressing conservation of ice mass.

For models in which the growth and decay of massive ice sheets are considered (which are likely to involve relatively large departures from radiative equilibrium), it may be a worthwhile approach to use Eq. (C.9a) rather than Eq. (C.9b), thereby trading the difficulty of finding relations for $\tilde{P}_i$ and $\tilde{F}$ for what are probably comparable difficulties of finding relations for $\tilde{N}^{\downarrow}$ and $\tilde{H}_S^{\uparrow}$. Most previous studies of ice-sheet formation, however, have been based on Eq. (C.9b) (e.g., Weertman, 1964; Birchfield, 1977; Källén et al., 1979; Sergin, 1979; Oerlemans, 1980; Pollard et al., 1980).

For climatic-mean conditions on continental *land* or *ice* surface layers it is clear from Table II that $\epsilon_T \leq \delta_c = 10$ yr $\approx 10^8$ sec, so that

$$(Mc)_L \frac{\partial \hat{T}_L}{\partial t} \approx \sigma_L(\tilde{H}_{SL}^{\downarrow} + \tilde{G}_L^{\uparrow}) \approx 0$$

i.e., assuming the geothermal flux is small, the net surface flux of heat into or out of the continental portions of the earth (land or ice sheet) is negligible. Since from Eq. (C.1)

$$\tilde{H}_S^\downarrow = \left(\frac{\sigma_L}{\sigma}\right)\tilde{H}_{SL}^\downarrow + \left(\frac{\sigma_W}{\sigma}\right)\tilde{H}_{SW}^\downarrow$$

we have

$$\tilde{H}_S^\downarrow \approx \left(\frac{\sigma_W}{\sigma}\right)\tilde{H}_{SW}^\downarrow = \left(\frac{\sigma_W}{\sigma}\right)\tilde{H}_{Sw}^\downarrow + \left(\frac{\sigma_{wi}}{\sigma}\right)\tilde{H}_{Swi}^\downarrow$$

i.e., the area average flux of heat into or out of the surface of the earth appearing in Eq. (C.9a) is due almost entirely to the flux across the *ocean* part of the surface. Thus, Eq. (C.9a) can be written in the form

$$\frac{dM_i}{dt} = -\frac{1}{L_f}(\sigma\tilde{N}^\downarrow + \sigma_w\tilde{H}_{Sw}^\downarrow + \sigma_{wi}\tilde{H}_{Swi}^\downarrow) \tag{C.9a'}$$

C.2.1. Ice-Sheet Mass and Extent. The total mass of ice contained in a continental ice sheet extending from $\xi = \zeta$ to $\xi = \gamma$, assuming a uniform depth profile with longitude, $I(\xi)$, is given by

$$M_i = 2\pi a^2 \rho_i \int_\zeta^\gamma \alpha(\xi)I(\xi)\,d\xi \tag{C.10}$$

where $\alpha(\xi)$ is the fraction of the latitude circle occupied by continent.

For simplicity, let us assume further that the continent occupies a fixed longitude sector, and let us adopt an idealized ice-sheet profile similar to that proposed by Weertman (1964) and schematically illustrated in Fig. 18, i.e.,

$$\alpha(\xi) = \alpha \qquad \text{(a constant)} \tag{C.11}$$

$$I(\xi) = \left(\frac{2\tau_0 a}{\rho_i g}\right)^{1/2}(1 + \epsilon)(\xi - \zeta)^{1/2} \tag{C.12}$$

where τ_0 is the shear stress at the bed of the ice sheet and $\epsilon/(1 + \epsilon)$ is the fraction of the total ice depth that is submerged below ground level due to the ice-load effect.

If we introduce Eqs. (C.11) and (C.12) into Eq. (C.10), we obtain

$$M_i = \tfrac{4}{3}\pi a^2 \alpha \rho_i \left(\frac{\tau_0 a}{\rho_i g}\right)^{1/2}(1 + \epsilon)(\gamma - \zeta)^{3/2} \tag{C.13}$$

from which we have

$$\frac{dM_i}{dt} = -2\pi a^2 \alpha \left(\frac{\rho_i \tau_0 a}{g}\right)^{1/2}(1 + \epsilon)(\gamma - \zeta)^{1/2}\frac{d\zeta}{dt} \tag{C.14}$$

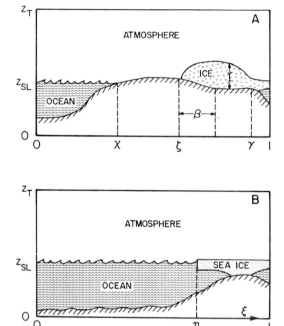

FIG. 18. Schematic vertical cross sections at $\Lambda = A$ and B of the climatic system portrayed in Fig. 17.

If we now introduce Eq. (C.14) into Eq. (C.9a'), we obtain the following equation governing the ice-sheet extent over a continent confined between two fixed meridians (α = constant):

$$\frac{d\zeta}{dt} = \left[2\pi a^2 \alpha \left(\frac{\rho_i \tau_0 a}{g} \right)^{1/2} (1 + \epsilon) L_f \right]^{-1} (\gamma - \zeta)^{-1/2}$$

$$\times [\sigma \tilde{N}^\downarrow + \sigma_w \tilde{H}^\uparrow_{Sw} + \sigma_{wi} \tilde{H}^\uparrow_{Swi}] \qquad (C.15)$$

As discussed in Section C.1, the sea-ice extent η is to be prescribed diagnostically as a function of ζ, i.e.,

$$\eta = \eta(\zeta) \qquad (C.16)$$

C.2.2. Sea-Ice Mass and Extent as a Free Variable. For an idealized all-ocean model ($\alpha = 0$) in which only sea ice in both shelf and pack form is considered, a formula for the "free" variations of η has been derived previously (Saltzman, 1978). This formula is based on the assumption that $M_i = \sigma \rho_i I_w (1 - \eta)$, $P_i = 0$, and $\bar{F} = \Lambda F(\eta)/2$, where ρ_i and I_w are prescribed constant values of the sea-ice density and thickness, respectively, and Λ is a parameter representing the distance (in units of ξ) over which freezing or melting is reduced linearly to zero from an assumed maximum value at the ice edge, $F(\eta)$. Thus, from Eqs. (C.9a) and (C.9b), and using the "surface heat balance condition" (3.10) to evaluate $F(\eta)$, we can write the formula for free variations only of sea ice in the alternate forms

$$\frac{d\eta}{dt} = \frac{1}{\rho_i I_w L_f} (\tilde{N}^{\downarrow} + \bar{H}_S^{\uparrow}) \tag{C.17a}$$

$$= \frac{1}{2\rho_i L_f (I_w/\Lambda)} (H_S^{(1)\downarrow} - H_S^{(2)\uparrow} - H_S^{(3)\uparrow} - H_S^{(4E)\uparrow} + H_S^{(5)\uparrow})_{\xi=\eta} \tag{C.17b}$$

where (I_w/Λ) plays the role of an "inertia" parameter for sea-ice mass change at its edge. From a comparison of Eqs. (C.17a) and (C.17b), we find that for this model,

$$\tilde{N}^{\downarrow} = \bar{H}_S^{\downarrow} + \frac{\Lambda}{2} (H_S^{(1)\downarrow} - H_S^{(2)\uparrow} - H_S^{(3)\uparrow} - H_S^{(4E)\uparrow} + H_S^{(5)\uparrow})_{\xi=\eta} \tag{C.18}$$

A cross-sectional view of this all-ocean model is given in Fig. 18B.

C.3. Mean Ocean Temperature Equation

Equation (3.12) is the fundamental relationship governing the variations of θ ($\equiv \hat{T}_w$). If we neglect the geothermal flux into the ocean from below ($\bar{G}_w^{\uparrow} = 0$), this equation takes the form

$$\frac{d\theta}{dt} = \frac{1}{C_w M_w} (\sigma_w \bar{H}_{Sw}^{\downarrow} - \sigma_{wi} \bar{H}_{Swi}^{\uparrow}) \tag{C.19}$$

No new variables are introduced in this equation, but we remain with the need to find closure expressions for $\bar{H}_S^{\uparrow}$ and $\tilde{N}^{\downarrow}$.

C.4. Vertical Heat Flux at Surface

A complete set of parameterization formulas for the components of $H_{Sw}^{\uparrow}(\xi; \eta, \theta)$ [i.e., $H_{Sw}^{(1)\uparrow}$, $H_{Sw}^{(2)\uparrow}$, $H_{Sw}^{(3)\uparrow}$, and $H_S^{(4E)\uparrow}$] and for $H_S^{(5)\uparrow}(\eta; \theta)$, has been presented in Saltzman and Moritz (1980). These formulas represent the results of an ongoing program of development of such parameterizations over a number of years (see Saltzman, 1967, 1968, 1973, 1980; Saltzman and Ashe, 1976a,b; Saltzman and Vernekar, 1968, 1971a, 1972, 1975, 1982). In the formulations of Saltzman and Moritz (1980) the possibility is allowed that the longwave radiative flux, $H_S^{(2)\uparrow}$, is affected explicitly by CO_2 changes in the atmosphere. It was assumed that such CO_2 changes can be associated with mean temperature changes of the entire ocean, $d\theta/$

dt, a conjecture that is superficially plausible because of the temperature-dependent dissolution properties of CO_2 in water, but which is debatable on many other grounds owing to the extreme complexity of the carbon cycle (see Bolin, 1981). In any event, it is at least possible that the bulk ocean temperature θ will affect $H_S^{\uparrow}$, and hence M_i, through this CO_2 effect as well as through its possible direct influence on the equatorial surface temperature [see (C.4) and (C.5)].

We now summarize the formulas and approximations for the climatic mean values of $H_{Sw}^{(1)\downarrow}$, $H_{Sw}^{(2)\uparrow}$, $H_{Sw}^{(3)\uparrow}$, $H_{Sw}^{(4E)\uparrow}$, $H_{Sw}^{(5)\uparrow}(\eta)$, and $H_{Swi}^{\uparrow}$ that were developed in Saltzman and Moritz (1980).

C.4.1. The Shortwave (Solar) Radiative Flux.

$$H_{Sw}^{(1)\downarrow}(\xi; \eta, \theta) = [1 - r - \chi(\xi; \eta, \theta) - \alpha_4 n(\xi; \eta)]A(\xi) \qquad (C.20)$$

where r is the reflectivity due to backscatter of the cloudiness atmosphere; $\chi(\xi; \eta, \theta) = [\alpha_1 + \chi_d(0) + \alpha_2 T_S(\xi; \eta, \theta)]\mathcal{P}(\xi)$ is the shortwave absorptivity with no clouds; $\chi_d(0)$ is the absorptivity due to all gases other than water vapor, measured at the equator; $\mathcal{P}(\xi) = 1 + \alpha_7\xi + \alpha_8\xi^2 + \alpha_9\xi^3$ is the path length; $A(\xi) = [1 - r_{Sw}(\xi)]R(\xi) = A(0)(1 + \alpha_5\xi + \alpha_6\xi^2)$ is the surface absorption of *undepleted* solar radiation; $r_{Sw}(\xi)$ is the ocean surface reflectivity; $R(\xi)$ is the annual mean solar radiation incident at the top of the atmosphere; $\alpha_4 = \chi_n + r_T + r_n - r$; χ_n is the cloud drop absorptivity; r_T is the reflectivity of the atmosphere above clouds; r_n is the reflectivity of cloud tops; α_1, α_2, α_5, α_6, α_7, α_8 are constants (see Saltzman and Moritz, 1980).

C.4.2. The Longwave (Terrestrial) Radiative Flux.

$$H_{Sw}^{(2)\uparrow}(\xi; \eta, \theta) = \sigma T_S^4(\xi; \eta, \theta)[1 - \beta_1 n(\xi; \eta)][1 - \epsilon(\xi; \eta, \theta)] \quad (C.21)$$

where σ is the Stefan–Boltzmann constant; $\epsilon = \epsilon_v + \epsilon_c - \epsilon_* + \epsilon_z$ is the emissivity of atmosphere; $\epsilon_v = \epsilon_v^* + \beta_2[T_S(\xi; \eta, \theta) - T_S^*]$ is the emissivity of water vapor; T_S^* is the standard temperature (286 K); $\epsilon_c = b_8 \ln[CO_2] + b_9 = \epsilon_c^* + b_{10}\vartheta$ is the emissivity of carbon dioxide; $[CO_2]$ is the carbon dioxide concentration (ppm); ϵ_c^* is the value of ϵ_c for the present value, $[CO_2]^* = 330$ ppm, which is taken as a standard; ϵ_* is the emissivity correction due to CO_2 and H_2O overlap; ϵ_z is the emissivity of ozone; β_1, β_2, b_8, b_9, b_{10} are constants (see Saltzman and Moritz, 1980).

C.4.3. The Sensible Heat Flux.

$$H_{Sw}^{(3)\uparrow}(\xi; \eta, \theta) = \gamma_1 + \gamma_2\Gamma Z + \gamma_5 \frac{(1 - \xi^2)}{a^2}\left(\frac{\partial T_Z}{\partial\xi}\right)^2 \qquad (C.22)$$

where γ_1, γ_2, and γ_5 are constants (see Saltzman and Moritz, 1980).

C.4.4. The Latent Heat Flux.

$$H_{Sw}^{(4E)\uparrow}(\xi; \eta, \theta) = \lambda_0 De_{SAT}[T_S(\xi; \eta, \theta)]$$

$$+ \lambda_1 \frac{\partial e_{SAT}[T_S(\xi; \eta, \theta)]}{\partial T} H_{Sw}^{(3)\uparrow}(\xi; \eta, \theta) \qquad (C.23)$$

where $\lambda_0 = 0.622 p_S^{-1} L_v \rho (1 - h_v)$; h_v is the relative humidity; $D = D_0 + D_2 \xi^2/\eta^2$ is the turbulent transfer coefficient; $\lambda_1 = 0.622 L_v (c_p p_S)^{-1}$; $e_{SAT}(T_S) = \lambda_2 + \lambda_3 T_S + \lambda_4 T_S^2$ is the saturation vapor pressure; $\partial e_{SAT}(T_S)/\partial T = \lambda_5 + \lambda_6 T_S + \lambda_7 T_S^2$; $D_0, D_2, \lambda_2, \ldots, \lambda_7$ are constants (see Saltzman and Moritz, 1980).

Together with the formulas for $T_S(\xi; \eta, \theta)$ and $T_Z(\xi; \eta, \theta)$ given in Saltzman (1982) [i.e., Eqs. (C.4), (C.5), (C.6)] and the formula for $n(\xi)$ [i.e., Eq. (C.8)], Eqs. (C.20)–(C.23) combine to yield our parameterization of $\tilde{H}_{Sw}^{\downarrow}$ in terms of the two variables η and θ, assuming $\tilde{H}_{Sw}^{(4F)}$ can be neglected.

In Saltzman (1978) we introduced the simplest possible assumption regarding the form of $H_S^{(5)\uparrow}(\eta; \theta)$, i.e.,

$$H_S^{(5)\uparrow}(\eta; \theta) = v(\theta - \tau_\eta) \qquad (C.24)$$

where v is a constant estimated from the present-day surface heat balance near the ice edge. At this time there appears to be no physical basis for developing a more rigorous parameterization of this flux.

The mean upward flux of heat from the ocean to the atmosphere through pack sea ice ($H_{Swi}^{\downarrow}$) depends on the atmospheric temperature and the thickness and areal density of fissures and leads in the sea ice. As suggested by Maykut and Untersteiner (1969), we can assume that in the first approximation a balance tends to be struck between these factors such that a fairly uniform, near-constant flux prevails in the long-term average. The value of 3 W m^{-2} is suggested by Maykut and Untersteiner (1969) and used by Saltzman and Moritz (1980), though somewhat higher values have more recently been suggested by Parkinson and Washington (1979).

C.5. Net Radiative Flux at Top of Atmosphere

Accurate parameterizations of the net downward flux of solar radiation and the net upward flux of terrestrial radiation at the top of the atmosphere ($H_T^{(1)\downarrow}$ and $H_T^{(2)\uparrow}$, respectively) are no less difficult to formulate than those for the surface of the earth ($H_S^{(1)\downarrow}$ and $H_T^{(2)\uparrow}$, respectively), for which errors of the order of $\pm 5\%$ are to be expected. Since we are now interested in computing a relatively small difference between these two quantities averaged over the globe,

$$\bar{N}^\downarrow = \bar{H}_T^{(1)\downarrow} - \bar{H}_T^{(2)\uparrow}$$

we are clearly stretching the limits of our capabilities. Nonetheless, we are obliged to try to make such an estimate if we are ever to fully explore the nature of long-term climatic change. We may entertain the hope that, although the absolute values have large errors, the evolutionary *changes* in the values may be at least qualitatively correct.

The simplest statements we can make are those embodied in the one-dimensional Budyko–Sellers type EBMs, in which expressions of the following form are postulated:

$$\bar{H}_T^{(1)\downarrow} = \{1 - \alpha_P[\Xi(T_S)]\}\bar{R}$$
$$\bar{H}_T^{(2)\uparrow} = a + b\bar{T}_S$$

where $\alpha_P[\Xi(T_S)]$ is a specified function; a and b are constants. These parameterizations, however, do not match the detail in which $H_S^{(1)\downarrow}$ and $H_S^{(2)\uparrow}$ are parameterized in Eqs. (C.20) and (C.21); for example, they contain no explicit functional dependence of the radiative fluxes on water vapor and CO_2 content. More appropriate formulas for $H_T^{(1)\downarrow}$ and $H_T^{(2)\uparrow}$, consistent with Eqs. (C.20) and (C.21), are of the following form (see Paltridge and Platt, 1976):

$$H_T^{(1)\downarrow}(\xi; \Xi, \theta) = [1 - A(\xi; \Xi, \theta)]R(\xi) \tag{C.25}$$

where

$$A(\xi; \Xi, \theta) = r - r_S\mu(\mu - r) + n[r - r_S\mu(\mu - r) - r_T - r_n(1 - r_T)]$$
$$\mu = \mu(\xi; \Xi, \theta) = 1 - \chi[T_S(\xi; \Xi, \theta)]$$
$$r_S = r_S(l, l_i, w, wi; \xi)$$
$$n = n(\xi; \Xi)$$
$$H_T^{(2)\uparrow}(\xi; \Xi, \theta) = \sigma T_S^4(\xi; \Xi, \theta)\{1 + g - \epsilon(\xi; \Xi, \theta)$$
$$- n(\xi; \Xi)[1 + G - \epsilon(\xi; \Xi, \theta) - \beta_{10}\epsilon']\} \tag{C.26}$$

where g and ϵ' are constants given by Paltridge and Platt (1976) representing the upward clear-sky emissivity of the atmosphere and above-cloud longwave transmissivity, respectively; β_{10} is a constant relating cloud-top temperature to surface temperature.

C.6. Complete Deterministic System

The ice-sheet equation (C.15) for α = const [or, alternatively, the sea-ice equation (C.17b) for $\alpha = 0$] and the mean ocean temperature equation (C.19), combined with the modeling approximations (C.3)–(C.8) and the heat flux parameterizations (C.20)–(C.26), comprise a closed deterministic dynamical system in two dependent variables, Ξ and θ, that can be written in the form,

$$\frac{d\Xi}{dt} = \Phi(\Xi, \theta) \tag{C.27}$$

$$\frac{d\theta}{dt} = \Psi(\Xi, \theta) \tag{C.28}$$

where Φ and Ψ are autonomous, nonlinear functions of high degree in Ξ and θ.

A useful representation of the physical processes governed by this coupled system of equations, showing the feedbacks among all the parameters that affect Ξ and θ, is the flow diagram shown in Fig. 19. A similar diagram for the special case of an all-ocean model where $\Xi = \eta$ (sea-ice

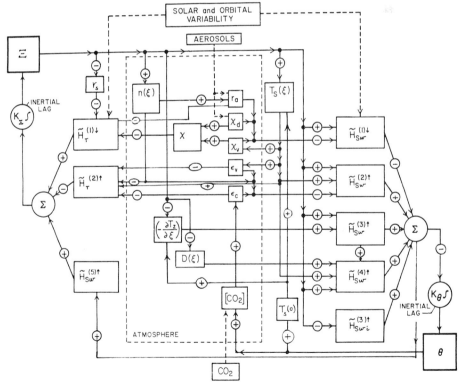

FIG. 19. Flow diagram showing feedback loops contained in the proposed dynamical system for ice-mass (Ξ) and ocean temperature (θ) variations.

extent) was given in Saltzman and Moritz (1980, Fig. 9), wherein a full description of the symbols used in this type of feedback diagram is given. Suffice it to note here that any closed set of links, following the direction of the arrows, represents a feedback loop that is *positive* if there is an even number of negative links and *negative* if there is an odd number of negative links. Two examples of positive feedbacks in our model, portrayed in Fig. 8, are $\{\Xi \rightarrow r_S \rightarrow \bar{H}_T^{(1)\downarrow} \rightarrow \Xi\}$ (the ice-albedo feedback) and $\{[CO_2] \rightarrow \epsilon_c \rightarrow \bar{H}_{Sw}^{(2)\uparrow} \rightarrow \theta \rightarrow [CO_2]\}$ (the hypothesized CO_2–ocean temperature greenhouse feedback), and an example of a negative feedback is $\{\Xi \rightarrow \bar{H}_{Sw}^{(2,3,4E)\uparrow} \rightarrow \theta \rightarrow (T_S, \partial T_Z/\partial \xi) \rightarrow \bar{H}_{Sw}^{(2,3,4E)\uparrow} \rightarrow \bar{H}_{Sw}^{(5)\uparrow} \rightarrow \Xi\}$ (ice-insulator effect).

ACKNOWLEDGMENTS

Support for preparing this review was provided by the National Science Foundation under grant ATM-7925013 of the Climate Dynamics Program, Division of Atmospheric Sciences. I am grateful to Dr. Alfonso Sutera for many useful and ongoing discussions of the questions treated.

REFERENCES

Benzi, R., Parisi, G., Sutera, A., and Vulpiani, A. (1982). Stochastic resonance in climatic change. *Tellus* **34**, 10–16.

Berner, W., Oeschger, H., and Stauffer, B. (1980). Information on the CO_2 cycle from ice core studies. *Radiocarbon* **22**, 227–235.

Birchfield, G. E. (1977). A study of the stability of a model continental ice sheet subject to periodic variations in heat input. *JGR, J. Geophys. Res.* **82**, 4909–4913.

Bolin, B., ed. (1981). "Carbon Cycling Modelling." Wiley, New York.

Boyce, W. E., and DiPrima, R. C. (1977). "Elementary Differential Equations and Boundary Value Problems." Wiley, New York.

Budyko, M. I. (1969). The effect of solar radiation variations on the climate of the earth. *Tellus* **21**, 611–619.

Dansgaard, W., Johnson, S. J., Clausen, H. B., and Langway, C. C., Jr. (1971). Climatic recored revealed by the Camp Century ice core. *In* "The Late Cenozoic Glacial Ages" (K. K. Turekian, ed.), pp. 37–56. Yale Univ. Press, New Haven, Connecticut.

Delmas, R. J., Ascencio, J.-M., and Legrand, M. (1980). Polar ice evidence that atmospheric CO_2 20,000 yr BP was 50% of present. *Nature (London)* **284**, 155–157.

Denton, G. H., and Hughes, T. J. (1981). The Arctic ice sheet: An outrageous hypothesis. *In* "The Last Great Ice Sheets" (G. H. Denton and T. J. Hughes, eds.), pp. 437–467. Wiley, New York.

Dietrich, G. (1963). "General Oceanography." Wiley, New York.

Duplessy, J. C., Chenouard, L., and Vila, F. (1975). Weyl's theory of glaciation supported by isotopic study of Norwegian core K 11. *Science* **188**, 1208–1209.

Eriksson, E. (1968). Air–ocean–icecap interactions in relation to climatic fluctuations and glaciation cycles. *Meteorol. Monogr.* **8** (30), 68–92.

Flint, R. E. (1971). "Glacial and Quaternary Geology." Wiley, New York.

Fraedrich, K. (1979). Catastrophes and resilience of a zero-dimensional climate system with ice-albedo and greenhouse feedback. *Q. J. R. Meteorol. Soc.* **105**, 147–167.

Gruber, A. (1977). Satellite estimates of the earth–atmosphere radiation balance. *In* "Radiation in the Atmosphere" (H.-J. Bolle, ed.), pp. 477–480. Science Press, Princeton, New Jersey.

Hasselmann, K. (1976). Stochastic climate models. Part 1. Theory. *Tellus* **28**, 473–485.

Hays, J. D., Imbrie, J., and Shackleton, N. J. (1976). Variations in the earth's orbit: Pacemaker of the ice ages. *Science* **194**, 1121–1132.

Hoffert, M. I., Callegari, A. J., and Hsieh, C.-T. (1980). The role of deep sea heat storage in the secular response to climatic forcing. *JGR, J. Geophys. Res.* **85**, 6667–6679.

Imbrie, J., and Imbrie, J. Z. (1980). Modeling the climatic response to orbital variations. *Science* **207**, 943–953.

Källén, E., Crafoord, C., and Ghil, M. (1979). Free oscillations in a climate model with ice-sheet dynamics. *J. Atmos. Sci.* **36**, 2292–2303.

Kellogg, T. B., Duplessy, J. C., and Shackleton, N. J. (1978). Planktonic foraminiferal and oxygen isotopic stratigraphy and paleoclimatology of Norwegian Sea deep-sea cores. *Boreas* **7**, 61–73.

Kim, J.-W., Chang, J.-T., Baker, N. L., and Gates, W. L. (1981). "The Climate Inversion Problem: Determination of the Relationship between Local and Large-scale Climate," Rep. No. 22. Climate Research Institute, Oregon State University, Corralis.

Knauss, J. A. (1978). "Introduction to Physical Oceanography." Prentice-Hall, Englewood Cliffs, New Jersey.

Lamb, H. H. (1977). "Climate, Present, Past and Future," Vol. 2. Methuen, London.

Lohmann, G. P. (1978). Abyssal benthonic foraminifera as hydrographic indicators in the western South Atlantic Ocean. *J. Foraminiferal Res.* **8**, 6–34.

Lorenz, E. N. (1963). Deterministic non-periodic flow. *J. Atmos. Sci.* **20**, 130–141.

Lorenz, E. N. (1968). Climatic determinism. *Meteorol. Monogr.* **5**, 1–3.

Lorenz, E. N. (1975). Climatic predictability. *GARP Publ. Ser.* **16**, 132–136.

Lorenz, E. N. (1979). Forced and free variations of weather and climate. *J. Atmos. Sci.* **36**, 1367–1376.

Marsden, J. E., and McCracken, M. (1976). "The Hopf Bifurcation and Its Applications." Springer-Verlag, Berlin and New York.

Mason, B. J. (1976). Towards the understanding and prediction of climatic variations. *Q. J. R. Meteorol. Soc.* **102**, 473–498.

Maykut, G. A., and Untersteiner, N. (1969). "Numerical Prediction of the Thermodynamic Response of Arctic Sea Ice to Environmental Changes," RM-6093-PR. Rand Corporation.

Mitchell, J. M. (1976). An overview of climatic variability and its causal mechanisms. *Quat. Res. (N.Y.)* **6**, 481–493.

Monin, A. S., and Yaglom, A. M. (1971). "Statistical Fluid Mechanics." MIT Press, Cambridge, Massachusetts.

Moritz, R. E. (1979). Nonlinear analysis of a simple sea-ice ocean temperature oscillator model. *JGR, J. Geophys. Res.* **84**, 4916–4920.

Nace, R. L. (1969). World water inventory and control, *In* "Water, Earth and Man" (R. J. Chorley, ed.), pp. 31–42. Methuen & Co., London.

Newell, R. E. (1974). Changes in the poleward energy flux by the atmosphere and ocean as a possible cause for ice ages. *Quat. Res. (N.Y.)* **4**, 117–127.

Nicolis, C. (1982). Stochastic aspects of climatic transitions—Response to a periodic forcing. *Tellus* **34**, 1–9.

North, G. R., Cahalan, R. F., and Coakley, J. A., Jr. (1981). Energy balance climate models. *Rev. Geophys. Space Phys.* **19**, 91–121.

Oerlemans, J. (1980). Continental ice sheets and the planetary radiation budget. *Quat. Res. (N.Y.)* **14**, 349–359.

Paltridge, G. W., and Platt, C. M. R. (1976). "Radiative Processes in Meteorology and Climatology." Elsevier, Amsterdam.

Parkinson, C. L., and Washington, W. M. (1979). A large-scale numerical model of sea ice. *JGR, J. Geophys. Res.* **84**, 311–337.

Pedlosky, J. (1971). Geophysical fluid dynamics. "Mathematical Problems in the Geophysical Sciences," Lect. Appl. Math., Vol. 13, pp. 1–60. Am. Math. Soc., Providence, Rhode Island.

Pollard, D., Ingersoll, A. P., and Lockwood, J. G. (1980). Response of a zonal climate–ice sheet model to the orbital perturbations during the Quaternary ice ages. *Tellus* **32**, 301–319.

Priestley, C. H. B. (1959). "Turbulent Transfer in the Lower Atmosphere." Univ. of Chicago Press, Chicago, Illinois.

Rhoads, D. C., and Morse, J. W. (1971). Evolutionary and ecologic significance of oxygen-deficient marine basins. *Lethaia* **4**, 413–428.

Robock, A. (1978). Internally and externally caused climate change. *J. Atmos. Sci.* **35**, 1111–1122.

Saltzman, B. (1964). On the theory of the axially-symmetric, time-average, state of the atmosphere. *Pure Appl. Geophys.* **57**, 153–160.

Saltzman, B. (1967). On the theory of the mean temperature of the earth's surface. *Tellus* **19**, 219–229.

Saltzman, B. (1968). Steady state solutions for axially-symmetric climatic variables. *Pure Appl. Geophys.* **69**, 237–259.

Saltzman, B. (1973). Parameterization of hemispheric heating and temperature variance fields in the lower troposphere. *Pure Appl. Geophys.* **105**, 890–899.

Saltzman, B. (1977). Global mass and energy requirements for glacial oscillations and their implications for mean ocean temperature oscillations. *Tellus* **29**, 205–212.

Saltzman, B. (1978). A survey of statistical-dynamical models of the terrestrial climate. *Adv. Geophys.* **20**, 183–304.

Saltzman, B. (1979). Equilibrium climatic zonation deduced from a statistical-dynamical model. *GARP Publ. Ser.* **22**, 803–841.

Saltzman, B. (1980). Parameterization of the vertical flux, of latent heat at the earth's surface for use in statistical-dynamical climate models. *Arch. Meteorol., Geophys. Bioklimatol., Ser. A* **29**, 41–53.

Saltzman, B. (1982). Stochastically-driven climatic fluctuations in the sea-ice, ocean temperature, CO_2 feedback system. *Tellus* **34**, 97–112.

Saltzman, B., and Ashe, S. (1976a). The variance of surface temperature due to diurnal and cyclone-scale forcing. *Tellus* **28**, 307–322.

Saltzman, B., and Ashe, S. (1976b). Parameterization of the monthly mean vertical heat transfer at the earth's surface. *Tellus* **28**, 323–332.

Saltzman, B., and Moritz, R. E. (1980). A time-dependent climatic feedback system involving sea-ice extent, ocean temperature, and CO_2. *Tellus* **32**, 93–118.

Saltzman, B., and Pollack, J. A. (1977). Sensitivity of the diurnal surface temperature range to changes in physical parameters. *J. Appl. Meteorol.* **16**, 614–619.

Saltzman, B., and Vernekar, A. D. (1968). A parameterization of the large-scale eddy flux of relative angular momemtum. *Mon. Weather Rev.* **96**, 854–857.

Saltzman, B., and Vernekar, A. D. (1971a). An equilibrium solution for the axially-symmetric component of the earth's macroclimate. *JGR, J. Geophys. Res.* **76**, 1498–1524.

Saltzman, B., and Vernekar, A. D. (1971b). Note on the effect of earth orbital radiation variations on climate. *JGR, J. Geophys. Res.* **76**, 4195–4197.

Saltzman, B., and Vernekar, A. D. (1972). Global equilibrium solutions for the zonally-averaged macroclimate. *JGR, J. Geophys. Res.* **77**, 3936–3945.

Saltzman, B., and Vernekar, A. D. (1975). A solution for the northern hemisphere climatic zonation during a glacial maximum. *Quat. Res. (N.Y.)* **5**, 307–320.

Saltzman, B., and Vernekar, A. D. (1983). The influence of poloidal motions and latent heat release on the equilibrium ice extent in a simple climate model. *J. Atmos. Sci.* (submitted).

Saltzman, B., Sutera, A., and Evenson, A. (1981). Structural stochastic stability of a simple auto-oscillatory climatic feedback system. *J. Atmos. Sci.* **38**, 494–503.

Schneider, S. H., and Dickinson, R. E. (1974). Climate modeling. *Rev. Geophys. Space Phys.* **12**, 447–493.

Schnitker, D. (1974). West Atlantic abyssal circulation during the past 120,000 years. *Nature (London)* **248**, 385–387.

Schuss, Z. (1980). Singular perturbation methods in stochastic differential equations of mathematical physics. *SIAM Rev.* **22**, 119–155.

Sellers, W. D. (1969). A global climatic model based on the energy balance of the earth–atmosphere system. *J. Appl. Meteorol.* **8**, 392–400.

Sergin, V. Ya. (1979). Numerical modeling of the glaciers–ocean–atmosphere global system. *JGR, J. Geophys. Res.* **84**, 3191–3204.

Sergin, V. Y., and Sergin, S. Y. (1976). Paper 1 *in* "The Simulation of the 'Glaciers–Ocean–Atmosphere' Planetary System" (S. Y. Sergin, ed.), pp. 5–51. Far East Sci. Cent., USSR Acad. Sci., Vladivostok (in Russ.).

Shackleton, N. J., and Kennett, J. (1975). "Initial Reports of the Deep Sea Drilling Project XXIX," pp. 743–755. U.S. Govt. Printing Office, Washington D.C.

Shackleton, N. J., and Opdyke, N. D. (1973). Oxygen isotope and paleomagnetic stratigraphy of equatorial Pacific core V28-238: oxygen isotope temperature and ice volumes on a 10^5 and 10^8 year scale. *Quat. Res. (N.Y.)* **3**, 39–55.

Smagorinsky, J. (1974). Global atmospheric modeling and the numerical simulation of climate. *In* "Weather and Climate Modification" (W. H. Hess, ed.), pp. 633–686. Wiley, New York.

Smith, W. L., Hickey, J., Howell, H. B., Jacobowitz, H., Hilleary, D. T., and Drummond, A. J. (1977). Numbus-6 earth radiation budget experiment. *Appl. Opt.* **16**, 306–318.

Streeter, S. S. (1973). Bottom water and benthonic foraminifera in the North Atlantic glacial–interglacial contrasts. *Quat. Res. (N.Y.)* **3**, 131–141.

Suarez, M. J., and Held, I. M. (1976). Modelling climatic response to orbital parameter variations. *Nature (London)* **263**, 46–47.

Sutera, A. (1980). Stochastic perturbation of a pure convective motion. *J. Atmos. Sci.* **37**, 245–249.

Sutera, A. (1981). On stochastic perturbation and long-term climate behavior. *Q. J. R. Meteorol. Soc.* **107**, 137–153.

Weertman, J. (1964). Rate of growth or shrinkage of nonequilibrium ice sheets. *J. Glaciol.* **6**, 145–158.

Wyrtki, K. (1981). An estimate of equatorial upwelling in the Pacific. *J. Phys. Oceanogr.* **11**, 1205–1214.

Part III

RADIATIVE, SURFICIAL, AND DYNAMICAL PROPERTIES OF THE EARTH–ATMOSPHERE SYSTEM

SATELLITE RADIATION OBSERVATIONS AND CLIMATE THEORY

GEORGE OHRING[1]

Department of Geophysics and Planetary Sciences
Tel-Aviv University, Tel-Aviv, Israel
and
National Earth Satellite Service, NOAA
Washington, D.C.

AND

ARNOLD GRUBER

Earth Sciences Laboratory
National Earth Satellite Service, NOAA
Washington, D.C.

1. INTRODUCTION

The purpose of this article is to illustrate some of the uses of satellite observations of the earth's radiation budget in studies of the earth's climate. The article is not meant to be a complete survey of the field but

[1] Present address: Laboratory for Atmospheric Sciences, Climate and Radiation Branch, Code 915, NASA Goddard Space Flight Center, Greenbelt, Maryland 20771.

237

ADVANCES IN GEOPHYSICS, VOLUME 25
ISBN 0-12-018825-2

rather an exposition of those topics of interest to the authors. The works of others on these topics are included in a natural fashion.

The radiation budget determines the sources and sinks of radiative energy for the earth–atmosphere system. It is these sources and sinks that drive the general circulation of the atmosphere and thus control the climate. Prior to the satellite era, our knowledge of the earth's radiation budget was based upon radiative transfer calculations (see, for example, the classical study of London, 1957). Because of their importance to studies of the climate, radiation budget observations from satellites were initiated as early as 1962, only 2 yr after the launching of the first weather satellite. Such observational programs have intensified during the last few years. One of these programs obtains estimates of the components of the radiation budget from the visible and infrared window scanning radiometers on the operational National Oceanic and Atmospheric Administration (NOAA) weather satellites. In Section 2, the results of these observations are discussed and compared with the results of other satellite observations of the earth's radiation budget.

One of the important characteristics of the earth's climate is its sensitivity to changes in boundary or external conditions. A typical measure of this sensitivity is the response of the mean surface temperature to a change in the solar constant. In turn, this response depends on the sensitivity to changes in surface temperature of the longwave radiation emitted by the earth–atmosphere system and of the earth's planetary albedo. The longwave radiation and planetary albedo are components of the radiation budget that are measured from satellites. Section 3 discusses the application of satellite radiation observations to such sensitivity studies.

Significant advances have been made in recent years in the use of numerical climate models to simulate the earth's climate and its sensitivity to changes in the external forcing. To validate the simulations, the distributions of the computed climatic variables are compared to the observed distributions of these elements. Because of the complexity of these models and the possibility of compensating errors in the treatment of various physical processes, good simulations of the observed climate are not a complete test of a climate model. Satellite observations of the radiation budget offer a means of validating model radiation calculations and of diagnosing possible causes of error in the simulations. Section 4 discusses the application of satellite radiation observations to the validation of a particular climate model of the statistical-dynamical type.

The seasonal cycle of the climatic variables at the surface of the earth is controlled, to a large extent, by the seasonal cycle in the surface heat and water budgets. In turn, these are related to the seasonal cycle of the components of the radiation budget of the earth–atmosphere system. Section 5 discusses the annual cycle of radiation budget components as observed from satellites for several climatic regions.

There are a number of climate-related applications of satellite radiation observations that are not dealt with in this article. Among these are the calculations of meridional heat transports by the oceans (Oort and Vonder Haar, 1976), the monitoring of the earth's climate, and the checking of various theories of climatic change. Together with the topics discussed in this article, they indicate the wide variety of contributions that satellite observations can make toward advancing our knowledge of climatic processes and climatic change.

2. The Earth's Radiation Budget

2.1. Introduction

Measurements by the two-channel scanning radiometer aboard the NOAA operational satellites were used between June 1974 and March 1978 to determine the components of the earth–atmosphere radiation budget over the entire globe. These radiation values were computed daily by the National Earth Satellite Service (NESS), NOAA. The components consist of the outgoing longwave radiation, the albedo and the related absorbed solar radiation, and the net radiation, which is defined as the absorbed solar radiation minus the outgoing longwave radiation. These quantities are suitable for meteorological and climatic studies over all areas of the globe on a variety of time and space scales. Several studies with these data are those of Gruber (1977, 1981), Jensenius et al. (1978), Winston and Krueger (1977), Ohring and Clapp (1980), Hartmann and Short (1980), and Heddinghaus and Krueger (1981). An atlas summarizing mean monthly, seasonal, and annual average conditions has been produced by Winston et al. (1979). Details of the derivation of the flux values from the operational data, the processing of the data to form suitable outputs such as maps and averages, and the creation of an archive are given by Gruber (1977). We will briefly describe some of the key points of that report to familiarize the reader with these data.

The measuring instrument is a scanning radiometer (SR) flown aboard NOAA operational polar-orbiting satellites. These satellites are sun synchronous and orbit the earth at $\sim$1500 km with a period of $\sim$117 min. Equator crossing times are at 2100 local standard time (LST) northbound and 0900 LST southbound. The radiometer senses energy in the visible (0.5–0.7 μm) portion of the spectrum and in the infrared window region (10.5–12.5 μm). The spatial resolutions of the sensors are $\sim$4 km for the visible and $\sim$8 km for the infrared at the subsatellite point. As part of the NESS operational processing, the two infrared observations per day are corrected to a nadir view at the top of the atmosphere, and the visible observations are divided by cos ζ, where ζ is the solar zenith angle. These data, expressed in terms of digital counts, are mapped and archived by the NESS (Conlan, 1973) and are the basic data base from which the albedo and outgoing longwave radiation are determined.

In estimating the albedo and outgoing longwave radiation, subsets of high-resolution data are averaged and two global data sets are produced. One set is aligned with the National Meteorological Center (NMC) polar stereographic map base and is composed of a 125 × 125 square array per hemisphere. Each element of the array represents an average of original data points (256 points up to September 15, 1976 and 65 points thereafter because of reduced data sampling in the operational archive) and has a size that varies from 102 km at the equator to 204 km near the poles. The other global set is a 2.5° latitude–longitude array derived by a bilinear interpolation of the polar stereographic square arrays.

The estimate of total flux of outgoing longwave radiation from radiance measurements in the window region is made by use of a regression model that was derived from radiation calculations made for 99 different atmospheres for cloudy and clear skies covering a broad range of temperatures. The original regression equation was determined to be linear and explained 98% of the total variance. However, subsequent research has established that a nonlinear regression gives a superior fit, especially at extreme flux values. All previous longwave fluxes based on the linear regression equation have been corrected, and the archive and mapped fields include only those corrected values (Abel and Gruber, 1979).

The albedo is determined from the visible channel by assuming that the reflectance in the 0.5–0.7 μm region is a good estimate of the full spectral reflectance (0.2–4.0 μm). Also, it is assumed that the observed reflectance is isotropic and independent of solar zenith angle, that there is no diurnal variation of the reflecting surface, and that the solar constant is known (a value of 1353 W m^{-2} was used). The diurnal assumption is common to all estimates of reflected energy based on polar-orbiting satellites, which generally obtain only one view per day of a given area.

Recent work suggests that the assumption that the reflectance within 0.5–0.7 μm is a good estimate of the full spectral reflectance (0.2–4.0 μm) may not be overly restrictive. Ramanathan and Brieglieb (1980) show from model calculations that the 0.5–0.7 μm albedo provides a good estimate of total albedo under clear skies for the zonal average case. We are currently investigating this assumption for other sky conditions through model calculations, as well as for a variety of land surfaces.

Another important assumption made in obtaining the albedo from the narrow field of view data is that the reflected radiance is isotropic. There were several reasons for making this assumption. First, no well-defined angular models existed which could be reliably applied to the data. For example, Raschke *et al.* (1973a,b) applied simplified angular models to Nimbus 3 data, whereas Ruff *et al.* (1968) suggested that for zonal average conditions there were only small differences between reflectance obtained isotropically and by application of angular models. Also, the data base used for the computations did not have the necessary angular information available for application of any angular model. This, combined with the uncertainty of the existing angular models, persuaded us to make the isotropic assumption. Recent experimental data on the angular distribution patterns obtained from Nimbus 7 (Stowe *et al.*, 1980a) have clarified the importance of the anisotropic behavior of surfaces. Nevertheless, a comparison of reflected flux computed from direct flux measurements (no angular adjustments required) and angularly corrected data and isotropic assumptions all based on Nimbus 7 experimental data suggest that the isotropic assumption provides excellent zonal averages, except at high solar zenith angles, particularly over ice and snow (Jacobowitz, 1981).

Two other assumptions require discussion. Both involve diurnal processes of reflected energy. It is assumed that there is no dependence of the reflected energy on the solar zenith angle and that the reflecting surface exhibits no physical diurnal variation, e.g., a cloud in the field of view would remain there all day. These assumptions were based in part on the lack of reliable information on the diurnal variability and on the zenith angle dependence of the complex surfaces we were computing radiation budget for, and the belief that for large time and space scales, these assumptions would not be too restrictive, since small time and space scale variability would be less of a factor.

The net effect of these assumptions is to emphasize the large space (zonal) and time (monthly) scales. Comparisons with other independent satellite measurements of the planetary radiation budget, e.g., Ellis and Vonder Haar (1976), denoted EV, and Campbell and Vonder Haar (1980a), denoted CV, tend to support the representativeness of the large-scale averages. They will be discussed later.

2.2. Time and Space Average Radiation Budget

In this section we present the time series of the global annual average, seasonal maps, and time–latitude sections of albedo, absorbed solar radiation, outgoing longwave radiation, and net radiation.

2.2.1. Annual Cycle. Figure 1 shows the annual cycle in albedo, outgoing longwave radiation, absorbed solar radiation, and the net radiation, for the sample of 45 months. The albedo exhibits a pronounced annual cycle with a maximum in the Northern Hemisphere winter and a minimum during the summer. The outgoing longwave radiation also exhibits a pronounced annual variation; however, its maximum is in Northern Hemisphere summer. This out-of-phase relationship is a result of the large annual cycle of temperature experienced over continents, as well as the effects of clouds and snow. Thus, the maximum in outgoing longwave radiation during the Northern Hemisphere summer is most likely due to the large annual variation of temperature over the continents which results in a global mean temperature that is higher in July than in January. (This will be seen later when we examine time–latitude sections of the outgoing longwave radiation.) During the Northern Hemisphere winter, cloud and snow cover over the continents contribute to the high global albedo. These effects will be seen more clearly when we examine time–latitude sections of the albedo and outgoing longwave radiation.

The amplitude of the annual cycle of absorbed solar radiation is about 5 W m^{-2}, which is small compared to the outgoing longwave radiation and much smaller than that implied by the variation in albedo. The phase of the variation is also inconsistent with the phase of the albedo, both exhib-

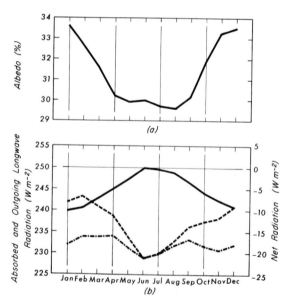

FIG. 1. Annual cycle of (a) albedo and (b) outgoing longwave radiation (——), absorbed solar radiation (-----), and the net radiation (----) based on 45-month SR data set.

iting minimum values during the summer months. This is because the minimum albedo occurs at the time when the sun is further from the earth, and the amount of incoming solar radiation is less than during the winter months. This relative difference of incoming solar energy ($\pm 3.4\%$) is enough to dampen the amplitude and reverse the phase of the annual cycle in absorbed solar energy from that implied by the albedo. As a result the amplitude and phase of the net radiation are essentially determined by the variations in outgoing longwave radiation. The net radiation exhibits a deficit in every month, with the largest deficit occurring in June. Since it is reasonable to expect a net radiation of close to zero for the long-term average, this condition highlights a probable bias in the estimates. The bias appears to be in the outgoing longwave radiation since the global average albedo is similar to that obtained from other observations. On the average, the outgoing longwave radiation appears to be overestimated by 13 W m^{-2}. If we subtract the annual mean deficit, the months May to October would exhibit deficits and the remaining months a surplus.

The global average value of albedo is 0.314, of absorbed solar energy is 232 W m^{-2}, and of net radiation is -13 W m^{-2}. If we were to adjust these values to a solar constant of 1376 W m^{-2} as measured by the earth radiation budget (ERB) experiment on Nimbus 7 (Hickey *et al.*, 1980) the albedo would be 0.306, the absorbed solar energy 239 W m^{-2}, and the net radiation -5 W m^{-2}.

2.2.2. Mapped Quantities. In this section the mean quantities based on the 45-month scanning radiometer data set will be presented and discussed. Because of the large volumes of data and maps we will limit ourselves to the summer (June, July, and August) and winter (December, January, and February) seasons and to the albedo, outgoing longwave radiation, and net radiation.

2.2.2.1. Winter. Maps of the mean winter albedo, outgoing longwave radiation, and net radiation are presented in Fig. 2a–c. Each field is shown in a mercator projection from 60° N to 60° S and then in a polar stereographic projection from 50° latitude poleward for the Northern (left side) and Southern (right side) hemispheres.

A good way of examining these maps is to look at them in the following zones: 30° N–30° S, 30°–60° N, S, and the polar regions. In the 30° N–30° S zone there is considerable structure evident in both latitude and longitude. The prominent features are the broad areas of high albedo over South America, Africa, and Indonesia at about 10°–15° S, which are associated with active convective rain zones in those areas. Other prominent high albedo areas are associated with clouds over the oceans. The Intertropical Convergence Zone (ITCZ) is clearly evident in the Atlantic

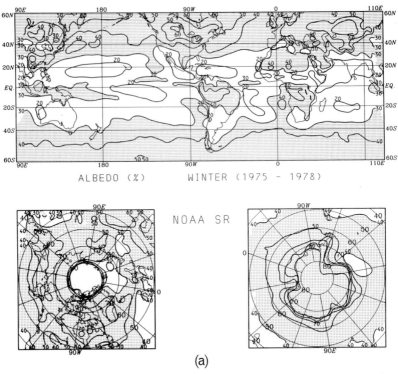

FIG. 2. Average winter (December–February) radiation budget maps: (a) albedo, (b) outgoing longwave radiation, and (c) net radiation.

and Pacific oceans at about 5°–10° N, and the high albedo regions in the eastern Pacific and eastern Atlantic are associated with the extensive stratocumulus clouds which are usually present there. The desert areas of North Africa also exhibit high albedo, some areas slightly in excess of 40%. The low albedo values north and south of the ITCZ represent regions of little or no cloudiness associated with the subtropical high-pressure zones. This entire region undergoes large seasonal shifts which will be evident when looking at the mean maps for the summer.

The outgoing longwave radiation exhibits similar distributions in this zone, except there is an out-of-phase relationship to the albedo. That is, the high-albedo convective rain areas and the ITCZ exhibit low outgoing longwave radiation (low-temperature emitting surface) and the low-albedo clear areas exhibit high outgoing longwave radiation (high-temperature emitting surface). The exception to this out-of-phase relationship occurs over the high-albedo stratocumulus clouds and the North African deserts. In the case of the cloudy area the closeness of the clouds

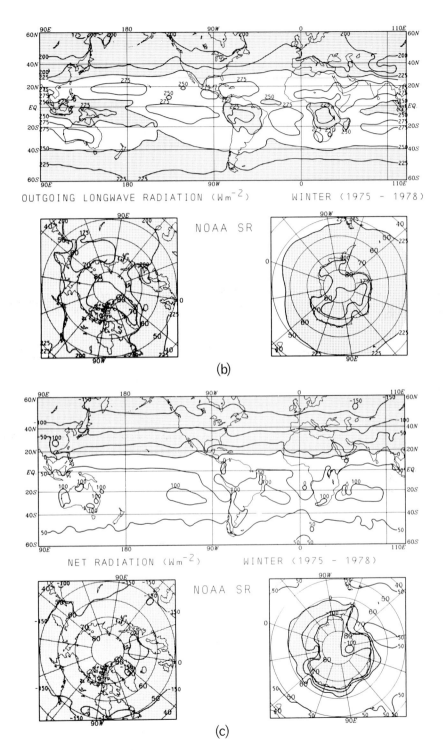

OUTGOING LONGWAVE RADIATION (Wm^{-2})   WINTER (1975 – 1978)

NOAA SR

(b)

NET RADIATION (Wm^{-2})   WINTER (1975 – 1978)

NOAA SR

(c)

Fig. 2b and c.

to the ocean surface gives them a high emitting temperature, thus resulting in large outgoing longwave radiation. The desert areas are also a high-temperature surface, thus emitting relatively large amounts of longwave radiation. This accounts for the minimum or slightly negative net radiation (Fig. 2c) calculated for those areas.

In the 30° N–30° S zone the net radiation achieves its greatest surplus, in excess of 100 W m^{-2} over the subtropical high-pressure zones of the Southern Hemisphere.

30°–60° N, S. In this zone there are distinct differences between the Northern and Southern Hemispheres, particularly in albedo, which can be related to differences in the distributions of oceans and continents. Thus, there is considerable east–west variability in the 30°–60° N zone with higher albedo over the continents than over the oceans. This is attributed to the effects of snow cover and perhaps more cloudiness over the continents. Conversely, the 30°–60° S zone exhibits more of a zonally oriented albedo pattern, broken only by intrusions of South America, Australia, and to a lesser extent Africa.

The outgoing longwave pattern, which is controlled by the temperature of the emitting surface, exhibits much more of a zonal character in the Northern Hemisphere and thus has more similarity to the Southern Hemisphere than did the albedo. From this pattern we can infer that there is considerable cloud cover over the Atlantic and Pacific oceans, with equivalent radiating temperatures of about 250 K, suggesting middle clouds.

The net radiation in these zones is also zonally oriented, however, the Northern Hemisphere is everywhere in a radiative deficit, whereas there is a surplus in the Southern Hemisphere.

60°–90° N, S. In the Northern Hemisphere the region around the pole is not illuminated during the winter months and albedo determinations cannot be made. In the region where it was possible to make an albedo determination there is considerable east–west variability associated with land and ocean differences as for the 30°–60° N zone.

In the Southern Hemisphere the high-albedo region is dominated by Antarctica. The area between 60° and 70° S at about 10° W represents relatively cloud-free conditions over what is normally ice-free water.

The outgoing longwave radiation, as would be expected, is low in both polar regions. Interestingly, both polar areas exhibit similar values, somewhat less than 175 W m^{-2}.

Finally, there is a net radiation deficit in the Northern Hemisphere 60°–90° N region, whereas the deficit in the Southern Hemisphere is concentrated over Antarctica and slightly to the southwest with the remainder of the zone exhibiting a radiation surplus.

2.2.2.2. Summer. The albedo, outgoing longwave radiation, and net radiation for the months of June, July, and August are displayed in Fig. 3a–c.

30° N–30° S. In the 30° N–30° S zone the seasonal shift of the cloud-free high-pressure zones and the cloudy rainy zones is quite evident in the albedo (Fig. 3a) and the outgoing longwave radiation (Fig. 3b) (compare to Fig. 2a and b). Thus, the major cloud regions observed south of the equator over Africa, South America, and Indonesia shifted northward from winter to summer as has the ITCZ. The clear areas associated with the subtropical high-pressure zones have also shifted northward. These seasonal shifts, both in pressure and in cloudiness of the ITCZ, have been well documented, e.g., by Riehl (1954) and Gruber (1972). Clearly the radiation budget in this zone can be characterized as being dominated by the seasonal movement of cloud/cloud-free areas.

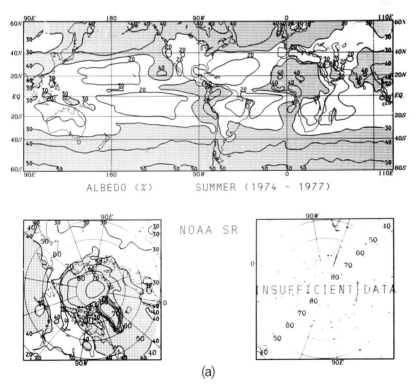

FIG. 3. Average summer (June–August) radiation budget maps: (a) albedo, (b) outgoing longwave radiation, and (c) net radiation.

(*continued*)

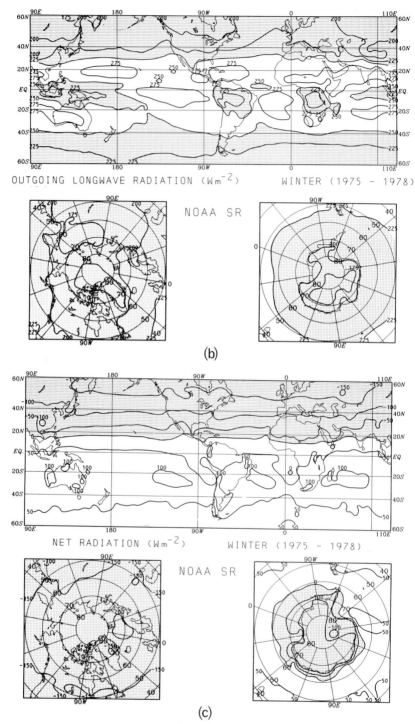

OUTGOING LONGWAVE RADIATION (Wm^{-2})   WINTER (1975 – 1978)

NOAA SR

(b)

NET RADIATION (Wm^{-2})   WINTER (1975 – 1978)

NOAA SR

(c)

FIG. 3b and c.

The low-level stratocumulus clouds on the eastern boundaries of the oceans, which were so prominent during the winter season, are also present during these months. They are now more evident in the eastern North Pacific Ocean.

The net radiation (Fig. 3c) is almost the reverse of that found during the winter; that is, north of 10° S there is a net radiation surplus, except for the stratocumulus cloud and desert regions; and between 10° and 30° S there is a deficit of net radiation.

30°–60° N, S. In the Northern Hemisphere there is appreciable east–west structure in the albedo (Fig. 3a), although in contrast to the winter situation, the albedo is higher over the oceans than over the land. This is attributed to more cloudiness over the ocean than over the land. A similar interpretation can explain the pattern of outgoing longwave radiation, which exhibits lower values over the oceans than over land areas. This is, of course, complicated by the land–sea temperature differences, which would tend to produce the same pattern although not with the same magnitude. In the 30°–60° S latitude zone both the albedo and outgoing longwave radiation exhibit a zonally oriented pattern much as they did during the winter months.

There is generally a surplus of net radiation in the 30°–60° N zone except for two small areas located at about 180° and 50° N and about 45° W and 50° N. In the Southern Hemisphere there is a deficit of net radiation everywhere.

60°–90° N, S. During this season the solar and viewing geometry made a reliable determination of albedo impossible in the Southern Hemisphere (Gruber, 1977). The albedo in the Northern Hemisphere exhibits high values associated with the snow and ice fields still present at high latitudes during these months. Notice the high albedo (greater than 70%) outlining the Greenland ice sheet. Also notice that the albedo pattern is asymmetrical with respect to the pole, due to the high albedo determined over the North Atlantic Ocean.

The outgoing longwave radiation in the 60°–90° N zone exhibits low values where the albedo was highest, consistent with our expectations for ice/snow covered surface. In the 60°–90° S zone extremely low values of outgoing longwave are measured, less than 125 W m^{-2} over Antarctica.

With no illumination over the south polar region, the net radiation essentially follows the pattern of the outgoing longwave radiation. Over the north polar region there is a deficit in net radiation that more or less follows the pattern of the albedo map. The high albedos result in little absorbed energy—thus the large deficits in net radiation in those regions.

2.2.3. Time–Latitude Variations. Further details about the temporal variation of the radiation balance components are shown in Figs. 4–7, which show the time–latitude distribution of zonally averaged monthly mean albedo, outgoing longwave radiation flux, absorbed solar radiation, and net radiation for the period June 1974 through February 1978.

The major features of the annual course of albedo are qualitatively similar to an earlier study by Winston (1971), who examined an entire year of ESSA 3 and 5 brightness data, and to the temporally limited study by Raschke *et al.* (1973b), who studied Nimbus 3 radiation data. They are also similar to the results of Jacobowitz *et al.* (1979), who studied wide-angle ERB data from Nimbus 6, and of Stephens *et al.* (1981), who composited 48 months of wide- and medium-field of view data from six different satellites spanning the years 1964–1977. The current results have the advantage of high spatial resolution, thus allowing a more detailed examination of small-scale features such as the ITCZ. In the middle latitudes of both hemispheres the maximum albedo occurs during the respective winter seasons—a result of increased snow cover (especially in the Northern Hemisphere), clouds, and solar zenith angle. In the subtropics (10°–20° N,

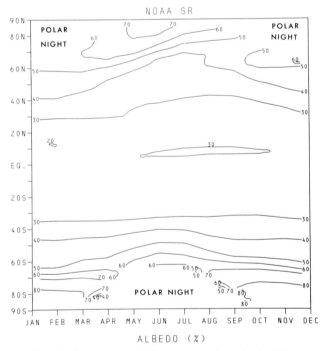

Fig. 4. Time–latitude section of the mean monthly albedo. EQ, equator.

S) the phase reverses, with a minimum in winter months and maximum during the summer months. In the belt 5°–10° N (ITCZ region) an annual cycle is not so clearly evident. However, the albedo appears to reach a maximum during the June–October period.

The outgoing longwave radiation (Fig. 5) exhibits many of the same characteristics of the albedo but in an inverse sense. This is not surprising since much of the cloudiness over the globe extends into the middle and upper troposphere, giving rise to the condition of high albedo and low outgoing longwave radiation. Snow and ice will also yield the same relationship. Thus, there are lower values of outgoing longwave radiation toward the poles, higher values of outgoing longwave radiation in the subtropical zones, which exhibit low amounts of cloud, and a secondary minimum in the equatorial zone, primarily north of the equator associated with the ITCZ. The excursion of the minimum in outgoing longwave radiation in the ITCZ region into the Southern Hemisphere during January and February is principally a result of the southward shift of cloudiness over Africa, South America, and the Indonesian–Melanesian area.

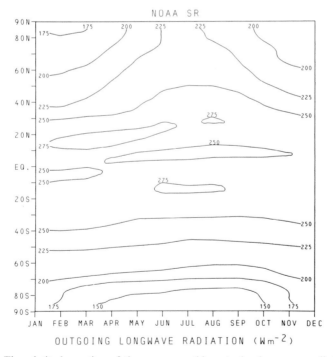

FIG. 5. Time–latitude section of the mean monthly outgoing longwave radiation. EQ, equator.

Over the open ocean the cloudiness associated with the ITCZ generally remains north of the equator (Gruber, 1972). As in the case of the albedo, the annual cycle is more pronounced in the Northern Hemisphere than in the Southern Hemisphere. Maximum values occur in the middle and polar latitudes during the summer months in both hemispheres mainly in response to increased temperature. Also, phase reversals take place in the 5°–20° N and 0°–20° S regions, that is, minima during the summer months and maxima during the winter months of each hemisphere.

The annual variation of absorbed solar energy is shown in Fig. 6. Outside the 5°–10° N latitude belt (ITCZ region), a pronounced annual variation is observed with the maximum of absorbed energy occurring during the summer months of each hemisphere, regardless of latitude. The phase of the variation clearly follows the course of the sun during the year, as seen by following the solar declination plotted in Fig. 6. The reason for this variation is that the annual variation of the incoming solar energy is much greater than the annual variation of the albedo; thus, the phase of the absorbed solar energy is essentially determined by the phase of incom-

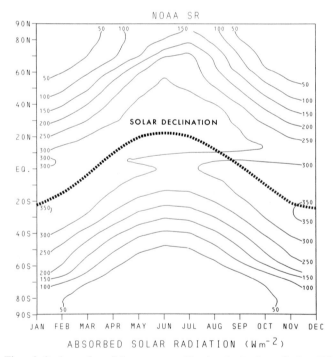

FIG. 6. Time–latitude section of the mean monthly absorbed solar radiation. EQ, equator.

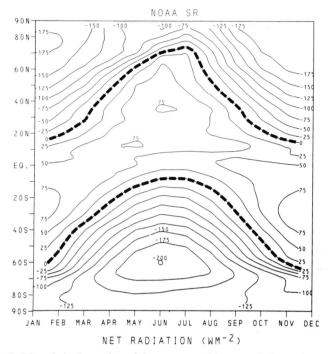

FIG. 7. Time–latitude section of the mean monthly net radiation. EQ, equator.

ing solar energy. This is consistent with the time series of the global average conditions presented earlier.

The net energy (Fig. 7) also exhibits a pronounced annual variation with phase relationships similar to the absorbed solar energy, i.e., maximum of net radiation during the summer months of both hemispheres. The maximum and surplus of energy also follow the solar declination. The reason for this is that the annual variation of net radiation is dominated by the variation in absorbed solar energy, which has a much greater amplitude than the variation of outgoing longwave radiation. The exception to this, of course, is in and near the polar night latitudes, where outgoing longwave radiation is the only component and thus determines the sign and magnitude of the net radiation.

2.3. Intercomparison of Satellite Radiation Budget Observations

Compilation of previous estimates of the planetary radiation budget obtained from other satellites has been performed by Ellis and Vonder

Haar (1976) (EV), Jacobowitz *et al.* (1979) (J), Campbell and Vonder Haar (1980a) (CV), and Stephens *et al.* (1981) (S). Since those compilations do not include any data from the NOAA operational spacecraft, they provide an excellent source for comparing our results. However, the spectral intervals of measurement, continuity and length of record, time of observation, and spatial revolution differ significantly from our data set.

The CV and J data sets both treat the Nimbus 6 wide-angle ERB observations. However, CV analyzed the observations to the top of the atmosphere differently and extended the data set to 24 months.

The EV data set is a composite of 29 months for the years 1964–1971, from seven different satellites. The S data set extends the EV set with the addition of the Nimbus data for a total of 48 months for the years 1964–1977.

We have selected the EV and CV data sets for comparison to the SR results. That choice was made because the EV and CV data represent two different eras in the estimation of radiation budget parameters from space, even though they share some common characteristics (e.g., spectral intervals of measuring instruments and spatial resolution). Table I compares characteristics of the three systems.

2.3.1. Zonal Average Comparison. The SR high-resolution data have been averaged over 10° latitude belts and all albedo values have been scaled to 1353 W m^{-2}, the value used in the SR computations, in order to facilitate comparisons. Figure 8 compares annual average meridional profiles of albedo. All three estimates compare quite favorably between 65° N and 65° S. Poleward of those latitudes differences are much larger, with EV exhibiting the lowest values. The low estimates of EV are partly due to the fact that at latitude 65° and greater, EV assumes an albedo ranging from 50 to 64% for low insolation months of June, July, August, September, October, and November. The differences between the SR and CV in the polar regions are probably related to difficulties in obtaining reliable estimates of albedo in polar regions from the SR data. The NOAA polar orbiting satellite has an 0830 LT descending node. Thus, when scanning over the North Pole, it observes reflected energy from a backscattered azimuth, and when scanning over the south pole, from a forward-scattered direction. At low solar zenith angles, as encountered at the equinox, isotropic albedos are generally underestimates over the North Pole and overestimates over the South Pole (L. L. Stowe, personal communication, 1982). A rough computation for average conditions indicates that the expected ratio of forward to backward isotropic albedo is about 1.12. The ratio based on SR observations of the estimates of Southern Hemisphere to Northern Hemisphere albedo at 85° N is 1.26. Thus, this effect may

TABLE I. COMPARISONS OF CHARACTERISTICS OF DATA SETS[a]

	SR	EV	CV
Radiometer type	High-resolution scanner	Flat plate medium-resolution scanner	Wide angle
Spectral range (μm)			
Shortwave	0.5–0.7	0.2–4	0.2–3.8
Longwave	10.5–12.5	4–50	3.8–>50
Years covered	1974–1978	1964–1971	1975–1977
Total months of data	45	29	24
Satellites	NOAA 2, 3, 4, 5	Experimental Nimbus 2, 3, ESSA-7, ITOS 1, NOAA 1	Nimbus 6
Local equator crossing time	0900	0830–1500 (depending on satellite)	1200
Resolution			
Original	4 km—shortwave 8 km—longwave	Full disk (~10°– 20°, half power)— medium resolution ~100 km	Full disk (~10°, half power)
Data set	2.5° latitude/ longitude	Zonal average— 10°–20° latitude	Zonal average— 10° latitude

[a] SR, scanning radiometer; EV, Ellis and Vonder Haar (1976); CV, Campbell and Vonder Haar (1980a).

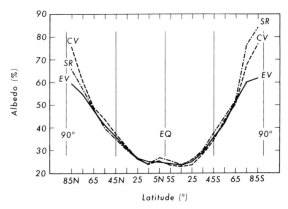

FIG. 8. Comparison of SR (-----), EV (——), and CV (----) estimates of the annual average zonal albedo. EQ, equator (see text for meaning of other abbreviations). Global averages are 31% for SR, 30% for EV, and 31% for CV.

account for a significant portion of the increased albedo in the Southern Hemisphere polar regions especially since the effect is systematic, i.e., the north polar regions are always observed in a backscattered direction and the south polar regions in a forward direction.

Also, we would not expect this difference in viewing geometry to account for the entire difference in albedo because it is reasonable, on the average, that the south polar region, which could be considered a snow-covered region, would exhibit a somewhat higher albedo than the north polar region, mainly an ice-covered region, based on the characteristics of the surface (L. L. Stowe, personal communication, 1982).

The comparison of profiles for the outgoing longwave radiation is shown in Fig. 9. The SR values are systematically higher than the CV values, with a difference in the global average of 14 W m^{-2}. The EV data are in good agreement with the CV observations in the Northern Hemisphere, and agree well with the SR data from 5° N to 45° S. The difference in the global average between SR and EV is 8 W m^{-2}. We originally thought a diurnal effect may have been a contributing factor to these differences, because the EV data set is composed from different local times and the SR data are all at a fixed local time, but different from the CV (Nimbus 6) data, which is also at a fixed local time. However, a comparison of annual zonal average flux from TIROS N (1530 LT equator crossing) and NOAA 6 (0730 LT equator crossing) agree with the SR values (Fig. 10), indicating that this is not the cause of the differences.

Since the SR flux computations are based on a regression model relat-

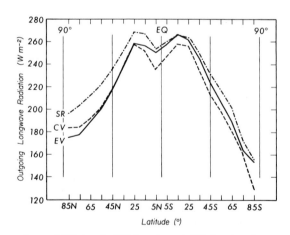

FIG. 9. Comparison of SR (-·-·-), EV (———), and CV (- - - -) estimates of the outgoing longwave radiation. EQ, equator. Global averages are 244 W m^{-2} for SR, 230 W m^{-2} for CV, and 236 W m^{-2} for EV.

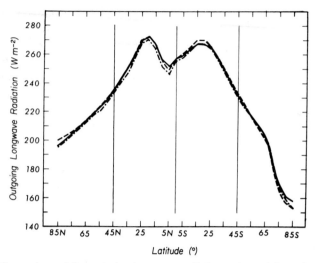

FIG. 10. Comparison of the outgoing longwave radiation estimated from the SR (———),
TIROS N (- - - -) and NOAA 6 (- - - -) data sets. Global averages are 244 W m^{-2} for SR, 245 W
m^{-2} for TIROS N, and 243 W m^{-2} for NOAA 6 data sets.

ing window channel radiance to broad-band fluxes (Abel and Gruber,
1979), we suspect that the systematic difference may be related to the
procedures used in deriving the regression coefficients. Possible causes
are:

(1) Model calculations of the flux are done for clear and completely
    cloudy conditions only.
(2) There is no stratification of regression equations according to sea-
    son or latitude zone.

Some preliminary calculations (R. Ellingson, personal communication,
1981) suggest that incorporation of partly cloudy conditions and spectrally
varying cirrus emissivity can account for about $\frac{1}{3}$ the difference between
SR and CV. An investigation is also planned which will study the possible
latitudinal and seasonal effects by extending the original 99 soundings that
were used for the model calculations to 1200 soundings.

Despite the difference in absolute magnitude, the profiles exhibit similar shapes, suggesting that calculations of changes or departures will be
represented with fidelity.

2.3.2. Time–Latitude Comparisons.   In order to gain some indication
of the spatial and temporal distribution of differences between the SR,

EV, and CV data sets we examined the time–latitude distribution of the differences. As in the previous comparisons, we averaged the high-resolution SR data to 10° zones, and scaled the EV and CV albedo values to the solar constant used in the SR computations: 1353 W m^{-2}.

The time–latitude differences in outgoing longwave radiation (SR–EV and SR–CV) are shown in Fig. 11a and b. As might be expected from the previous discussion on the comparison of the annual profiles, the SR outgoing longwave radiation is everywhere larger than the CV data and is also larger than the EV data nearly everywhere except for a small region in the tropics between April and October and a small area in the south polar region between November and May.

Interestingly the largest differences occur in the polar regions of both hemispheres and tend to be concentrated during the winter half year, suggesting a certain dependency on emitting temperature. This dependency on emitting temperature is also suggested by the relatively large differences found in the CV data set between 5° N and 15° N, roughly corresponding to the location of the ITCZ, which because of deep convection represents a low-temperature emitting region. This feature, however, is not found in the comparison with the EV data. In fact, differences with EV are more narrowly confined in time than the CV data set.

Time–latitude differences in albedo are shown in Fig. 12a and b. Within the area bounded by 65° N and 65° S differences are generally small, mostly between 0 and 5%. In the comparison with EV there is more of a tendency toward compensating positive and negative values within a latitude zone, whereas the comparison with CV indicates a tendency for the differences to exhibit the same sign within broad latitude zones with compensation occurring between zones. The regions poleward of 65° latitude exhibit some of the largest differences, much as we saw on the annual profile. These regions represent a relatively small area and therefore do not significantly impact the global average value, but naturally raise questions relative to the appropriate values of the radiation budget in polar latitudes.

Finally, the time–latitude differences in the net radiation estimates are displayed in Fig. 13a and b. The comparison with CV exhibits negative differences everywhere between 65° N and 65° S. The comparison with EV exhibits a strong tendency for some amount of compensation within latitude zones between 65° N and 65° S much the same as the albedo and to a slight extent with the outgoing longwave. The differences in the polar regions in both comparisons tends to be large. However, they appear to be concentrated in the summer half year when the outgoing longwave differences were smallest, therefore indicating the influence of the albedo on the net radiation pattern in this area. This is physically reasonable

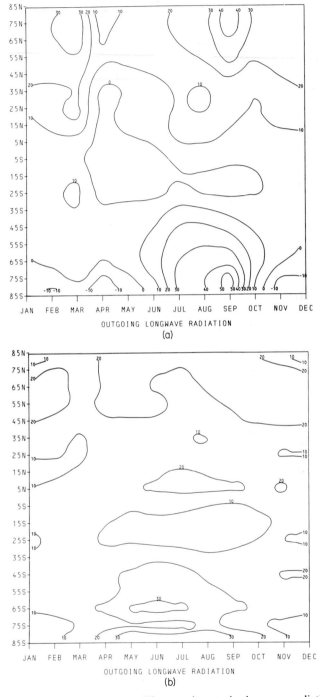

Fig. 11. Time–latitude sections of the differences in outgoing longwave radiation: (a) SR minus EV; (b) SR minus CV.

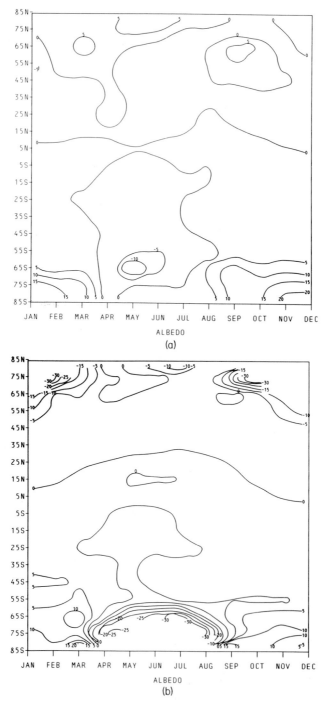

FIG. 12. Time–latitude sections of the differences in albedo: (a) SR minus EV; (b) SR minus CV.

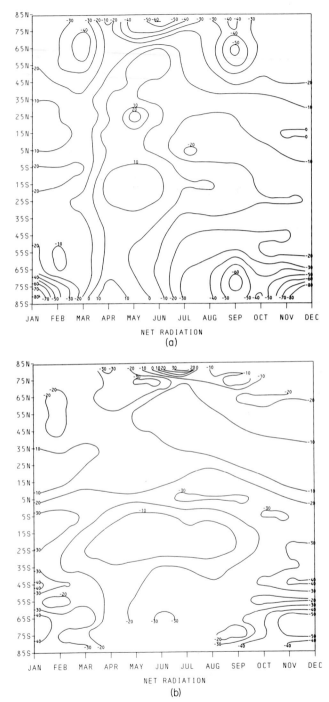

FIG. 13. Time–latitude sections of the differences in net radiation: (a) SR minus EV; (b) SR minus CV.

since there is a maximum of incoming solar energy during the summer months and even a small difference in albedo will have a large impact on the absorbed solar energy and thus the net radiation.

2.4. Additional Remarks

The comparison of zonal average monthly mean and annual radiation budget components from the SR data set with two other independent radiation budget data sets provides some measure of the reliability of the SR observations. For example, the global average albedo is, for practical purposes, the same for all three estimates, about 31%. The SR outgoing longwave estimate of 244 W m^{-2} is positively biased compared to EV and CV, which gave values of 8 and 14 W m^{-2}, respectively. The net radiation, calculated using a solar constant of 1376 W m^{-2}—determined by the ERB experiment (Hickey et al., 1980)—is -5, 5, and 11 W m^{-2} for the SR, EV, and CV data sets, respectively. The net radiation, being the small difference between two large somewhat uncertain quantities, is naturally a difficult number to obtain accurately.

The distributions of the mean monthly radiation budget, given by the time–latitude sections, are also comparable. Outside the polar regions the differences between the SR and CV and between the SR and EV data sets are about 2–3% in albedo and mostly less than 10 W m^{-2} for the outgoing longwave radiation, when accounting for the bias. Differences in the polar regions are greater for the reasons previously discussed.

The fact that there are differences between SR and EV and between SR and CV as well as between EV and CV should not be surprising, especially when differences in spectral resolution, spatial resolution, period of record, time of observations, simplifying assumptions, and instrumentation are considered. Perhaps the most surprising element is the reasonably good agreement between estimates.

The potential for more reliable measurements of the mean monthly radiation budget exists with the Earth Radiation Budget Experiment (ERBE), a NASA experiment consisting of a three-satellite system designed to measure the planetary radiation budget and assess its diurnal variability from regional to global scales. It is anticipated that this experiment will reduce the many sources of uncertainty associated with the current radiation budget estimates through improved instruments, measurements of the diurnal variability of the radiation budget, and application of the most recent angular distributions of reflected and emitted radiation obtained from the Nimbus 7 ERB experiment (Stowe and Taylor, 1981).

3. Sensitivity Studies

3.1. Introduction

It is generally assumed in meteorology and climatology that over a sufficiently long period (>1 yr) there is a balance between the solar radiation absorbed and the longwave radiation emitted by the global earth–atmosphere system. This balance is expressed by the equation

$$F = (S/4)(1 - \alpha) \qquad (3.1)$$

where F is the longwave radiation, S is the solar constant, and α is the planetary albedo.

The validity of this assumption has not been strictly proven. However, there are several reasons for believing that it is at least approximately correct: (1) observations indicate that global mean temperatures are approximately constant from year to year, implying that F is approximately constant and that F balances the solar input to the system, and (2) the heat capacity of the atmosphere and the upper mixed layer of the oceans is small enough so that they would respond relatively rapidly to changes in the solar input—within times of the order of a month for the atmosphere and years for the upper layer of the ocean. Satellite observations would represent the best way to validate the equation. But, currently available observations are not accurate enough to validate it to better than about a few percent.

Equation (3.1) represents a convenient starting point for examining the sensitivity of the earth's·climate to changes in the solar input to the system. Schneider and Mass (1975) have defined a global sensitivity parameter

$$\beta = S_0 \, dT_s/dS \qquad (3.2)$$

where S_0 is the current value of the solar constant and T_s is the global mean surface temperature. $\beta/100$ may be interpreted as the change in the global mean surface temperature that would be produced by a 1% change in the solar constant. For the current solar constant, Eq. (3.1) may be written

$$F_0 = (S_0/4)(1 - \alpha_0) \qquad (3.3)$$

Differentiating Eq. (3.1) with respect to mean global surface temperature and substituting current values for the solar constant and planetary albedo where they appear undifferentiated, we obtain

$$\frac{dF}{dT_s} = \frac{(1 - \alpha_0)}{4} \frac{dS}{dT_s} - \frac{S_0}{4} \frac{d\alpha}{dT_s} \qquad (3.4)$$

Substituting for $(1 - \alpha_0)$ from Eq. (3.3), rearranging, and solving for $S_0 \, dT_s/dS$, we obtain

$$\beta = \frac{F_0}{dF/dT_s + (S_0/4)(d\alpha/dT_s)} \tag{3.5}$$

This equation has been discussed by Cess (1976), among others. Important feedback mechanisms are included in the denominator of the right-hand side of Eq. (3.5). The basic longwave radiation sensitivity to changes in surface temperature, dF/dT_s, is a positive quantity; thus, it represents a negative feedback mechanism, whose strength depends on its magnitude. As surface temperature increases due to an increase in the solar constant, the longwave radiation loss to space will also increase, by an amount depending on the magnitude of dF/dT_s, tending to constrain the initial surface temperature increase. The sensitivity of the absorbed solar radiation (or shortwave sensitivity) to changes in surface temperature is related to the second term in the denominator of Eq. (3.1). It is generally believed that this is a negative quantity, the main idea being that an increase in surface temperature could lead to reduced ice and snow cover and hence, a reduced planetary albedo (ice-albedo feedback). This term would then represent a positive feedback mechanism, acting to amplify any initial temperature change due to a change in the solar constant. Within each of the terms in the denominator are hidden other potentially important feedback mechanisms; these will be discussed later. What is of importance to the present article is that the denominator contains two quantities, F and α, that are ideally suited for satellite observation. It is the purpose of this section to review the use of satellite observations in determining some of the explicit and implicit sensitivity coefficients contained in Eq. (3.5). Such determinations are not only useful from a theoretical point of view but are also important components of the energy balance type of climate model, in which all processes are parameterized in terms of surface temperature.

3.2. Longwave Radiation and Surface Temperature

If the earth had no atmosphere, it would radiate to space as a blackbody with temperature T_s. Assuming its planetary albedo was the same as the present one and did not vary with temperature, Eq. (3.5) would reduce to

$$\beta = \frac{F_0}{dF/dT_s} = \frac{\sigma T_s^4}{4\sigma T_s^3} = \frac{T_s}{4} \tag{3.6}$$

This yields $dF / dT_s = 3.76$ and $\beta = 61°C$. Thus, under such conditions, a 1% change in the solar input to the system would cause a change of about 0.6°C in the mean surface temperature of the earth. This value of β serves as a handy reference value.

The actual situation is much more complex. The outgoing radiation of the earth–atmosphere system is a function of the vertical distributions of several variables:

$$F = f(T(Z), w_i(Z), A_c(Z)) \tag{3.7}$$

where Z is height, w_i is the mixing ratio of the ith absorbing constituent, and A_c is the cloud amount. The major absorbing constituents are H_2O, CO_2, O_3, and aerosols. A number of investigators have performed simulations (model calculations) to examine the sensitivity of the outgoing radiation to changes in one or more of these variables (e.g., Manabe and Wetherald, 1967; Budyko, 1969; Ramanathan et al., 1976; Coakley, 1977). These simulations suggest that although F depends on all the variables listed in Eq. (3.7), a large part of the variance of F can be explained by variations in surface temperature, at least for highly averaged conditions (e.g., zonal averages). For example, from simulations based upon monthly mean vertical profiles at 260 stations around the world, Budyko (1969) found that

$$F = 226 + 2.26T_s - (48 + 1.62T_s)A_c \tag{3.8}$$

with an rms errors of less than 5%. The units are watts per meter squared for F and degrees centigrade for T_s. Since zonal average cloudiness does not vary too much from a value of 0.5, Eq. (3.8) may be approximated by

$$F = a + bT_s = 202 + 1.45T_s \tag{3.9}$$

As a result of these findings, formulas such as Eq. (3.9) have been used to parameterize the outgoing radiation in energy balance climate models, leading to fairly realistic simulations of the earth's zonal average surface temperature climate. The coefficient b represents the longwave sensitivity parameter, dF/dT_s.

The value of dF/dT_s obtained from such studies is much less than the value 3.76 for a blackbody at the earth's distance from the sun. This difference is usually explained as follows. The earth's atmosphere behaves in such a manner as to maintain constant relative humidity rather than constant absolute humidity as the surface temperature changes. As a result, with an increase of surface temperature there is an increase in the absolute amount of water vapor in the atmosphere. The effective emission-to-space layer of the atmosphere rises and, since temperature decreases with altitude, occurs at a lower effective temperature. This combination of events reduces the longwave sensitivity of the actual earth–atmosphere system from that of a blackbody.

Estimates of the values of b have been obtained in a variety of ways that are not always consistent with each other. Method 1 of obtaining b is by simple linear regression between simulated or observed values of F and observed surface temperatures. In this case, $b = dF/dT_s$ and includes within it all possible feedbacks between surface temperature and other variables affecting F. Method 2 of obtaining b is by multiple linear regressions of the form $F = a + bT_s + cA_c$. In this case, the value of $b = (\partial F/\partial T_s)_{A_c}$. This is a partial derivative and is to be interpreted as the change in F with a change in surface temperature, with cloud amount held constant. Method 3 is based upon computing F for two model atmospheres which differ only in surface temperature. In such computations the lapse rate, relative humidity, cloudiness, and other absorber distributions are generally held fixed. The value of b obtained in this way may be interpreted as a partial derivative with all other variables held fixed, i.e., $(\partial F/\partial T_s)_{all}$. The values of b obtained by these methods may be different; nevertheless, they have all been used as estimates of dF/dT_s for use in energy balance climate models.

There is some doubt about the appropriate method for determining b for use in energy balance climate models. In such models, all processes are parameterized in terms of the surface temperature, which is the only climatic variable that is predicted. Warren and Schneider (1979) argue that the b of Method 1, dF/dT_s, should be used since a total derivative is required in order to include (implicitly) changes in other variables that affect F and which are correlated with surface temperature. Coakley (1979) suggests that implicit inclusion of the effect of cloud amount on b determined from applying Method 1 to zonally averaged data may be incorrect. There may be no real physical relationship between cloud amount and surface temperature and the correlation between them may be due only to distributions with latitude.

Values of b determined from simulations range from 1.28 W m^{-2} °C^{-1} (Coakley and Wielicki, 1979) to 2.25 W m^{-2} °C^{-1} (Ramanathan et al., 1976). The lower value was obtained by applying Method 2 to the zonal means of a general circulation model (GCM) simulation (Wetherad and Manabe, 1975); the higher value was obtained from applying Method 3 to a case in which cloud-top altitude (as opposed to cloud-top temperature) is held fixed as the surface temperature is changed. The upper part of Table II shows estimates of b from simulations.

Warren and Schneider (1979) have examined the significance of uncertainties in the value of b on the results of an energy balance climate

TABLE II. SUMMARY OF ESTIMATES OF THE LONGWAVE SENSITIVITY PARAMETER b^a

Method	Investigator	b
Simulation		
M2, 260 stations, monthly means	Budyko (1969)	1.45
M3, 1-D global average model,	Ramanathan et al. (1976)	
fixed cloud-top altitude,		2.25
fixed cloud-top temperature		1.37
M2, 3-D GCM, zonal means	Coakley and Wielicki (1979)	1.28
Observation		
M1, zonal means, global	Budyko (1975)	1.67
M2, annual zonal means, Northern Hemisphere	Cess (1976)	1.57
M1, annual zonal means, global	Warren and Schneider (1979)	1.78
excluding 70°–90° S		2.17
M1, annual zonal means, global	Oerlemans and Van den Pool 1978 (1978)	2.23
M1, monthly means, varying latitude, global	Warren and Schneider (1979)	1.83
M2, seasonal means, varying latitude, Northern Hemisphere	Ohring and Clapp (1980)	1.8

^{a} $b = dF/dT_s$ (W m^{-2} °C^{-1}); M1, M2, and M3 refer to methods 1, 2, and 3, respectively, described in the text.

model. Such models include the ice-albedo effect through a parameterization of the shortwave sensitivity parameter, which, from Eq. (3.5), is proportional to $d\alpha/dT_s$. They find significant dependence of the results on b. Figure 14 (after Warren and Schneider, 1979) shows the latitude of the polar ice/snow cap as a function of the change in solar constant for different values of b and two different values of the shortwave sensitivity. For example, for $b = 1.37$, a decrease of the solar constant by 6.5% is required to produce an ice-covered earth; for $b = 2.25$, a decrease of 21% is required.

Since the exact value of b is important, and since it cannot be determined from simulations due to a number of uncertainties, especially the treatment of the clouds, and since satellites are capable of observing F on a global basis, several attempts have been made to estimate b observationally, i.e., from observed F and T_s values.

Apparently, the first attempt to determine b from satellite observations was that of Budyko (1975). He applied Method 1 to the zonal averages of the Nimbus 3 medium-resolution radiometer observations (Raschke et al., 1973a) of F and a corresponding set of mean surface temperatures. He

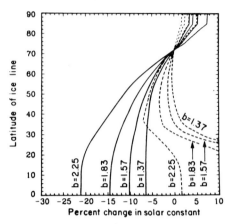

FIG. 14. Sensitivity of the extent of the polar snow/ice cap to the value of b, as determined from an energy balance climate model: $f = 0.004$ K^{-1} (———) and 0.009 K^{-1} (- - - -). f is the albedo–temperature feedback coefficient, which controls the shortwave sensitivity. (After Warren and Schneider, 1979.)

obtained $b = 1.67$ W m^{-2} °C^{-1} from this data set, and he states that the early satellite observations of the earth's radiation budget (1962–1966), summarized by Vonder Haar and Suomi (1971), yielded almost the same value for b.

A satellite data set that has been used to determine b by a number of investigators is that of Ellis and Vonder Haar (1976). This data set consists of zonally averaged monthly means of the earth–atmosphere system radiation budget.

Cess (1976) applied Method 2 to the annual averages of F from this data set, the annual zonal averages of surface temperature from Crutcher and Meserve (1970), and the annual zonal cloud amounts from London (1957). For the Northern Hemisphere he obtained

$$F = 257 + 1.6T_s - 91A_c \tag{3.10}$$

A similar expression with similar coefficients was derived for the Southern Hemisphere. For fixed cloud amount, this yields $b = 1.6$.

The lower part of Table II summarizes past determinations of b from satellite observations. Except for the Budyko (1975) determination, the observed values are based on the Ellis and Vonder Haar (1976) satellite data set for F.

Using the regression methods, T. S. Chen and G. Ohring (personal communication, 1982) are examining the dependence of b on satellite radiation data sets, and some results are shown in Table III. Pre-1972

TABLE III. DEPENDENCE OF b AND r ON SATELLITE DATA SETS[a,b]

| Area/time | Satellite data set | b (W m^{-2} °C^{-1}) and r from | |
		$F = a + bT_s$	$F = a + bT_s + cA_c$
Global, zonal, monthly	Pre-1972	1.78 (0.93)	1.74 (0.94)
(18 latitude zones, 12 months)	Nimbus 6	1.74 (0.95)	1.70 (0.96)
	NOAA SR	1.61 (0.98)	1.58 (0.98)
Oceans, zonal, monthly	Nimbus 6	1.70 (0.91)	1.46 (0.94)
	NOAA SR	1.58 (0.95)	1.45 (0.96)
Land, zonal, monthly	Nimbus 6	1.81 (0.93)	1.69 (0.97)
	NOAA SR	1.57 (0.95)	1.49 (0.98)

[a] From T. S. Chen and G. Ohring (personal communication, 1982).
[b] r is the correlation coefficient for the linear regressions.

refers to the Ellis and Vonder Haar (1976) data set, Nimbus 6 refers to 2 yr of wide-angle longwave flux observations from the Nimbus ERB experiment (Campbell and Vonder Haar, 1980a), and NOAA SR refers to the Gruber and Winston (1978) data set. The cloudiness data are from Berlyand et al. (1980) and the surface temperature data are from Crutcher and Meserve (1970) for the Northern Hemisphere and Taljaard et al. (1969) for the Southern Hemisphere. For the globe as a whole, the NOAA SR values of F are less sensitive to surface temperatures than the F values of the other data sets—$b = 1.6$ versus $b = 1.7$–1.8—and such differences would affect the results of energy balance climate models (see Fig. 14). [The value of 1.78 for the pre-1972 data set is slightly different than that obtained by Warren and Schneider (1979) ($b = 1.83$) because of slightly different methods of obtaining 10° latitudinal averages of surface temperature.] Another interesting difference between the data sets is the value of b for oceans and for land. The NOAA SR values of b are approximately the same for land and ocean, whereas the Nimbus values are higher for land than for ocean. The cloud amounts for these regressions are from Berlyand et al. (1980) and the surface temperatures for land and ocean separately are from A. Robock (1980; personal communication, 1981). It should be pointed out that some of the dependence on satellite data set may be due to the fact that the satellite data sets are not simultaneous with one another and cover only a few years each. Also, the surface temperature and cloud amounts are based upon climatological averages and are not concurrent with the satellite data sets.

Table IV (T. S. Chen and G. Ohring, personal communication, 1982) shows the dependence of b on latitude for the Northern Hemisphere. Because of the small annual variation of temperature at low latitudes and

TABLE IV. DEPENDENCE OF b ON LATITUDE USING NOAA SR MONTHLY DATA SET[a,b]

Latitude	b (W m^{-2} °C^{-1}) and r from	
	$F = a + bT_s$	$F = a + bT_s + cA_c$
80°–90° N	1.62 (0.99)	1.92 (0.99)
70°–80° N	1.70 (0.99)	2.20 (0.99)
60°–70° N	1.70 (0.99)	1.98 (0.996)
50°–60° N	1.79 (0.99)	1.89 (0.99)
40°–50° N	2.25 (0.996)	
30°–40° N	2.56 (0.99)	2.35 (0.99)
20°–30° N	1.33 (0.90)	

[a] From T. S. Chen and G. Ohring (personal communication, 1982).
[b] r is the correlation coefficient for the linear regressions.

uncertainties in the data, it is not possible to derive b with any confidence using this method at these latitudes. Also, at 20°-30° N and 40°–50° N the addition of cloud amount (Method M2) does not improve the specification of F so that at these latitudes only the results for Method M1 are shown. It is quite clear that b is not constant but has significant meridional variations, generally decreasing from equator to pole. Similar variations of b with latitude occur in the Southern Hemisphere. Such variations have also been found by Warren and Schneider (1979), who used the Ellis and Vondar Haar (1976) radiation data set, and by Ramanathan (1977) from simulations. Such a dependence on latitude might be expected from the meridional variation of effective emission temperature of the earth–atmosphere system. If the outgoing longwave flux is represented by σT_e^4, then $b = dF/dT_e$ would decrease with increasing latitude from a value ~ 4 in the equatorial zone to ~ 3 at the poles. The feedback effect of water vapor in an atmosphere with constant relative humidity (see discussion at beginning of this section) would reduce both of the above values, but a meridional gradient in b would remain. The above argument also implies a seasonal variation in b, especially where there is a large annual temperature variation.

The value of b may vary for other reasons. After all, as indicated earlier, the outgoing radiation depends upon the entire vertical structure and composition of the atmosphere. In view of the variability of b, the validity of sensitivity studies with energy balance climate models using a constant value of b is open to question. [See Coakley and Wielicki (1979) for some interesting work on this subject.] Satellite observations will help to better define this variability and thus help to answer this question, but first better consistency between different observing systems must be obtained.

3.3. The Radiation Budget and Cloud Amount

It is quite possible that because of changes in surface temperature resulting from a change in the solar constant (or in any of the other boundary conditions controlling the earth's climate), there will be a change in the amount of cloud cover, A_c. Clouds have two important effects on the radiation budget of the earth–atmosphere system. Because of their scattering properties at solar radiation wavelengths, clouds increase the albedo of the system, and an increased albedo means a reduction in the amount of absorbed radiation. Every weather forecaster knows that, other things being equal, the maximum temperature on a cloudy day will be less than that on a clear day. On the other hand, because of their absorption properties at large wavelengths, clouds decrease the longwave radiation loss to space. Forecasters will predict higher nighttime maximum temperature in the presence of cloudiness. Because of the potential for cloud amount to act as a feedback mechanism in the earth's climatic system, and thus amplify or damp any externally caused climatic change, it is important to analyze its effect on the radiation budget of the system. The term dF/dT_s may be written as

$$\frac{dF}{dT_s} = \frac{\partial F}{\partial T_s} + \frac{\partial F}{\partial A_c}\frac{dA_c}{dT_s} \qquad (3.11)$$

The interpretation of the terms on the right-hand side of Eq. (3.11) is as follows. The term $\partial F/\partial T_s$ represents the change in F with a change in surface temperature and includes all possible feedback mechanisms (e.g., water vapor amount, temperature profile, and cloud height feedbacks) except for cloud amount; the possibility of cloud amount feedback on F is included in the second term, where dA_c/dT_s represents the possibility of a change in cloud amount as the surface temperature changes. Similarly, $d\alpha/dT_s$ may be written as

$$\frac{d\alpha}{dT_s} = \frac{\partial \alpha}{\partial T_s} + \frac{\partial \alpha}{\partial A_c}\frac{dA_c}{dT_s} \qquad (3.12)$$

where $\partial \alpha/\partial T_s$ includes all possible feedback effects on albedo (e.g., ice-albedo and surface-type-albedo feedbacks) resulting from changing surface temperature, except for cloud amount feedback, which is represented by the second term on the right-hand side of Eq. (3.12).

The net radiation passing through each horizontal unit area at the top of the earth–atmosphere system may be written as

$$\text{Net} = Q - F \qquad (3.13)$$

where Q is the solar radiation absorbed by the system. Schneider (1972) has introduced a cloud sensitivity parameter δ for determining the effect of a change in cloud amount on the net radiation:

$$\delta = \frac{\partial(\text{Net})}{\partial A_c} = \frac{\partial Q}{\partial A_c} - \frac{\partial F}{\partial A_c} \tag{3.14}$$

where $\partial Q/\partial A_c$ may be called the albedo or shortwave effect of the clouds, and $\partial F/\partial A_c$ may be called the greenhouse or longwave effect of the clouds.

Substituting Eqs. (3.11) and (3.12) into Eq. (3.5) and making use of Eq. (3.14), one may obtain

$$\beta = \frac{F_0}{(\partial F/\partial T_s) - \delta(dA_c/dT_s) + (S_0/4)(\partial\alpha/\partial T_s)} \tag{3.15}$$

Equation (3.15) shows that if cloud amount does change with a change in surface temperature, then δ will determine both the sign and the magnitude of the cloud feedback mechanism.

Early calculations with an analytical model (Ohring and Mariano, 1964) and the classic thermal equilibrium calculations of Manabe and Wetherald (1967) suggested that the global mean value of δ is negative—i.e., the cloud-albedo effect is greater than the greenhouse effect. A recent finding by Cess (1976) has stimulated renewed interest in the sign and magnitude of δ. He found that the global mean of δ is approximately zero. If true, this would mean that the greenhouse effect of the clouds completely cancels their albedo effect. In this case the net radiation at the top of the atmosphere would be insensitive to changes in cloud amount, which would rule out cloud amount as a potential feedback mechanism. If true, this finding has important implications for global climate studies. For some simple climate models, it means that the global mean surface temperature is insensitive to changes in cloud amount. There are also implications for global climate models of the general circulation type. Because of difficulties in generating or predicting cloudiness, much of the climate sensitivity work being performed with general circulation models is done with prescribed rather than predicted cloudiness (Manabe and Wetherald, 1975; Wetherald and Manabe, 1975). If the net radiation of the system is insensitive to cloud amount, this deficiency of the models may not be that critical, and it may be possible to obtain approximate estimates of climate sensitivity even with models that use prescribed cloudiness. It is therefore of some importance to determine whether the opposite effects of clouds on the radiation budget of the earth–atmosphere system do indeed cancel each other.

TABLE V. SUMMARY OF ESTIMATES OF $\partial F/\partial A_c$ and $\partial F/\partial Q$

Method	Investigator	$\partial F/\partial A_c$ (W m^{-2})	$\partial F/\partial Q$
Simulation			
1-D, global average model			
Single cloud	Schneider (1972)	−75	
	Ramanathan (1976)	−71	
	Coakley (1977)	−51	
Multiple clouds	Cess (1974)	−68	
	Wang and Domoto (1974)	−61	
Global average of zonal energy budget, multiple clouds	Hoyt (1976)	−34	
2-D, zonal average model, single cloud	Ohring and Adler (1978)	−33	
3-D, GCM, from $F(T_s, A_c)$ zonal regression	Coakley and Wielicki (1979)	−71	
260 representative points on earth, monthly means $F(T_s, A_c)$ regression	Budyko (1969)	−73	
Observation			
$F(T_s, A_c)$ regression			
Annual, zonal means	Cess (1976)	−91	
Seasonal, zonal means	Ohring and Clapp (1980)	−60	
Monthly, zonal means	Warren and Schneider (1979)	−58	
Comparison of clear and cloudy sky values of F	Ellis (1978)	−40	
$F(\alpha(A_c))$ regressions, monthly means, regional	Ohring *et al.* (1981)	−35	0.33
$F(\alpha)$ regressions, daily values, regional	Hartmann and Short (1980)		0.4
$F(Q)$ regressions, monthly means, low latitudes	Cess *et al.* (1982)		
Ellis and Vonder Haar (1976) sat. data			1.1
Campbell and Vonder Haar (1980b) sat. data			1.0
Gruber and Winston sat. data			0.5
$F(Q)$ regressions, daily values, regional Nimbus 7, scanning radiometer (preliminary)	G. Ohring and H. Ganot (personal communication, 1982)		0.51

values of δ suggest almost complete cancellation of the albedo effect of clouds by their greenhouse effect.

To explain why the value of $\partial F/\partial A_c$ deduced from the satellite observation is so much larger than that obtained in model calculations, Cess and Ramanathan (1978) suggest that with an increase in total cloud amount, there is typically an increase in the high-cloud fraction and a decrease in the low-cloud fraction as observed from above. Since higher clouds have a greater greenhouse effect, the value of $\partial F/\partial A_c$ for the real atmosphere is larger than is obtained in model calculations, in which it is generally assumed that the percentage change in cloud amount is the same for all cloud layers and is equal to the percentage change in total cloud amount, or in the notation of Cess and Ramanathan (1978), that $(dA_{c_i}/A_{c_i}) = (dA_c/A_c)$, where the subscript i refers to a nonoverlapped cloud layer at a particular height. Combining the seasonal cloud-cover fractions of London (1957) into hemispheric averages of high, middle, and low clouds, Cess and Ramanathan do indeed find that with an increase in total cloud amount the high-cloud fraction increases and the low-cloud fraction decreases. Using this information to calculate a hemispheric average value of $\partial F/\partial A_c$, they obtain a value of -104 W m^{-2} as opposed to a value of -49 W m^{-2} for a calculation in which the percentage change in cloud amount is the same for all cloud layers. These results tend to confirm Cess's (1976) original conclusion that the albedo effect of a change in cloud amount is essentially completely cancelled by the greenhouse effect of a change in cloud amount, i.e., $\delta \approx 0$. They also point to the desirability of arriving at estimates of δ, and particularly $\partial F/\partial A_c$, from observations.

Ohring and Clapp (1980) and Ohring et al. (1981) have estimated $\partial F/\partial A_c$ from radiation budget data obtained from the NOAA SR data set (Gruber and Winston, 1978). The method is based upon the use of albedo variations at specific geographical areas as a measure of cloud amount variations. The basic data consist of the monthly mean values of longwave radiation and albedo for the 45-month period from June 1974 to February 1978 inclusive, and for the $2\frac{1}{2}°$ latitude by $2\frac{1}{2}°$ longitude grid. They examined the changes in longwave radiation and albedo for the same month in successive years at selected geographical locations. The albedo change is interpreted as being due to a change in mean cloudiness characteristics and, in particular, to a change in the mean cloud amount for the month [see Eq. (3.19)]. Unless the selected geographical area is susceptible to albedo variations due to snow-cover variations, an interpretation in terms of a change in cloudiness characteristics appears to be valid. For oceanic areas this is almost certainly true, and even for land areas this seems like a reasonable assumption, since there is no reason to expect the clear-sky albedo to vary significantly from one month of one year to the same month the following year. Thus, one may associate an increase (decrease) in albedo with an increase (decrease) in cloud amount. By plotting change in longwave radiation against the change in albedo, one can examine the relationship between changes in longwave radiation and cloud amount.

Figures 15 and 16 (from Ohring and Clapp, 1980) show such graphs for two different oceanic regions. ΔF and $\Delta \alpha$ are the interannual changes in

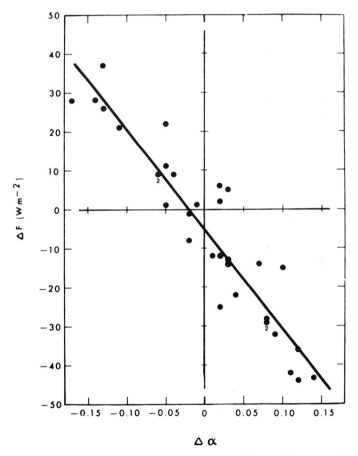

FIG. 15. Scatter diagram showing interannual changes in monthly means of longwave radiation ΔF plotted against interannual changes of albedo $\Delta \alpha$ for area of interest at the equator, 180°. Two data points at the same location are indicated by the number 2. The sloping line is the least squares regression line through the points. $r(\Delta F, \Delta \alpha) = -0.93$ is the correlation coefficient between ΔF and $\Delta \alpha$. The slope of the regression line (-254 W m^{-2}) gives the best estimate of the average value of $(\Delta F/\Delta \alpha)$, which is interpreted as $(\partial F/\partial \alpha)$. (After Ohring and Clapp, 1980.)

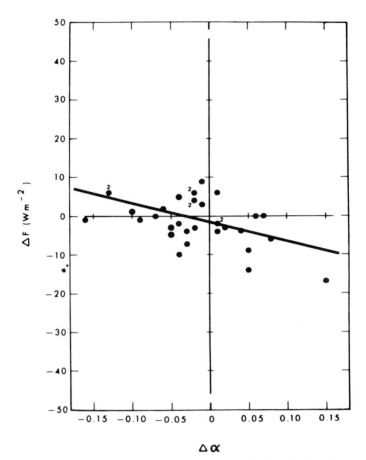

FIG. 16. Same as Fig. 2 except for area centered at 35° N, 125° W off the coast of California. $r(\Delta F, \Delta\alpha) = -0.49$, $(\Delta F/\Delta\alpha) = -45$ W m^{-2}. (After Ohring and Clapp, 1980.)

the monthly means of F and α, respectively. It is assumed that the slope (b) of the least-square regression line

$$\Delta F = a + b \, \Delta\alpha \qquad (3.21)$$

is the best estimate of the average value of $\Delta F/\Delta\alpha$ for the area, which is interpreted as $\partial F/\partial\alpha$. Ohring and Clapp (1980) discuss how this procedure for determining $\partial F/\partial\alpha$ suppresses long-period trends—whether real or instrumental—in the data. Figure 15 represents a region in the equatorial Pacific that is subject to incursions of deep convective cloudiness from the south and west, whereas Fig. 16 is representative of the stratus cloud regime off the coast of California. The explanation for the large difference in the $\partial F/\partial \alpha$ values for the two locations most likely lies in the different cloud regimes. The low clouds off the California coast have only a small effect on the longwave radiation, whereas the deep convective clouds of the equatorial Pacific, with their higher cloud-top heights, have a much greater effect.

A determination of the relative importance of the greenhouse and albedo effect of clouds is obtained as follows. From Eqs. (3.13) and (3.16), it follows that

$$\partial (\text{Net})/\partial \alpha = -Q_0 - \partial F/\partial \alpha \qquad (3.22)$$

If $\partial (\text{Net})/\partial \alpha$ is positive, the greenhouse effect is greater than the albedo effect; if it is negative, the albedo effect is greater; and if it is zero, there is complete cancellation of the two effects. Ohring and Clapp (1980) and Ohring et al. (1981) show that $\partial (\text{Net})/\partial \alpha$ is negative on a regional and global basis. Similar results have been obtained by Hartmann and Short (1980), who examined the day-to-day variations of the NOAA scanning radiometer observations.

Estimates of $\partial F/\partial A_c$ may be obtained from the $\partial F/\partial \alpha$ values by using Eq. (3.19), writing

$$\frac{\partial F}{\partial A_c} = \frac{\partial F}{\partial \alpha} \frac{\partial \alpha}{\partial A_c} = (\alpha_c - \alpha_s) \frac{\partial F}{\partial \alpha} \qquad (3.23)$$

and estimating $(\alpha_c - \alpha_s)$ for each geographical area. This has been done by Ohring and Clapp (1980) for a small number of geographical areas and by Ohring et al. (1981) on a quasi-global basis. In the work by Ohring et al. (1981), a 45-month time series of longwave radiation and albedo was extracted from the data archive for each $10°$ latitude–longitude gridpoint, from $60°$ N to $60°$ S. The annual and semiannual cycles were removed from the time series at each grid point by subtracting out these components using harmonic analysis. Long-term trends in albedo and longwave radiation were also removed by the use of least squares linear fits to the time series at each grid point. These modified values of albedo and longwave radiation are used to determine $\partial F/\partial \alpha$ from the slope (b) of the least squares regression relating F to α at each grid point

$$F = a + b\alpha \qquad (3.24)$$

Estimates of $(\alpha_c - \alpha_s)$ for use in Eq. (3.23) were obtained as follows. The clear-sky albedo values α_s are based on processing the data of Posey and Clapp (1964), relating surface albedo to land-use type, together with a correction for atmospheric effects on the radiation. The cloud albedo values α_c are based on the meridional distribution of α_c from Ohring and

Adler (1978). In a similar manner, the net sensitivity parameter

$$\delta = \frac{\partial(\text{Net})}{\partial A_c} = \frac{\partial(\text{Net})}{\partial \alpha}\frac{\partial \alpha}{\partial A_c} = (\alpha_c - \alpha_s)\frac{\partial(\text{Net})}{\partial \alpha} \quad (3.25)$$

can be evaluated.

Figures 17 and 18 show the distributions of $\partial F/\partial A_c$ and $\partial(\text{Net})/\partial A_c$, respectively, from 60° N to 60° S obtained by Ohring et al. (1981). Figure 17 shows that the largest (in absolute value) cloud greenhouse effects are generally at low latitudes, associated with the typically high cloud heights in the equatorial region. There are significant deviations from a purely zonal pattern. Of particular interest are the regions of reduced cloud greenhouse effect located in the eastern portions of the tropical and sub-tropical oceans; these are associated with the extensive regions of low cloudiness typical of these regions. The Northern Hemisphere average (0° to 60° N) of $\partial F/\partial A_c$ is -35 W m^{-2}. Figure 18 shows that δ, the net sensitivity, is negative over the entire map, indicating that the cloud albedo effect dominates. The largest absolute values of δ, values of -100 W m^{-2} or greater, are generally found over those low-latitude oceanic regions where the cloud greenhouse effect is small, the available solar radiation is large, and the surface albedo is low.

Thus, the studies of Ohring and Clapp (1980), Hartmann and Short (1980), and Ohring et al. (1981) with the NOAA scanning radiometer observations all indicate that the cloud albedo effect is greater than the cloud greenhouse effect, and, hence, that clouds are potentially an important radiative feedback mechanism in the earth's climatic system.

These studies contradict the finding by Cess (1976) that the greenhouse and albedo effects of the clouds compensate one another. In an attempt to understand these differences, Cess et al. (1982) evaluated the relative magnitudes of the cloud albedo and greenhouse effects using three satellite data sets: Ellis and Vonder Haar (1976), Campbell and Vonder Haar (1980b), and Gruber and Winston (1978). Cess et al. (1982) define a quantity

$$\epsilon = \frac{\partial F/\partial A_c}{\partial Q/\partial A_c} = \frac{\partial F}{\partial Q} \quad (3.26)$$

where the partial derivatives imply that changes in F and Q are due solely to variations in cloud amount. Thus, $\epsilon = 1$ corresponds to cloud greenhouse–albedo compensation ($\delta = 0$), whereas for $\epsilon < 1$, cloud albedo dominates, and for $\epsilon > 1$ the cloud greenhouse effect dominates. Using long-term monthly means, they examined the seasonal variations of

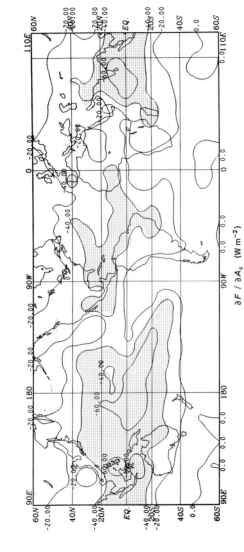

FIG. 17. Quasi-global distribution of the greenhouse effect of the clouds, $\partial F/\partial A_c$, with annual and semiannual cycles and trend removed. Values less than -40 W m^{-2} are stippled. EQ, equator. (After Ohring *et al.*, 1981.)

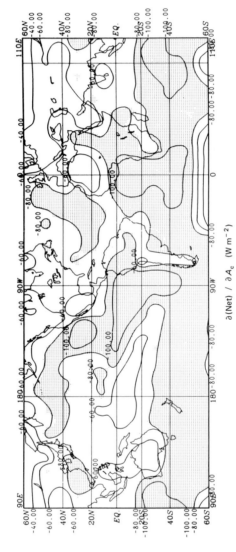

Fig. 18. Quasi-global distributions of sensitivity of net radiation at top of earth–atmosphere system to cloud amount, $\partial(\text{Net})/\partial A_c$, with annual and semiannual cycles and trend removed. Values less than $-80\ \text{W m}^{-2}$ are stippled. (After Ohring *et al.*, 1981.)

F and Q at low latitudes, where there is a large seasonal variation of cloudiness. By removing the effect of seasonal variations of surface temperature on F (by assuming a reasonable value of $\partial F/\partial T_s$) and the effect of seasonal variations of solar zenith angle on Q [by adopting the zenith angle dependence of albedo described by Lian and Cess (1977)], they are left with variations in F and Q that are presumably related to variations in cloud amount. They show that $\epsilon \simeq 1$ for both the Ellis and Vonder Haar (1976) and Campbell and Vonder Haar (1980b) data sets, and that $\epsilon \simeq 0.5$ for the Gruber and Winston (1978) data set. It should be pointed out, however, that the Ellis and Vonder Haar (1976) and Campbell and Vonder Haar (1980b) data sets are not independent: the latter includes the former, and it would have been better to examine these two data sets separately. As a possible cause of these differences, Cess *et al.* (1982; see also Ramanathan and Briegleb, 1980) suggest that the Gruber and Winston (1978) data are biased toward low values of ϵ because they are based upon narrow spectral interval observations. They suggest that (a) the effect of cirrus clouds on the longwave radiation is underestimated because cirrus is more transparent in the infrared window region than in the remainder of the longwave radiation spectrum, and (b) the contrast between cloudy-sky and clear-sky albedos is enhanced in the visible window, thus enhancing the effect of clouds on absorbed solar radiation. But with regard to (a), the calculations of Platt and Stephens (1980) indicate that the effective flux emittance of cirrus in the IR window is approximately equal to the total IR flux emittance.

To overcome the limitations of the NOAA scanning radiometer observations, G. Ohring and H. Ganot (personal communication, 1982) have begun a study to determine ϵ from the flux and albedo observations of the broad-band, narrow-field of view (NFOV) radiometers that are part of the Nimbus 7 ERB experiment (Jacobowitz *et al.*, 1978). Aside from the use of broad-band radiometers, the Nimbus 7 satellite differs from the NOAA satellites in that it is in a noon orbit. Values of the flux and albedo are obtained for 2070 target areas covering the globe, each approximately 500×500 km. Using preliminary daily data for the last half of November 1978, Ohring and Ganot calculated ϵ from a linear regression of Q on F, for each of the 2070 target areas. The resulting values of ϵ were then averaged zonally, and these zonal means are shown in Fig. 19. At almost all latitudes $\epsilon < 1$, and the global average value (area weighted) is approximately 0.5. Thus, these preliminary data from the Nimbus ERB experiment support the idea that the cloud albedo effect dominates.

The lower half of Table V summarizes some of the determinations of $\partial F/\partial A_c$ and $\epsilon = \partial F/\partial Q$ from satellite observations, as discussed above.

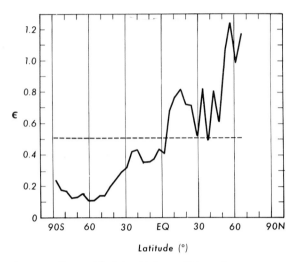

FIG. 19. Zonal means of $\epsilon = \partial F/\partial Q$ for 11 days of November 1978 as determined from Nimbus 7 ERB observations (preliminary) (G. Ohring and H. Ganot, personal communication, 1982). EQ, equator. Dashed line indicates global average of 0.51.

3.4. Absorbed Solar Radiation and Surface Temperature

The second term in the denominator of the right-hand side of Eq. (3.5) when multiplied by -1 represents the sensitivity of the absorbed solar radiation (shortwave sensitivity) to changes in surface temperature. It can be seen that this term depends upon the sensitivity of the planetary albedo to surface temperature, $d\alpha/dT_s$. We have already discussed (Section 3.3) how variations in cloud amount affect the shortwave sensitivity. We shall now discuss other effects.

One of the major feedback mechanisms introduced by the first energy balance climate models was the ice-albedo feedback. As surface temperature decreases, the snow/ice cover increases, leading to an increase in albedo, a decrease in absorbed solar radiation, and a further decrease in temperature. Thus, this mechanism amplifies any temperature perturbation caused by a change in the external or boundary conditions controlling the climate. Budyko (1969) introduced the mechanism in a very simple way. If the temperature is less than a certain critical value, the albedo of the earth–atmosphere system is set equal to that of a snow-covered area: $\alpha = 0.62$. If the temperature is greater than the critical value, the albedo is set equal to that of a snow-free area: $\alpha = 0.32$. The critical temperature is taken as the mean annual temperature at $72°$ N, the latitude to which the earth's north polar ice cap extends. Sellers (1969) assumed that for temperatures greater than 10°C the albedo is not temperature dependent but is equal to the presently observed albedo as a function of latitude. For temperatures less than 10°C, the albedo is a function of latitude and temperature with $\partial\alpha/\partial T_s = -C_t$, where C_t is assumed to be 0.009 °C^{-1}.

Cess (1976) and Lian and Cess (1977) pointed out the importance of the zenith angle dependence of albedo. The albedos of various surface types (e.g., ocean, land, clouds) increase with increasing solar zenith angle. This effect is mitigated to some extent by increased atmospheric absorption, but the net result is an increased planetary albedo with increasing solar zenith angle. Since both temperature and solar zenith angle decrease poleward, and since they both influence albedo in the same direction, it is important to separate their effects in any parameterization of the albedo as a function of surface temperature. Using satellite observations, Cess and his co-workers have separated these effects.

We may differentiate Eq. (3.18) with respect to surface temperature, holding cloud amount constant

$$\frac{\partial\alpha}{\partial T_s} = A_c \frac{\partial\alpha_c}{\partial T_s} + (1 - A_c) \frac{\partial\alpha_s}{\partial T_s} \qquad (3.27)$$

Assuming that the clear-sky albedo is a function of only surface temperature and solar zenith angle, we can obtain

$$\frac{\partial\alpha_s}{\partial T_s} = \frac{d\alpha_s}{dT_s} - \frac{\partial\alpha_s}{\partial\mu} \frac{d\mu}{dT_s} \qquad (3.28)$$

where μ is the cosine of the zenith angle. From the minimum albedos derived by Vonder Haar and Ellis (1975) for zonal, annual means and from corresponding surface temperatures, Lian and Cess (1977) determined $d\alpha_s/dT_s$. From the mean annual variation of solar zenith angle with latitude, they were able to determine $d\mu/dT_s$. To evaluate $\partial\alpha_s/\partial\mu$, they first estimated $\partial\alpha_g/\partial\mu$, where α_g is the surface albedo, from available summaries, and then used Lacis and Hansen's (1974) parameterization of atmospheric absorption and scattering to calculate α_s from α_g, and thus obtain estimates of $\partial\alpha_s/\partial\mu$. With the use of Eq. (3.28), $\partial\alpha_s/\partial T_s$ could now be computed. To determine $\partial\alpha_c/\partial T_s$, they assumed that the cloud albedo is a function only of solar zenith angle and clear-sky albedo. Using Eq. (3.19), they solved for α_c as a function of latitude by inserting climatological values for zonal, annual cloudiness (London, 1957), clear-sky albedo (Vonder Haar and Ellis, 1975), and planetary albedo (Ellis and Vonder Haar, 1976). Applying linear regression analysis, they obtained α_c as a function of α_s and μ, from which they estimated $\partial\alpha_c/\partial T_s$ by differentiating with respect to surface temperature. They were now able to estimate

$\partial\alpha/\partial T_s$ from Eq. (3.27). Inspection of Eq. (3.28) shows that with temperature and μ decreasing with latitude, the ice-albedo feedback mechanism, as represented by $\delta\alpha/\delta T_s$, is weaker than the combination of ice-albedo and zenith angle-albedo effects, as represented by $d\alpha/dT_s$. The early energy balance models (Budyko, 1969; Sellers, 1969) essentially used $d\alpha/dT_s$ and hence overestimated β, the earth's climatic response to a change in solar constant. Using more realistic values of the ice-albedo feedback parameter that are based upon satellite albedo observations, $\beta/100$ is reduced by a factor of 2—from 4° to 2°C (Lian and Cess, 1977; Coakley, 1979; Warren and Schneider, 1979).

Warren and Schneider (1979) have examined the effect of different values of $\partial\alpha/\partial T_s$ on their energy balance model. Figure 20 (after Warren and Schneider, 1979) shows the percent reduction in solar constant required to produce an ice-covered earth, as a function of $f = -\partial\alpha/\partial T_s$. The values of f used in the early energy balance climate models were ~ 0.009, whereas those corrected for zenith angle dependence are ~ 0.005. Inspection of Fig. 20 shows that for $f = 0.009$, a 2% reduction in solar constant is required, whereas for $f = 0.005$, a decrease of more than 6% is necessary.

More direct information on the solar zenith angle dependence of albedo comes from the Nimbus-7 ERB experiment (Jacobowitz et al., 1978). Figures 21 and 22 show albedos for different surface types (land, ocean, snow, and ice) and cloud types (low, middle, high water, and high ice) as a function of the cosine of the solar zenith angle. The method of calculating the albedos from the NFOV observations is described by Stowe et al. (1980b). The graphs are based upon data for a 59-day period (L. L. Stowe,

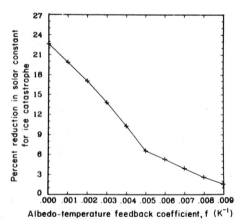

FIG. 20. Percent reduction in solar constant required to produce an ice-covered earth, as a function of $f = -\partial\alpha/\partial T_s$, the ice-albedo–temperature feedback coefficient. (After Warren and Schneider, 1979.)

personal communication, 1982). Scene identification for the albedo observations is based upon U.S. Air Force data for land/water separation and the presence of snow or ice (Stowe *et al.*, 1980a) and Nimbus 7 temperature humidity infrared radiometer (THIR) observations for cloud discrimination (Chen *et al.*, 1980). Data such as that shown in Figs. 21 and 22 should be very useful for climate modelers.

Aside from furnishing information on albedos, satellite radiation observations furnish information on snow and ice cover. Such information has been used by Ohring and Adler (1978) and Chou *et al.* (1981) to parameterize the fraction of a latitude belt covered by snow/ice as a function of surface temperature, in a zonally averaged statistical dynamical model. And Robock (1980) has used these observations to calculate the seasonal variation of the earth's zonally averaged surface albedo.

The zenith angle-albedo effect, although not important for solar constant variations, does come into play in evaluating Milankovitch type climatic changes in which the orbital elements of the earth vary (Cess and Wronka, 1979).

Aside from the ice-albedo feedback mechanism, another kind of surface type-albedo feedback is possible: biosphere-albedo feedback. The possibility of such a feedback mechanism has been discussed by Charney (1975) in connection with the process of desertification. The idea is that a decrease in surface vegetation (e.g., as a result of overgrazing) would lead to an increased albedo, leading to downward vertical motion in the atmo-

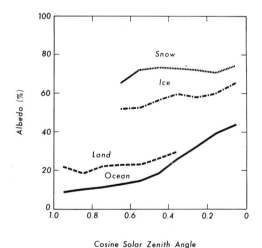

Fig. 21. Albedo as a function of the cosine of the solar zenith angle for different surface types [L. L. Stowe (personal communication, 1982), from 59 days of preliminary data from the Nimbus 7 ERB experiment]: snow ($\cdots$), ice ($-\cdot-\cdot-$), land ($----$), and ocean ($———$).

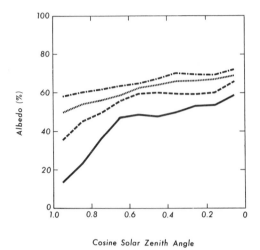

FIG. 22. Albedo as a function of the cosine of the solar zenith angle for different cloud types [L. L. Stowe (personal communication, 1982), from 59 days of preliminary data from the Nimbus 7 ERB experiment]: high (ice) (-·-·-), high (water) (·····), middle (----), and low (——).

sphere, leading to reduced rainfall, which would amplify the original decrease in surface vegetative cover. Satellite albedo observations have been (Norton *et al.*, 1979) and are currently (Kandel *et al.*, 1981) being used to monitor secular changes in clear-sky albedo of regions such as the Sahel that may be subject to desertification processes.

4. Validation of Climate Models

4.1. Introduction

The first test of a climate model is its ability to reproduce the present climatic conditions. In such a test the simulated climatic variables—temperature, pressure, wind, precipitation—are compared to the observed fields. But, it may be possible to obtain good simulations as a result of compensating errors in the treatment of various physical and dynamical processes. It is desirable, therefore, to also validate the simulations of the important physical and dynamical processes included in the model. One of the critical processes is radiative transfer, and satellite observations offer a means for global verification of model radiation simulations. Although not particularly appropriate for energy balance model validation (since the radiation parameterizations of these simple models are largely based on satellite radiation observations), such an application of the satellite observations is particularly suitable for verifying climate models of the general circulation or statistical-dynamical type. Also, to the extent that errors in the simulation of the climatic elements may be due to incorrect prescriptions of inputs for calculation of radiative processes (e.g., cloudiness, snow/ice cover), the differences between the observed and simulated radiation budget should be useful in diagnosing the causes of such errors. In this section, we show how satellite observations are being used to assist in the validation of a particular model of the statistical-dynamical type.

4.2. Description of the Model

The model is a seasonal hemispheric zonal average model (SHZAM). It is an extension and further development of the mean annual model of Ohring and Adler (1978) to permit studies of the annual climatic cycle; some sensitivity results have been reported by Ohring and Adler (1980). The primary output of the model consists of the meridional and seasonal variations of oceanic and continental surface temperatures, 500-mbar temperature, and zonal wind at 250 and 750 mbar.

The model is based upon the zonally averaged form of the two-level quasi-geostrophic potential vorticity system of equations, including diabatic heating and frictional dissipation, and surface energy budget equations. Solar radiative processes included are absorption by water vapor, ozone, and cloud particles; scattering by air molecules and cloud particles; and reflection by the surface of the earth. Longwave radiative processes include absorption and emission by water vapor, carbon dioxide, and clouds. Heat transfers by evaporation at the surface, convection, condensation in the atmosphere, and ocean currents are parameterized. All diabatic heating processes are as in the work by Ohring and Adler (1978) except for a new evaporation parameterization (Saltzman, 1980).

The main modification of the Ohring and Adler (1978) model required for simulating the annual cycle is the treatment of the surface energy budget. Because of different heat storage characteristics, oceanic and continental temperatures go through different annual cycles, and it is important, in a model that attempts to simulate the annual variation, to be able to simulate these differences. We solve this problem by resolving the model surface at each latitude belt into oceanic and continental portions and applying separate surface energy budget equations to each portion. Thus, separate surface temperatures are computed for the continents and oceans of each latitude belt. The method used is based upon the force–

restore method of computing surface temperatures (see, for example, Taylor, 1976). The relevant equation is of the form

$$\frac{pk}{2\pi} \frac{\partial T_s}{\partial t} = -k(T_s - \bar{T}_s) + \sum_{i=1}^{n} H_s(i) \tag{4.1}$$

where p is the period of the annual cycle, k is the heat storage coefficient, which depends on the nature—continent or ocean—of the subsurface layer participating in the annual cycle, T_s is the surface temperature, $\bar{T}_s$ is the temperature at the depth below the surface where the annual variation is negligible, and $\sum_{i=1}^{n} H_s(i)$ is the sum of the heat fluxes at the earth's surface. In the integration with respect to time that is used to solve the equations, the temperatures at the depth of no seasonal variation are assumed to be equal to the previous year's computed mean annual surface temperature (once a steady state is reached, these temperatures approach constant values). Because its heat storage characteristics are closer to those of land than ocean, sea ice is incorporated into the continental portion of the latitude belt. The 500-mbar temperature is assumed to be the same over oceans and land, and the atmospheric heating term is a weighted average of the contributions of the continental and oceanic portions at each latitude belt.

Starting from isothermal conditions and a state of rest, the model is run for 20 yr, using 3-day time steps. This is a sufficient period for obtaining an annual cycle that reproduces itself from one year to the next.

The prescribed inputs to the model include the following: (1) the mean annual meridional distributions of surface relative humidity, water vapor profile shape, eddy mixing coefficients, and height and thickness of a single effective cloud layer (Ohring and Adler, 1978); (2) mean monthly meridional distributions of cloud amount (Berlyand et al., 1980) and snow and sea ice amounts (Robock, 1980); (3) a constant cloud optical depth ($\tau_c = 6.1$), which is obtained by tuning the model to reproduce the observed mean annual hemispheric surface temperature of 288 K; and (4) two subsurface heat storage coefficients—one for land (3.5 W m^{-2} K^{-1}) and one for oceans (25 W m^{-2} K^{-1})—that are obtained by tuning the model's annual temperature range.

4.3. Validation

For purposes of illustrating the use of satellite observations in validation, we shall focus on latitude–month cross sections of the differences between computed and observed surface temperatures, 500-mbar temperatures, albedo, and longwave flux. The observed temperatures are from Oort and Rasmussen (1971) and extend only to 75° N; the observed radiation values are the NOAA SR data set of Gruber and Winston (1978).

Since the observed temperatures and radiation values have uncertainties associated with them, we shall adopt the following strategy in comparing the simulations with the observations. We shall focus on the large differences and we shall call these model errors. The question arises as to the meaning of "large" differences. In the present context, large differences are those that are greater than the uncertainties in the observed values. For temperatures, we would estimate the uncertainties to be 1°–2°C for monthly means for 10° latitude belts, with the larger values pertaining to higher latitudes. For radiation budget quantities, we may estimate the uncertainties from the differences of two observed radiation budget data sets as shown and discussed in Section 2. We would estimate the uncertainties to be 5–10 W m^{-2} for the outgoing flux and 2–3% for the albedo (albedo measured in units of percent), with even larger values for the polar regions. We shall also focus on the patterns of the difference fields, which should also be useful in analyzing model errors.

If we analyze the surface temperature differences (Fig. 23) using the above strategy, we find that there is good agreement between simulated and observed temperatures over most of the time–space domain covered.

Model minus Observed Surface Temperature (°C)

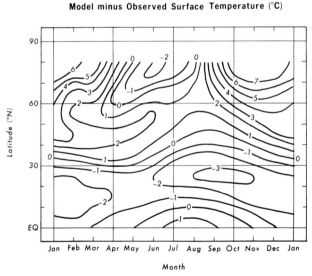

FIG. 23. Latitude–month cross section of model minus observed (Oort and Rasmussen, 1971) surface temperatures (°C). EQ, equator.

There appears to be a belt of modest negative errors (2°–3°C) at subtropical latitudes and a region of large positive errors (3°–7°C) in polar regions during the fall and winter. In addition, the pattern of differences suggests that the amplitude of the model's annual cycle is too large at high latitudes. The 500-mbar temperature difference field (Fig. 24) suggests that the amplitude of the annual variation is overestimated at tropical and polar latitudes and that there appears to be a positive bias at equatorial latitudes. Below 45° N the pattern of 500-mbar temperature differences is similar to that of the surface temperature differences. Above 45° N, they appear to be inversely related for the most part.

We next turn to the radiation budget components. The albedo differences are shown in Fig. 25 and the longwave flux differences in Fig. 26. In analyzing the radiation budget components, one should keep in mind an important difference between them. Since cloudiness, snow/ice cover, and surface albedos are prescribed in the model, feedback effects on model albedo are limited to the small effects of water vapor changes. This means that errors in model albedo can be largely attributed to inaccurate prescriptions or parameterizations. On the other hand, feedback effects on model longwave flux, although not including cloud feedback effects, *do* include the important effects of temperature and water vapor. Thus,

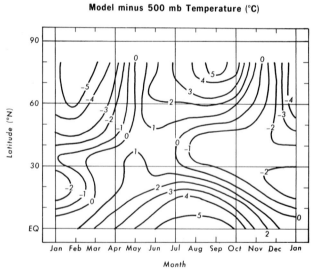

Model minus 500 mb Temperature (°C)

Fig. 24. Latitude–month cross section of model minus observed (Oort and Rasmussen, 1971) 500-mbar temperatures (°C). EQ, equator.

Model minus Observed Albedo (%)

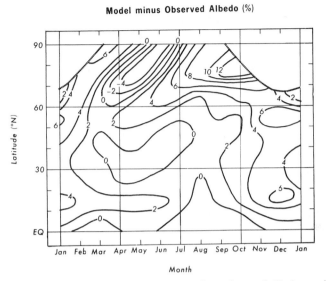

FIG. 25. Latitude–month cross section of model minus observed (Gruber and Winston, 1978) albedos (%). EQ, equator.

Model minus Observed Longwave Flux

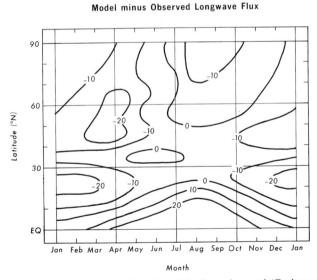

FIG. 26. Latitude–month cross section of model minus observed (Gruber and Winston, 1978) longwave radiation fluxes (W m⁻²). 10 W m⁻² has been subtracted from the observed fluxes to correct for an apparent positive bias (see text). EQ, equator.

errors in model longwave fluxes are not so easily interpreted since they can be caused by errors in the simulated climate as well as by inaccurate prescriptions (cloudiness) and parameterizations.

In the albedo difference cross section there is a suggestion of a belt of modest positive albedo errors (2 or 4%) at subtropical latitudes. These albedo errors appear to be negatively correlated with the surface temperature errors in this same latitude belt. This is the kind of relationship one would expect between albedo and surface temperature anomalies and suggests that in this latitude belt the temperature errors may be due, at least in part, to albedo errors resulting from incorrect prescriptions. In polar regions, there is an annual variation of albedo difference with large positive values in the fall ($>12\%$) to large negative values in the late spring ($<-4\%$). However, it is not at all certain that these are due to model errors. Examination of the cross sections in Section 2, showing the differences between the NOAA SR albedos and albedos derived from broad spectral band observations, reveals large differences at high latitudes with a pattern somewhat similar to the albedo differences of Fig. 25. Thus, we are reluctant to draw any conclusions with respect to model albedo errors at high latitudes.

The longwave flux difference cross section is shown in Fig. 26. As indicated in Section 2, there appears to be a positive bias of a little more than 10 W m^{-2} in this observed longwave flux data set. For purposes of comparison with the model output, we have subtracted 10 W m^{-2} from the observed values. (The original difference field can be obtained from the corrected difference field in Fig. 26 by subtracting 10 from each of the isopleth labels.) Model longwave fluxes appear to be too low in the subtropical dry zone. From the discussion of the albedos, it will be recalled that this is also a region of excessive model albedos that leads to negative surface temperature errors. The lower surface temperatures can explain some but not all of the reduced longwave flux values. The discussion on longwave sensitivity in Section 3 indicates that surface temperature errors of only ~ 2 K would lead to reductions of only ~ 5 W m^{-2} in the longwave flux. The negative longwave flux errors in this subtropical belt are much larger than $|5|$ W m^{-2}. A possible explanation for these relatively large negative longwave flux differences that is also consistent with the positive albedo differences in this belt is overestimated cloudiness. The prescribed cloudiness is after Berlyand et al. (1980) and although their cloud amounts are ~ 0.10 greater than those of London (1957) at these latitude belts, they are also greater by a similar amount at all latitude belts. Thus, it is difficult to verify this possible explanation on the basis of currently available cloud climatologies.

At tropical latitudes there are positive longwave flux errors with a maximum in late summer (>20 W m^{-2}). The model albedos appear to be correct at these latitudes (see Fig. 25), as do the model surface temperatures (Fig. 23). Some, but not all, of this positive longwave flux error is due to positive temperature errors at 500 mbar at these latitudes, which also peak during the summer. Thus, although the model surface temperatures are approximately correct, the entire model atmosphere is much too warm, resulting in enhanced emission to space. A possible cause of the excessive 500-mbar temperatures is lack of sufficient meridional transport in tropical latitudes, possibly as a result of the quasi-geostrophic nature of the model.

At high latitudes, the model longwave fluxes appear to be approximately correct. On the basis of such a comparison alone one might conclude that the model temperatures are reasonably correct at these latitudes. However, inspection of the surface temperature differences (Fig. 23) shows large errors at these latitudes. One might conclude at this point that there is something wrong with the longwave radiation computation scheme since the wrong surface temperatures yield correct fluxes. However, inspection of the 500-mbar temperature errors (Fig. 24) shows that, for the most part, they vary inversely as the surface temperature errors. Because of this compensation, the effective radiating temperature of the earth–atmosphere system at these latitudes is probably correct, resulting in good longwave flux values. The problem remains of determining the cause of the temperature errors at the surface and at 500 mbar. The inverse relationship between the two suggests that the errors may be due to difficulties with the surface–atmosphere sensible and latent heat fluxes at these latitudes.

4.4. Summary

It was not intended to review all of the results from SHZAM in this section but rather to illustrate the application of satellite radiation observations to the diagnosis and validation of climate models. Because of present uncertainties in the satellite measurements of the earth's radiation budget (see Section 2), such an application can be frustrating at times, but improved observations should reduce these uncertainties. Climate models of the general circulation type can be validated in a similar fashion, i.e., using zonal averages, but what may be just as important for such models is a diagnostic validation against regional averages. A very useful diagnostic tool might be validation of the annual cycles of longwave radiation and albedo for different geographic/climatic regions against satellite observations of the type shown in the following section.

5. Climatology from Satellites

5.1. Introduction

The annual cycle of climatic variables at the surface of the earth in different climatic regions has always been one of the key elements of descriptive climatology. Because of their effects on human life and agriculture, temperature and precipitation are the main variables that have been studied. However, the annual cycle of these variables is controlled by radiative and dynamic factors, i.e., by the annual cycles of the energy budget and the general circulations of the atmosphere and the oceans. Budyko (1974) has developed a climatology of the energy budget at the surface of the earth. Included in this climatology are examples of the annual march of the energy budget components at locations representative of the different climatic regimes prevailing on the earth. The satellite observations of the earth's radiation budget permit similar information to be obtained from the upper boundary of the earth–atmosphere system. In this section we present the annual cycle of the energy budget at the top of the atmosphere for some of the same areas discussed by Budyko. The data are from the NOAA scanning radiometer observations (Gruber and Winston, 1978).

5.2. Annual Cycle of Energy Budget at Top of Atmosphere for Selected Climatic Regions

The satellite radiation observations are from $2\frac{1}{2}° \times 2\frac{1}{2}°$ areas whose centers are as close as possible to the climatic locations chosen by Budyko. The nomenclature on climatic type is also after Budyko. To assist in the interpretation of the observations, information on the annual cycle of cloudiness at these locations has been obtained from Volumes 7, 10, and 12 of the "World Survey of Climatology" (Landsberg, 1972–1977). In cases where no cloudiness data were available for a particular station, an attempt was made to find a nearby station (within 100–200 km) with cloud data or to obtain the annual cycle of precipitation (as a surrogate measure of the annual variation of cloudiness) from "Climates of the World" (U.S. Department of Commerce, 1969).

5.2.1. *Equatorial Continental Climate (São Gabriel, Brazil)—Fig. 27.* The net radiation curve shows two maxima—in September and in February—in qualitative agreement with the solar insolation cycle for an equatorial location. Theoretically, the second maximum should be in March when the sun is over the equator, but the albedo curve shows a

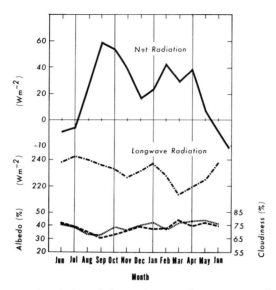

FIG. 27. The annual variation of the earth–atmosphere system radiation budget and cloudiness for an equatorial continental location (equator, 67.5° W–São Gabriel, Brazil): net radiation (——), longwave radiation (-·-·-), albedo (----), and cloudiness (·····).

maximum in March, causing a depression in the net radiation. The albedo maximum is probably due to the peak in cloudiness at this time. This interpretation is consistent with the longwave radiation curve, which shows a minimum in March. Because of the small annual variation of temperature at this equatorial location, longwave radiation variation can safely be related to cloudiness variations. That the maximum net radiation occurs in February, and not in March when both insolation and cloudiness are larger than in February, implies that the albedo effect of the clouds is greater than their greenhouse effect.

The annual variation of cloudiness suggests that the ITCZ is furthest north in September (minimum cloudiness and albedo at São Gabriel) and furthest south in March (maximum cloudiness and albedo at São Gabriel). This is supported by both cloudiness and precipitation data. As one would expect for an equatorial location, the annual variation of net radiation is small. Of interest is the difference of about 25 W m^{-2} in net radiation between June and December—two months with the same solar declination angle. The albedos are similar for the two months. A small part ($\sim$7 W m^{-2}) of the net radiation difference can be explained by the longwave radiation difference between the two months. The largest part of this difference, however, is due to the variation in earth–sun distance between June and December.

5.2.2. Equatorial Monsoon Climate (Saigon, Vietnam)—Fig. 28. The effects of the monsoon circulation are most evident in the annual marches of albedo and longwave radiation. The late winter/early spring minimum in albedo and maximum in longwave flux are associated with the dry tropical air, with little cloudiness, that covers this area at this time. The radiation data suggest that the monsoon reaches its climax in September (maximum albedo, minimum longwave flux); the precipitation data confirm this observation.

5.2.3. Tropical Continental Climate (Aswan, Egypt)—Fig. 29. With little cloudiness or precipitation, this desert area's radiation budget is controlled by the annual march of insolation. Both the net radiation and the longwave emission are at a maximum in summer and at a minimum in winter. In contrast with the equatorial regions discussed above, the longwave emission curve reflects the annual variation of temperature rather than cloudiness. The net radiation curve shows a deficit for most of the year. This is a result of the combination of clear skies, which permit escape of longwave surface radiation to space, and relatively high albedo (30–35%) despite the absence of cloudiness. For some other desert areas (e.g., in Libya and in Saudi Arabia) the net radiation is negative and less than -20 W m^{-2} all year round.

5.2.4. Subtropical Continental (Aidin, USSR)—Fig. 30. The higher the latitude the greater the annual variation of insolation; as a result, the amplitude of the annual cycle of net radiation also increases with latitude as can be seen by comparing this figure with the preceding ones. The large annual cycle in cloudiness with a maximum in winter is reflected in the albedo curve (maximum in winter) and in the longwave radiation curve (minimum in winter), although the longwave flux is also influenced by the annual temperature cycle and the albedo may be influenced by winter snow.

5.2.5. Midlatitude Continental (Barnaul, USSR)—Fig. 31. The winter peak in albedo (60%) is probably associated with surface snow cover rather than cloudiness, which is at a minimum during this season. Typical of midlatitude continental areas, there is a very large annual range of longwave emission—in this case over 70 W m^{-2}—reflecting the large annual temperature range.

5.2.6. Equatorial Monsoon (15° N, 70° E, Indian Ocean)—Fig. 32. Since the annual range of sea surface and atmospheric temperatures is small for this low-latitude oceanic area, the large annual variations of

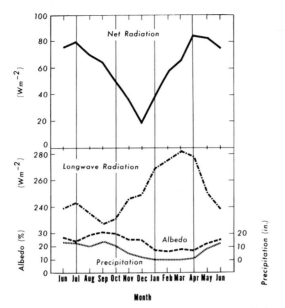

FIG. 28. The annual variation of the earth–atmosphere system radiation budget and precipitation for an equatorial monsoon location (10° N, 107.5° E—Saigon, Vietnam): net radiation (——), longwave radiation (-·-·-), albedo (----), and precipitation (·····).

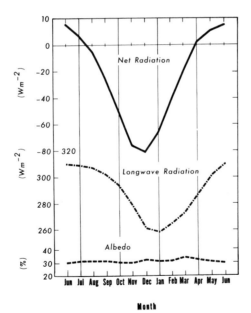

FIG. 29. The annual variation of the earth–atmosphere system radiation budget for a tropical continental location (25° N, 32.5° E—Aswan, Egypt): net radiation (——), longwave radiation (-·-·-), and albedo (----).

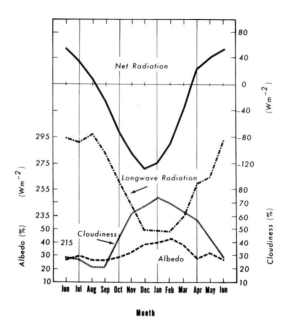

FIG. 30. The annual variation of the earth–atmosphere system radiation budget and cloudiness for a subtropical continental location (40° N, 55° E—Aidin, USSR): net radiation (——), longwave radiation (-·-·-), albedo (----), and cloudiness (·····).

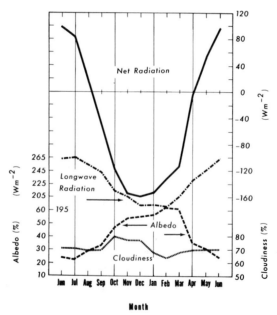

FIG. 31. The annual variation of the earth–atmosphere radiation budget and cloudiness for a midlatitude continental location (52.5° N, 82.5° E—Barnaul, USSR): net radiation (——), longwave radiation (-·-·-), albedo (----), and cloudiness (·····).

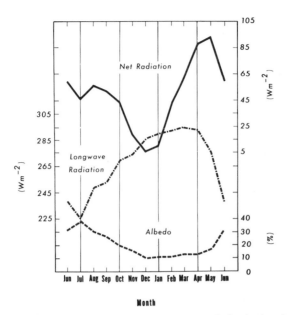

Fig. 32. The annual variation of the earth–atmosphere radiation budget for an equatorial monsoon location (15° N, 70° E—Indian Ocean): net radiation (——), longwave radiation (-·-·-), and albedo (----).

longwave emission and albedo can be attributed to cloudiness variations. The albedo maximum and longwave minimum in July are associated with the peak cloudiness of the Indian monsoon at this time. The inverse relationship between longwave flux and albedo, resulting from the cloud greenhouse effect and the cloud albedo effect, are clearly defined in these annual variations. The dominance of the cloud albedo effect over the cloud greenhouse effect can be inferred from the May maximum of net radiation. May and June have about the same insolation totals at this latitude. Both the longwave radiation and the albedo values indicate that May cloudiness is less than June cloudiness. Therefore, the fact that May net radiation is greater than the June value implies that the cloud albedo effect dominates.

Acknowledgments

We wish to acknowledge the many individuals whose help was necessary to complete the manuscript: Marilyn Varnadore and Shoshana Adler for computer programming support, Bob Ryan for drafting the figures, Gene Dunlap for his photographic work, and Carol

Maunder and Regina Woodward for typing various portions of the manuscript. We also wish to acknowledge the fruitful discussions with our colleagues, Mark Chen and Larry Stowe. The work at Tel-Aviv University was supported by NOAA Grants NA81AA-H-00010 and NA80AA-H-00064 and by the Israel Academy of Sciences and the U.S.–Israel Binational Science Foundation, Jerusalem, Israel.

REFERENCES

Abel, P., and Gruber, A. (1979). *NOAA Tech. Memo., NESS* **106**, 1–24.

Berlyand, T. G., Strokina, L. A., and Greshnikova, L. Ye. (1980). *Meteorol. Gidrol.* No. 3, pp. 15–23.

Budyko, M. I. (1969). *Tellus* **6**, 611–619.

Budyko, M. I. (1974). "Climate and Life." Academic Press, New York.

Budyko, M. I. (1975). *Meteorol. Hydrol.* **10**, 1–10.

Campbell, G. G., and Vonder Haar, T. H. (1980a). "An Analysis of Two Years of Nimbus 6 Earth Radiation Budget Observations: July 1975 to June 1977," Atmos. Sci. Pap. No. 320. Dept. Atmos. Sci., Colorado State University, Fort Collins.

Campbell, G. G., and Vonder Haar, T. H. (1980b). "Climatology of Radiation Budget Measurements from Satellites," Atmos. Sci. Pap. No. 323. Dept. Atmos. Sci., Colorado State University, Fort Collins.

Cess, R. D. (1974). *J. Quant. Spectrosc. Radiat. Transfer* **14**, 861–872.

Cess, R. D. (1976). *J. Atmos. Sci.* **33**, 1831–1843.

Cess, R. D., and Ramanathan, V. (1978). *J. Atmos. Sci.* **35**, 919–922.

Cess, R. D., and Wronka, J. C. (1979). *Tellus* **31**, 185–192.

Cess, R. D., Briegleb, B. P., and Lian, M. S. (1982). *J. Atmos. Sci.* **39**, 53–59.

Charney, J. (1975). *Q. J. Roy. Meteorol. Soc.* **101**, 193–202.

Chen, T. S., Stowe, L. L., Taylor, V. R., and Clap, P. F. (1980). *Ext. Abstr., Int. Radiat. Symp., 1980* pp. 315–317.

Chou, S.-H., Curran, R. J., and Ohring, G. (1981). *J. Atmos. Sci.* **38**, 931–938.

Coakley, J. A., Jr. (1977). *J. Atmos. Sci.* **34**, 465–470.

Coakley, J. A., Jr. (1979). *J. Atmos. Sci.* **36**, 260–269.

Coakley, J. A., Jr., and Wielicki, B. A. (1979). *J. Atmos. Sci.* **36**, 2031–2039.

Conlan, E. F. (1973). *NOAA Tech. Memo., NESS* **52**, 1–57.

Crutcher, H. L., and Meserve, J. M. (1970). "Selected-Level Heights, Temperatures and Dew Point Temperatures for the Northern Hemisphere," U. S. Navy, NAVAIR 50-IC-52. Washington, D.C. (available from Chief, Naval Operations).

Ellis, J. S. (1978). Ph.D. Thesis, Colorado State University, Fort Collins.

Ellis, J. S., and Vonder Haar, T. H. (1976). "Zonal Average Earth Radiation Budget Measurements from Satellites for Climate Studies," Atmos. Sci. Pap. No. 240. Dept. Atmos. Sci., Colorado State University, Fort Collins.

Gruber, A. (1972). *J. Atmos. Sci.* **29**, 193–197.

Gruber, A. (1977). *NOAA Tech. Rep., NESS* **76**, 1–28.

Gruber, A. (1981). *Repr., Conf. Atmos. Radiat., 4th, 1981* pp. 237–239.

Gruber, A., and Winston, J. S. (1978). *Bull. Am. Meteorol. Soc.* **59**, 1570–1573.

Hartmann, D. L., and Short, D. A. (1980). *J. Atmos. Sci.* **38**, 1233–1250.

Heddinghaus, T. R., and Krueger, A. F. (1981). *Mon. Weather Rev.* **109**, 1208–1218.

Hickey, J. R., Stowe, L. L., Jacobowitz, H., Pelligrino, P., Maschoff, R. H., House, F., and Vonder Haar, T. H. (1980). *Science* **208**, 281–283.

Hoyt, D. V. (1976). *NOAA Tech. Rep.* **ERL 362-ARL4**, 1–124.

Jacobowitz, H. (1981). *Repr., Conf. Atmos. Radiat., 4th, 1981* pp. 146–148.
Jacobowitz, H., Stowe, L. L., and Hickey, J. R. (1978). "In the Nimbus 7 User's Guide," pp. 33–69. NASA Goddard Space Flight Center, Greenbelt, Md.
Jacobowitz, H., Smith, W. L., Howell, H. B., Nagle, F. W., and Hickey, J. R. (1979). *J. Atmos. Sci.* **36,** 501–507.
Jensenius, J. A., Cahir, J. J., and Panofsky, H. A. (1978). *Q. J. R. Meteorol. Soc.* **104,** 119–130.
Kandel, R. S., Courel, M. F., Cisse, E. H., and Rasool, S. I. (1981). *Pap., IAMAP Gen. Assembly, 1981.*
Lacis, A. A., and Hansen, J. E. (1974). *J. Atmos. Sci.* **31,** 118–133.
Landsberg, H. E. (1972–1977). "World Survey of Climatology," Vols. 7, 10, and 12. Elsevier, Amsterdam.
Lian, M. S., and Cess, R. D. (1977). *J. Atmos. Sci.* **34,** 1058–1062.
London, J. (1957). "A Study of the Atmospheric Heat Balance," Final Report, Contract No. AF 19(122)-165. New York University, New York.
Manabe, S., and Wetherald, R. T. (1967). *J. Atmos. Sci.* **24,** 241–259.
Manabe, S., and Wetherald, R. T. (1975). *J. Atmos. Sci.* **32,** 3–15.
Norton, C. C., Mosher, F. R., and Hinton, B. (1979). *J. Appl. Meteorol.* **18,** 1252–1262.
Oerlemans, J., and Van den Dool, H. M. (1978). *J. Atmos. Sci.* **35,** 371–381.
Ohring, G., and Adler, S. (1978). *J. Atmos. Sci.* **35,** 186–205.
Ohring, G., and Adler, S. (1980). *Ext. Abstr., Int. Radiat. Symp., 1980* pp. 277–279.
Ohring, G., and Clapp, P. (1980). *J. Atmos. Sci.* **37,** 447–454.
Ohring, G., and Mariano, J. (1964). *J. Atmos. Sci.* **21,** 448–450.
Ohring, G., Clapp, P. F., Heddinghaus, T. R., and Krueger, A. F. (1981). *J. Atmos. Sci.* **38,** 2539–2541.
Oort, A. H., and Rasmussen, E. M. (1971). *NOAA Prof. Pap.* **5,** 323 pp.
Oort, A. H., and Vonder Haar, T. H. (1976). *J. Phys. Oceanogr.* **6,** 781–800.
Platt, C. M. R., and Stephens, G. L. (1980). *J. Atmos. Sci.* **37,** 2314–2322.
Posey, J. W., and Clapp, P. F. (1964). *Geofis. Int.* **4,** 33–48.
Ramanathan, V. (1977). *J. Atmos. Sci.* **34,** 1885–1897.
Ramanathan, V., and Briegleb, B. P. (1980). *Ext. Abstr., Int. Radiat. Symp., 1980* pp. 355–357.
Ramanathan, V., Callis, L. B., and Boughner, R. E. (1976). *J. Atmos. Sci.* **33,** 1092–1112.
Raschke, E., Vonder Haar, T. H., Pasternak, M., and Bandeen, W. R. (1973a). *NASA Tech. Note* **D-7249,** 1–73.
Raschke, E., Vonder Haar, T. H., Bandeen, W. R., and Pasternak, M. (1973b). *J. Atmos. Sci.* **30,** 341–364.
Riehl, H. (1954). "Tropical Meteorology." McGraw-Hill, New York.
Robock, A. (1980). *Mon. Weather Rev.* **108,** 267–285.
Ruff, I., Koffler, R., Fritz, S., Winston, J. S., and Rao, P. K. (1968). *J. Atmos. Sci.* **25,** 323–332.
Saltzman, B. (1980). *Arch. Meteorol., Geophys. Bioklimatol., Ser. A* **29,** 41–53.
Schneider, S. H. (1972). *J. Atmos. Sci.* **29,** 1413–1422.
Schneider, S. H., and Mass, C. (1975). *Science* **190,** 741–746.
Sellers, W. O. (1969). *J. Appl. Meteorol.* **8,** 392–400.
Stephens, G. L., Campbell, G. G., and Vonder Haar, T. H. (1981). *JGR, J. Geophys. Res.* **86,** 9739–9760.
Stowe, L. L., and Taylor, V. R. (1981). *Prepr., Conf. Atmos. Radiat., 4th, 1981* pp. 124–127.
Stowe, L. L., Jacobowitz, H., Ruff, I., Chen, M., Baldwin, E., Van Cleef, F., Hill, M., and Coleman, D. (1980a). *Ext. Abstr., Int. Radiat. Symp., 1980* pp. 413–415.

Stowe, L. L., Jacobowitz, H., and Taylor, V. R. (1980b). *Ext. Abstr., Int. Radiat. Symp., 1980* pp. 430–432.

Taljaard, J. J., Van Loon, H., Crutcher, H. L., and Jenne, R. L. (1969). "Climate of the Upper Air: Southern Hemisphere. Vol. I. Temperatures, Dew Points, and Heights at Selected Pressure Levels," U. S. Navy NAVAIR 50-IC-55. Washington, D.C. (available from Chief, Naval Operations).

Taylor, K. (1976). *J. Appl. Meteorol.* **15**, 1129–1138.

U.S. Department of Commerce (1969). "Climates of the World." U.S. Govt. Printing Office, Washington, D.C.

Vonder Haar, T. H., and Ellis, J. E. (1975). *Abstr., Conf. Atmos. Radiat., 2nd, 1975* pp. 107–110.

Vonder Haar, T. H., and Suomi, V. E. (1971). *J. Atmos. Sci.* **28**, 305–314.

Wang, W. C., and Domoto, G. A. (1974). *J. Appl. Meteorol.* **13**, 521–534.

Warren, S. G., and Schneider, S. H. (1979). *J. Atmos. Sci.* **36**, 1377–1391.

Wetherald, R. T., and Manabe, S. (1975). *J. Atmos. Sci.* **32**, 2044–2059.

Winston, J. S. (1971). *Mon. Weather Rev.* **99**, 818–827.

Winston, J. S., and Krueger, A. F. (1977). *Pure Appl. Geophys.* **115**, 1131–1144.

Winston, J. S., Gruber, A., Gray, T. I., Jr., Varnadore, M. S., Earnest, C. L., and Manello, L. P. (1979). "Earth–Atmosphere Radiation Budget Analyses Derived from NOAA Satellite Data, June 1974–Feb. 1978," Vols. 1 and 2. U.S. Dept. of Commerce, NOAA, NESS, Washington, D.C.

LAND SURFACE PROCESSES AND CLIMATE—SURFACE ALBEDOS AND ENERGY BALANCE

ROBERT E. DICKINSON

National Center for Atmospheric Research[1]
Boulder, Colorado

1. INTRODUCTION

Land covers about 150 million km² (30%) of the earth's surface. Soil forms its underlying surface, but over much of the land lies a vegetative cover. At high latitudes and altitudes glaciers and permanent snow fields (covering about 16 million km² of land) are found. During January, snow covers about 40 million km² of land in the Northern Hemisphere.

The land's vegetative cover includes a wide range of ecosystems and human uses as summarized in Table I. Approximately ¼ of the land is desert, ¼ forests, ¼ grasslands, ⅛ farmlands and cities, and the remainder largely tundra but also swamps, marshes, lakes, and rivers.

Popular concepts of climate largely involve the interaction of seasonally varying atmospheric temperature, precipitation, and surface radiation

[1] The National Center for Atmospheric Research is sponsored by the National Science Foundation.

ADVANCES IN GEOPHYSICS, VOLUME 25

TABLE I. LAND USE AS PERCENTAGE OF TOTAL LAND AREA

Land use	%
Arable–mixed farming and human areas	10–13
Grazing land	20–25
Extratropical forests (mostly conifer)	10–15
Tropical forests and woodlands	13–18
Deserts	25–30
Tundra/high altitude	6–9
Swamp and marshes, lakes and streams	2–3

with these various land types. Many scientific disciplines also study various aspects of the interaction between climate and land surface processes. Hydrologists concern themselves with the time-varying flows of streams and rivers, the consequent availability of water resources for irrigation, power generation, industrial and urban use, and the possible threats of flooding due to excess supply. Glaciologists study the response of small mountain glaciers and continental ice sheets to snow supply in accumulation zones and melting in ablation zones.

Agronomers, foresters, ecologists, and range managers consider the radiative fluxes within plant canopies as the energy source both for photosynthesis and for maintaining the temperature and transpiration of the canopy. Plant temperature controls the rates of various biochemical processes. If temperature is too low or too high, crop production suffers. Transpiration occurs as a by-product of photosynthesis. As plants open stomata on their leaves to catch carbon dioxide, they lose water vapor that has evaporated within the leaf. If roots are unable to restore this lost water, the stomata close to the point where a water balance is maintained, and in so doing reduce their intake of carbon dioxide. Hence, the supply of water within the rooting zone of plants is also of considerable concern.

Geologists study the role of temperature, wind, and rainfall in the weathering of rocks, formation of soils, and movement of sediments to streams, and physical geographers study the general question of the influence of climate on the land's surface.

I come from yet another discipline, with a strong interest in climatic processes—dynamic meteorology. My educational training and professional associations have largely been concerned with developing better scientific understanding and predictive capabilities for the motions of the atmosphere. Perhaps the major practical achievement of dynamic meteorology has been development over the last 35 years of real-time data acquisition systems and computer models of the atmosphere sufficiently realistic to provide numerical weather forecasts, now useful up to a week in advance.

The same upper-air soundings that provided initial conditions for the forecast models have also been used to obtain climatologies of atmospheric motion. My own entry into meteorology was under Victor Starr at MIT, who pioneered the use of atmospheric statistics for studying the conservation of momentum and conversions of energy within the global atmosphere. Numerical forecast models were also applied to simulation of these and other statistics of the global atmospheric circulation. Numerical models constructed to carry out such studies were referred to as general circulation models (GCMs). J. Smagorinsky pioneered much of this work and developed the Geophysical Fluid Dynamics Laboratory (now under NOAA in Princeton) which continues in these activities.

The large-scale motions of the atmosphere and their role in transport and energy conversions have been the primary climate variables of concern to dynamic meteorologists. In the past, everything else occurring in the atmosphere, e.g., radiation, clouds, small-scale turbulence, and rainfall, were lumped together as "physics" and considerable intellectual effort was devoted to showing these terms were less important than the dynamics of motions (for those questions of interest to dynamic meteorologists). Nevertheless, dynamic meteorologists had to pay some attention to heating of the atmosphere by other than adiabatic redistribution of energy. In particular, it has long been necessary to recognize the latitudinal variation of the net radiative heating–cooling of the atmosphere–earth system in order to account for the atmosphere's well-known function of moving excess heat from equatorial to polar latitudes.

Just as dynamic meteorologists assumed all but motions were "given," so the disciplines concerned with other parts of the climate system assumed the atmosphere was given. Oceanographers drove their models with measured winds and surface temperatures, glaciologists assumed measured ablation and accumulation rates for their models, and those disciplines concerned with micrometeorology measured net radiative fluxes, winds, and atmospheric stability as inputs for models of evapotranspiration and soil moisture.

There has been a renaissance in climate studies over the last decade. Scientists in the different disciplines concerned with the climate system have grown increasingly appreciative of the connections between the various components of the climate system and the hazards of overly narrow viewpoints. Of course, some interdisciplinary leaps have led to blunders, but on the whole, these efforts have been stimulating.

Up to now, the disciplines concerned with land microclimatology have been less integrated into the mainstream of current efforts to understand the climate system as an integrated whole than have some other disciplines, such as oceanography. They have been largely lumped together by meteorologists into the category of "climate impacts" and recognized more as consequences of the climate system than as an integral part. The lack of appreciation of the importance of processes over land for the climate system as a whole is due in part to inadequate knowledge on the part of many atmospheric scientists as to the physical functioning of land processes and in part to the narrow conception of the climate system held by many scientists concerned primarily with the functioning of land processes.

It is the intent of the present article to help bridge the gap between disciplinary studies of atmospheric versus surface climate. It is primarily directed toward describing physical processes involving the land surface that should be of concern to scientists interested in modeling the global climate system. Considerable emphasis is given to the physics of land surface albedos; modeling surface energy balance processes is considered somewhat more briefly.

2. Surface Albedos

This section is intended to provide a conceptual basis for understanding land surface albedos. The processes that determine surface albedos are complicated and it is tempting in developing climate models to rely on summaries of observational data. However, existing data sets can be quite misleading in conveying the impression that albedos are fixed constants. Rather, albedos are generally dependent on both the wavelength and incidence angle of incoming solar radiation. Because of this dependence, it is necessary to consider the modification of the incoming solar beam by the atmosphere as done in the following subsection. Surface albedos, furthermore, are dependent on surface textures and structure as well as composition; the primary intent of this section is to use physical reasoning and simple mathematical models to summarize how these factors determine surface albedos.

2.1. Incoming Solar Flux

The solar flux $F_{st}^{\downarrow}$ reaching the top of the atmosphere normal to a surface is given by

$$F_{st}^{\downarrow} = 1370(1 + 0.034 \cos \gamma_\epsilon) \quad \text{W m}^{-2} \tag{2.1}$$

to three-digit accuracy, where $\gamma_\epsilon = 2\pi(\text{day of year} -3)/365$ days. Globally averaged, about half of the flux reaching the top of the atmosphere is absorbed at the surface, the rest being absorbed or reflected by atmospheric constituents or reflected by the surface. Both surface and atmospheric radiative properties are strongly wavelength-dependent over the spectrum of solar energy. Practically all solar radiation of wavelength shorter than 310 nm is absorbed by ozone and other absorbers in the stratosphere and above. About 2% of the incident beam is removed from the ultraviolet solar flux by these absorbers, somewhat independently of latitude and solar zenith angle (as a result of the saturation of the absorption).

Due to the smooth variation of atmospheric and surface spectral properties at wavelengths reaching the lower atmosphere and surface, considerable spectral averaging is possible for calculation of solar energy fluxes. The minimum practical division is into "visible" (including relatively long-wavelength ultraviolet) and into near-infrared. A convenient division between these two categories is at wavelengths of about 700 nm. This division is chosen for two reasons. First, approximately half the solar energy incident at the top of the atmosphere is at shorter and half at longer wavelengths than this value. Second, the chlorophyll absorption bands occur at 700 nm and shorter wavelengths, so plant canopies reflect a much larger fraction of the incident radiation in the near-infrared part of the solar spectrum than they do in the visible. Other surfaces also have greatly different reflectances (albedos) for near-infrared than for the visible solar radiation, but the spectral variation of the reflectances for these surfaces is not as discontinuous as it is for green leaves. In particular, soils typically reflect about twice as much near-infrared as visible radiation, and conversely, snow surfaces reflect about half as much near-infrared as visible radiation. Typical spectral reflectances of a plant canopy and soil are shown in Fig. 1.

Solar fluxes at the surface are obtained from the fluxes at the top of the atmosphere by subtracting the energy absorbed or reflected within the

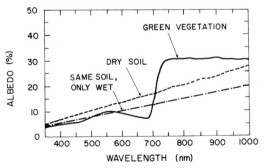

FIG. 1. Spectral reflectances for dry soil, wet soil, and a plot of blue grama grass. (From Tucker and Miller, 1977.)

atmosphere, i.e., the downward flux at the bottom of the atmosphere is, approximately,

$$F_{sb}^{\downarrow} = \cos \theta \, F_{st}^{\downarrow}(1 - A_{H_2O} - A_{O_3} - R_{dust} - R_{air} - R_{clouds}) + F_{sr}^{\downarrow} \quad (2.2)$$

where A_{H_2O} and A_{O_3} are, respectively, the fractional absorption due to water (vapor or liquid) and ozone; R_{dust}, R_{air}, and R_{clouds} are the reflectivities (and any assumed absorption) due to dust, air (Rayleigh scattering), and clouds, respectively; $F_{sr}^{\downarrow}$ is the solar radiation backscattered from the sky after one or more reflections from the surface. It can be as large as the first term in Eq. (2.2) for a bright snow surface under thick clouds, but for clear sky and a nonsnow surface would be typically about 1% of the first term. The term $\cos \theta$ gives the projection of the solar beam in the direction of the local vertical, where θ is the solar zenith angle (angle between a solar beam and the local vertical), and is conventionally obtained from

$$\cos \theta = \cos \phi \cos h \cos \delta + \sin \phi \sin \delta \quad (2.3)$$

where ϕ is the latitude, h is the hour angle (from noon), and δ is the solar declination (tilt of earth's axis from a right angle to the direction to the sun).

Solar radiation is "scattered" into other directions from the direction of the direct solar beam if it interacts with matter in the atmosphere. The reflected radiation consists of that radiation which has been scattered in an upward direction. The solar radiation incident at the surface consists of that fraction of the solar beam that has not been absorbed or scattered plus the radiation that has been scattered in the downward direction. The latter is referred to as "diffuse" radiation. Numerical procedures are available to describe the directional details of scattered radiation to considerable accuracy. However, the limiting factor in calculating radiative transfer through the atmosphere in climate studies is generally inadequate information as to the distribution and optical properties of atmospheric dust and clouds. The diffuse radiation is often assumed to be isotropic and for many purposes it can be approximated by a single beam with a 60° solar zenith angle. The fraction of incident radiation that is diffuse varies from about 0.1 for an overhead sun and cloudless sky to more than 0.9 below thick clouds.

In summary, for global climate models, radiative fluxes at the surface can be subdivided into the following four categories: direct visible beam, diffuse visible beam, direct near-infrared beam, and diffuse near-infrared beam. The depletion terms in Eq. (2.2) are different for each of these beams and depend on the solar zenith angle. In particular, the Rayleigh scattering term, because it is proportional to the fourth power of inverse wave number, affects significantly only the visible and ultraviolet fluxes.

The water vapor absorption term applies almost entirely to the near-infrared fluxes, whereas the ozone absorption term, which includes the Chappius bands at 600 nm as well as absorption of ultraviolet radiation, removes on the average about 3% of the incident solar radiation. Absorption by other species, in particular, O_2, CO_2, and NO_2, is in total less than 2% of the incident beam but should be included to ensure adequate accuracy.

Absorption of solar radiation by water vapor involves a large number of individual lines which are broadened by pressure; vertical absorption paths are usually defined by the pressure-weighted amount of water in a column, converted to units of liquid water, and are typically about 1–2 cm, but vary from 0.1 cm in the Antarctic and winter Arctic to as large as 5 cm for extremely warm moist tropical conditions. The corresponding absorption averaged over a day is typically about 10% but ranges from about 6% to about 16% of the solar beam. For a nearly overhead sun and summer conditions, this absorption would be in the range 10–13%. The additional liquid and vapor water in clouds can increase absorption by up to another several percent.

Rayleigh scattering over clear sky reflects about 6% of the incident beam upward, on the average, but globally averaged, about $\frac{1}{4}$ of this upscatter is masked by clouds. Aerosols in a clear sky scatter typically 10% of the incoming beam, but only about 1% is scattered upward and so lost.

In summary, between 15 and 50% of a clear-sky solar beam is lost due to absorption or upscatter; this loss varies primarily with water vapor concentrations and solar zenith angle, and would be about 20–25% for a nearly overhead sun and a clear summer day. Cloudy skies, on the average, reflect about 50% of the incident solar radiation, but this reflection can range from a few percent for high thin cirrus to more than 90% of the otherwise unabsorbed radiation for thick low clouds.

2.2. Important Factors When Considering Surface Albedos

About three-fourths of the incident solar beam reaches the surface under a clear sky and nearly overhead sun and typically half that much under cloudy conditions. Under clear sky conditions, nearly all the solar radiation reflected from the surface returns to space. Under cloudy conditions typically half the reflected radiation may return through the clouds to space, with the remainder returned to the surface as the $F_{sr}^{\downarrow}$ term in Eq. (2.2).

The *surface albedo* is the ratio of reflected to incident solar radiation. It is evident from the above discussion that modifications in surface albedo significantly change the fraction of solar radiation that is reflected from the earth–atmosphere system, that is, the *planetary albedo*. The temperature of the earth–atmosphere system is quite sensitive to small changes in its energy balance, such as are possible from large-scale changes in surface albedos. What are the processes that enter into determining surface albedos over land?

It might seem that the albedo of a surface should simply depend on the chemical composition of the surface. Unfortunately, albedo depends even more on the physical structure of a surface. For example, the albedo of water in a pond will be less than about 0.04 (overhead sun), but the albedo of water in a low thick fog layer can be nearly 1.0. The primary reason for this difference is that light incident on a pond has essentially only one opportunity to be reflected at the air–water interface; nearly all the light entering the pond stays there. For fog or cloud, on the other hand, almost all the incident solar beam that is not up-scattered the first time it enters a droplet is down-scattered out the bottom of the droplet and has many further opportunities to be up-scattered. Only a very small fraction of the light intercepted by a droplet is actually absorbed. The thicker the cloud, the larger is the fraction of the incident beam that eventually leaks out the top of the cloud.

Since vegetation covers at least half of the total land surface, the determination of surface albedo in the presence of vegetation is an important question. A novice modeling the albedo of leaf canopies might expect that their albedos should correspond to the albedos of single leaves. However, measured canopy albedos are typically not much more than half the albedos measured for individual leaves. Why is there this difference? Leaves, unlike water droplets, absorb approximately half the incident solar radiation, mostly in the visible, due to their chlorophyll. Normally, most of the light transmitted through the leaf is absorbed by lower leaves or by the ground. If all the transmitted light were subsequently absorbed and all the light reflected from the upper leaf surfaces were to escape the canopy, the canopy albedo would correspond to the leaf albedo. However, much of the light reflected from leaves situated below the top surface of the canopy is shadowed by other leaves and so further attenuated; this light trapping reduces the canopy albedos to values considerably lower than that of individual leaf surfaces as illustrated in Fig. 2.

Many other geophysical examples of such light trapping are easily imagined. For example, a deep canyon is generally quite dark compared to surrounding terrain even if locally the canyon surface is of the same albedo as that of the surrounding terrain. The only difference, then, is that much of the light reflected from the canyon surface strikes other parts of the canyon surface and the effective albedo is correspondingly reduced.

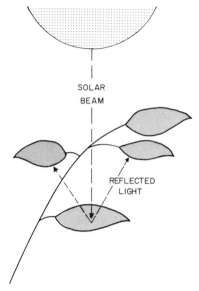

FIG. 2. Sketch of the partial trapping of light reflected from a canopy leaf by overlying leaves.

Just as holes of all sizes at the surface reduce the effective albedo, so do upward-protruding objects. For example, a vertical pole that is a diffuse reflector will send half the light it intercepts downward to the ground, which will in part absorb the light. Hence, the effective albedo of the pole and surface it shades is correspondingly half the reflectivity of the pole multiplied by one plus the albedo of the underlying surface (neglecting further trapping). Leafless hardwood canopies over a snow surface can consequently greatly reduce the effective surface albedo (see, e.g., Federer, 1971). Plant canopies increase their trapping of light not only by arranging their leaves randomly in space but also by inclining their leaves somewhat vertically. Ordering of leaves into clusters can further reduce albedos.

2.3. Mathematical Theory of Plant Canopy Albedos

The optical depth of the canopy is measured by the leaf area index distribution function $L(z)$, which is defined as the area of one side of the leaves (projected onto a flat surface, if the leaves are not flat), per unit area of space, above level z. If the leaves are randomly distributed in space, and the direct solar beam at the top of the canopy is $F_s(\infty)$, the expected value of the vertical flux of the incident solar beam at level z and for a given solar zenith angle whose cosine is μ, is

$$F_s^{\downarrow}(z, \mu) = F_s^{\downarrow}(\infty) \, \exp[-\bar{L}(z, \mu)] \qquad (2.4)$$

where the optical path $\bar{L}(z, \mu)$ is given by

$$\bar{L}(z, \mu) = \frac{1}{\mu} \int_z^{\infty} G(z', \mu) \, dL(z') \qquad (2.5)$$

$\mu = \cos\theta$, and $G(z, \mu)$ when multiplied by $L(z)$ is the projection of the leaf area in the direction of the solar beam (see, e.g., Ross, 1975). For example, $G = \mu$ for horizontal leaves and $G = 0.5$ for leaves with all orientations of equal probability. The assumption is made here and in the following analysis that G is independent of the azimuthal angle, i.e., as would be obtained if all leaf azimuths were of equal probability. The geometry of the leaf scattering is sketched in Fig. 3. Note that any radiation that has interacted with leaves is removed from the direct beam as described by Eq. (2.4). Diffuse sky radiation is likewise attenuated as is the direct beam, except that it is incident at all angles so that integration over the angles of incidence becomes necessary.

When solar radiation strikes a single leaf, a fraction of the radiation is absorbed, and the rest is scattered in all directions and emerges on both

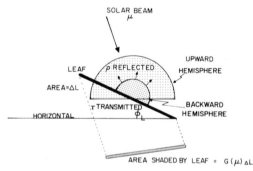

FIG. 3. Sketch of the geometry of leaf scattering. The solar beam is incident at an angle whose cosine with the vertical is μ. The leaf's orientation is defined by the direction of its normal, which is determined by an inclination angle ϕ_L and an azimuth relative to the azimuth of the solar beam. The term $G(\mu)$ is the cosine of the angle between the leaf normal and the direction of solar beam. The term ΔL is the one-sided leaf area. The product of $G(\mu)$ and ΔL gives the area shaded by the leaf. Radiation scattered into the backward hemisphere is referred to as "reflected" and radiation scattered into the forward hemisphere as "transmitted." The net radiation scattered by all leaves into the *upward* hemisphere is required to determine the canopy albedo.

sides of the leaf. The general formalism of scattering from a leaf corresponds to that for scattering from a nonspherical aerosol. In other words, the scattered radiation depends not only on the angle between the incident and scattered beams and on the wavelength of the light but also on the orientation of the leaf relative to the direction of the incident light. It is usually assumed for theoretical calculation that the scattering is isotropic (that is, "diffuse") on the near and far side of the leaf. However, in practice, the radiation reflected from the near side of the leaf has a specular as well as diffuse component that can be quite pronounced in the case of glaucous (i.e., shiny) leaves (Gates, 1980).

The fraction of incident light transmitted through a leaf is denoted τ and the fraction of light reflected from the leaf is denoted ρ; $\omega = \tau + \rho$ is the fraction of light scattered, and $1 - \omega$ is the light absorbed. Typical values are indicated in Table II. Note that visible radiation is largely absorbed and near-infrared radiation is largely reflected or transmitted. The reflection is usually somewhat larger than the transmission, as might be expected for scattering from an object, such as a thick cloud, that consists of a large number of individual scattering centers (i.e., the leaf mesophyll cells and chloroplasts within these cells). Also, the presence of the multiple scattering centers within the leaf implies that its reflectivity and transmissivity should depend on the orientation of the leaf relative to the incident beam. Some observational evidence for this dependence has been discussed by Gates (1980). The dependence of leaf transmissivity on leaf orientation apparently only becomes significant if the angle between leaf normal and solar beam is more than 60°; the fraction of intercepted beam absorbed is decreased at near-glancing incidence.

A precise calculation of canopy albedos is not practical because of the complex geometry of leaf and stem orientation and insufficient information as to the optical properties of the leaves and stems of most plants as functions of scattering angles and leaf and stem orientations.

Fortunately, approximate treatments for canopy albedos can be formu-

TABLE II. TYPICAL SCATTERING PARAMETERS FOR LEAVES[a] FOR THE VISIBLE AND NEAR-INFRARED COMPONENTS OF A SOLAR BEAM

	Visible	Near-infrared
Scattering parameter ω	0.15–0.20	0.80–0.85
Reflection ρ	0.6ω	0.7ω
Transmission τ	0.4ω	0.3ω

[a] Based on Ross (1975) and Goudriaan (1977).

lated that are compatible with likely observational detail. For the following discussion, the canopy is assumed to consist only of leaves, the properties of these leaves are independent of the horizontal coordinates, and their orientations are independent of the vertical coordinates in the canopy. The albedo for visible solar radiation a_{vis}^{c} can be adequately determined by assuming that all scattered light is absorbed upon encountering another leaf.

The albedo of the canopy with leaves which are constant in orientation but randomly distributed in space and of infinite optical depth is obtained in this "single scattering" approximation by

$$a_c^{ss}(\mu) = \omega \int_0^1 \mu' \Gamma(\mu, \mu') f(\mu, \mu')\, d\mu' \qquad (2.6)$$

where

$$f(\mu, \mu') = [\mu G(\mu') + \mu' G(\mu)]^{-1} \qquad (2.7)$$

The parameter $\Gamma(\mu, \mu') = G(\mu)\, G(\mu')\, P(\mu, \mu')$ where $P(\mu, \mu')$ is the scattering phase function giving the relative fraction of scattered flux in direction μ', per projected leaf area in that direction; it is normalized such that $\int_{-1}^{1} P(\mu, \mu')\, G(\mu')\, d\mu' = 1$. The various terms entering this calculation are sketched in Fig. 4. The integration over all upward scattering angles gives the relative upward flux.

All the terms whose product gives $\Gamma(\mu, \mu')$ depend on leaf orientation. For the more realistic case where leaf orientation is given by a distribution

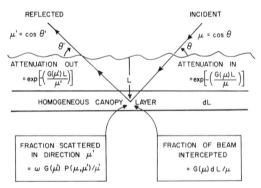

FIG. 4. Sketch of the terms entering the single-scatter approximation; a single layer of unit area scatters $\omega G(\mu)\, G(\mu')\, P(\mu, \mu')\, dL$ of the radiation incident on it from the direction defined by μ into the direction defined by μ'; multiplication by the attenuation of the incident and scattered beam and integration over all μ' and all layers gives the total single-scatter canopy albedo.

function, it is necessary to integrate these factors over the leaf orientation distribution function. Ross (1981) provides a detailed formulation for this situation and derives $\Gamma(\mu, \mu')$, the "generalized phase function," for several examples.

For a leaf whose backscatter and forward scatter are isotropic, the scattering phase function is independent of solar zenith angle and depends only on what fraction of backscattered or forward-scattered flux has been up-scattered versus down-scattered.

Equation (2.6) simply illustrates the profound role of light trapping by a canopy. The term $f(\mu, \mu')$ determines the relative reduction of the scattered beam due to absorption over the optical path in leaving the canopy. Light penetrates holes in the canopy, is reflected from lower leaves, and then is absorbed along its upward path. If the leaves were instead arranged in layers with no holes, the canopy albedo would decrease to that of a single leaf. This limit is also reached in Eq. (2.6) for a glancing sun, i.e., $\mu = 0$, incident on horizontal leaves, i.e., $G(\mu') = \mu'$.

In considering the canopy albedo for near-infrared fluxes, it is necessary to allow for multiple scattering of light intercepted by leaves. For doing this, it is useful to introduce a two-stream type of approximation. The commonly used models of this kind have been summarized by Meador and Weaver (1980). We assume that the diffuse fluxes are isotropic in the upward and downward directions (see, e.g., Coakley and Chylek, 1975). Let $I^\uparrow$ and $I^\downarrow$ be the upward and downward diffuse solar flux divided by the incident solar flux, respectively. Leaves are taken to be flat plates with no difference between the optical properties of their tops and bottoms. Appropriate equations in the two-stream approximation are then

$$-\bar{\mu}\frac{dI^\uparrow}{dL} + [1 - (1 - \beta)\omega]I^\uparrow - \omega\beta I^\downarrow = \omega\bar{\mu}k\beta_0 \exp(-kL) \qquad (2.8)$$

$$\bar{\mu}\frac{dI^\downarrow}{dL} + [1 - (1 - \beta)\omega]I^\downarrow - \omega\beta I^\uparrow = \omega\bar{\mu}k(1 - \beta_0) \exp(-kL) \qquad (2.9)$$

These expressions use k for the optical depth of the direct beam per unit leaf area

$$k = G(\mu)/\mu \qquad (2.10a)$$

whereas $\bar{\mu}$ is the average inverse diffuse optical depth per unit leaf area

$$\bar{\mu} = \int_0^1 [\mu'/G(\mu')] \, d\mu' \qquad (2.10b)$$

and β and β_0 are the up-scatter parameters for the diffuse and incident beams, respectively, to be defined below. An alternative formulation differing in some details is given in Ross and Nilson (1975) and Ross (1981).

A diffuse up-scatter parameter β can be obtained for leaves whose forward scatter and backscatter are isotropic. If the normal to the leaf makes an angle ϕ_L relative to the local vertical, the top surface of the leaf absorbs downward flux and emits upward flux into a hemisphere with a wedge (between the leaf bottom and the horizontal plane through the point of solar incidence) removed, as illustrated in Fig. 3. Within the wedge, upward flux is emitted and downward flux absorbed by the bottom surface. As inferred from the analysis in the appendix of Norman and Jarvis (1975)

$$\omega\beta = 0.5[\rho + \tau + (\rho - \tau)\cos^2\phi_L)] \qquad (2.11)$$

where again ρ is the reflectivity and τ is the transmissivity of a leaf. Only for horizontal leaves or if $\rho = \tau$ is it strictly correct to use $\omega\beta = \rho$. For example, for leaves with all orientations of equal probability, $\omega\beta = \frac{2}{3}\rho + \frac{1}{3}\tau$. For nearly vertical leaves, the leaf diffuse up-scatter parameter is approximately an average of the transmissivity and reflectivity.

Equations (2.8) and (2.9) solved in the $\omega \to 0$ limit (single-scatter approximation and semi-infinite canopy) give for single-scattering canopy albedo,

$$I^\uparrow(L = 0, \omega \to 0) = \frac{\omega\bar{\mu}k\beta_0}{1 + \bar{\mu}k} \qquad (2.12)$$

This expression can be made equal to the "exact" single-scatter limit by defining the direct-beam up-scatter parameter to be given by

$$\beta_0 = \frac{1 + \bar{\mu}k}{\omega\bar{\mu}k}\, a_c^{ss}(\mu) \qquad (2.13)$$

where $a_c^{ss}(\mu)$ is defined by Eq. (2.6). With this substitution, solutions to Eqs. (2.8) and (2.9) should give excellent results for the visible solar albedo and acceptable results for the near-infrared albedo, for homogeneous leaf canopies. Even for completely isotropic scattering $\omega\beta_0$ will be slightly different than ρ if $\mu/G(\mu)$ is not independent of solar zenith.

Note that $\bar{\mu} = 1$ provided either the leaves are horizontal (i.e., $G = \mu'$) or all orientations are of equal probability (i.e., $G = 0.5$, $\int_0^1 \mu'\, d\mu' = 0.5$). However, $\bar{\mu}$ can be significantly larger (e.g., 1.2) for predominantly vertical leaves. Largely vertical leaves increase the average path of upward-diffusing light. The precise definition of $\bar{\mu}$ is somewhat arbitrary and is expressed differently by Ross (1981).

Equations (2.8) and (2.9) can be solved straightforwardly as constant-coefficient differential equations [reflection and transmission coefficients may be obtained from Eqs. (21) and (22) of Coakley and Chylek (1975) or Eqs. (14) and (15) of Meador and Weaver (1980)]. Of special interest is the albedo for a semi-infinite canopy, which may be written

$$a_c(\mu) = \frac{\omega \bar{\mu} k}{(\alpha + \bar{\mu} k)(1 + c)} [1 + (2\beta_0 - 1)c] \qquad (2.14)$$

where

$$\alpha = (1 - \omega)^{1/2}(1 - \omega + 2\beta\omega)^{1/2}$$

$$c = (1 - \omega)/\alpha$$

As $\omega \rightarrow 0$, $\alpha \rightarrow 1$, $c \rightarrow 1$, and Eq. (2.14) reduces to Eq. (2.12). Equation (2.14) is useful for analyzing the albedo of the canopy to solar near-infrared radiation. For rough calculations of diurnal average canopy albedo, we can take $\bar{\mu} = k = 2\beta_0 = 2\beta = 1$, so that Eq. (2.14) simply becomes

$$a_c \simeq \frac{\omega}{[1 + (1 - \omega)^{1/2}]^2} \qquad (2.15)$$

For example, if $\omega = 0.84$, appropriate to the near-infrared beam, $a_c = 0.43$ compared to 0.21 for the single-scattering estimate for an optically thick canopy as obtained from Eq. (2.15) by assuming $\omega = 0$. The simplest expression for zenith angle dependence is obtained by taking $\bar{\mu} = 2\beta_0 = 2\beta = 1$ but allowing k to vary. The consequent expression is discussed in Ross (1981, p. 327).

To see under what conditions the finite thickness of a canopy need be considered, note that a reflected light ray has the average optical path per unit leaf area $k + \bar{\mu}^{-1} \approx 2$. Hence, corrections due to a finite canopy are roughly proportional to $\exp[-2L(0)]$, which are negligible for an integrated leaf area parameter $L(0) > 2$. Full canopies generally have leaf area parameters roughly between 5 and 10.

The preceding analysis has assumed a random distribution of leaf positions; in other words, leaves in any infinitesimal layer ΔL are uncorrelated with any other such layer. However, on some plants leaves tend to organize themselves into spatially correlated layers; in the extreme all the leaves in some finite leaf area increment ΔL may all be located at the same level. This situation is treated most simply for a semi-infinite homogeneous canopy by considering the effect of adding one such layer to the existing canopy. This does not change the canopy albedo but gives a relationship between the canopy albedo a_c and layer reflectivity and transmissivity. If we denote the albedo of this layer to a direct and a diffuse beam by $\rho_{l,0}$ and ρ_l and the transmissivity of this layer to a direct and a diffuse beam by $\tau_{l,0}$ and τ_l, respectively, then summation of all the multiple reflections between the canopy and added layer gives

$$a_c = \rho_{l,0} + \frac{\tau_{l,0} \, a_c \tau_l}{1 - \rho_l a_c} \qquad (2.16)$$

This gives a quadratic expression for a_c for any sort of layer. If all leaves are at the same level in each layer, $\rho_{l,0} = \beta_0 \omega k \, \Delta L$, $\rho_l = \beta \omega \bar{\mu}^{-1} \, \Delta L$, $\tau_{l,0} = 1 - (1 - \omega + \beta_0 \omega) k \, \Delta L$, $\tau_l = 1 - (1 - \omega + \beta \omega) \bar{\mu}^{-1} \, \Delta L$. For example, (a) if $\Delta L \to 0$, an expression equivalent to Eq. (2.14) is obtained. On the other hand, if we take $\Delta L = 1$, $\beta = \beta_0 = 0.5$, $k = \bar{\mu} = 1$,

$$a_c = \frac{\omega}{1 + (1 - \omega^2)^{1/2}} \qquad (2.17)$$

which was first derived by deWitt [cf. Eq. 2.10 of Goudriaan (1977)].

Equation (2.17) for leaves in incremental layers can be compared with Eq. (2.15) for randomly spaced leaves. The albedo of a canopy with random leaf position is seen to be not much more than half as large as that of the layered canopy for small ω; for the layered canopy example, light trapping by shading of reflected radiation is small.

Some plant canopies depart from random leaf positioning by organization of leaves into clusters or clumps. Clustering is characteristic of conifers where the needles are organized into "shoots" and the shoots into whorls (Norman and Jarvis, 1974, 1975). With clustering, the internal shading is larger than that for random leaf positions. The effect of clustering on canopy albedo can be considered by treating each cluster as a single leaf unit with unequal reflection and transmission coefficients (Goudriaan, 1977). Since these units will generally have lower albedos than a single leaf, a clustered canopy will generally have a lower albedo than it would if the clusters were individual leaves.

For the above discussion, it has been assumed that leaves were uniformly distributed in the horizontal dimension. However, many canopies have open spaces between individual plant elements with lower vegetation or ground beneath and most have some individual plants that extend to greater heights than others. Such organization corresponds to clumping of canopy elements and further increases the internal shading. A single isolated cylindrical plant shades not only its own area, but also an additional area $hw \tan \theta$, where w is the diameter of the plant and h its height (Brown and Pandolfo, 1969). If such cylindrical plants are randomly distributed in space, the average fractional shading S due to their vertical extent is given by $S = \exp[-P \tan \theta]$, where P is the average vertical cross-sectional area per unit horizontal area (Federer, 1971). If the canopy has holes, P should be corrected for these, e.g., for random leaves $P = P_0[1 - \exp(-L')]$, where L' is the effective leaf area index through a plant element in the direction of the solar beam, and P_0 is the vertical cross-sectional density for opaque plants. Vertical elements will diffusely reflect and transmit a fraction a_v and τ_v, respectively, of the radiation removed from the shaded area. Only half of the diffuse radiation will be upward, so for small ground albedos, a combination of vertical elements with dense leaves ($\tau_v \rightarrow 0$) and their underlying shaded area will have but half the albedo of a homogeneous canopy. Also note that the area shaded by vertical elements increases greatly with a lower sun. Hence, canopy inhomogeneities tend to reduce albedos more with a larger solar zenith angle. This dependence is the opposite of that for a homogeneous canopy as given by Eq. (2.14).

The above discussion provides a basis for interpreting the wide range of plant canopy albedos that have been measured in nature.

2.4. Summary of Parameters Determining Plant Canopy Albedos

The following parameters together determine the albedo of a plant canopy:

(1) *The spectral composition of the incident solar radiation.* For a clear day, the ratio of visible to near-infrared solar flux typically varies diurnally from 0.9 for a low sun to 1.1 for a nearly overhead sun, and albedo correspondingly decreases by about 0.02–0.03. This variation is due primarily to the greater solar zenith dependence of the Rayleigh scatter and ozone (Chappius band) absorption in the visible compared to that of the water vapor absorption in the near-infrared.

On the other hand, near the summer solstice the solar zenith and, hence, visible fluxes change little with latitude, but less infrared radiation is incident on tropical surfaces due to larger amounts of water vapor in the tropical atmosphere. This relative depletion of near-infrared radiation lowers tropical albedos by 0.01–0.02 compared to midlatitudes.

(2) *The zenith angle of the incident solar beam.* For a homogeneous canopy with equal probability for all leaf positions and orientations, Eq. (2.14) applies with $\bar{\mu} = 1$, $k = 0.5/\cos\theta$, so the albedo is proportional to $(1 + 2\alpha \cos\theta)^{-1}$. For visible radiation, $\alpha \simeq 1$, giving a decrease in albedo from sunrise to noon by a factor of 3, for near-infrared radiation, $\alpha \simeq 0.4$, giving a variation in albedo from sunrise to noon by nearly a factor of 2. Figure 5 (Idso *et al.*, 1978), showing the albedo variation with solar zenith

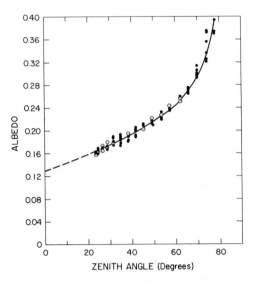

FIG. 5. Variation of wheat field total albedo with solar zenith angle (Idso *et al.*, 1978).

of a wheat field, is typical of such variation for near-uniform crop canopies. It indicates a variation with solar zenith by a factor of 2–3 as suggested by Eq. (2.14). Rougher surfaces have less diurnal variation due to increased shadowing by the vertical faces of roughness elements as the solar zenith angle increases. Hence, conifers have a smaller variation with solar zenith than suggested by Eq. (2.14); pines have an increase of less than a factor of 2 and spruces practically no increase from sunrise to noon (Jarvis *et al.*, 1975). Spruces with their spire-shaped crowns depart more drastically from the assumption of a homogeneous canopy than do pines with their more rounded crowns. A solar beam normal to the spire-shaped spruce crown will penetrate farther into the foliage than a beam from an overhead sun; hence, in the vicinity of the top of the crown the back reflection will be greater at overhead than at low sun. The trapping of solar radiation by a spruce–fir forest at low sun is illustrated in Fig. 6.

The architecture of spruce–fir forests, which nearly maximizes the possible trapping of solar radiation from a sun near the horizon, largely explains why the average albedo of such forests is so low. Measured overhead sun albedos of spruce–fir forests are not much lower than the noontime albedos of other plant canopies (i.e., 0.1–0.15). Tropical forests with their multiple strata and isolated giant trees also have relatively weak variation of albedo with solar zenith angle. Pinker *et al.* (1980) discussed the diurnal variation of a tropical dry evergreen forest in Thailand. Their results as reproduced in Fig. 7 show about a 30% decrease of albedo from

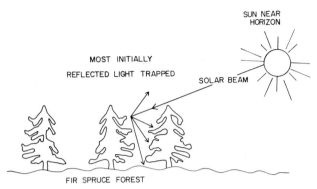

FIG. 6. Sketch of the efficient trapping of a nearly horizontal solar beam scattered by a tree in a spruce–fir forest. As a consequence, albedos of this canopy may be somewhat lower for low sun than high sun in contrast to the solar zenith dependence of the albedo of a smooth homogeneous canopy. Roughening of other surfaces (e.g., plowing a field) can likewise lower albedos and weaken or reverse the increase of albedo with lower sun forward for homogeneous surfaces.

sunrise (sunset) to noon, for both the tropical forest and an adjacent clearing.

(3) *The optical properties of individual leaves.* Gates (1980) has listed mean reflectances and absorptances for a large number of deciduous plant leaves for both low and high sun. Most of his leaf reflectances with overhead sun lie between 0.2 and 0.3, but a few are as small as 0.15 or as large as 0.4. Low sun reflectances are typically about 0.05 larger, consistent with the expected diurnal variation of the ratio of visible to near-infrared radiation already discussed. Norman and Jarvis (1975) found a value of about 0.2 for spruce needle reflectances. Table II, showing the spectral

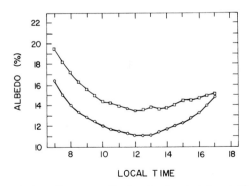

FIG. 7. Diurnal variation of average albedo of a tropical forest (○) and clearing (□) (Pinker *et al.*, 1980).

variation of leaf reflectances, has already been discussed. Albedos of homogeneous canopies at low sun are close to the albedos of individual leaf elements. The visible albedos of a homogeneous canopy are reduced relatively more at high sun than are the near-infrared albedos, since upward light leakage after multiple scattering is significant for the latter. Albedos of woody branches may be somewhat lower than those of green leaves, and dry nongreen leaves have higher albedos in the visible than do green ones due to their lack of chlorophyll but they also have lower albedos in the near-infrared (see, e.g., Gausman *et al.*, 1976).

(4) *The structure of the canopy.* As already discussed, inhomogeneity in the horizontal distribution of canopy elements lowers albedo, especially for a low sun. Clumping of leaves or needles within an individual plant does likewise. According to Goudriaan (1977), differences in clustering are

> probably one of the main reasons why heather, gorse, and different types of woodland reflect less (about 0.16) than pastures and farm crops such as grains (about 0.23). It must be noted that broad-leafed species with a more regular leaf arrangement like sugar beet, cucumber, and bracken score even higher (about 0.26).

It would be expected that taller vegetation would usually expose rougher surfaces to the incident solar radiation and so would have lower albedos. Such a correlation between vegetation height and albedo has been reported by Stanhill (1970) and Oguntoyinbo (1970) and is reproduced in Fig. 8.

Vegetative cover over drier terrain is also more in patches and clumps and hence should have lower albedos due to this structure compared to vegetation of the same average height in moist areas.

(5) *The orientation of plant leaves.* As indicated by Eq. (2.14), a homogeneous canopy with random leaf distribution and horizontal leaves has no solar zenith dependence. The greatest solar zenith dependence is expected if the leaf orientations are largely vertical, i.e., $k = G(\mu)/\mu$ in Eq. (2.14) decreases more rapidly than $1/\mu$ as $\mu \to 1$ at high sun, since the leaf projection factor $G(\mu)$ also decreases as μ increases. Many plants arrange their upper leaves with nearly vertical inclinations to help capture the incident sunlight. Largely vertical needles at the end of conifer twigs help to reduce the albedo of a conifer forest at high sun.

2.5. Snow Albedos

Snow surfaces generally have much higher albedos than do other surface materials. Furthermore, since the extent of snow cover would vary with climate change, it provides an important and extensively discussed

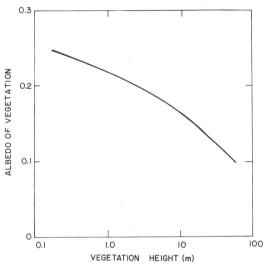

F1G. 8. Variation in the albedo of plant canopies with height of the vegetation (based on Oguntoyinbo, 1970).

feedback on surface temperatures in high latitudes, which couples to atmospheric temperatures by vertical and latitudinal transport. It is hence desirable that the descriptions of snow albedo in climate models be improved.

Semiempirical models of snow albedo have been developed in the past to interpret existing data sets. The first entirely theoretical calculation of snow albedos has been presented by Wiscombe and Warren (1980). For the purpose of the radiative calculations, they assume that a snow field is equivalent to a homogeneous cloud layer of ice spheres of uniform radius. They consider radii ranging from 50 to 1000 μm, corresponding to extremely fresh snow and old recrystallized snow, respectively. Besides the dependence on grain radii, they consider dependence on snow depth and solar zenith angle. Particles of the size range considered strongly forward-scatter the incident light.

To see why this is so, note that the reflection from a particle large enough for geometric optics to apply is given by Fresnel's formula, which for normal incidence depends only on the index of refraction n and is simply $(n - 1)^2/(n + 1)^2 \simeq 0.02$ (for ice and water).

Most of the light entering a large sphere is not of normal incidence, so the average reflectivity is increased somewhat and further doubled by the additional reflection as the light leaves the sphere. The net reflection for large water or ice spheres is 11% or less of the light intercepted by the sphere, depending on how much internal absorption occurs. The remainder of the intercepted light ray, plus an amount equal to the intercepted light which is diffracted by the edge of the sphere, is bent somewhat in direction. This is the forward-scattered component.

Light rays that have interacted with a large number of such spheres may be reversed in direction with a consequent large increase in the reflectivity of the assemblage of spheres compared to that of a single interface. This effect can be contrasted with the effect of canopies of leaves that have albedos lower than those of individual leaves. Due to their relatively large absorptances, leaves more effectively trap light than turn it around.

Wiscombe and Warren (1980) use Mie scattering theory to determine the angular distribution of scattered light from a single snow grain and the "delta-Eddington" approximation for obtaining the snow albedo. The delta-Eddington approximation was developed particularly to solve analytically radiative transfer problems with a strongly forward angular distribution of scattered light. Their solution for snow albedo depends on the snow depth in liquid water equivalent, the radius of the snow particles, the albedo of whatever surface underlies the snow, the solar zenith angle, and the wavelength of the incident solar radiation. The optical cross section of the snow particles, hence optical depth for given snow depth, is approximately twice the geometrical cross section. The wavelength of the solar radiation determines the precise optical cross section and the parameters ω and g, where ω as before is the fraction of light that is scattered rather than absorbed and g is the average of the cosine of the angle that the scattered light makes relative to the incident beam. The delta-Eddington approximation obtains solutions in terms of the parameters

$$\omega^* = (1 - g^2)\omega/(1 - g^2\omega) \qquad (2.18)$$

$$g^* = g/(1 + g) \qquad (2.19)$$

The absorption of weakly absorbing large spheres is proportional to their radii but their reflection is nearly independent of radii. Hence, the absorption per encounter, i.e., $1 - \omega$, is proportional to radius, and large snow grains are more highly absorbing than small. Optical depth per unit mass is inversely proportional to grain radius. Hence, as the individual grains of a snow surface grow in size, absorption per unit mass remains about the same but back-reflection decreases, light penetrates more deeply, and more light reaches the underlying surface. Light also penetrates more deeply in the visible than in the more strongly absorbing near-infrared. These dependencies are illustrated in Fig. 9 taken from Wiscombe and Warren (1980). It shows theoretical calculations of albedo for

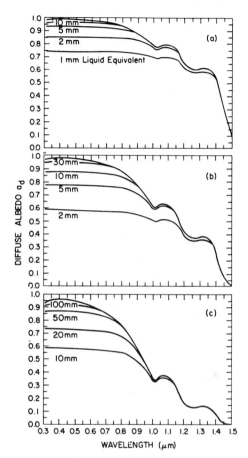

FIG. 9. Values of diffuse albedo versus wavelength for a variety of snow depths expressed in liquid water equivalent and three values of grain radius: (a) 50, (b) 200, (c) 1000 μm (Wiscombe and Warren, 1980).

snow surfaces of three different grain radii and for several values of snow liquid water equivalent. For example, the surface albedo for grains of snow of 50-μm radius and with 1 cm liquid water appears nearly the same as that for snow of infinite depth, but if the same amount of snow occurs in 1000-μm grains, nearly half of the incident visible light penetrates it, and the albedo is correspondingly reduced.

In the limit of deep snow, the delta-Eddington albedo $a_s(\mu)$ becomes

$$a_s(\mu) = \frac{\omega^*}{1 + \frac{2}{3}s} \frac{1 - g^*s\mu}{1 + fs\mu} \tag{2.20}$$

where

$$f = 1 - \omega^* g^*, \qquad s = [3f^{-1}(1 - \omega^*)]^{1/2}$$

Equation (2.20) is functionally similar to Eq. (2.14) for the two-stream albedo for plant canopies. The term $g^* s \mu$ in the numerator corresponds to the $(2\beta_0 - 1)c$ term in Eq. (2.14). For symmetric scattering, $g = 0$ and the delta-Eddington formulation reduces to the more well-known Eddington approximation.

Wiscombe and Warren discuss extensively many of the currently available observations of snow albedo and compare these with their theoretical results. They find good agreement in the near-infrared part of the spectrum but predict albedos higher than those indicated by most observations in the visible. They consider possible errors in their analysis and conclude that the only likely explanation is the usual presence of small impurities in the snow. This possibility is examined in detail in a companion article (Warren and Wiscombe, 1980). They suggest 0.1-μm soot particles present at concentrations of about 0.1–1 ppm could reduce visible albedos to those usually observed.

It is interesting to contrast snow albedos with those of sea ice. Ice without air bubbles has an albedo as low as that of a water surface. Sea-ice albedos are determined primarily by the size and distribution of air spheres in the ice. Snow, on the other hand, consists of ice "spheres" in air. In both cases, reflections occur at the air–ice interface whereas absorption occurs within the ice.

It would appear that there is now a good theoretical basis for including snow albedos in climate models. However, several problems arise in attempting accurate determinations of surface albedo in the presence of snow. First, in the context of the theory of Wiscombe and Warren (1980) it is necessary to determine the decrease of snow albedo due to the passage of time. Snow albedos decrease because of growth of grain size, which depends on snow temperature and possible melting. They also decrease as more impurities collect on the snow surface. City dwellers know how rapidly the surface of snow darkens due to accumulation of soot. However, there seems to be little observational evidence for the rate at which impurities darken old snow in rural areas. Such darkening could be important for modeling spring snow melt over continental areas.

Also difficult to correctly parameterize are the effects of inhomogeneities in snow coverage. As a result of drifting and other small-scale circulations, snow coverage is not of uniform depth even over a completely level surface. Microrelief further contributes to irregularities in the depth of the snow coverage.

Finally, vegetative cover, from dry grass stems to conifer trees, may greatly reduce surface albedo if these obstacles are not completely covered by snow. This point was nicely illustrated in the study by Federer (1971) of the effects of a leafless hardwood forest on surface albedo. Figure 10, taken from that study, shows that the stems and branches of the hardwood forest readily lower surface albedos by a factor of 2 over that of a bare snow surface. It is evidently important for climate models with snow prediction to prescribe other surface features that may shade the snow and only use albedos for deep snow when the surface features, vegetative or otherwise, are buried by the snow. Shading by dead vegetation can also significantly darken desert surfaces (Otterman, 1981).

2.6. Soil Albedos

Soil and sand surfaces have a wide range of albedos from less than 0.1 for black organic soils to greater than 0.5 for white sands. If soil occurs as small grains without small-scale surface corrugations and the near-field interference of radiation scattered by different particles can be ignored, its albedo can be calculated from Eq. (2.20) provided the optical parameters g and ω are determined for the given particles. To my knowledge, such a calculation has only been attempted for Mars. This was done by Zurek (1978) in order to use measured surface albedos to help establish the optical properties of Martian dust clouds, presumed to consist of the same material as that lying on the surface.

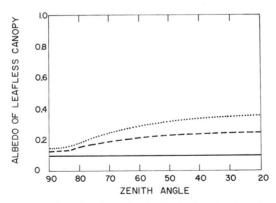

FIG. 10. Modeled variation of surface albedo of a leafless hardwood canopy for various values of solar zenith angle and ground albedo (due to degree of snow cover): ($\cdots\cdots$) 0.75, (----) 0.50, (——) 0.16 (Federer, 1971).

Due to their larger index of refraction, grains of mineral dust are about twice as reflective as ice or water droplets. The fraction of radiation that interacts with a sphere in air and is backscattered is given by $0.5(1 - g)$. For large, nearly nonabsorbing ice and water spheres, $g \simeq 0.9$, and for comparable rock spheres, $g \simeq 0.8$. The scattering fraction parameter ω for mineral grains is likely to be larger than 0.9 for largely quartz grains, depending on wavelength and particle size, but ω can drop to between 0.7 and 0.8 or less for particles made of darker rocks or with a ferric oxide coating. (The lowest possible value for ω is 0.5—appropriate to soot; it is not zero since the optical cross section is about twice the physical cross section due to inclusion of radiation diffracted by the particle's edge.) These remarks regarding likely values of ω are necessarily rather speculative because of the near-linear dependence of absorption on particle size and the wide range of surface particle sizes. Patterson (1981) has recently reviewed data on the optical properties of crustal aerosol.

As an example of the application of Eq. (2.20) to a flat soil surface, Fig. 11 plots its calculated albedo versus zenith angle for $g = 0.8$, $\omega = 0.91$. The dashed line illustrates the albedo expected from a mixture of direct and diffuse solar radiation. The relative fraction of diffuse radiation was assumed to be $0.1/(0.1 + \mu)$, varying from 0.1 at overhead sun to 1 at sunset, as might result from a mixture of Rayleigh and aerosol downward scattering of sunlight. The curve including diffuse sunlight resembles

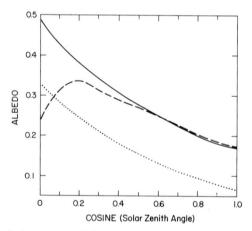

FIG. 11. Theoretical variation of albedo with solar zenith for a flat sand surface. The curves represent albedos for the direct solar beam and are obtained from Eq. (2.20) by assuming $\omega = 0.91$ and $g = 0.8$ for dry sand (——) and $g = 0.92$ for wet sand (·····). The dashed curve is the same as the solid curve except it includes the diffuse light contribution to the albedo.

some observations of desert albedos (see, e.g., Ashburn and Weldon, 1956). However, a simple theory based on Eq. (2.20) would only be expected to apply under special conditions of a flat surface composed of large distinctive particles. The albedo of most soil surfaces varies much less with solar zenith angle than would be predicted by Eq. (2.20).

Observational studies of soil albedo have been reviewed by Myers and Allen (1968) and Heilman *et al.* (1978). According to Myers and Allen, with reference to laboratory studies of soil albedo:

> These studies conclude that increasing particle diameter results in a decrease in reflectivity. This conclusion is correct only for the laboratory case of dispersed soils . . . fine textured soil materials usually have a darker tone than coarse soils. . . . Differences in soil moisture and humus content, in general, overshadow differences in soil texture . . . In the undisturbed case, fine-textured soils generally have structure, which gives them the characteristics of aggregates coarser than sand. . . . Structureless soils reflect 15% to 20% more light than soils with well-defined structure.

Sands are perhaps better approximated than soils by a model of simple grains [e.g., Leu (1977) reports on correlation between reflectance and grain size of iron-stained beach sand].

Clumping of soil particles into aggregates is analogous to the previously discussed organization of plant canopies into clusters and such clumping should not only lower albedos overall, but also flatten the zenith angle variation inferred from Eq. (2.20). Coating of soil particles by humic and fluvic acids reduces the fractional scattering parameter of individual particles. Presence of ferric oxide reduces scattering over the visible spectrum at wavelengths shorter than red.

Soil moisture has long been known to lower soil albedos by typically a factor of 2. To see the likely reason for this, consider a soil surface that has been totally saturated and covered by water so that there are no interfaces between soil particles and air. Reflection would occur only from the air–water interface at the top of the soil and from soil–water interfaces. Since the index of refraction of water is 1.33 and that of rock particles is about 1.6, the reflectivity of soil particles in water is only about 0.4 that of soil particles in air. In other words, the asymmetry factor g changes from about 0.8 to about 0.92. The possible reduction of soil albedo due to water is illustrated in Fig. 11 by the curve calculated for $g = 0.92$. As air penetrates the soil, there should be increased reflections from air bubbles. However, observations indicate no significant increase in the albedos of soils until their water content reaches field capacity, i.e., the level at which water is no longer freely draining through them.

As moist soil becomes aerated, the problem of determining the transfer of radiation through it becomes complicated because of the mixture of air, water, and soil surfaces. However, it seems plausible that the dominant effect of the water remains the reduction of the backscatter of light rays incident on soil particles. A similar explanation proposed by Angström (1925), involving total internal reflections in the water film covering a soil particle, has been discussed by Planet (1969).

Another way to lower the albedo of a surface is by roughening it. For example, raking a dry soil will reduce its albedo and plowing it will lower the albedo even more. The basic principle involved is the same as that discussed for plant canopies (e.g., as illustrated in Fig. 6) and applies to any surface of low reflectivity. Some of the radiation reflected upward by a rough surface strikes other parts of the surface and is further attenuated. Furthermore, the radiation reflected from slanted surfaces will have a downward component that will also again strike other parts of the surface and be further attenuated. By contrast, all radiation leaving a flat surface is in the upward direction and none is returned to the surface (except by reflections within the atmosphere).

The effect on albedos of large departures from flat planar surfaces has been studied most extensively in the context of clouds (e.g., as reviewed by Welch et al., 1980). Davies (1978) has derived a three-dimensional delta-Eddington formalism that could be applied to the question of the albedo of rough surfaces.

Coulson and Reynolds (1971) made a detailed study of the variation of albedo between 0.3 and 0.8 μm for various surfaces. One particular soil they examined was "Yolo loam." The initial surface had been "disked." They first obtained the albedo of the disked surface. They then established a flat surface by sprinkling water over it ("puddling it"). They obtained the albedo of this surface while wet and after it had dried. Figure

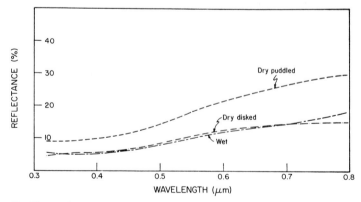

Fig. 12. Observed variation with wavelength of the albedo of three soil surfaces, all derived from "Yolo loam" (Coulson and Reynolds, 1971).

12 shows their comparison of the albedos of the dry disked, wet puddled, and dry puddled surfaces. Remarkably enough, the albedo of a dry disked surface was about as low as that of the wet flat surface, whereas the albedo of the dry puddled surface was nearly double that of the other two surfaces. The albedo of these surfaces varied only slightly with solar zenith angle, presumably due to their surface roughness and organization of soil particles into clods. A model of the effects of tillage on soil albedo has been described by Cruse et al. (1980).

In summary, the primary factors establishing soil albedo are the following: (a) soil composition, especially the amounts of highly absorbing organic and iron compounds, (b) particle size, (c) degree of aggregation of soil particles into larger soil units, (d) soil moisture, and (e) surface roughness.

2.7. Surface Albedos for Climate Models

As the preceding discussion has indicated, the processes determining land surface albedos are complex and far from completely understood in quantitative terms. Yet there will always be immediate requirements for some kind of albedo prescription to put into climate models. Table III summarizes the prescriptions adopted by Dickinson et al. (1981) as a tentative basis for assignment of albedos in a climate model over snow-

TABLE III. VEGETATION AND GROUND ALBEDOS[a]

Surface type		Albedos	
		Visible ($\lambda < 0.7$ μm)	Near-infrared ($\lambda > 0.7$ μm)
Bare nondesert	dry	0.13	0.26
soil	wet	0.05	0.10
Desert soil	dry	0.26	0.52
or sand	wet	0.10	0.20
Short vegetation (crops or grass-land)		0.10	0.30
Conifer forest		0.04	0.20
Midlatitude deciduous forest		0.08	0.28
Tropical forest		0.04	0.20
Tundra		0.15	0.40

[a] As adopted by Dickinson et al. (1981). Over partially vegetated areas, albedos of bare ground and those of vegetative canopies are averaged.

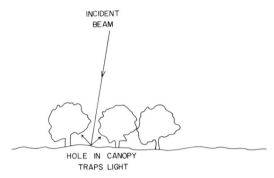

INCIDENT
BEAM

HOLE IN CANOPY
TRAPS LIGHT

FIG. 13. Sketch illustrating trapping of light by a hole in a forest canopy. Due to this mechanism, limited removal of trees from a forest with an otherwise near smooth canopy can lower the net surface albedo even though the underlying surface may have a higher albedo than that of the canopy.

free surfaces. In considering the question of albedo change due to surface change, it is important to remember that it is not just the amount of vegetative cover that determines the albedo, but that the surface texture and structure are also quite important. For example, the common belief is that deforestation increases surface albedo. This is usually true in the case of complete tree removal, but selective cutting of trees would most likely increase the light trapping capability of a canopy and so reduce its albedo, as illustrated in Fig. 13. Further observational studies of surface albedo would be especially helpful if carried out with a theoretical framework in mind and if directed to the question of surface albedo changes due to land use change.

It is important in calculating snow albedos in a climate model to account for the effects on albedo of overlying vegetative canopies and of variations in the snow depth. Decrease of snow albedos due to crystal growth and accumulation of impurities appears to be quite important, but there is yet little observational basis for incorporating these effects into climate models.

3. SURFACE ENERGY BALANCE

Principles of energy balance are fundamental to the functioning of land surfaces as a component of the climate system. They determine (a) the temperature of surface objects, (b) the rate at which the surface loses water by evapotranspiration, and (c) the fluxes of energy between the surface and atmosphere. The basic statement of energy conservation is simply that the net flux of energy into a surface object must equal the rate at which the object is storing energy or, stated as an equation,

$$F_{\text{STORAGE}} = F_{\text{Rn}} - L_v F_{\text{E}} - F_{\text{H}} \tag{3.1}$$

where F_{STORAGE} is the rate at which energy is stored, F_{Rn} is the net radiation absorbed by the surface, F_{H} is the flux of sensible heat away from the surface, F_{E} is the evaporative water loss from the surface, and L_v is the coefficient for latent heat of evaporation (or sublimation if the surface water is in the form of ice or snow). The storage term includes the thermal energy required to melt snow or ice, or the negative energy required to freeze it. Equation (3.1) applies to the surface as a whole or to its individual components, in particular, to the ground and to plant canopies or parts of the plant canopy (e.g., individual leaves). The terms of Eq. (3.1) generally depend on the temperature of the surface; if their relationship to temperature can be expressed mathematically, Eq. (3.1) may be used as an equation for determining surface temperature or the temperature of individual surface elements, given values for the other parameters that enter into the relationships.

The intent of this section is to summarize some of the aspects of surface energy balance considerations that are important for global climate models. Another review from a somewhat different perspective has recently been given by Monteith (1981).

After determining temperature(s) by solution of Eq. (3.1), the individual flux terms may be evaluated to obtain the energy exchange between the surface and atmosphere. In observational studies, temperature is generally obtained by measurement, and Eq. (3.1) is used directly to establish fluxes. The moisture flux is usually emphasized in such studies because of its importance for soil water usage.

If the surface is characterized by a single temperature, the net radiative fluxes can be decomposed into the following individual terms:

$$F_{\text{Rn}} = F_{\text{sb}}^{\downarrow}(1 - a_s) + \epsilon(\overline{F_{\text{IR}}^{\downarrow}} - \overline{\sigma T_s^4}) - F'_{\text{IRn}} \tag{3.2}$$

where $F_{\text{sb}}^{\downarrow}$ is the incident solar flux obtained from an atmospheric model as expressed by Eq. (2.2), a_s is the surface albedo. Bars over quantities are used to indicate a diurnal average; $F_{\text{IR}}^{\downarrow}$ is the downflux of thermal infrared radiation at the surface, a fraction $(1 - \epsilon)$ of which is reflected, and $\epsilon \sigma T_s^4$ is the surface thermal emission, where σ is the Stefan–Boltzmann constant and ϵ is the thermal emissivity. The term F'_{IRn} is the diurnal variation of net infrared cooling at the surface. The accuracy of Eq. (3.2) is improved if, as previously discussed, the dependence of a_s on wavelength and incidence angle are accounted for by including separate terms for visible and near-infrared fluxes and for the direct and diffuse components of the solar beam. The emissivity ϵ is significantly wavelength-dependent at least for those surfaces for which it departs greatly from unity (e.g., quartz sand). If this spectral dependence is not to be accounted for in detail, it has been suggested by V. Ramanathan that at least the emissivity of the 8- to 12-μm window region be treated separately; at other thermal infrared wavelengths the downward radiation is emitted from near-surface atmospheric water vapor and its reflection nearly compensates for the reduction in surface emission due to ϵ being less than unity. The term F'_{IRn} can be approximated by

$$F'_{\text{IRn}} = 4\sigma\epsilon \bar{T}_s^3 (T_s - \bar{T}_s) \tag{3.3}$$

For considering more than one type of surface, Eq. (3.2) is further decomposed into an appropriately weighted average over the surfaces involved.

The outward fluxes of sensible heat and water vapor from surfaces are usually obtained from the aerodynamic transfer formulas

$$F_H = \rho_a C_p C_{\text{DH}} V(T_s - T_a) \tag{3.4}$$

$$F_E = \rho_a C_{\text{DE}} V(q_s - q_a) \tag{3.5}$$

where ρ_a is the density of surface air; C_p is the specific heat of air; C_{DH} and C_{DE} are aerodynamic transfer coefficients for heat and moisture, respectively; V is the near-surface wind velocity; T_a is near-surface air temperature; q_s is the water vapor concentration of air immediately adjacent to the surface; and q_a is the near-surface water vapor concentration. The quantities C_{DH}, C_{DE}, q_a, and T_a are all evaluated at some reference level. This level is usually assumed in climate models to be 10 m above the surface (or 10 m above the zero displacement level of the near-surface logarithmic wind profile for tall vegetation), but for observational studies over land the reference level is often taken to be 2 m above the surface (anemometer level).

The coefficients C_{DH} and C_{DE} may be nearly equal and generally depend on surface roughness and also on temperature lapse rate and wind shear (e.g., as combined into a "Richardson number") as well as on the chosen reference level. In a comprehensive climate model, T_a and q_a and the information required to obtain the aerodynamic transfer coefficients (except surface roughness) would be determined from an atmospheric boundary-layer parameterization. Consideration of surface energy balance processes is required to obtain T_s and q_s.

To show how surface energy balance [Eq. (3.1)] and the energy transfer relationships [Eqs. (3.2)–(3.5)] are applied in practice, a number of specific examples are now discussed.

3.1. Uniform Single Surface with No Heat Storage and Saturated Air

The assumption $F_{STORAGE} = 0$ is frequently used in climate models because of its simplicity. It is approximately satisfied for land surfaces averaged over 24 hr but leads to large errors in determining surface temperature, sensible heat flux, and latent heat flux instantaneously over a diurnal cycle.

For this example, the surface is assumed to be moist. The concentration of water vapor adjacent to a wet surface is a known function of its temperature (equal to that of the surface), i.e.,

$$q_s = q_s^{SAT}(T_s) \tag{3.6}$$

where q_s^{SAT} is the saturated water vapor concentration. For this example, the overlying air is also assumed to be saturated, hence $q_a = q_a^{SAT}(T_a)$. With no storage, Eq. (3.1) becomes

$$L_v F_E + F_H = F_{Rn} \tag{3.7}$$

where all the terms in Eq. (3.7) are functions of T_s as found from Eqs. (3.2)–(3.6). Hence, given atmospheric parameters, Eq. (3.7) is solved for T_s. As a simple numerical example, take $V = 6$ m sec^{-1}—a somewhat strong surface wind, and $C_{DH} = C_{DV} = 2.75 \times 10^{-3}$—a typical value for the aerodynamic drag coefficient evaluated at the 10-m level over a smooth bare ground surface under conditions of neutral stability; then $\rho_a C_p C_{DH} V \simeq 20$ W m^{-2} K^{-1}. If q_a is also saturated at T_a, $L_v(q_s - q_a) \simeq B_e^{-1} C_p(T_s - T_a)$, where $B_e^{-1} = L_v(\partial q^{SAT}/\partial T)/C_p$. Implicit in the assumption of q_a being saturated is the formation of dew when T_s drops below T_a. The term B_e is referred to as the equilibrium Bowen ratio; $B_e = 1$ at $T = 7°C$. Its inverse B_e^{-1} can range from 5 for a warm tropical or summer day ($T = 35°C$) to 0.1 in polar winter ($T = -30°C$).

For the sake of argument, assume $B_e^{-1} = 3$, appropriate to $T = 25°C$. Equation (3.7) then reduces to

$$\lambda_s(T_s - T_a) = F_{Rn} \tag{3.8}$$

where $\lambda_s = \rho_a C_p C_{DH} V(1 + B_e^{-1}) = 80$ W m^{-2} K^{-1}. Typical maximum clear-sky values for F_{Rn} with nearly overhead sun (e.g., June in midlatitudes) is 600 W m^{-2}. During the night, the value of F_{Rn} depends on T_s, atmospheric moisture, and cloudiness; during midlatitude summer, a typical value would be $F_{Rn}(night) \simeq -0.2\sigma T_s^4 \simeq -80$ W m^{-2}.

Given the above values of F_{Rn},

$$T_s - T_a = 7.5°C \quad \text{(daytime)}$$
$$T_s - T_a = -1°C \quad \text{(night)}$$

An atmospheric boundary-layer model, in turn, determines T_a from T_s and atmospheric temperature at higher levels; for example, a simple model parameterization used in past NCAR GCMs (Washington and Williamson, 1977) of the balance between surface sensible heat transfer and convection of that heat into the atmosphere gives

$$T_a = \eta T_r + (1 - \eta)T_s \qquad (3.9)$$

where η is a weight factor between 0 and 1 depending on the ratio of the vertical eddy diffusion coefficient to the aerodynamic drag coefficient and T_r is the temperature the near-surface air would reach by atmospheric mixing if there were no energy exchanged with the surface (i.e., it is the temperature of the lowest model layer, e.g., centered at 1.5 km for past NCAR GCMs, and extended down to the surface by some prescribed lapse rate). Equation (3.9), together with Eq. (3.8), gives T_s and T_a in terms of T_r and F_{Rn}, i.e.,

$$\begin{aligned} T_a &= (\eta^{-1} - 1)F_{Rn}/\lambda_s + T_r \\ T_s &= F_{Rn}/(\eta\lambda_s) + T_r \end{aligned} \qquad (3.10)$$

As typical values, assume $\eta = 0.8$ for midday and $\eta = 0.2$ for night. During the day, T_a is strongly coupled to the atmosphere above by convection; the example gives

$$T_a = 1.9°C + T_r, \qquad T_s = 9.4°C + T_r \quad \text{(daytime)}$$

while during the night there is little vertical coupling in the atmosphere; the numerical example gives

$$T_a = -4°C + T_r, \qquad T_s = -5°C + T_r \quad \text{(nighttime)}$$

As the numerical example indicates, atmospheric temperatures near the surface commonly are about as much out of balance with the free atmosphere at night as they are during the day. The much weaker cooling of the surface at night compared to its heating during the day is compensated by the much weaker nighttime convective coupling to the free atmosphere.

The above numerical example gives a 6° diurnal range of near-surface air temperature, which is typical for moist areas and for latitudes and seasons appropriate to the solar heating assumed here. There are many possible inaccuracies in this example; the major ones involving surface processes are the assumptions that there is no subsurface energy storage, that atmospheric air is saturated, and that the surface soil remains moist at midday. For a diurnal cycle of heating, a column of moist soil has a thermal inertia which is typically about 10 W m^{-2}. In this example, the thermal inertia reduces the difference between T_s and T_a by about 10%.

Under conditions of weak surface winds (i.e., $V < 1$ m sec^{-1}) or dry soil, the soil thermal inertia is a major factor in determining the amplitude of the diurnal cycle of surface air temperature in nonvegetated areas.

Also, even if the nighttime air is saturated with respect to water vapor, it will not generally be saturated during the day. A 2.0°C increase of temperature increases the amount of water vapor required for saturation by about 12%. However, daytime convection prevents q_a from increasing much over its nighttime values, and daytime relative humidity in a moist region is likely to be about 60–80%. Consequently, if the surface soil remains moist, midday evaporation will proceed at a greater rate (e.g., 50% faster) than that deduced here and nighttime dew formation will be slower or nonexistent, all other parameters being the same. Both inclusion of thermal inertia and lower air relative humidity act to reduce the daytime temperature maximum from that inferred here. The surface transfer coefficients would increase over their neutral values during the day and decrease at night, which would also reduce the daytime maximum temperature but amplify the nighttime temperature decrease. These errors were partially compensated for in the example by the assumption of a surface wind V_s somewhat larger than typical. Also note that it is common for the surface soil to be dry at midday even if moist at night, and this dryness may reduce the evaporative flux and so allow the surface to be warmer. Since surface winds can range anywhere between 1 and 10 m sec^{-1}, the inferred diurnal temperature could assume a wide range of values.

If the ground were so dry that no daytime evaporation or nighttime dew formation could occur but all other parameters were the same, then $\lambda_s = 20$ W m^{-2} K^{-1} and we would have

$$T_a = 7.5°C + T_r, \qquad T_s = 37.5°C + T_r \quad \text{(daytime)}$$
$$T_a = -16°C + T_r, \qquad T_s = -20°C + T_r \quad \text{(nighttime)}$$

For this example of dry soil, heat storage by conduction into the soil could reduce the inferred daytime T_s by as much as 15°C, and return of this heat during the night would also reduce the nighttime temperature drop from that inferred here.

3.2. Thermal Capacity of Soil and Snow Surfaces

Here the effect of $F_{STORAGE}$ on surface temperature is evaluated as the rate of heat conduction into the underlying surface, which for soil or snow can be written

$$F_{STORAGE} = -\rho_s C_s k_s \left. \frac{\partial T(z)}{\partial z} \right|_{z=0} \tag{3.11}$$

where ρ_s is the surface density, C_s is the specific heat of surface material per unit mass, k_s is the surface thermal diffusivity (units of $m^2 \, sec^{-1}$), z is the depth into the soil, and $T(z)$ is the subsurface temperature. The heat capacity per unit volume $\rho_s C_s$ for soil varies primarily with soil water content and may be evaluated from

$$\rho_s C_s = (0.23 + \rho_w) C_w \tag{3.12}$$

for a typical soil half-filled by voids (see, e.g., deVries, 1963). The term ρ_w is the volume of liquid water per volume of soil and C_w is the specific heat of water per unit volume. The thermal diffusivity k_s of soil depends on soil moisture, and fractional concentrations of organic material and quartz; k_s decreases with decreasing moisture, decreasing quartz content, and increasing organic content. For a typical loam soil, $k_s \approx 5$ for soil moisture varying from saturated to nearly dry but drops to 2 for oven-dry soil (in units of $10^{-7} \, m^2 \, sec^{-1}$) (see deVries, 1975, Fig. 3.1). For peat, k_s can be as low as 1 and for moist quartz sand as large as 9 in the same units. For snow, k_s and $\rho_s C_s$ depend on ρ_{sw}, the snow density relative to water, i.e.,

$$k_s = 14 \times 10^{-7} \rho_{sw} \; m^2 \, sec^{-1} \tag{3.13}$$
$$\rho_s C_s = 0.49 \rho_{sw} C_w \tag{3.14}$$

Typical values of ρ_{sw} for surface snow range from 0.1 (freshly fallen snow) to 0.3 (old snow) (see, e.g., Anderson, 1976). Hence, snow thermal diffusivity near the surface varies between 2×10^{-7} and $5 \times 10^{-7} \, m^2 \, sec^{-1}$; it has about the same range of values as loam. However, the heat capacity of fresh snow is an order of magnitude less than that of moist soil, primarily because of its large air content. Due to its low heat capacity, a snow surface is much more insulating than is a soil surface.

The temperature gradient term $\partial T / \partial z$ in Eq. (3.11) is evaluated by solution of the equation of thermal conductivity:

$$\frac{\partial T}{\partial t} - k_s \frac{\partial^2 T}{\partial z^2} = 0 \tag{3.15}$$

[Equation (3.15) is derived by neglecting the variation of $\rho_s C_s k_s$ with z.]

For periodic forcing proportional to $\exp(i v t)$, this equation has the solution

$$T(z) = \bar{T}_s + (T_s - \bar{T}_s) \exp[-(i v / k_s)^{1/2} z] \tag{3.16}$$

where $\bar{T}_s$ is the diurnal average temperature. Hence

$$F_{STORAGE} = \rho_s C_s (k_s v)^{1/2} (T_s - \bar{T}_s) \exp(i\pi/4) \tag{3.17}$$

The significance of the $\exp(i\pi/4) = i^{1/2}$ term is that the rate of heat storage leads maximum surface temperature by $\frac{1}{8}$ period (3 hr). For numerical models of the diurnal cycle, it is convenient to approximate F_{STORAGE} by

$$F_{\text{STORAGE}} = C_{\text{SOIL}}\left[v^{-1}\frac{\partial T_s}{\partial t} + (T_s - \bar{T}_s)\right] \tag{3.18}$$

hence simulating the storage phase lag. With the soil "thermal inertia"

$$C_{\text{SOIL}} = \rho_s C_s (0.5 k_s v)^{1/2} \tag{3.19}$$

Equation (3.18) reduces to the "exact" solution for diurnal periodic forcing, since $(0.5)^{1/2}(1 + i) = i^{1/2}$. This approximation is referred to as the "force–restore" procedure; its accuracy for realistic diurnal forcing has been demonstrated by Deardorff (1978) and again by Lin (1980). For moist soil corresponding to $\rho_w = 0.33$, Eq. (3.12) gives $\rho_s C_s = 2.34 \times 10^6$ J m^{-3}. With a loam soil value of $k_s = 5 \times 10^{-7}$ m^2 sec^{-1}, and diurnal forcing $v = 7.3 \times 10^{-5}$ sec^{-1}, the soil thermal inertia becomes $C_{\text{SOIL}} = 10$ W m^{-2} K^{-1}. As extreme cases, a very dry soil might have $C_{\text{SOIL}} \simeq 4$ W m^{-2} K^{-1} and a saturated soil, $C_{\text{SOIL}} \simeq 20$ W m^{-2} K^{-1}.

The soil heat storage as indicated by Eqs. (3.17) and (3.18) has a term in phase with T_s, hence with the sensible and latent fluxes, and a term equal in magnitude to the first term but leading T_s by a quarter-day. The sum of these two terms gives the 3-hr phase advance previously noted.

3.3. Nonsaturated Air over Bare Ground

Now both q_s and q_a are assumed to be less than their saturated value; in general, they then can be written

$$q_s = \gamma q_s^{\text{SAT}} + (1 - \gamma)q_a \tag{3.20}$$
$$q_a = h q_a^{\text{SAT}} \tag{3.21}$$

where γ and h are parameters lying between 0 and 1. The parameter h is simply the relative humidity at the reference level at which q_a is evaluated. The parameter γ is determined by the diffusion of water upward through the soil in response to evaporative demand. It is usually near 1 at night when demand is low and drops to its lowest value in early afternoon when evaporative demand is greatest; this minimum value depends on the maximum rate water can diffuse upward through the soil from its storage in underlying layers. The relative humidity h will also generally be smallest in early afternoon.

From Eqs. (3.20) and (3.21),

$$q_s - q_a = \gamma(q_s^{\text{SAT}} - h q_a^{\text{SAT}}) \tag{3.22}$$

The latent heat flux due to this humidity difference is obtained from

$$L_v(q_s - q_a) \simeq \gamma h B_e^{-1}(T_s - T_a)$$
$$+ \gamma(1 - h)[B_e^{-1}(T_s - \bar{T}_s) + L_v \bar{q}_s^{SAT} \quad (3.23)$$

Using Eq. (3.23) to evaluate latent heat fluxes and Eq. (3.18) to evaluate soil heat storage, we can write the ground energy balance, Eq. (3.1), as

$$\frac{C_{SOIL}}{v} \frac{\partial T_s}{\partial t} + \lambda_1(T_s - T_a) + \lambda_2(T_s - \bar{T}_s) = F \quad (3.24)$$

where

$$F = F_{Rn} + F'_{IRn} - \rho_a L_v C_{DE} V\gamma(1 - h)\bar{q}_s^{SAT}$$
$$\lambda_1 = \rho_a C_p C_{DH} V[1 + \gamma h(C_{DE}/C_{DH})B_e^{-1}]$$
$$\lambda_2 = 4\epsilon\sigma T_s^3 + C_{SOIL} + \gamma(1 - h)\rho_a C_p C_{DE} VB_e^{-1}$$

At this level of complexity, detailed numerical computations are best done by computers. The objective here is to reveal the general structure of the response of surface temperature to the diurnal cycle of solar heating.

According to Eq. (3.24), two "restore" or "feedback" factors act to limit the daily range of T_s. First, sensible and latent heat fluxes as summarized in λ_1 act to restore T_s back to the temperature of overlying air. The ratio of sensible to latent fluxes in this term, denoted B_a, is given by

$$B_a = \frac{C_{DH}}{C_{DE}} (\gamma h)^{-1} B_e \quad (3.25)$$

Both dryness of the soil ($\gamma < 1$) and dryness of the air at the reference level ($h < 1$) act to increase B_a and hence reduce λ_1 over that inferred for saturated conditions. Second, three terms act to restore surface temperature to its diurnal average value. These terms are infrared thermal damping ($4\epsilon\sigma T_s^3 \simeq 6$ W m^{-2} K^{-1}), the soil thermal inertial ($C_{SOIL} \simeq 10$ W m^{-2} K^{-1}), and the diurnal variation of evaporation resulting from departures of the reference level air from saturation. This last term is quite variable, but would typically be in the range 10–30 W m^{-2} K^{-1} and it is unlikely to be much larger than 100 W m^{-2} K^{-1} except under conditions of extremely dry air, extremely strong surface winds, and moist soil. It would, for example, be a dominant term in the heat balance of a desert oasis.

It is useful to note that since λ_1 and λ_2 are usually considerably larger than C_{SOIL}, the balance between the diurnal variation of F and the restore terms is primarily responsible for determining the magnitude of the diurnal range of T_s.

With neglect of $\partial T_s/\lambda t$ in Eq. (3.24) and with the assumed coupling between T_a and equilibrium air temperature T_r given by Eq. (3.9), Eq. (3.24) can simply be solved for T_s and T_a.

The solution for T_a is

$$T_a \simeq (1 - \delta)T_r + \delta\bar{T}_s + \frac{(1 - \eta)F}{\eta\lambda_1 + \lambda_2} \tag{3.26}$$

where

$$\delta = \frac{(1 - \eta)\lambda_2}{\eta\lambda_1 + \lambda_2}$$

This expression may be compared to Eq. (3.10) derived assuming saturated conditions and neglecting contributions to λ_2. It shows, for example, that for weak enough coupling to the overlying atmosphere ($\eta \ll 1$), $\delta \approx 1$ and

$$T_s \approx T_a \approx \bar{T}_s + F/\lambda_2$$

In other words, the restoration to $\bar{T}_s$ dominates the diurnal variation of T_s and T_a during conditions of extreme vertical stability.

In summary, the above analysis reveals the role of various parameters which determine the diurnal cycle of surface soil and air temperature. The diurnal cycle of evaporation and sensible heat fluxes also follows from T_s, T_a, and $\bar{T}_s$ by use of Eqs. (3.23), (3.4), and (3.5). However, obtaining more quantitative results requires modeling the soil moisture fluxes to obtain the parameter γ used in Eq. (3.20). A numerical model of diurnally varying surface energy balance and evaporation which includes a realistic soil moisture flux parameterization but with prescribed atmospheric temperatures and humidity has been developed by van Bavel and Hillel (1976).

The analysis given above has also suggested the importance of the convective coupling of the surface to the free atmosphere for determining surface temperatures. A realistic treatment of this coupling requires a model of the planetary boundary layer.

In summary, the instantaneous temperature of near-surface air over bare ground, for given solar fluxes, is determined by some combination of the temperature of air aloft, diurnal average ground temperature, and radiative forcing. The relative weight given to these factors depends on the strength of vertical atmospheric mixing, atmospheric humidity, surface winds and roughness, and soil moisture. Diurnal variation of these factors influences mean as well as diurnal temperature variations. On the one hand, peak vertical convection at times of maximum solar heating and little vertical coupling during the night both lower the average surface temperature from that which would be inferred by using diurnal average heating and convection. On the other hand, the diurnal cycle of surface soil moisture acts to raise daytime surface temperature over that which would be inferred from diurnal average moisture conditions.

3.4. Evapotranspiration from Vegetative Cover

Most land surfaces are at least partially shaded by vegetation; hence, consideration of the energy balance of vegetative canopies is quite important for global climate models. The most general models of the micrometeorology within vegetative canopies approximate the canopy by a sufficient number of vertical layers to adequately resolve vertical variations of temperature and humidity within the canopy. Such a model has recently been used to calculate the water usage of different plant covers (Sellers and Lockwood, 1981a,b). Vertical-layered canopy models calculate net energy exchange by each layer, depending on the vertical distribution of absorption of solar radiation within the canopy, and the winds and eddy diffusion within the canopy.

A convenient simplified approach that appears to simulate at least qualitatively the role of plant canopies with regard to surface energy exchanges is to model the canopy with a single layer of uniform properties. This approach has been referred to as the "big leaf" model by Sinclair *et al.* (1976), who made detailed comparisons between the transpiration and photosynthesis calculated from such a simple model and from a detailed layer model. They found remarkably good agreement between the numerical results inferred from these two approaches.

In the simplest models, the presence of any underlying surface is neglected. The heat storage of plant canopies is usually also considered negligible, implying a balance between radiative, convective, and latent heat fluxes as expressed by Eq. (3.7), with the fluxes obtained from the aerodynamic transfer formulae, Eqs. (3.4) and (3.5), written in the form

$$F_H = \rho_a C_p \hat{r}_H^{-1}(T_l - T_a) \tag{3.27}$$

$$F_E = \rho_a \hat{r}_E^{-1}(q_l - q_a) \tag{3.28}$$

where $\hat{r}_H^{-1} = C_{DH} V$ and $\hat{r}_E^{-1} = C_{DE} V$; T_l and q_l are the temperature and water vapor concentration, respectively, on the surface of the canopy leaves; T_a and q_a refer to temperature and humidity, respectively, at some reference level above the canopy. Observational studies evaluate these latter quantities at the zero level of the logarithmic wind profile, neglecting additional foliage boundary layer resistance (cf. Thom, 1975).

If the relative area of transpiring vegetation equals L_{AI}, then the flux of water out of the leaves, which must balance F_E, is given by

$$F_E = \rho_a r_s^{-1} L_{AI}[q^{SAT}(T_l) - q_l] \tag{3.29}$$

The term r_s represents the resistance of the leaf stomata to water diffusion through the leaf. The first term in the brackets is the water vapor concentration inside a leaf, which is generally assumed to be the saturated concentration at the temperature of the leaf. If the leaf is wetted by rain or dew, q_l is also at this saturated concentration. The flux is generally still nonzero, implying $r_s = 0$.

Elimination of q_l in Eqs. (3.28) and (3.29) gives

$$F_E = \rho_a L_{AI} \frac{[q^{SAT}(T_l) - q_a]}{r_E + r_s} \qquad (3.30)$$

where $r_E = L_{AI}\hat{r}_E$. It is convenient to decompose the humidity difference in Eq. (3.30) into a term due to temperature difference and a term due to the relative dryness of the overlying air, i.e., by taking

$$q^{SAT}(T_l) = B_e^{-1}L_v^{-1}C_p(T_l - T_a) + q^{SAT}(T_a) \qquad (3.31)$$

where again $B_e^{-1} = L_v C_p^{-1}(\partial q/\partial T)$. It has been found convenient in observational studies of transpiration, to eliminate $(T_l - T_a)$ in Eq. (3.27) in terms of F_E using Eqs. (3.30) and (3.31), and from the assumption of energy balance to obtain a single expression for F_E. If we denote as the mixing ratio deficit, $q^{SAT}(T_a) - q_a = \Delta q$; and $r_H = L_{AI}\hat{r}_H$, this expression may be written

$$L_v F_E = \frac{F_{Rn} + \hat{r}_H^{-1}B_e L_v \rho_a \Delta q}{1 + \tilde{B}_e} \qquad (3.32)$$

where $\tilde{B}_e = r_H^{-1}B_e(r_E + r_s)$. Equation (3.32) is equivalent to Eq. (134) of Thom (1975), and its original derivation is credited to J. Monteith.

Equation (3.32) shows that transpiration is driven by net radiative heating and ventilation by dry atmospheric air. The term $\tilde{B}_e$ is the canopy Bowen ratio for saturated atmospheric air. In other words, if atmospheric air is saturated, the stomatal resistance increases the Bowen ratio over that for a moist surface. Daytime values of stomatal resistance r_s for green leaves are usually found to be in the range 100–1000 sec m^{-1}. Tall grass or crops would typically have transfer coefficients C_{DH}, C_{DV} about four times as large as was appropriate for the smooth surface previously considered. With these larger coefficients but other parameters the same as before, $r_H^{-1}L_{AI}\rho_a C_p = 80$ W m^{-2} K^{-1}, and the aerodynamic resistance to heat transfer is $r_H = 15L_{AI}$ sec m^{-1}, or for typical values of $L_{AI} = 5$ for a well-developed canopy, $r_H = 75$ sec m^{-1}, which is the same magnitude or somewhat smaller than r_s. Short grass prairies would have C_{DH} decrease by a factor of 2 and r_H increased by a factor of 2. On the other hand, C_{DH} for forests is an order of magnitude larger than for crops and grassland, but winds are somewhat weaker and so for forests r_H typically assumes ʻvalues of the order of 10–30 sec m^{-1}.

The water vapor transfer coefficient r_E^{-1} is usually assumed to have the same value as the heat transfer coefficient r_H^{-1}. If a significant fraction of the canopy surface area consists of stems, branches, trunks, dead leaves, or other nontranspiring surfaces, then effective stomatal resistance r_s should be increased by the ratio of total surface to transpiring surface. More generally, if the canopy consists of j nonwetted surfaces with relative area A_j and stomatal resistance r_s^j, then the average stomatal resistance for these surfaces is given by

$$r_s^{-1} = \sum (A_j / r_s^j) \tag{3.33}$$

In this expression, the fraction of dry nontranspiring surface has $r_s^j = \infty$. The evaporation from wetted surfaces is calculated separately.

The net stomatal resistance of a dry forest canopy evidently can greatly influence $\bar{B}_e$ and thus the sensible heat fluxes relative to latent fluxes for nonsaturated air. However, under conditions of weak radiative fluxes or large mixing ratio deficit, F_E for a forest still exceeds that for a grassland because of the larger roughness of the forest and hence the larger transfer coefficient C_{DV}. For example, if we neglect F_{Rn} in Eq. (3.32) and assume $\bar{B}_e \gg 1$, then

$$F_E \simeq \frac{L_{Al}\rho_a \, \Delta q}{r_E + r_s} \tag{3.34}$$

which for r_E not much smaller than r_s decreases significantly for increasing surface smoothness. In the limit described by Eq. (3.34), the canopy functions as a wet bulb and collects sensible heat from the atmosphere for canopy evapotranspiration. This limit with $r_s = 0$ qualitatively describes the large interception losses (reevaporation from leaves) that can occur in forests during precipitation events and nonsaturated surface air [e.g., as measured by Stewart (1977) and modeled by Murphy and Knoerr (1975)].

In arid regions, the stomatal resistance r_s may increase greatly at times due to scarcity of soil moisture. Recent studies of the evaporation from natural prairies include those of Ripley and Saugier (1978) and of Parton *et al.* (1981).

3.5. Temperature of a Nontranspiring Canopy

Monteith's formula, Eq. (3.32), has been widely used in studies of water budgets of plant canopies, when the net radiation F_{Rn} is obtained from observations. In climate models, on the other hand, F_{Rn} needs to be calculated from model parameters. It depends in part on the exchange of thermal infrared radiation between the canopy, ground, and overlying atmosphere, which in turn depends on the temperature T_l of the canopy elements. Hence, it is appropriate to formulate the energy balance equation in terms of T_l as the primary dependent variable [e.g., as done by Deardorff (1978)]. Consider now the simplest example of such a formulation—a nontranspiring surface, such as provided by branches and dead leaves, or a slatted roof. For simplicity, exchange with any underlying surface is neglected. The absorbed radiation F_{Rn} is decomposed into two parts:

$$F_{Rn} \simeq \bar{F}_{Rn} - \epsilon \sigma T_l^4$$

where $\bar{F}_{Rn}$ is the part of the absorbed radiation that does not depend on surface temperature, and $\epsilon \sigma T_l^4$ is the canopy (leaf) blackbody emission. Surface energy balance then requires

$$\rho_a C_p \hat{r}_H^{-1}(T_l - T_a) + \epsilon \sigma T_l^4 = \bar{F}_{Rn} \tag{3.35}$$

which is easily solved for T_l in terms of $\bar{F}_{Rn}$ and T_a. For example, if we subtract diurnal average terms from Eq. (3.35), and take F'_{Rn}, T'_l, and T'_a to represent departures from the diurnal average of $\bar{F}_{Rn}$, T_l, and T_a, respectively, then Eq. (3.35) can be solved approximately for T_l:

$$T'_l = \frac{\lambda_1 T'_a + F'_{Rn}}{\lambda_1 + \lambda_2} \tag{3.36}$$

where for the assumed dry surface $\lambda_1 = \rho_a C_p \hat{r}_H^{-1}$ and $\lambda_2 = 4\epsilon \sigma T_l^3$ are assumed to have no diurnal variation. If, as before, $\lambda_1 = 80$ W m^{-2} K^{-1} (appropriate to crops or tall grass), $\lambda_2 = 6$ W m^{-2} K^{-1}, and $F_{Rn} = 320$ W m^{-2}, then $T'_l = 0.93\, T'_a + 3.7°$. Jackson et al. (1981) have found similar temperature differentials in observing a dry wheat canopy. Again, T_a and T_l are also related by the requirement of continuity of heat flux into the atmosphere, as expressed by Eq. (3.9), which could be linearized to give another relationship between T'_s and T'_a. It is more realistic to include exchange with an underlying surface (e.g., the ground) as well. If that surface has temperature T_g and also emissivity ϵ, and the canopy has the same temperature on top and bottom, then a term $\epsilon \sigma (T_g^4 - T_l^4)$ is added to the left side of Eq. (3.35), and Eq. (3.36) becomes

$$T'_l = \frac{\lambda_1 T'_a + \lambda_2 T'_g + F'_{Rn}}{\lambda_1 + 2\lambda_2} \tag{3.37}$$

where T'_g represents the departure of T_g from its diurnal average.

Note that if $T'_g = T'_l$, Eq. (3.37) reduces to Eq. (3.36). The net radiative exchange between canopy and underlying ground has a magnitude of at most about 10 W m^{-2}. It is usually neglected in micrometeorological studies. However, it can be the primary term heating the ground under a well-developed plant canopy, and hence helps to force the ground temperature to approach that of the plant canopy.

3.6. Modeling the Temperature of a Transpiring Canopy

If the stomatal resistance of the canopy is finite F_E must be included in evaluating the energy balance. Its dependence on leaf temperature is obtained from Eqs. (3.30) and (3.31), which can be used, for example, to relate the diurnal fluctuation in leaf temperature to a diurnal forcing term, i.e., define F^* as

$$F^* = F'_{Rn} - \rho_a L_v r_v^{-1} (q_a^{SAT} - q_a)' \qquad (3.38)$$

where ( )' denotes departure from diurnal average. Equation (3.37) generalizes to include transpiration if F^* from Eq. (3.38) is used in place of F'_{Rn} in Eq. (3.37) and $\rho_a C_p B_e^{-1} (r_s + r_v)^{-1} L_{Al}$ is added to λ_1.

The difficult problem in including evapotranspiration in evaluating the energy balance of a canopy is the specification of the stomatal resistance r_v. The following three questions arise:

(1)   How is the fraction of surface area covered by water from rain or dew and the corresponding canopy water storage determined? Deardorff (1978) suggested a simple parameterization for the fractional wetted area; considerable research in forest hydrology has been devoted to the question of canopy water storage because of its importance in determining how much rainfall is reevaporated from the canopy (see, e.g., Rutter, 1975).

(2)   If the plants have an ample soil water supply, what environmental parameters determine the stomatal resistance r_s? The stomata of most nondesert plants respond to solar radiation and are closed at night. Their resistance varies, in addition, with atmospheric temperature and with the difference between water vapor concentration and its saturated value [see, for example, the discussions of Jarvis (1976) and Hinckley et al. (1978)].

(3)   How is plant transpiration regulated under conditions of dry soil? Under moisture stress, stomata control water flow to match that available from the soil.

Plant physiologists have studied extensively the flow of water through plants as driven by "potentials" along the pathway of water from soil, through roots, stems, and leaves. The mesophyllic cells in leaves develop negative osmotic potentials by assimilating dissolved ions, and these potentials extract water from roots and soil when needed. The lowest osmotic potentials that can be maintained are generally between -10 and -30 atm of pressure. If the soil water potential drops below the minimum potential of plants, the plants "wilt," for the leaf cells can no longer be supplied by soil water to remain turgid. The stomata would be fully closed at this point, but some leakage of water through leaf walls continues to desiccate the leaf cells. Under less dry conditions, roots may still not be able to extract water from the soil at the rate that water would be required for normal transpiration; if so, some stomatal closure is necessary.

Models of plant transpiration relate stomatal resistance to leaf water potential; the leaf water potential in turn depends on leaf water storage, which is increased by water flow from roots and decreased by transpiration. The water flow into the roots depends on the resistance of the soil to moisture diffusion to the roots (a function of soil water content) and on the difference between soil and plant water potential. The roots also provide a significant resistance to the flow of water to the leaves. A parameterization of these processes simple enough to be used in climate models has been developed by Federer (1979).

It has been observed that the stomatal resistance of many plants is not altered by plant water potential until the plant potential is nearly as low as the wilting potential. This suggests that as a useful approximation, the plant stomatal resistance r_s may be assumed to remain at its minimum value until water demand exceeds the maximum possible rate at which water can be supplied from the soil through the roots. At this point, r_s adjusts to match the net canopy transpiration to the maximum possible rate of water supply. For example, Denmead and Millar (1976) find a maximum transpiration rate of 0.6 mm hr^{-1} for wheat in Australia. Such a parameterization for plant water flow has been developed for use with global climate models (Dickinson et al., 1981).

3.7. Modeling Energy Exchange over a Snow Surface

Snow fields reflect a large fraction of the incident solar radiation and, if widespread in spring and early summer, significantly reduce the albedo of the earth–atmosphere system. It is thus important for climate models to correctly calculate the timing and degree of spring snow melt in temperate and high latitudes of the Northern Hemisphere. Modeling snow melt is also required for the simulation of stream flows as needed for developing water resources or protecting against floods. This subject has recently been reviewed by Male and Granger (1981). A framework for considering this topic has already been developed earlier in this article. It is worth repeating that the shading effects of vegetative canopy over a snow surface can greatly modify surface albedo compared to that of open snow. Furthermore, solar radiation absorbed by needles, branches, dead grass, etc. is not available directly to the snow surface; the exchange of sensible heat and thermal infrared radiation between the canopy and the snow surface needs to be considered. Petzold (1981) has recently reported observations of the radiation balance of melting snow in an open boreal forest.

In high latitudes, freezing and thawing of soil water contribute significantly to surface energy balance; furthermore, the presence of frozen water can greatly reduce the permeability of soil to infiltration of water, and so contribute to spring flooding.

4. FURTHER REMARKS ON LAND SURFACE PROCESSES AS A COMPONENT OF CLIMATE MODELS

Although this study has emphasized surface processes, it is well worth reiterating the point that surface and atmospheric climate are coupled in many ways. The modifications of solar fluxes reaching the surface by atmospheric constituents have been mentioned. Likewise, the thermal infrared fluxes incident at the surface depend on cloud and water vapor concentrations and atmospheric temperature structure, especially in the lowest 1 or 2 km. Fluxes of sensible and latent heat are even more dependent on atmospheric structure.

In considering sensitivity of surface climate to changes in energy balance components, it is important to account for atmospheric changes, which in turn depend on the spatial scale of the disturbance. For example, if the albedo of a plant canopy were reduced such that the canopy absorbed more solar radiation, its immediate temperature change would be given by an expression equivalent to Eq. (3.36) averaged over the diurnal cycle, i.e., $T'_l = \lambda_l^{-1} F'_{Rn}$, where primes refer to the assumed perturbation. For example, with $F'_{Rn} = 8 \text{ W m}^{-2}$ and $\lambda_l = 80 \text{ W m}^{-2} \text{ K}^{-1}$, $T'_l = 0.1°C$. In other words, evaporative and sensible fluxes to the atmosphere would at first allow the canopy temperature to change by only a small amount. If the canopy were of small horizontal extent, e.g., a few trees, the atmospheric response to its increased heating could be negligible. On the other hand, if the albedo of the land surface were changed to give 8 W m^{-2}

more absorption over all land, the earth as a whole would absorb about 2 W m^{-2} more solar radiation; it is known from models of the climate system that the eventual combined change of atmosphere, land, and ocean surfaces would be an increase in temperature by about 1°C. In other words, over the plant canopy, the initial 8 W m^{-2} would lead to an eventual perturbation heating of 80 W m^{-2}. The additional 72 W m^{-2} of radiation would correspond to increased downward flux of thermal infrared radiation and the decreased sensible and latent heat fluxes due to a warmer atmosphere containing more water vapor.

In this study, I have considered what factors are important for determining the albedos and energy balance of land surfaces. The concepts of albedo and energy balance are central to the inclusion of land surface processes in climate models. Most climate models to date as based on atmospheric GCMs have used very simple descriptions of land surfaces which, however, have been adequate for simulation of gross features of global atmospheric climate. Their success has been, in part, due to the relative importance of oceans for atmospheric climate and their prescription of observed ocean temperatures.

However, in modeling details of surface climate over land, it becomes important to include realistic descriptions of vegetation cover, soil types, and soil moisture to adequately model surface roughness, surface albedos, and surface water budgets and energy exchange. If these processes are included, a climate model should be able to simulate reasonable diurnal variations of temperature not only of the air, but also of the ground and plant canopies. It is also feasible to improve the realism of the parameterizations of surface hydrology in global climate models, and in doing so to better model not only evapotranspiration but also time-dependent stream flow. Only with such improvements will climate models be able to plausibly answer such questions as to the effects of climate change on water resources and flood hazards.

One of the primary obstacles to improving model descriptions of surface processes over the earth is the great heterogeneity in surface structure over most land areas. The minimum horizontal spatial elements of global climate models are generally rectangular surfaces with sides at least several hundred kilometers in dimension. Over such a surface, there can be thousands of individual land elements as characterized by particular vegetative cover, soil type, and terrain. The question as how to properly characterize averages over these individual elements within a model grid square is still largely unresolved. Before such complexities are addressed, it is perhaps important to first better establish the sensitivity of different climate parameters to various aspects of simplified but still somewhat realistic average descriptions of land surfaces.

Acknowledgments

Past discussions with T. Federer and S. Warren and a seminar by J. Otterman introduced to the author some of the material in this text. Helpful comments were made on a draft version by J. Coakley, R. Chervin, S. Thompson, and S. Warren.

References

Anderson, E. A. (1976). "A Point Energy and Mass Balance Model of Snow Cover." Off. Hydrol., U.S. Natl. Weather Serv., Washington, D.C.

Angström, A. (1925). *Geogr. Ann.* **7**, 323–342.

Ashburn, E. V., and Weldon, R. G. (1956). *J. Opt. Soc. Am.* **46**, 583–586.

Brown, P. S., and Pandolfo, J. P. (1969). *Agric. Meteorol.* **6**, 407–421.

Coakley, J. A., Jr., and Chylek, P. (1975). *J. Atmos. Sci.* **32**, 409–418.

Coulson, K. L., and Reynolds, D. W. (1971). *J. Appl. Meteorol.* **10**, 1285–1295.

Cruse, R. M., Linden, D. R., Radke, J. K., Larson, W. E., and Larntz, K. (1980). *Soil Sci. Soc. Am. J.* **44**, 378–383.

Davies, R. (1978). *J. Atmos. Sci.* **35**, 1712–1725.

Deardorff, J. W. (1978). *J. Geophys. Res.* **83**, 1889–1903.

Denmead, O. T., and Millar, B. D. (1976). *Agron. J.* **68**, 307–311.

deVries, D. A. (1963). "Physics of Plant Environment," pp. 210–235. North-Holland Publ., Amsterdam.

deVries, D. A. (1975). In "Heat and Mass Transfer in the Biosphere" (D. A. deVries and N. H. Afgan, eds.), Part 1, pp. 5–28. Wiley, New York.

Dickinson, R. E., Jäger, J., Washington, W. M., and Wolski, R. (1981). "Boundary Subroutine for the NCAR Global Climate Model," NCAR Tech. Note TN-173+1A.

Federer, C. A. (1971). *Agric. Meteorol.* **9**, 3–20.

Federer, C. A. (1979). *Water Resour. Res.* **15**, 555–562.

Gates, D. M. (1980). "Biophysical Ecology." Springer-Verlag, Berlin and New York.

Gausman, H. W., Rodriguez, R. R., and Richardson, A. J. (1976). *Agron. J.* **68**, 295–296.

Goudriaan, J. (1977). "Crop Micrometeorology: A Simulation Study." Pudoc, Wageningen.

Heilman, J. L., Myers, V. I., Moore, D. G., Schmugge, T. J., and Friedman, D. B., eds. (1978). *NASA Conf. Publ.* **2073**.

Hinckley, T. M., Lassoie, J. P., and Running, S. W. (1978). *For. Sci. Monogr.* **20**, 72.

Idso, S. B., Hatfield, J. L., Reginato, R. J., and Jackson, R. D. (1978). *Remote Sens. Environ.* **7**, 273–276.

Jackson, R. D., Idso, S. B., Reginato, R. J., and Pinter, P. J., Jr. (1981). *Water Resour. Res.* **17**, 1133–1138.

Jarvis, P. G. (1976). *Philos. Trans. R. Soc. London B Ser.* **273**, 593–610.

Jarvis, P. G., James, G. B., and Landsberg, J. J. (1975). In "Vegetation and the Atmosphere" (J. L. Monteith, ed.), Vol. 2, pp. 171–240. Academic Press, New York.

Leu, D. J. (1977). *Remote Sens. Environ.* **6**, 169–182.

Lin, J. D. (1980). *JGR, J. Geophys. Res.* **85**, 3251–3254.

Male, D. H., and Granger, R. J. (1981). *Water Resour. Res.* **17**, 609–627.

Meador, W. E., and Weaver, W. R. (1980). *J. Atmos. Sci.* **37**, 630–643.

Monteith, J. L. (1981). *Q. J. R. Meteorol. Soc.* **107**, 1–27.

Murphy, C. E., Jr., and Knoerr, K. R. (1975). *Water Resour. Res.* **11**, 273–280.

Myers, V. I., and Allen, W. A. (1968). *Appl. Opt.* **7**, 1819–1838.

Norman, J. M., and Jarvis, P. G. (1974). *J. Appl. Ecol.* **11,** 375–398.

Norman, J. M., and Jarvis, P. G. (1975). *J. Appl. Ecol.* **12,** 839–878.

Oguntoyinbo, J. S. (1970). *Q. J. R. Meteorol. Soc.* **96,** 430–441.

Otterman, J. (1981). *Tellus* **33,** 68–77.

Parton, W. S., Laurenroth, W. K., and Smith, F. M. (1981). *Agric. Meteorol.* **24,** 97–109.

Patterson, E. M. (1981). *JGR, J. Geophys. Res.* **86,** 3236–3246.

Petzold, D. E. (1981). *Arct. Alp. Res.* **13,** 287–293.

Pinker, R. T., Thompson, O. E., and Eck, T. F. (1980). *Q. J. R. Meteorol. Soc.* **106,** 551–558.

Planet, W. G. (1969). *Remote Sens. Environ.* **1,** 127–129.

Ripley, E. A., and Saugier, B. (1978). *J. Appl. Ecol.* **15,** 459–479.

Ross, J. (1975). *In* "Vegetation and the Atmosphere" (J. Monteith, ed.), Vol. 1, pp. 13–56. Academic Press, New York.

Ross, J. (1981). "The Radiation Regime and Architecture of Plant Stands." Junk, The Hague.

Ross, J., and Nilson, T. (1975). *In* "Heat and Mass Transfer in the Biosphere" (D. A. deVries and N. H. Afgan, eds.), Part 1, pp. 327–336. Wiley, New York.

Rutter, A. J. (1975). *In* "Vegetation and the Atmosphere" (J. Monteith, ed.), Vol. 1, pp. 111–154. Academic Press, New York.

Sellers, P. J., and Lockwood, J. G. (1981a). *Clim. Change* **3,** 121–136.

Sellers, P. J., and Lockwood, J. G. (1981b). *Q. J. R. Meteorol. Soc.* **107,** 395–414.

Sinclair, T. R., Murphy, C. E., Jr., and Knoerr, K. R. (1976). *J. Appl. Ecol.* **13,** 813–829.

Stanhill, C. J. (1970). *Sol. Energy* **13,** 59–66.

Stewart, J. B. (1977). *Water Resour. Res.* **13,** 915–921.

Thom, A. S. (1975). *In* "Vegetation and the Atmosphere" (J. Monteith, ed.), Vol. 1, pp. 57–110. Academic Press, New York.

Tucker, C. J., and Miller, L. D. (1977). *Photogramm. Eng. Remote Sens.* **43,** 721–726.

van Bavel, C. H. M., and Hillel, D. I. (1976). *Agric. Meteorol.* **17,** 453–476.

Warren, S. G., and Wiscombe, W. J. (1980). *J. Atmos. Sci.* **37,** 2734–2745.

Washington, W. M., and Williamson, D. L. (1977). *Methods Comput. Phys.* **17,** 111–172.

Welch, R. M., Cox, S. K., and Davis, J. M. (1980). *Meteorol. Monogr.* **17,** 1–96.

Wiscombe, W. J., and Warren, S. G. (1980). *J. Atmos. Sci.* **37,** 2712–2733.

Zurek, R. W. (1978). *Icarus* **35,** 196–208.

GLOBAL ANGULAR MOMENTUM AND ENERGY BALANCE REQUIREMENTS FROM OBSERVATIONS

ABRAHAM H. OORT

Geophysical Fluid Dynamics Laboratory/NOAA
Princeton University
Princeton, New Jersey

AND

JOSÉ P. PEIXÓTO

Geophysical Institute
University of Lisbon
Lisbon, Portugal

1. INTRODUCTION

The objectives of this article are to present a global picture of the general circulation of the atmosphere and its thermal structure, to inter-

ADVANCES IN GEOPHYSICS, VOLUME 25

pret the results in the framework of the balance requirements of the general circulation, and to show the inferred role of the oceans. The main emphasis will be on the balances of angular momentum and energy. Certain aspects of the global balance requirements for water vapor will be presented here, but an extensive treatment will be given in another article (Peixóto and Oort, 1983). The main purpose of our studies is to obtain a deeper comprehension of certain physical mechanisms responsible for maintaining the present climatic conditions. Our results are also essential for the formulation of general circulation models and for their verification since they constitute the constraints that must be fulfilled.

The basic theoretical framework for the present work was laid, for a large part, by the late Victor P. Starr, Edward N. Lorenz, and their co-workers at the Massachusetts Institute of Technology (MIT). During the last three decades the MIT group has contributed significantly to our present knowledge of the general atmospheric and oceanic circulations by painstaking analyses of direct observations. The results are shown in a long list of scientific publications. As highlights we may select the articles by Starr (1948, 1951, 1953) on the nature of the large-scale eddies in the atmosphere, by Starr and White (1954) on the balance requirements of the general circulation, by Lorenz (1955, 1963) on the concept of available potential energy and on atmospheric predictability, by Phillips (1956) on a first numerical experiment of the general circulation, by Saltzman (1957) on the breakdown of the eddies in wave-number domain, and by Peixóto (1973) on the hydrological cycle. Much of this information was summarized in the monographs by Lorenz (1967), Starr (1968), Oort and Rasmusson (1971), and Newell et al. (1972, 1974).

Of course, many basic contributions were made by investigators from other research groups, such as by Priestley (1951) on the wind stress on the oceans, by Bjerknes (1953) on the theory of the general circulation, by Bjerknes (1966) and Namias (1979) on air–sea interaction, by Mintz (1954) on the structure of the zonal flow, and by Palmén et al. (1958) on the mean meridional circulation.

The early research mainly considered the Northern Hemisphere, but during the last decade a beginning was made in the analysis of the Southern Hemisphere. However, the construction of a comprehensive, global picture of the climate is only now possible since the analysis of a 10-yr truly global, homogeneous data set was completed at the Geophysical Fluid Dynamics Laboratory (GFDL) (Oort, 1983).

The present analyses are based on real observations. No indirectly derived data, such as geostrophic or balanced winds or model-derived relationships, are used, with only one exception. The exception is the mean meridional circulation which cannot, at present, be objectively determined south of about 20° S. The basic data set used is fairly homogeneous and consists of (1) all available upper air data from the global rawinsonde network for the 10-yr period May 1963 through April 1973, and (2) all available surface ship reports for the same period. The typical data coverage in space and time is shown in Figs. 1 and 2 as the number of years that each station had sufficient reports during January to be accepted in our analyses. The coverage and other characteristics of the atmospheric data used in calculating the statistics were carefully considered in order to arrive at a product with the best overall quality. Because of space limitations, we will restrict our discussion to annual mean conditions and to those of the extreme seasons, December–February (DJF) and June–August (JJA).

In addition to the atmospheric and oceanographic data, we will show some results based on satellite measurements of the radiation balance at the top of the atmosphere provided by our colleagues at Colorado State University (Campbell and Vonder Haar, 1980).

Based on the conservation of angular momentum, we will present a comprehensive picture of the angular momentum cycle in the atmosphere–ocean–solid earth system. After elaborating on the implications of the balance of angular momentum, some global aspects of the energetics are discussed in order to explain how the radiation imbalance drives the atmosphere and oceans, and how the required energy balance over the globe is maintained. The concept of the energy cycle is finally used to get a better understanding, not only of the various forms of energy involved, but also to assess the energy conversions that take place in the climatic system. The difference between the largely land-covered Northern and the ocean-covered Southern Hemispheres will be of particular interest.

The present, diagnostic approach to the study of climate can lead to insights on how the system works. However, it cannot provide answers to questions such as: Why does the climate operate as observed? or What mechanisms are responsible for climatic change? In addition, other approaches are clearly needed. Analytical, stochastic, and deterministic models are valuable in this respect, but perhaps most promising are the large-scale numerical, general circulation models. Some of these models have already shown remarkable success in simulating the actual climate, and promise to become reliable tools in testing theories of climatic change (see, e.g., Manabe, this volume). Undoubtedly the most rapid progress in climate research will come about if equal emphasis is given to observational–diagnostic research and theoretical–numerical experiments.

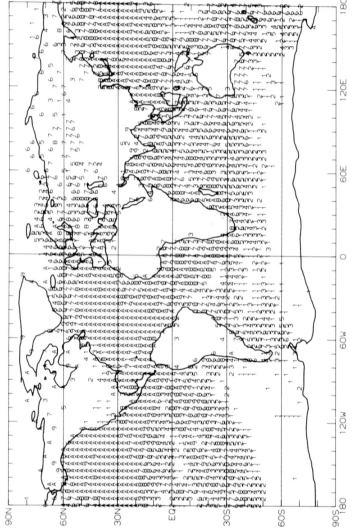

Fig. 1. Distribution of input data used in the mean January analyses at the 1000-mbar/surface level, and number of years of observations available, ranging between 1 and 10 (= A) yr. Data over land are from rawin-sonde stations (1000 mbar) and data over ocean from 2° × 2° averaged surface ship reports (only plotted for every 5° longitude) for the 10-yr period 1963–1973. (EQ, equator.)

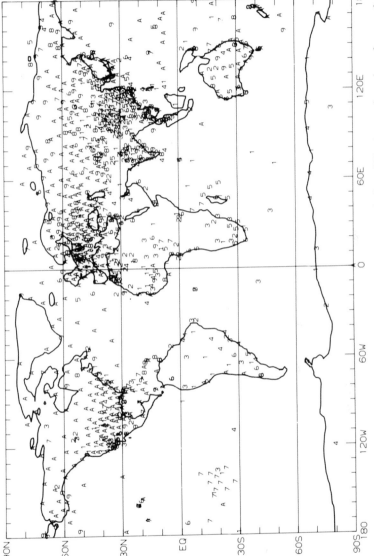

Fig. 2. Distribution of rawinsonde data used in the mean January analysis at 500 mbar, and number of years of observations available, ranging between 1 and 10 (= A) yr, for the 10-yr period 1963–1973 (the total number of stations over the globe for January was 1093).

2. DATA HANDLING AND ANALYSIS PROCEDURES

The basic input data are contained in two sets. Set 1 contains all available daily (twice daily for the 1968–1973 period) reports from the global rawinsonde network for the 10-yr period May 1963 through April 1973. This sample contains the bulk of the information for the present study. Set 2 supplements the first set over the oceans at the lowest level of analysis (1000 mbar/surface); it contains all available daily surface ship reports for the same 10-yr period.

Both data sets were carefully checked for erroneous reports at several phases in the data processing scheme. The most important checks were an initial, gross test for unreasonable meteorological values and, at a later stage, the application of a cutoff criterion for those values of x which fell outside the range $(\bar{x} - 4\sigma(x), \bar{x} + 4\sigma(x))$. The seasonal average $\bar{x}$ and the seasonal standard deviation $\sigma(x)$ were computed from the 10-yr time series at each station and for each isobaric level and season. In the case of the surface ship data, all reports within a grid square of 2° latitude × 2° longitude were considered to represent values for a hypothetical station located in the center of the square. From the checked time series monthly mean statistics were computed at each station for each of the 120 months under study. These statistics include the usual time-mean, variance, and covariance estimates. Only those stations with monthly values based on 10 or more days for the upper air and 3 or more days for the surface were considered acceptable and therefore used in the further analysis.

Next, for the present study of long-term mean conditions, n-yr average station values, where $1 \leq n \leq 10$, were calculated for each calendar month. Interannual, i.e., year-to-year, variations were not included in the transient eddy statistics. Then, the n-yr average statistics served as input for the global objective analysis scheme applied. This scheme uses the zonal averages of the data as a first-guess field. For further information on the analysis procedures, see Oort and Rasmusson (1971), Oort (1978), or Lau and Oort (1981). Typical examples of the distribution of the input data and the number of good years available for the surface (ship data)/ 1000-mbar (rawinsonde stations) and 500-mbar levels are found in Figs. 1 and 2, respectively. The lack of upper air data over the southern oceans is clear. Thus, some of the results between 30° S and 70° S latitude must be regarded as tentative. In the tropics of the Northern Hemisphere there are also some large data gaps over the eastern North Pacific and the Atlantic oceans.

The various maps, zonal mean cross sections, and profiles shown in this article are based on the point values of a regular 73 × 73 grid (2.5° latitude × 5° longitude). Most of the horizontal maps shown in this article were

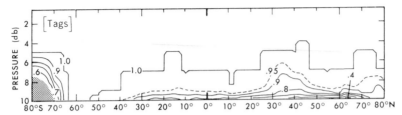

FIG. 3. Latitude–pressure diagram of the zonal average of the number of points above the earth's surface. A value of 1.0 indicates that all points along the latitude circle lie above the earth's surface.

somewhat smoothed before drafting by twice applying a two-dimensional 5-point median smoother to the 73 × 73 grid (resolution 2.5° latitude, 5° longitude) followed by a weak Laplacian smoother as proposed by Tukey (see, e.g., Rabiner *et al.*, 1975). The required mean seasonal and annual grid point values were calculated from the 12 monthly analyses. One should note that all fluctuations within a year are considered as transient eddies, so that the annual mean statistics include the contributions from both synoptic eddies and the "eddies" associated with the normal seasonal variation. If one is interested in annual mean statistics where only the synoptic systems are included as transient eddies and the effects of the annual cycle are removed, a good measure of the annual mean can be obtained by taking a straight average of the DJF and JJA transient eddy statistics.

The topography of the earth's surface was taken into account in all computations so that only grid points above the mountains contribute to the mean zonal and mean vertical estimates. To accomplish this, the topography was expressed in pressure coordinates, assuming that it is invariant throughout the year and equal to the mean annual value. To show the relative importance of the topography at various levels, we show in Fig. 3 a zonal cross section of the fraction of points above the earth's surface.

A description of the satellite radiation data and a critical discussion of their characteristics can be found in an article by Stephens *et al.* (1981).

3. ANGULAR MOMENTUM BALANCE OF THE CLIMATIC SYSTEM

The angular momentum with respect to the earth's axis of rotation is one of the fundamental parameters used to characterize the general circulation of the atmosphere and the climate. The fact that angular momentum for the earth as a whole, i.e., for the atmosphere–ocean–solid earth system, is practically invariant makes it an attractive and suitable property for geophysical research. Much of the historic development of modern meteorology is connected with the study of the transport of angular momentum in the atmosphere and of its exchange with the oceans and solid earth.

The total or absolute angular momentum (M) about the axis of rotation for a unit of atmospheric mass is composed of two terms: the Ω angular momentum (M_Ω) and the relative angular momentum (M_r). The first term represents the angular momentum if the atmosphere were in solid rotation with the earth, and the second term the angular momentum relative to the rotating earth:

$$M = M_\Omega + M_r \tag{3.1}$$

where

$$M_\Omega = \Omega R^2 \cos^2 \phi \tag{3.2a}$$

$$M_r = uR \cos \phi \tag{3.2b}$$

This is illustrated in Fig. 4.

Recently attention has also been given to the components of angular momentum about axes located in the equatorial plane at right angles to the earth's axis of rotation in order to study, e.g., the earth's wobble. The astronomical implications of exchanges of angular momentum between the fluid envelope and the solid earth itself can be seen clearly in the recorded changes in the length of day (see, e.g., Hide *et al.,* 1980; Lambeck, 1981). For the study of the atmosphere it is the angular momentum component about the polar axis that is of most interest and therefore will be the only component to be discussed here.

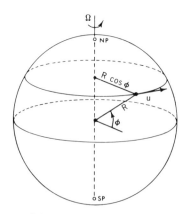

FIG. 4. Schematic diagram of the angular momentum component around the earth's axis of rotation. $M = M_\Omega + M_r = (\Omega R \cos \phi + u)R \cos \phi$. SP, South Pole; NP, North Pole.

TABLE I. HEMISPHERIC AND GLOBAL INTEGRALS OF THE RELATIVE ANGULAR
MOMENTUM M_r IN THE ATMOSPHERE[a]

	Year		DJF		JJA		DJF–JJA	
NH[b]	5.28	(5.11)	9.57	(9.78)	0.20	(−0.26)	9.37	(10.04)
SH[b]	7.58	(7.42)	4.84	(5.02)	9.48	(9.50)	−4.64	(−4.48)
GL[b]	12.88	(12.54)	14.41	(14.80)	9.71	(9.24)	4.70	(5.56)

[a] Given in units of 10^{25} kg m^2 sec^{-1}. Estimates by Newell et al. (1972) are given in parentheses.
[b] NH, Northern Hemisphere; SH, Southern Hemisphere; GL, globe.

Our estimates of the relative angular momentum M_r in the atmosphere are presented in Table I. The table shows that the annual mean value for the globe of 12.88×10^{25} kg m^2 sec^{-1} is, of course, small (about 1%) compared with the corresponding value for the Ω angular momentum, M_Ω, which is 10.1×10^{27} kg m^2 sec^{-1} (i.e., atmosphere in solid rotation with underlying surface). For comparison, we have added similar estimates of the various integrals by Newell et al. (1972) in Table I. Both data sets give about the same results.

The computed increase in M_r from JJA to DJF of 4.70×10^{25} kg m^2 sec^{-1} must be accompanied by a decrease in the rotation rate of the solid earth because the total angular momentum of the atmosphere–ocean–earth system must remain constant. In other terms, this implies a lengthening of the day from JJA to DJF of about 0.7 msec day^{-1}, which is about the value observed by astronomers. Hide et al. (1980) have reported a very interesting extension of these meteorological–astronomical comparisons to periods from 1 day to 1 yr. These analyses were based on daily global analyses of M_r and astronomical length-of-day observations during the year 1979, when the first GARP Global Experiment (FGGE) was taking place and the atmospheric data coverage was greatly improved. They found a very close correspondence between the day-to-day and month-to-month fluctuations in the length of day and those in the atmospheric angular momentum. This appears to justify the conclusion that fluctuations in the rotation rate of the solid earth with periods of a year or less can be explained solely by the exchange of angular momentum with the atmosphere. Interactions with the oceans or the liquid core are therefore probably only important at the longer, say, decadal time scale.

3.1. Description of the Basic Circulation

3.1.1. Global Distribution. The contribution of the zonal wind to the global integral of relative angular momentum M_r varies with location on the globe and also with the season as seen from Eq. (3.2b). The spatial patterns of the vertical mean zonal wind component $\hat{\bar{u}}$ are shown in Fig. 5 for the year, and the two extreme seasons DJF and JJA, respectively. These patterns show that in middle latitudes, between about 30° and 60° latitude, the dominant contribution comes from the subtropical and polar westerly jet streams. In the tropics the easterly trade winds are reflected in the negative values of $\hat{\bar{u}}$. The patterns in the Southern Hemisphere appear more zonally distributed than in the Northern Hemisphere, as would be expected from the greater homogeneity of the earth's surface in that hemisphere. We may mention that poor data coverage over the southern oceans is also a contributing, although less important, factor in making the patterns in the Southern Hemisphere more uniform. The major contributions to the angular momentum in the Northern Hemisphere stem from the midlatitude jets over eastern North America, Asia, and the adjacent oceans.

Comparing Fig. 5b and c, the seasonal shifts of about 10° latitude in the belt of westerlies and 10°–15° latitude in the belt of easterlies, in both cases toward the summer pole, become very clear. The westerlies are strongest in the winter hemisphere due to an increase in the north–south temperature gradient, whereas the tropical easterlies are strongest in the summer hemisphere. These seasonal shifts are perhaps more clearly seen in the zonal-mean profiles of Fig. 8, to be discussed in more detail later.

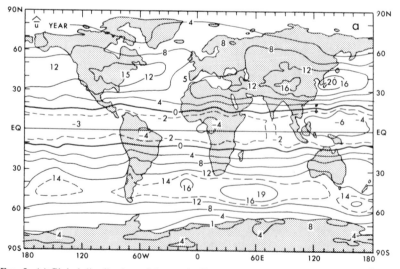

FIG. 5. (a) Global distribution of the vertically averaged zonal wind component $\hat{\bar{u}}$ for the 10-yr period. Positive values indicate eastward flow. Units are m sec^{-1}. (b) Same as (a) except for DJF season. (c) Same as (a) except for JJA season.

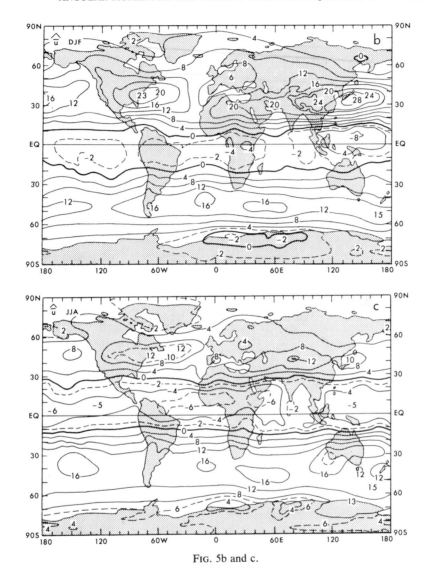

FIG. 5b and c.

For completeness, as well as for later reference, we also present the vertical-mean meridional wind component $\hat{\bar{v}}$ in Fig. 6. The fields of $\hat{\bar{u}}$ and $\hat{\bar{v}}$ taken together are the vector components of the vertically averaged horizontal flow over the globe. Obviously the circulation is mainly east–west due to the predominance of the u component. In fact, maximal meridional velocities are at most 6 m sec^{-1} during the northern winter, and are associated with the semipermanent troughs over the east coasts of North America and Asia. The zonal average value of $\hat{v}$ should practically vanish because the net flow of mass across latitude circles has to be negligible, even on a seasonal basis, as a consequence of the equation of continuity. To account for the accumulation of mass, as revealed by the observed pressure variations, mean meridional velocities on the order of 1 mm sec^{-1} or less would be sufficient. Yet the values for $[\hat{v}]$ calculated directly from the wind maps in Fig. 6 are two orders of magnitude too large (Rosen, 1976). Therefore, the directly obtained $[\hat{v}]$ values cannot be representative of the real conditions in the atmosphere.

3.1.2. Vertical Structure of the Circulation. The vertical structure of the zonal wind is shown in Fig. 7a for the year, and in Table A2 in Appendix A for the year and for the two seasons DJF and JJA. The similarities between the two hemispheres for the annual mean are quite striking. For example, the zonal circulation in both hemispheres is dominated by a westerly jet maximum of about 25 m sec^{-1} around 200 mbar. However, there are also differences since in the Southern Hemisphere the winds at all levels between 35° and 60° latitude are consistently stronger, by about 5 m sec^{-1}, than in the Northern Hemisphere, reflecting the greater strength of the zonal winds in that hemisphere. The lower surface

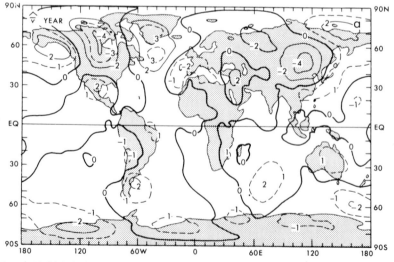

FIG. 6. (a) Global distribution of the vertically averaged meridional wind component $\hat{v}$ for the 10-yr period. Positive values indicate northward flow. Units are m sec^{-1}. (b) Same as (a) except for DJF season. (c) Same as (a) except for JJA season.

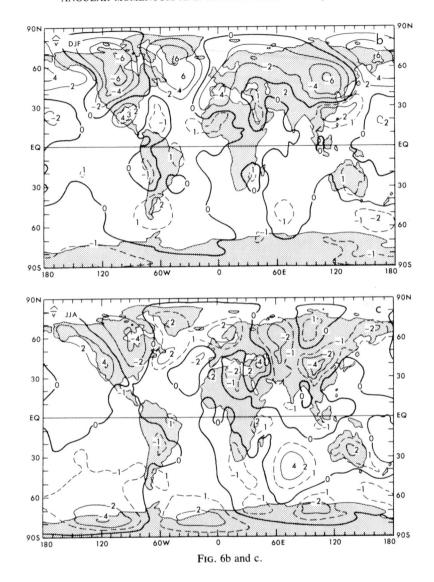

FIG. 6b and c.

values of the easterlies in the equatorial zone are clearly associated with the doldrums.

Vertical integration of the $[\bar{u}]$ values leads to the profiles shown in Fig. 8, which summarize the discussion of the mean zonal flow. To assess the intraannual variability, the profiles for the two extreme seasons are also included. Summer weakening and poleward shifts of the jet are most

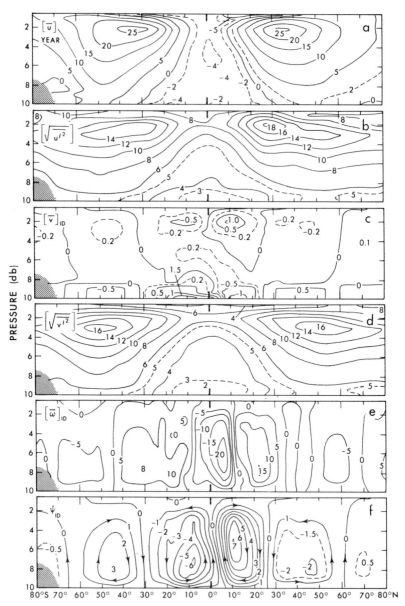

FIG. 7. Latitude–pressure diagrams of (a) the mean zonal winds in m sec^{-1}, (b) the mean day-to-day standard deviation $[\sigma(u)]$ in m sec^{-1}, (c) the mean meridional wind (computed by an indirect method poleward of 20° latitude) in m sec^{-1}, (d) the mean day-to-day standard deviation $[\sigma(v)]$ in m sec^{-1}, (e) the mean vertical velocity (computed from c using conservation of mass) in 10^{-5} mbar sec^{-1}, and (f) the stream function for mass in 10^{10} kg sec^{-1} for the 10-yr period 1963–1973.

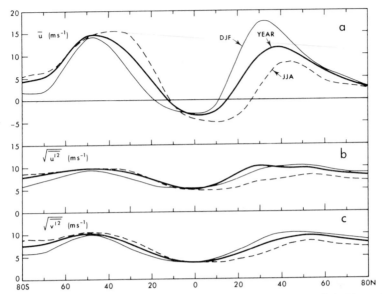

FIG. 8. Meridional profiles of the zonal and vertical mean (a) zonal wind, (b) standard deviation of the zonal wind, and (c) standard deviation of the meridional wind.

pronounced in the Northern Hemisphere but are also obvious in the Southern Hemisphere.

The meridional and vertical components of the mean cellular circulation are shown separately in Fig. 7c and e, and also in Fig. 7f in the form of a steam function. Outside the tropics the direct evaluation of the mean meridional circulation from the $\bar{v}$ field, which involves the averaging of positive and negative values of almost equal magnitude, becomes unreliable or even unusable in the Southern Hemisphere. In view of this difficulty, which is especially apparent in the Southern Hemisphere (see Table A3), we have used poleward of 15° latitude, in both hemispheres, indirectly computed values of $[\bar{v}]$ and $[\bar{\omega}]$. The method we followed was first developed by Kuo (1956) and later used by Gilman (1965), Holopainen (1967), among others, and most extensively by Newell *et al.* (1972). For the region between 15° S and 15° N, however, directly measured $[\bar{v}]$ values were used because there the meridional winds are more uniform and the observational network seems adequate to monitor the dominant Hadley circulation. Between 10° and 20° latitude the values derived from the two methods were averaged with weights varying between 0 and 1 to ensure a smooth transition between the directly and indirectly calculated $[\bar{v}]$. For the methodology utilized the reader is referred to Appendix B, where a detailed analysis is given of the actual procedure used and its implications for the various transport calculations.

Both the direct (uncorrected for net mass flow across latitude circles) and the indirect [$\bar{v}$] values are tabulated in Tables A3 and A4 for the year and the two seasons. As we have seen before, the direct [$\bar{v}$] are unusable in the extratropics for the Southern Hemisphere, giving a much too strong indirect Ferrel circulation (Oort, 1978).

The combined direct/indirect values show reasonable patterns with strong Hadley cells in the tropics, and weaker indirect circulations in middle latitudes. At high latitudes again weak direct circulations are found. The seasonal variation in the intensity and in extent of the tropical Hadley cells is remarkable, the winter Hadley cell being the dominant one. The Ferrel cells show little seasonal variation with only a slight tendency for weakening and poleward migration in summer.

The vertical velocity pattern illustrated in Fig. 7e shows strong rising motion of 2×10^{-4} mbar sec^{-1} (or about 3 mm sec^{-1}) centered at about 5° N, associated with the mean position of the Intertropical Convergence Zone (ITCZ). This equatorial belt is flanked in each hemisphere by sinking motion between about 10° and 40°, by rising motion between 50° and 70°, and again by sinking motion poleward of 70° latitude.

The two-dimensional stream function ψ is given in Fig. 7f. The stream function was evaluated from the [$\bar{v}$] values using the equations

$$[\bar{v}] = g \ \partial\psi/2\pi R \cos \phi \ \partial p \tag{3.3a}$$

$$[\bar{\omega}] = -g \ \partial\psi/2\pi R^2 \cos \phi \ \partial\phi \tag{3.3b}$$

assuming that $\psi = 0$ at the earth's surface and at the top of the atmosphere.

Inspection of Fig. 7f reveals clearly a three-cell structure in both hemispheres. The direct Hadley cells in the tropics are much stronger than the indirect Ferrel cells in middle latitudes. The direct polar cells are quite weak. The Hadley cell of the Southern Hemisphere penetrates across the equator, reinforcing the upward motions characteristic of the ITCZ. Our results agree very well with those published by Newell et al. (1972).

3.1.3. Variability of the Circulation. In order to study the variability of the wind components, the mean standard deviations for u and v, $\sigma(u)$ and $\sigma(v)$, were computed from all daily values available. The annual cross sections of the zonal-mean standard deviations for u and v are shown in Fig. 7b and d, and the seasonal values are given in Tables A12 and A13. We should point out that $\sigma^2(u)$ and $\sigma^2(v)$ represent the components of the transient eddy kinetic energy.

It is of interest to notice the close symmetry between the two hemispheres. Further, for both u and v the standard deviations are of the same magnitude as $[\bar{u}]$ or even larger, except near the subtropical jet stream between 30° and 40° latitude. This feature as well as the nearly identical patterns for the standard deviations of u and v clearly point out the turbulent character of the atmospheric general circulation. The structure of the maximum $\sigma(u)$ in the Northern Hemisphere reflects the larger variability of the northern jet stream. In terms of the transient eddy kinetic energy we find an equal partitioning of energy between the u and v components, i.e., $\sigma^2(u) \approx \sigma^2(v)$. The seasonal differences, as shown in Tables A12 and A13, are small, the winter standard deviations being somewhat larger than the summer ones, as one would expect.

The vertically integrated values are shown in the profiles of Fig. 8b and c. As mentioned above, the seasonal variations in the variances are relatively weak. The yearly curves sometimes lie above both seasonal profiles due to the extra variance resulting from the winter–summer variation included in the yearly standard deviation.

3.2. Meridional Transport of Angular Momentum

3.2.1. Balance Equation of Absolute Angular Momentum. In a rotating coordinate system Newton's second law of motion implies that the time rate of change of total angular momentum of a unit volume is equal to the sum of all torques acting on it. In the context of the atmosphere, the important torques are the pressure and the frictional torques. Using the (λ, ϕ, p, t) system, one can therefore write

$$dM/dt = -R \cos \phi \, g \, \partial z/R \cos \phi \, \partial \lambda + R \cos \phi \, F_\lambda \qquad (3.4)$$

where z is the geopotential height of a constant pressure surface and F_λ is the frictional force component in the λ direction. Expanding the total time derivative and using conservation of mass gives

$$\partial M/\partial t = -\text{div}_2 \, M\mathbf{v} - \partial(M\omega)/\partial p - g \, \partial z/\partial \lambda + R \cos \phi \, F_\lambda \qquad (3.5)$$

where

$$\text{div}_2 = \left(\frac{\partial}{R \cos \phi \, \partial \lambda} , \frac{\partial(\cos \phi)}{R \cos \phi \, \partial \phi} \right) \quad \text{and} \quad \mathbf{v} = (u, v)$$

Averaging Eq. (3.5) over time (denoted by an overbar) gives

$$\partial \bar{M}/\partial t = -\text{div}_2 \, \overline{M\mathbf{v}} - \partial \overline{M\omega}/\partial p - g \, \partial \bar{z}/\partial \lambda + R \cos \phi \, \bar{F}_\lambda \qquad (3.6)$$

For later applications it is useful to integrate Eq. (3.6) over the mass in a polar cap north of a "vertical wall" at a latitude of ϕ. After using Gauss' divergence theorem, one obtains

$$\frac{\partial}{\partial t} \iiint \bar{M}\, dm = \iint\limits_{\text{wall}} \overline{Mv}\, \frac{dx\, dp}{g} - g \iiint \frac{\partial \bar{z}}{\partial \lambda}\, dm + \iiint R \cos \phi\, \bar{F}_\lambda\, dm \quad (3.7)$$

where dm is a mass element. Equation (3.7) states that the time rate of change of angular momentum in the polar cap is balanced by three principal terms. The first term represents the horizontal in- or outflow of angular momentum through the vertical wall. The second term is the so-called mountain torque. It only contributes where mountains intersect the isobaric levels, so that we may write

$$-g \iiint \partial \bar{z}/\partial \lambda\, dm = \oint \left(\sum \bar{z}_E - \sum \bar{z}_W \right) R \cos \phi\, dy\, dp \quad (3.8)$$

where $\oint$ indicates that the integral is taken over the entire earth's surface north of the wall. This term can be calculated by summing the pressure (or geopotential height) differences between the east (subscript E) and west (subscript W) sides of all major mountain ranges in the polar cap. Finally, the third term represents the frictional torque. Below-grid-scale mountain effects not contained in the mountain torque are also included in the third term. Since

$$F_\lambda = -\partial \tau_{F\lambda}/\partial x - \partial \tau_{F\phi} \cos^2 \phi/\cos^2 \phi\, \partial y - \partial \tau_{Fp}/\partial p$$

where $(\tau_{F\lambda}, \tau_{F\phi}, \tau_{Fp})$ are the components of the friction stress in the λ direction, the mass integral can be reduced to a surface boundary integral

$$\iiint R \cos \phi\, \bar{F}_\lambda\, dm = \oint \bar{\tau}_0 R \cos \phi\, dx\, dy \quad (3.9)$$

where the $\tau_{F\phi}$ contributions are neglected at the wall. The surface stress τ_0 ($\tau_0 = -\tau_{Fp}/g$ at $p = p_0$) is counted positive when westerly angular momentum is transferred from the earth to the atmosphere, and vice versa.

Substituting Eqs. (3.8) and (3.9) in Eq. (3.7) yields the final equation

$$\frac{\partial}{\partial t} \iiint \bar{M}\, dm = \iint\limits_{\text{wall}} \overline{Mv}\, \frac{dx\, dp}{g} + \oint \left(\sum \bar{z}_E - \sum \bar{z}_W \right) R \cos \phi\, dy\, dp$$

$$+ \oint \bar{\tau}_0 R \cos \phi\, dx\, dy \quad (3.10)$$

which will be used in the following sections.

3.2.2. Horizontal Transfer Modes. In order to investigate the mechanisms of horizontal transport the first term on the right-hand side of Eq. (3.10) may be written as

$$\iint \overline{Mv} \, dx \, dp/g = 2\pi R \cos \phi \int [\overline{Mv}] \, dp/g$$

where the brackets indicate a zonal average.

For a unit pressure layer the total transport can be expanded as follows:

$$[\overline{Mv}] = [\overline{M'v'}] + [\bar{M}^* \bar{v}^*] + [\bar{M}][\bar{v}] \qquad (3.11)$$

where the prime and asterisk indicate departures from the time mean and zonal mean, respectively.

Writing M explicitly in terms of relative and Ω angular momentum, using Eq. (3.2), reduces Eq. (3.11) to

$$[\overline{Mv}] = ([\overline{u'v'}] + [\bar{u}^* \bar{v}^*] + [\bar{u}][\bar{v}] + \Omega R \cos \phi \, [\bar{v}]) R \cos \phi \qquad (3.12)$$

The terms on the right-hand side of Eq. (3.12) show the various mechanisms that contribute to the total northward flux of angular momentum. These mechanisms are the flux of relative angular momentum by transient eddies, by stationary or standing eddies, and by mean meridional circulations, and the flux of Ω angular momentum by the mean meridional circulation, respectively.

The vertical distribution of the zonal-mean flux of momentum and a breakdown in its various components is shown in the cross sections of Fig. 9 for the year. The first qualitative impressions are that (1) there is an overall symmetry with respect to the equator, except in the case of stationary eddies and (2) transient eddies dominate the total transport in the upper levels. The asymmetry in the case of standing eddy transport, $[\bar{v}^* \bar{u}^*]$, reflects the more zonal character of the time-mean flow south of 20° S, where the $[\bar{v}^* \bar{u}^*]$ values are, in fact, at the "noise" level. The mean meridional flux of momentum, $[\bar{v}][\bar{u}]$, illustrated in Fig. 9d, shows the effects of the three-cell structure in each hemisphere. Its contribution to the total transport, similar to that of the standing eddies, is relatively small compared with the contribution by the transient eddies, except in the surface boundary layer.

Looking more closely, some interesting quantitative differences between the two hemispheres can be pointed out. Thus, in spite of the fact that the stationary eddies are unimportant in the Southern Hemisphere, the total flux of angular momentum by transient, stationary, and mean meridional circulations together is somewhat greater in this hemisphere. Further, one finds the maximum poleward flux near 35° S at the 300-mbar level, whereas the Northern Hemisphere maximum occurs near 25° N at

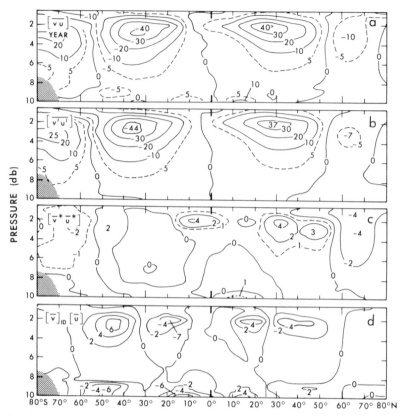

FIG. 9. Zonal mean cross sections of the northward flux of momentum by (a) all motions, (b) transient eddies, (c) stationary eddies, and (d) mean meridional circulations for the 10-yr period. Units are m² sec⁻².

the 200-mbar level. At polar latitudes the maximum equatorward flux appears stronger in the Southern Hemisphere than in the Northern Hemisphere. These differences are practically all due to the differences in the transient eddy flux.

Let us now consider the seasonal variations for the DJF and JJA seasons as tabulated in Tables A17–A19. Again the bulk of the momentum transport is generally accomplished by transient eddies (i.e., eddies with time scales less than 3 months). During each season the transient fluxes in the Southern Hemisphere are considerably stronger than those during the corresponding season in the Northern Hemisphere. Throughout the year the Antarctic appears to be an important source of momentum (i.e., there is a divergence of angular momentum), probably due to the effects of topography. There appears to be a tendency for cross-equatorial transports from the winter to the summer hemisphere.

Our transport patterns are very similar to those presented by Newell *et al.* (1972, pp. 168, 169) but tend to be somewhat weaker. In the case of stationary eddies (Table A18), the Southern Hemisphere contributes very little south of 20° S, as first documented by Obasi (1963). On the other hand, inside the tropics the stationary eddies for the seasons are important, and in the Northern Hemisphere winter they are actually major contributors to the total flux of momentum. Mak (1978) has called attention to possible uncertainties in the published $[\bar{v}^*\bar{u}^*]$ values associated with both real year-to-year variations and sampling and analysis problems due to spatial data gaps. Nevertheless, a comparison of our seasonal transports with those of Newell *et al.* (1972, pp. 166, 167), based on slightly different data and analysis techniques, shows reasonably good agreement. Finally, the mean meridional fluxes (Table A19) show some substantial seasonal contributions in the planetary boundary layer and in the upper troposphere. However, the total integrated values are small compared with the transient eddy transports, except in the tropical region where the Hadley cells are important.

Meridional profiles of the vertical-mean momentum fluxes for the year and for the DJF and JJA seasons are shown in Fig. 10. They condense the points already made in our previous discussion. To obtain the total flux of angular momentum, one must multiply the values of the momentum transport (as shown in Fig. 10a) by a factor $F = 2\pi R^2 \cos^2 \phi \, p_0/g = 2.56 \times 10^{18} \cos^2 \phi$ kg, assuming $p_0 = 1000$ mbar. The flux of earth's angular momentum, as given by the last term in expression (3.12), does not contribute to the net transport of angular momentum because

$$\int_0^{p_0} [\bar{v}] \, dp = 0$$

at all latitudes. The net meridional transports of angular momentum by atmospheric circulations are shown in Fig. 11. For the sake of comparison, we include previous results obtained by the pioneers in this field, Starr and White (1951, 1952, 1954), for the Northern Hemisphere, by Obasi (1963) for the Southern Hemisphere, and by Newell *et al.* (1972, p. 150) for the globe. In general, good agreement is found, which is surprising, especially in the Southern Hemisphere, in view of the large data gaps and the differences in analysis schemes and techniques (e.g., hand analysis in Obasi's study and semiobjective machine analysis in the case of Newell *et al.* and in our case). The differences during the Northern Hemisphere winter with Newell *et al.* (1972) are large, but can perhaps be explained by their use of largely geostrophic wind estimates based on

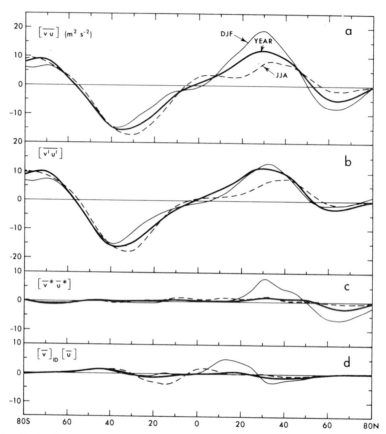

FIG. 10. Meridional profiles of the zonal and vertical mean northward transport of momentum by (a) all motions, (b) transient eddies, (c) stationary eddies, and (d) mean meridional circulations.

Crutcher's (1961) analyses. Most likely our present data set is more comprehensive and our analyses are more representative of present climatic conditions than those published previously.

3.3. Vertical Transport of Angular Momentum

Going back to the basic balance equation for total angular momentum, Eq. (3.6), and taking the zonal average, we obtain the following transformed equation:

$$\partial[\bar{M}]/\partial t = -\partial[\overline{Mv}] \cos \phi / R \cos \phi \, \partial \phi$$
$$- \partial[\overline{M\omega}]/\partial p + R \cos \phi [\bar{F}_\lambda] \qquad (3.13)$$

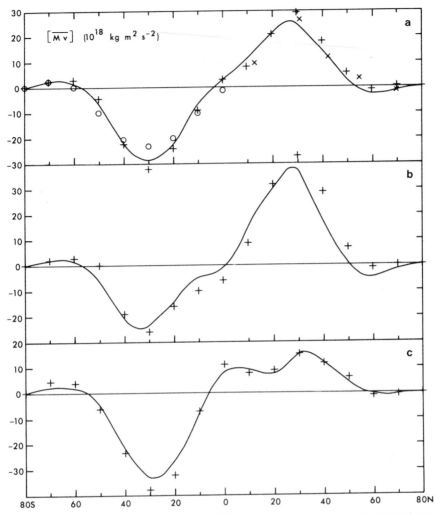

FIG. 11. A comparison of the net northward flow of angular momentum in the atmosphere by various investigators with the present results (solid line): (×) Starr and White (1954) for year 1950; (+) Newell *et al.* (1972); (○) Obasi (1963) for IGY. (a) For the 10-yr period, (b) for DJF season, and (c) for JJA season.

in which, for simplicity, the mountain torque [Eq. (3.8)] is included in a "gross" frictional force $[\bar{F}_\lambda]$. This force may therefore be written as

$$[\bar{F}_\lambda] = -\partial[\bar{\tau}_{F\phi}]R\cos^2\phi/R^2\cos^2\phi\ \partial\phi - \partial[\bar{\tau}_{Fp}]/\partial p \qquad (3.14)$$

where the stress components in the λ direction on the latitudinal wall ($\phi =$ constant), $\tau_{F\phi}$, and on the isobaric level ($p =$ constant), τ_{Fp}, are produced by at least three factors. These are friction, subgrid-scale momentum exchange, and pressure differences across topographic features ranging from the smallest hills to the largest mountains.

Substituting Eq. (3.14) into Eq. (3.13) gives

$$\partial[\bar{M}]/\partial t = -\partial\{[\overline{Mv}] \cos \phi + [\bar{\tau}_{F\phi}]R \cos^2 \phi\}/R \cos \phi \; \partial\phi$$
$$-\partial\{[\overline{M\omega}] + [\bar{\tau}_{Fp}]R \cos \phi\}/\partial p \qquad (3.15)$$

To clarify the important mechanisms involved in the vertical flux $[\overline{M\omega}]$, it will be decomposed into different components in a manner analogous to the decomposition of $[\overline{Mv}]$ presented before in Eq. (3.12). Thus, we have

$$[\overline{M\omega}] = ([\overline{u'\omega'}] + [\bar{u}^*\bar{\omega}^*] + [\bar{u}][\bar{\omega}] + \Omega R \cos \phi \; [\bar{\omega}])R \cos \phi \quad (3.16)$$

Since time variations of the angular momentum are usually small, the left-hand side of Eq. (3.15) can be set equal to zero. Under this condition we can define a stream function ψ_M for the angular momentum of a zonal ring of air, following Starr *et al.* (1970). We may write

$$2\pi R^2 \cos^2 \phi \; [\bar{\tau}_\phi] = -\partial\psi_M/\partial p \qquad (3.17a)$$

$$2\pi R^2 \cos^2 \phi \; [\bar{\tau}_p] = \partial\psi_M/R \; \partial\phi \qquad (3.17b)$$

where

$$[\bar{\tau}_\phi] = [\overline{Mv}]/R \cos \phi + [\bar{\tau}_{F\phi}] \qquad (3.18a)$$

$$[\bar{\tau}_p] = [\overline{M\omega}]/R \cos \phi + [\bar{\tau}_{Fp}] \qquad (3.18b)$$

Substitution of the expansions (3.12) and (3.16) in these equations gives for the total stress components, τ_ϕ and τ_p, the expressions

$$[\bar{\tau}_\phi] = [\overline{u'v'}] + [\bar{u}^*\bar{v}^*] + [\bar{u}][\bar{v}] + \Omega R \cos \phi \; [\bar{v}] + [\bar{\tau}_{F\phi}] \qquad (3.19a)$$

$$[\bar{\tau}_p] = [\overline{u'\omega'}] + [\bar{u}^*\bar{\omega}^*] + [\bar{u}][\bar{\omega}] + \Omega R \cos \phi \; [\bar{\omega}] + [\bar{\tau}_{Fp}] \qquad (3.19b)$$

These expressions show that the total stress results from the combined effects of large-scale transient eddies, stationary eddies, mean meridional circulations (associated with the advection of zonal momentum and the sphericity of the earth), and finally, mountains, subgrid-scale eddies, and friction.

Above the surface boundary layer in the free atmosphere $[\bar{\tau}_{F\phi}]$ can usually be neglected. This permits the evaluation of $[\bar{\tau}_\phi]$ from the vertical cross sections shown in Figs. 7c and 9a, and the corresponding tables. Finally the ψ_M distribution can be obtained for all points in the (ϕ, p) plane by integrating Eq. (3.17a) downward starting with $\psi_M = 0$ at the top level (here chosen as $p = 25$ mbar). The vertical transport $[\bar{\tau}_p]$ can then be determined using Eq. (3.17b).

The resulting patterns of the flow of total angular momentum for the year, DJF, and JJA are shown in Fig. 12. However, in order to better show the surface source and sink regions of angular momentum for the atmosphere, we also evaluated the nondivergent component of the flow of relative angular momentum. This component was obtained by neglecting the $\Omega R \cos \phi$ $[\bar{v}]$ term, i.e., the internal source (or sink) term of relative angular momentum, in integrating Eq. (3.17a) downward as described above. The final outcome is presented in Fig. 13. Through these operations the "free-wheeling" circulations of the earth's angular momentum associated with the mean meridional overturnings which do not yield a net transport across latitude circles (since $(1/g) \int_0^{p_0} [\bar{v}]\, dp = 0$) are eliminated. At the earth's surface Figs. 12 and 13 should be identical.

Figures 12 and 13 together give a clear picture of the overall cycle of angular momentum in the atmosphere. They depict the interactions with the earth's surface, defining very clearly the sources and sinks of angular momentum for the atmosphere.

In the yearly picture (Figs. 12a and 13a) the circulation of angular momentum is symmetric with respect to the equator with a dominant flow

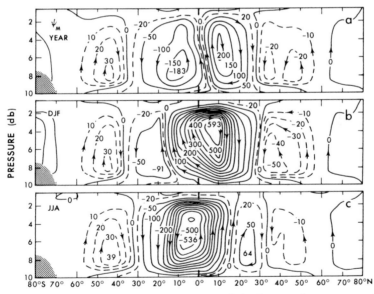

FIG. 12. Streamlines of the zonal mean transport of absolute angular momentum in the atmosphere for the 10-yr period (a) and the DJF (b) and JJA (c) seasons. Units are 10^{18} kg m^2 sec^{-2}.

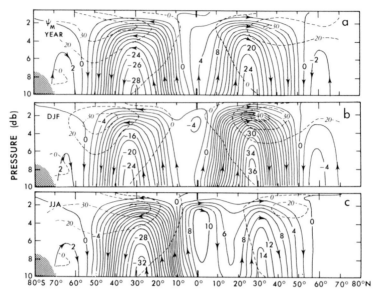

FIG. 13. Streamlines of the nondivergent component of the zonal mean transport of relative angular momentum in the atmosphere for the 10-year period (a) and the DJF (b) and JJA (c) seasons. Added are some dashed contours of $[\bar{u}]/\cos\phi$ in units of m sec^{-1}, which show the countergradient nature of the eddy transports. Units are 10^{18} kg m^2 sec^{-2}.

from the source regions in the subtropics to the sink regions in midlatitudes. The atmosphere gains westerly momentum in the region of surface easterlies between 30° S and 25° N with maxima near 15° S and 15° N, where the easterlies are strongest (see Fig. 7a). In the free atmosphere the vertical transports are mainly accomplished by mean meridional circulations as shown by a comparison of Figs. 7e and 12a. However, the meridional transport around 30° latitude is due to the eddies. Finally the downward transport into the sink regions, in the area of surface westerlies, is again accomplished by the mean meridional circulations. The polar surface easterlies also provide some westerly momentum for middle latitudes.

To show the implications of these transports for the zonal kinetic energy balance, dashed contours of the angular rotation of the atmosphere, given by $[\bar{u}]/R\cos\phi$, are added in Fig. 13. Note that for simplicity R is left out in labeling the contours. It is interesting to see that the meridional (mainly eddy) flux of relative angular momentum in the upper troposphere is directed toward higher values of angular rotation. Starr (1953, 1968) coined this a "negative viscosity" effect. It is a very important concept that has, to some extent, revolutionized our ideas of the operation of the atmospheric general circulation. As we shall discuss later, negative viscosity implies a conversion from eddy to zonal kinetic energy.

Let us consider next the seasonal diagrams (Figs. 12b, c and 13b, c). The main source of westerly angular momentum is found in the subtropics of the winter hemisphere, where the surface easterlies are strongest. Most of this angular momentum is transported toward midlatitudes of the same hemisphere, where it is returned to the earth's surface. Only a small fraction is transported across the equator into the summer hemisphere (Fig. 13b and c).

3.4. Angular Momentum Transfer between Earth and Atmosphere

3.4.1. Global Distribution. Integration of the balance equation of angular momentum [Eq. (3.10)] over the entire global atmosphere shows that in the long run the sum of the friction and mountain torques at the earth's surface must vanish, since

$$\frac{\partial}{\partial t} \iiint \bar{M} \, dm \approx 0$$

Thus we may write

$$\oint \bar{\tau}_0 R \cos \phi \, dx \, dy + \oint \left(\sum \bar{z}_E - \sum \bar{z}_W \right) R \cos \phi \, dy \, dp/g = 0$$

where the $\oint$ symbol indicates again that the integrals are taken over all points of the earth's surface. As will be seen later, the mountains generally reinforce the friction effects, although they are of less importance (White, 1949; Newton, 1971a). This equation leads to the general requirement that from a global perspective the surface area covered by easterly trade winds ($\tau_0 > 0$) should be roughly the same as the surface area covered by midlatitude westerlies ($\tau_0 < 0$) (Lorenz, 1967).

We can investigate these questions more deeply through a study of the horizontal distribution of the torque associated with the sum of the vertically integrated horizontal divergence of angular momentum and the pressure torque in Eq. (3.6), which in the long run ($\partial \bar{M}/\partial t \approx 0$) should be equal to the surface torque:

$$\bar{\tau}_0 R \cos \phi = \int_0^{p_0} \mathrm{div}_2 \, \overline{Mv} \, dp/g + \int_0^{p_0} (\partial \bar{z}/\partial \lambda) \, dp \qquad (3.20)$$

This expression can be rewritten by expanding the first term on the right-hand side as follows:

$$\bar{\tau}_0 R \cos \phi = \int_0^{p_0} \text{div}_2(\overline{u'v'}R \cos \phi) \, dp/g$$

$$+ \int_0^{p_0} \text{div}_2 (\bar{u}\bar{v} R \cos \phi) \, dp/g$$

$$- \int_0^{p_0} f\bar{v}_{ag} R \cos \phi \, dp/g \qquad (3.21)$$

where

$$\bar{v}_{ag} = \bar{v} - (g/f) \, \partial \bar{z}/R \cos \phi \, \partial \lambda.$$

The last term is difficult to measure because it represents the sensitive balance between wind and geopotential height. In order to expect reasonable results, as pointed out by Holopainen (1982), the ageostrophic wind component should be determined from samples of independent v and z data that are as complete as possible. There is some evidence, although indirect and tentative, for an approximate balance between the second and third terms on the right-hand side of Eq. (3.21) (Holopainen and Oort, 1981). This possible balance implies that the transient eddy term may provide the essential contribution to the surface stress. A map of this last quantity is shown in Fig. 14 for annual mean conditions. The distribution

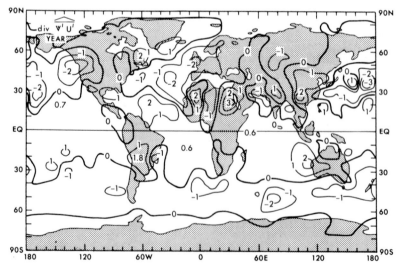

FIG. 14. Global distribution of the vertical mean divergence of momentum by transient eddies for the 10-yr period. Units are dyn cm^{-2}.

shows generally divergence in the tropical latitudes (source region) and convergence in the middle latitudes (sink region), as one would expect. One finds a tendency for the largest values (about 2 dyn cm^{-2}) to occur over the oceans. The narrow convergence zone over the central Sahara in between two divergence centers is probably unreliable and due to data problems.

3.4.2. Latitudinal Profiles.   To summarize the exchange of angular momentum between the earth and the atmosphere, zonal mean stress profiles are probably best suited. Figure 15a shows the latitudinal distribution of the zonal mean stress due to friction and mountains as inferred from the divergence of angular momentum using Eq. (3.20). For easy comparisons with other studies, the same profiles as given in Fig. 15a are redrawn in Fig. 15b in the form of total flux divergence profiles integrated over 5° latitude belts in Hadley units (10^{18} kg m^2 sec^{-2}). We can compare the first set of profiles with those presented by Priestley (1951) in his pioneering study of the surface stress over the oceans evaluated from surface ship reports. He used a bulk aerodynamic formulation,

$$\tau_0 = \rho c_D |\mathbf{v}| u$$

where the drag coefficient $c_D = 0.0013$. In spite of the uncertainties in his formulation and the fact that only ocean values were included, the quantitative agreement between our zonal mean land-plus-ocean values and his all-ocean values is very good both for the annual mean and the extreme seasons. The same agreement holds for results from later studies, notably those by Hellerman (1967, 1982), who used a more sophisticated, but still tentative, drag formulation from Bunker (1976), and by Newton (1971b).

In Fig. 16 Hellerman's (1967, 1982) ocean stress values are compared with our total stress values. A direct comparison is only valid if the (as yet unproven) assumption is made that in a latitude belt the average stress over the continents due to friction and mountains is equal to the average stress as computed over the ocean longitudes alone. Because of the small ocean–land fraction, one should disregard the comparison north of about 50° N. Hellerman's most recent results in the Southern Hemisphere agree to within 20–30% with our values, but his older stress values around 50° S were probably too large. In the tropics the agreement is, on the other hand, better with his older results. In this context we may point out that there is some evidence from recent measurements (S. D. Smith, personal communication) that Bunker's drag coefficients are too high and that therefore Hellerman's (1982) results represent overestimates.

To show the relative contribution by mountains, the large-scale mountain torques as evaluated by Newton (1971a) are plotted in Fig. 15c (see

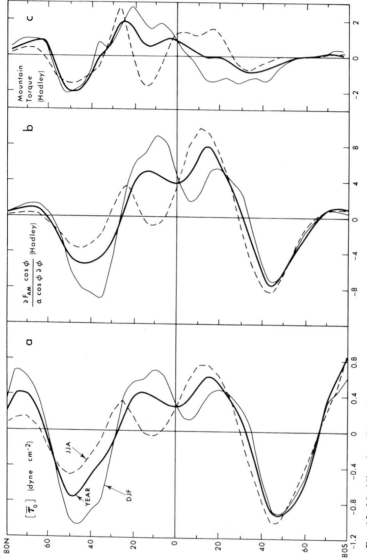

FIG. 15. Meridional profiles of (a) the mean surface stress (due to friction and mountains) per unit area in dyn cm^{-2}, (b) the mean surface torque (due to friction and mountains) integrated over 5° latitude belts in Hadley (= 10^{18} kg m^2 sec^{-2}) units, and (c) the mountain torque in Hadley units from Newton (1971a). The profiles (a) and (b) are computed using the present data for the atmospheric transports of angular momentum.

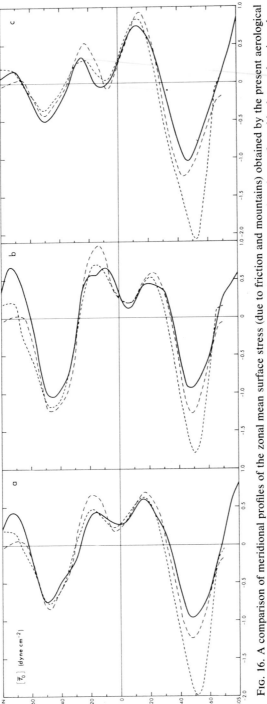

FIG. 16. A comparison of meridional profiles of the zonal mean surface stress (due to friction and mountains) obtained by the present aerological method (——) with similar stress profiles computed by Hellerman (- - -, 1967; ——, 1982) for the oceans only using surface ship data and various drag formulations; (a) the year; (b) DJF, (c) JJA.

also Oort and Bowman, 1974). Comparing Newton's mountain torque curves with the friction plus mountain torque curves in Fig. 15b, one sees qualitative similarity between the two patterns, at least in the Northern Hemisphere with its major mountain ranges. Interesting is the strong seasonal variation in the mountain torque in the Northern Hemisphere and Southern Hemisphere tropics, showing that it is an important term in the seasonal variation of the total torque, and that it generally works in the same sense as surface friction.

4. ENERGY BALANCE OF THE CLIMATIC SYSTEM

As is well known, the radiational energy of the sun constitutes the basic driving force for the atmospheric and oceanic general circulations. In this section we will discuss how the source regions of incoming solar radiation and the sink regions of outgoing terrestrial radiation are distributed over the globe, and how the atmosphere and oceans respond to these forcing factors.

4.1. Radiational Forcing

The most important scientific contribution of meteorological satellites has probably been the measurement of the radiation budget at the top of the atmosphere. This involves measuring both the solar shortwave input and the terrestrial longwave output components. Measurements of the planetary albedo combined with the known impinging solar radiation at the top of the atmosphere supply the necessary information on the first component of the radiation budget, the net solar input. Uncertainties in the present albedo values are related to inadequate diurnal sampling and incomplete knowledge of how to extrapolate from measurements at one zenith angle to full half-sphere values. The second component of the radiation budget involves measuring the longwave, terrestrial radiation. Less uncertainty seems to be involved in this determination.

Figure 17 shows the net radiation at the top of the atmosphere, F_{TA}, for the year and for the DJF and JJA seasons as determined by Campbell and Vonder Haar (1980) from several years of satellite observations. The annual picture shows basically a zonal pattern with energy input ranging from about 60 to 70 W m^{-2} near the equator and energy losses of about -100 W m^{-2} near the south pole and -120 W m^{-2} near the north pole. For the year as a whole, Fig. 17a shows that the ocean regions generally gain more energy than the land regions, requiring a net annual transport of energy from the oceans to the land. Of particular interest is the strong negative anomaly over the North African desert, requiring an appreciable influx of atmospheric energy or, in other words, requiring adiabatic heating by compression of the air to compensate for the radiational cooling. This is mainly a summer phenomenon.

Predominant zonality of the net incoming radiation is clear in the winter hemispheres in Fig. 17b and c. The summer hemisphere patterns are broken up in minima in net radiation over the continents and maxima over the oceans because the surface albedo is larger over land than over water, giving more reflection of solar radiation, and also because infrared radiation losses tend to be somewhat greater over the relatively warm continents than over the cool oceans. Much of the excess of summer radiation over the oceans is absorbed in the oceans themselves. By subtracting estimates of the seasonal heat storage in the oceans and in the overlying atmosphere from their net radiation values, Campbell and Vonder Haar (1982) were able to determine the need for a substantial atmospheric transport of energy from the land to the ocean regions during summer, and from the oceans to the land during winter. The strongest north–south energy fluxes in the atmosphere should occur during winter when the meridional gradients in radiation are greatest. It is also of interest to note that the maximum winter heat losses are found not at the poles but near 65° latitude, which is possibly connected with the areas of ice-free water, where heat losses can go on continuously.

Zonal mean radiation profiles again based on data from Campbell and Vonder Haar (1980) are shown in Figs. 18 and 19. The solar radiation available at the top of the atmosphere (shown in Fig. 18a) has a strong decrease in the winter hemisphere from a value of about 475 W m^{-2} in the subtropics of the summer hemisphere to a zero value at the winter pole, and only a weak decrease toward the summer pole. A considerable part of this radiation is reflected back to space as shown in Fig. 18b. This is especially the case at high latitudes, where the albedo (as illustrated in Fig. 19a) is very large ($>70\%$) because of the greater angle of incidence of the incoming radiation and, to some extent, because of snow and ice coverage. Finally, the absorbed solar radiation available for driving the earth's general circulation is given in Fig. 19b. The main difference with the original solar energy curves in Fig. 18a lies in the decrease of available radiation over the summer pole, leading to an appreciable north–south gradient of absorbed solar radiation even in the summer hemisphere. The annual curves in Figs. 18a and 19b have practically the same shape with an almost uniform translation of 100 W m^{-2} with respect to latitude. In analyzing the implications of these curves, it must be kept in mind that for the globe as a whole, assuming an albedo of 0.30 and a solar constant of

1360 W m^{-2}, the annual average absorbed solar radiation is 238 W m^{-2} with an expected seasonal variation of about 8 W m^{-2} amplitude due to the eccentricity of the earth's orbit around the sun (Ellis *et al.*, 1978).

The profiles of outgoing terrestrial radiation at the top of the atmosphere in Fig. 19c show a high plateau between 30° N and 30° S with a slight dip over the ITCZ, mainly due to extensive cloudiness, and low values at high latitudes. The atmosphere over the Antarctic seems to lose less infrared radiation than the Arctic atmosphere, presumably because of the high ice cap in the Antarctic (Bowman, 1982). However, the north–south gradients are always weak. The net meridional heating profiles (given in Fig. 19d) are obtained by subtraction of the absorbed and the emitted radiation profiles. They are not corrected for global radiation balance, and reflect largely the patterns of absorbed radiation shown in Fig. 19b. Variability in cloudiness and differences in surface albedo, such as those over deserts, forests, and oceans, also play a significant role in shaping the observed radiation profiles in Figs. 18 and 19. However, it is beyond the scope of the present article to further analyze these factors.

A synthesis of the hemispheric and global radiation components of the energy budget was prepared and the final outcome is presented in Table II. In Table IIA we give the results uncorrected for global annual balance. It shows a net imbalance of +9.2 W m^{-2} for the globe. This annual imbalance may be correct for periods of several years and possibly longer,

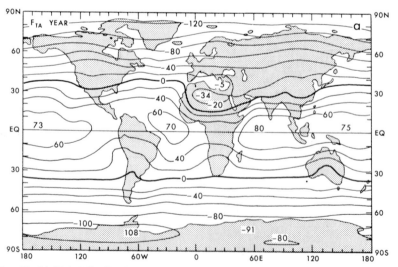

FIG. 17. (a) Global distribution of the net incoming radiation at the top of the atmosphere (F_{TA}) from Campbell and Vonder Haar (1980) for annual mean conditions. Units are W m^{-2}. (b) Same as (a) except for DJF season. (c) Same as (a) except for JJA season.

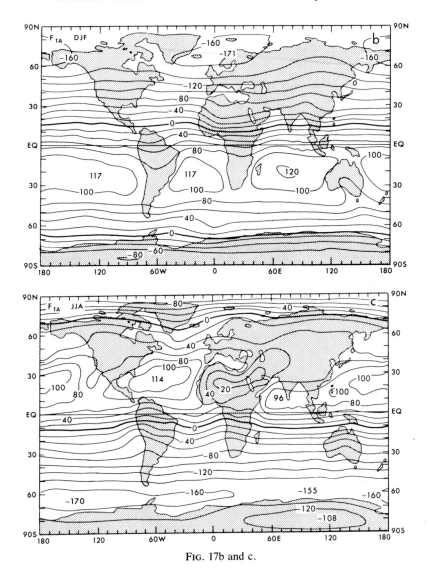

FIG. 17b and c.

during which energy may be stored in the oceans or cryosphere (Saltz-man, this volume). However, the imbalance must be spurious for long-term mean conditions. Therefore, in the present context we assume the global imbalance of about 9 W m^{-2} to be a measure of the error in the net flux calculations. It requires then some correction to preserve the climatic balance. Two different types of possible corrections were introduced as

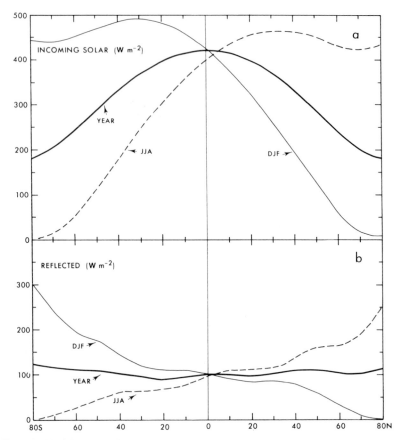

FIG. 18. Meridional profiles of (a) the zonal mean incoming and (b) the reflected solar radiation (Campbell and Vonder Haar, 1980). No corrections were made for global radiation balance.

shown in Table IIB and C. The first correction is the one recommended by Campbell and Vonder Haar (1980), in which the reflected solar and the terrestrial radiation are multiplied by a factor 1.025, which reduces the global imbalance to almost zero. The second correction is based on the assumption that the entire error is due to the albedo measurements. In this case, multiplying the reflected solar radiation by 1.09 ensures global annual balance. The spread in the estimates in the three sections of Table II gives some idea of the uncertainties involved in the various radiation terms.

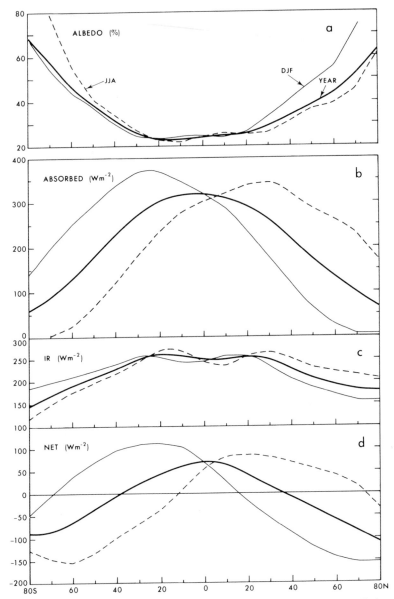

FIG. 19. Meridional profiles of the zonal mean (a) albedo, (b) absorbed solar radiation, (c) emitted infrared radiation (IR), and (d) net incoming radiation. No corrections were made for global radiation balance.

TABLE II. HEMISPHERIC AND GLOBAL RADIATION VALUES (W m^{-2}) AT TOP
OF THE ATMOSPHERE

	Year			DJF[a]			JJA[a]		
	NH[b]	SH[b]	GL[b]	NH	SH	GL	NH	SH	GL
A. No Correction									
Incoming solar	345	344	345	243	467	355	445	225	335
Reflected solar	105	104	103	74	146	110	137	70	98
Absorbed solar	240	243	241	169	321	245	308	165	237
Outgoing terrestrial	232	233	232	222	238	230	242	230	236
Net	8.2	10.1	9.2	−53	83	15.2	66	−64	0.9
Albedo (%)	30.5	29.5	30.0	30.3	31.2	30.9	30.8	26.6	29.4
B. Correction I[c]									
Reflected solar	108	104	106	76	149	112	140	61	101
Absorbed solar	237	240	239	167	317	242	304	164	234
Outgoing terrestrial	237	239	238	227	244	236	248	235	242
Net	−0.2	1.7	0.8	−60	74	6.7	56.5	−71.5	−7.5
Albedo (%)	31.2	30.2	30.7	31.1	32.0	31.7	31.6	27.2	30.1
C. Correction II[d]									
Reflected solar	115	111	113	80	159	120	149	65	107
Absorbed solar	230	234	232	163	308	235	295	160	228
Net	−1.2	1.0	−0.1	−59	70	5.3	54	−70	−8.0
Albedo (%)	33.2	32.2	32.7	33.1	34.0	33.7	33.6	28.9	32.0

[a] DJF, December–February; JJA, June–August.

[b] NH, Northern Hemisphere; SH, Southern Hemisphere; GL, global.

[c] Corrected for global balance by multiplying reflected solar and outgoing terrestrial radiation by 1.025.

[d] Corrected for global balance by multiplying reflected solar radiation only by 1.09.

The net radiation values in Fig. 19d for the annual mean enable us to evaluate the energy transports in the ocean–atmosphere system, $T_A + T_O$, as a function of latitude, needed to maintain the observed temperature structure. The actual calculations proceed by integrating over latitude the net radiation flux starting with $T_A + T_O = 0$ at the north pole:

$$T_A + T_O = - \int_\phi^{\pi/2} F_{TA} 2\pi R^2 \cos \phi' \, d\phi'$$

The results for the raw data uncorrected for global balance (Table IIA) are shown in Fig. 20 as a dashed curve, and for the corrected data (Table IIB, with an additional small correction to ensure exact global balance) as a solid curve. It is clear that a correction for global annual-mean radiation

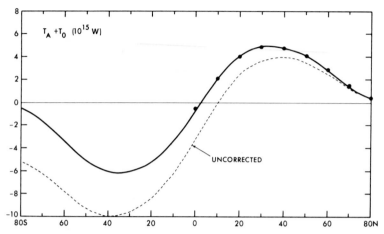

FIG. 20. Meridional profiles of the total transport of energy by the atmosphere and oceans, $T_A + T_O$, for annual mean conditions calculated from radiation requirements. The dashed curve is obtained from the uncorrected data (Table IIA), and the solid curve from the data after correction for global radiation balance (Table IIB). (●) Oort and Vonder Haar (1976).

balance is essential in order to arrive at reasonable values of the required energy transport. For example, a global imbalance of $+10$ W m^{-2} would lead to a fictitious southward energy flux of -2.6×10^{15} W at the equator and -5.1×10^{15} W at the south pole. This example shows that great care must be taken to obtain exact global balance.

The corrected curve in Fig. 20 shows almost antisymmetry with respect to the equator. The largest poleward transports calculated are about 5.0×10^{15} W near 30° N and about -6.0×10^{15} W near 35° S. A small cross-equatorial transport of energy from the Northern to the Southern Hemisphere seems to be necessary. In the seasonal case, the situation is somewhat different because considerable amounts of energy may be stored in the oceans, atmosphere, and continental surfaces, listed in decreasing order of importance. By subtracting the mean storage, $S_A + S_O$, at each latitude seasonal curves of $T_A + T_O$ could have been obtained. However, various corrections would have to be made for each storage component for global balance. Since this requires great care and much critical discussion, this will be left to a later publication (Oort et al., 1983).

The present estimates of $T_A + T_O$ for the Northern Hemisphere are quite similar to those presented before by Oort and Vonder Haar (1976), as can be seen by the comparison of the solid curve with the plotted values in Fig. 20.

4.2. Description of the Energy in the Atmosphere

Energy can be stored in the atmosphere in various forms, namely as internal energy, potential energy, latent heat, or kinetic energy. Thus, the total energy per unit mass, E, is given by the expression

$$E = c_v T + gz + Lq + \tfrac{1}{2}(u^2 + v^2) \tag{4.1}$$

4.2.1. Global Distribution of the Various Forms of Energy.

Expression (4.1) shows that the evaluation of the various forms of energy requires the three-dimensional distributions of the fields of T, z, q, u, and v. The various maps of u and v were already discussed in Section 3.1.1. Because the vertical-mean internal and potential energy are linearly related to each other through the relation $PE = (c_p/c_v - 1)IE$, only maps of the mean temperature $\hat{T}$ and of the specific humidity $\hat{q}$ are needed to describe the geographical distribution of the energy.

These maps are presented in Figs. 21 and 22 for the annual-mean conditions and for the DJF and JJA seasons. The mean values of $\hat{T}$ and $\hat{q}$ were evaluated for the layer between the surface and 25 mbar. Because of the strong decrease of T and q with height, values over mountainous regions are generally lower than over flat land and oceans. One of the reasons for presenting these maps is to show the response of the atmosphere to the radiational forcing. The seasonal changes in the gradients and intensity of

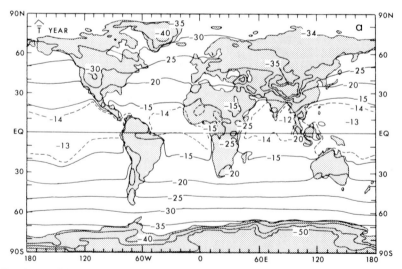

FIG. 21. (a) Global distribution of the vertical mean temperature in °C for the 10-yr period. (b) Same as (a) except for DJF season. (c) Same as (a) except for JJA season.

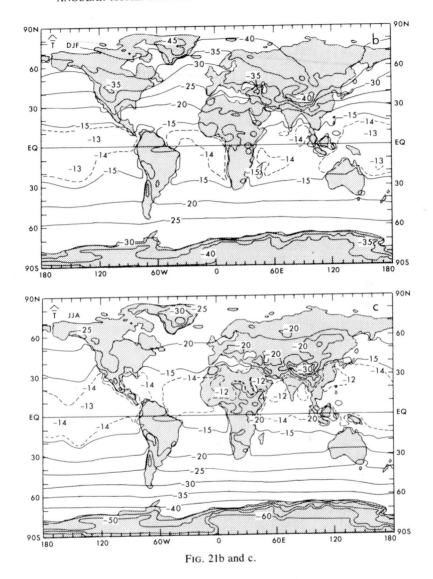

FIG. 21b and c.

the $\hat{\tilde{T}}$ and $\hat{\tilde{q}}$ fields must follow the changes in radiational forcing. Of course, their amplitude should be much smaller than those in the forcing, due to the moderating effects of heat storage in the oceans and to the advection by atmospheric and oceanic currents. The $\hat{\tilde{T}}$ maps in Fig. 21 show clearly the effects of the land–sea distribution, the nature of the land surface, and its topography. The influence of topography is evident in the configuration of the isolines over the mountainous regions and in the strong gradients over the Antarctic coast. The absence of continents in the middle latitudes of the Southern Hemisphere explains the characteristic zonality of the $\hat{T}$ in this hemisphere. Seasonal temperature changes tend to be most pronounced outside the tropics.

The $\hat{q}$ maps in Fig. 22 parallel the $\hat{T}$ maps, as is to be expected. However, the seasonal changes àre now more pronounced in the tropics. In both cases the role of the oceans is apparent. In fact, the highest values of $\hat{T}$ and $\hat{q}$ occur over the oceans, mainly in the winter season. For further discussion of the $\hat{q}$ fields see Peixóto and Oort (1983).

4.2.2. Vertical Structure of the Energy Distribution.

The vertical structure of the temperature (internal energy), geopotential height (potential energy), and specific humidity (latent heat) fields is shown in Fig. 23 for the annual mean, and in Tables A6–A8 of Appendix A for the year and for the DJF and JJA seasons. In order to show the structure of the geopotential height more clearly, the vertical profile of the NMC standard atmosphere (see Table III) was subtracted. The decrease in energy toward the poles for all three components, as well as the seasonal shifts in their gradients, result from the changes in radiational forcing. The annual mean pictures show predominant symmetry with respect to the "meteorological" equator at about 5° N. Starr *et al.* (1969) show global humidity cross

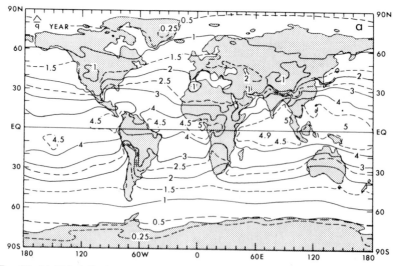

Fig. 22. (a) Global distribution of the vertical mean specific humidity in g kg⁻¹ for the 10-yr period. (b) Same as (a) except for the DJF season. (c) Same as (a) except for JJA season.

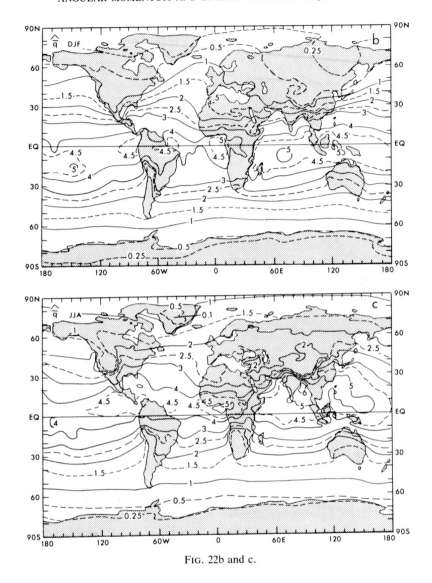

FIG. 22b and c.

sections which compare well with our Fig. 23e and f. Peixóto's (1974) Southern Hemisphere temperature cross sections are also in agreement with our Fig. 23a and b.

In Fig. 24 we present the meridional profiles of the vertically averaged $\bar{T}$, $\bar{z} - z_{SA}$, $\bar{q}$ and, finally, the total energy $\bar{E}$. These profiles synthesize the general behavior of the corresponding forms of energy. The contribution

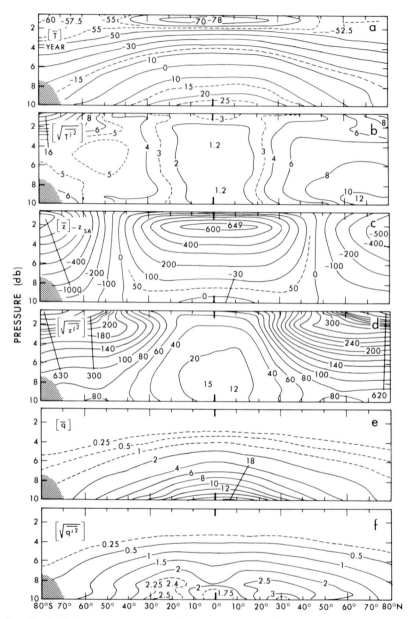

FIG. 23. Latitude–pressure diagrams of (a) the zonal mean temperature in °C, (b) the day-to-day standard deviation of temperature in °C, (c) the geopotential height departure from the NMC standard atmosphere in gpm, (d) the day-to-day standard deviation of geopotential height in gpm, (e) the specific humidity in g kg⁻¹, and (f) the day-to-day standard deviation of specific humidity in g kg⁻¹ for the 10-yr period 1963–1973.

TABLE III. GEOPOTENTIAL HEIGHT (z_{SA}) OF VARIOUS ISOBARIC LEVELS ACCORDING TO
THE NMC STANDARD ATMOSPHERE

p (mbar)	1000	850	700	500	300	200	100	50
z_{SA} (gpm)	113	1457	3011	5572	9159	11,784	16,206	20,632

from kinetic energy is negligible, as we will see later. The latent heat (Fig. 24c) shows a strong seasonal variation mainly in the lower latitudes, in contrast with the internal energy and the potential energy for which the seasonal variations occur mainly in the high latitudes. The profiles of total energy reflect these facts. The profiles also show that the Southern Hemisphere high latitudes contain less internal and potential energy, as well as less latent heat, than the Northern hemisphere high latitudes contain. It is

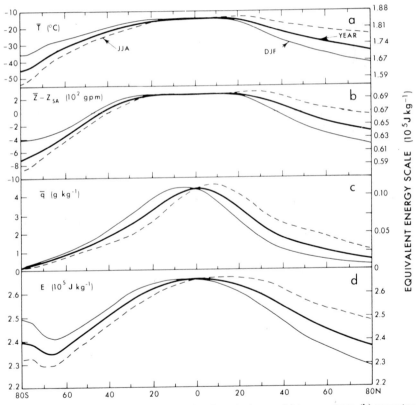

FIG. 24. Meridional profiles of the zonal and vertical mean (a) temperature, (b) geopotential height departure, (c) specific humidity, and (d) the total energy. An equivalent energy scale is given on the right-hand side margin of the figure.

worth mentioning that mainly due to the influence of topography on the value of the potential energy, the profiles of E show an artificial maximum poleward of 70° S.

Using our basic set of data, the mean values of the various forms of energy for the entire atmosphere were also evaluated. The corresponding values are presented in Table IV. The inspection of the table shows that the most important forms of energy are the internal energy (70.4% for global annual mean), potential energy (27.1%), and latent heat (2.5%). The kinetic energy is only a minute fraction (0.05%) of the total energy of the atmosphere. It plays, however, a very important role in the energetics of the general circulation in linking and redistributing the other forms of energy over the globe. The amplitude of the annual cycle in the Northern Hemisphere is almost twice as large as that in the Southern Hemisphere, which is mostly due to the differences in land–sea distribution in the two hemispheres.

Standard deviations of the daily values were also evaluated to assess the variability of the fields. The vertical distributions are shown in Fig. 23b, d, and f and in Tables A14–A16, whereas the vertical integrated profiles are given in Fig. 25. As expected, the larger day-to-day variability

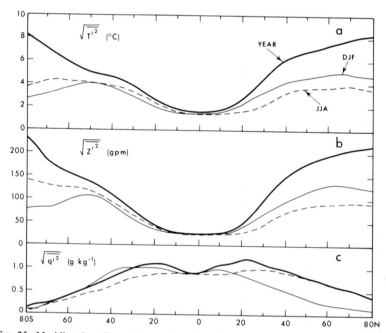

FIG. 25. Meridional profiles of the zonal and vertical mean day-to-day standard deviations of (a) temperature, (b) geopotential height, and (c) specific humidity.

TABLE IV. HEMISPHERIC AND GLOBAL INTEGRALS OF THE ATMOSPHERIC ENERGY (10^7 J m^{-2})

	Year			DJF[a]			JJA[a]			DJF-JJA		
	NH[b]	SH[b]	GL[b]	NH	SH	GL	NH	SH	GL	NH	SH	GL
PE[c]	70.0	68.7	69.3	69.0	69.3	69.2	71.0	68.3	69.6	-2.0	1.0	-0.4
IE[c]	180.6	180.0	180.3	178.2	181.6	179.9	183.4	178.4	180.9	-5.2	3.2	-1.0
LH[c]	6.48	6.28	6.38	5.15	7.16	6.16	8.07	5.38	6.72	-2.92	1.78	-0.56
K	0.116	0.131	0.123	0.168	0.100	0.134	0.072	0.156	0.114	0.096	-0.056	0.020
E	257.2	255.1	256.1	252.5	258.2	255.4	262.5	252.3	257.4	-10.0	5.9	-2.0
K/E (%)	0.05	0.05	0.05	0.07	0.04	0.05	0.03	0.06	0.04			
LH/E (%)	2.52	2.46	2.49	2.04	2.77	2.41	3.07	2.13	2.61			

[a] DJF, December–February; JJA, June–August.
[b] NH, Northern Hemisphere; SH, Southern Hemisphere; GL, global.
[c] PE, Potential energy per unit mass; IE, internal energy per unit mass; LH, latent heat per unit mass.

occurs outside the tropics in the middle and lower troposphere for the internal energy and latent heat, and in the upper atmosphere for the potential energy. The maxima are located at high latitudes in the case of internal and potential energy and in the subtropics in the case of latent heat. The general behavior of this variability can immediately be seen from the profiles. The curves are similar in both hemisheres. Nevertheless, minor differences occur, such as the slightly greater variability in the Northern Hemisphere except near 50° latitude during summer. The profiles in Fig. 25 reveal how much the seasonal cycle of the temperature (internal and potential energy) contributes to the annual variance in high latitudes. This can be seen by comparing the annual curve with the average of the DJF and JJA seasonal curves. We must point out again that all temporal eddies up to 1 year were included in the annual estimates of the variances (standard deviations) and covariances.

Finally, let us consider the kinetic energy. Using a similar breakdown as in Eq. (3.11), we write the total kinetic energy in its transient eddy, stationary eddy, and zonal mean components:

$$K = K_{TE} + K_{SE} + K_M \qquad (4.2)$$

where

$$
\begin{aligned}
K &= \tfrac{1}{2}[\overline{u^2} + \overline{v^2}] \\
K_{TE} &= \tfrac{1}{2}[\overline{u'^2} + \overline{v'^2}] \\
K_{SE} &= \tfrac{1}{2}[\bar{u}^{*2} + \bar{v}^{*2}] \\
K_M &= \tfrac{1}{2}([\bar{u}]^2 + [\bar{v}]^2)
\end{aligned}
\qquad (4.3)
$$

The corresponding values of the vertical structures of the mean transient eddy (K_{TE}), standing eddy (K_{SE}), and mean kinetic energy (K_M) together with the total kinetic energy (K) are displayed in Fig. 26 for the year and in Tables A9–A11 for the seasons. The mean vertically integrated values are shown in the meridional profiles in Fig. 27. The annual-mean total kinetic energy shows a similar pattern in both hemispheres. The 200-mbar maxima in the total kinetic energy K at 35° latitude are obviously associated with the subtropical jets. As seen from Fig. 27, the main components of the total kinetic energy are K_{TE} and K_M. The standing eddies only contribute significantly in the Northern Hemisphere. The broad midlatitude maxima in the K_{TE} cross sections are due to the daily meandering of the polar and subtropical jet streams and are associated with the seasonal shifts in latitude of the subtropical jet.

The profiles summarize the points already made. They show very clearly the differences between K_{SE}, and K_{TE} and K_M. The large seasonal

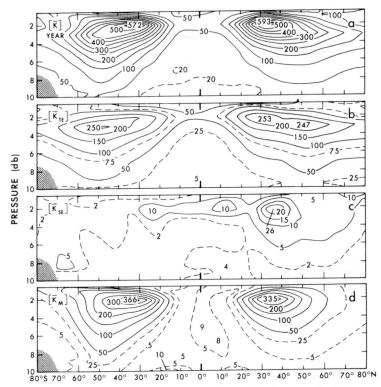

FIG. 26. Latitude–pressure diagrams of the zonal mean kinetic energy for (a) the total, (b) the transient eddy, (c) the stationary eddy, and (d) the mean component in m^2 sec^{-2} for the 10-yr period.

variations occur mainly in K_M. However, the variations are much more pronounced in the Northern than in the Southern Hemisphere, with a summer–winter change in the Northern Hemisphere peak value of almost a factor 3 and with high values in the Southern Hemisphere throughout the year. However, the variations in K_{TE} are also appreciable, in contrast with those in K_{SE}.

Kinetic energy and net radiation are the only energy components exhibiting a yearly maximum that is higher in the Southern than in the Northern Hemisphere.

As a final comment, we should mention that the midlatitude values of kinetic energy in the Southern Hemisphere probably represent considerable underestimates of the actual values. The occurrence of large data gaps and the bias due to a possible, selective loss of balloons during strong winds at some stations will tend to give too low values for the zonal winds

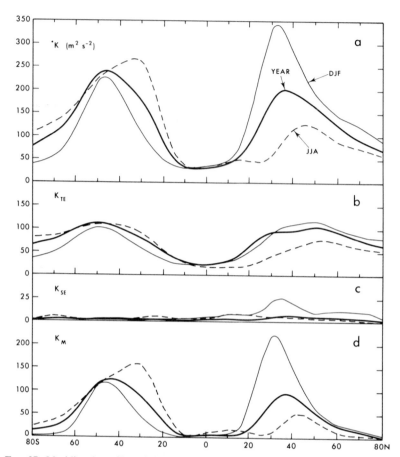

FIG. 27. Meridional profiles of the zonal and vertical mean (a) total kinetic energy, (b) transient eddy, (c) stationary eddy, and (d) mean kinetic energy in m² sec⁻².

in the Southern Hemisphere [see for comparison the geostrophic wind estimates by Van Loon *et al.* (1971) and Swanson and Trenberth (1981)].

4.3. Poleward Transport of Energy

It is well known that the motions in the atmosphere and oceans play an important role in carrying energy from the regions of net incoming radiation to the regions of net outgoing radiation. Thereby the atmospheric and oceanic currents have a moderating influence in the formation of the climate.

4.3.1. *Balance Equation of Energy.*   For the atmosphere the energy balance can be written in the (x, y, z, t) coordinate system as follows:

$$\rho \frac{dE}{dt} = -\text{div } \mathbf{J}_q \tag{4.4}$$

where $\mathbf{J}_q$ indicates the sum of the radiation flux ($\mathbf{F}_{rad}$), the energy flux by conduction plus subgrid-scale processes ($\mathbf{F}_{con}$), and the mechanical energy flux (i.e., pressure work term and frictional stress). Thus,

$$\mathbf{J}_q = \mathbf{F}_{rad} + \mathbf{F}_{con} = p\mathbf{c} + \boldsymbol{\tau} \cdot \mathbf{c} \tag{4.5}$$

Expanding the total time derivative and using conservation of mass, Eq. (4.4) becomes

$$\frac{\partial(\rho E)}{\partial t} = -\text{div } \rho E\mathbf{c} - \text{div } \mathbf{J}_q \tag{4.6}$$

Substituting Eq. (4.1) for E and Eq. (4.5) for $\mathbf{J}_q$ into Eq. (4.6), we obtain the general balance equation of energy for the atmosphere:

$$\frac{\partial}{\partial t} \rho(c_v T + gz + Lq + \tfrac{1}{2}|\mathbf{c}|^2) = -\text{div } \rho(c_v T + gz + Lq + \tfrac{1}{2}|\mathbf{c}|^2)\mathbf{c}$$
$$-\text{div}(\mathbf{F}_{rad} + \mathbf{F}_{con} + p\mathbf{c} + \boldsymbol{\tau} \cdot \mathbf{c}) \tag{4.7}$$

This equation can be rearranged by introducing the enthalpy ($c_v T + p/\rho = c_p T$), which represents the total potential energy content of the atmosphere. Furthermore, the kinetic energy and $\boldsymbol{\tau} \cdot \mathbf{c}$ contributions are very small and can be disregarded. If we integrate the resulting equation with respect to time, we obtain

$$\frac{\partial}{\partial t} \overline{\rho(c_v T + gz + Lq)}$$
$$= -\text{div } \overline{\rho(c_p T + gz + Lq)\mathbf{c}} - \text{div}(\bar{\mathbf{F}}_{rad} + \bar{\mathbf{F}}_{con}) \tag{4.8}$$

Integrating this equation over the volume of a polar cap leads to

$$\frac{\partial}{\partial t} \iiint \overline{\rho(c_v T + gz + Lq)} \, dV$$
$$= \iint_{\text{wall}} \overline{\rho(c_p T + gz + Lq)v} \, dx \, dz + F_{TA} - F_{BA} \tag{4.9}$$

where the horizontal component of the subgrid-scale flux ($\mathbf{F}_{con}$) is neglected since it is small compared with the large-scale flux. The term F_{TA} represents the net radiation flux at the top of the atmosphere, and F_{BA} the net flux of energy (radiation + sensible heat + latent heat) at the bottom, counted positive if directed from the atmosphere to the earth.

Equation (4.9) states that the change of energy in a polar cap is the net result of the exchanges of energy with the rest of the atmosphere, with outer space, and with the underlying surface. The relative importance of the fluxes of the various forms of energy across the three boundaries can be quite different. In fact, the exchange through the vertical, latitudinal wall takes place in all forms of energy, whereas across the upper boundary (F_{TA}) the exchange occurs in the form of radiation only, and through the bottom (F_{BA}) in the form of radiation, sensible and latent heat.

The flux across the vertical latitudinal wall in Eq. (4.9) can be rewritten in the (x, y, p) coordinate system as follows:

$$\iint_{\text{wall}} \overline{\rho(c_p T + gz + Lq)v}\ dx\ dz$$
$$= 2\pi R \cos \phi \int [\overline{(c_p T + gz + Lq)v}]\ dp/g \qquad (4.10)$$

Further, the integrand can be decomposed into the contributions by transient eddies, stationary eddies, and mean meridional circulations in order to get a better understanding of the physical mechanisms involved in the fluxes:

$$[\overline{(c_p T + gz + Lq)v}] = c_p[\overline{v'T'}] + c_p[\bar{v}^*\bar{T}^*] + c_p[\bar{v}][\bar{T}]$$
$$+ g[\overline{v'z'}] + g[\bar{v}^*\bar{z}^*] + g[\bar{v}][\bar{z}]$$
$$+ L[\overline{v'q'}] + L[\bar{v}^*\bar{q}^*] + L[\bar{v}][\bar{q}] \qquad (4.11)$$

4.3.2. Meridional Transport of Sensible Heat.

Due to the importance of large-scale turbulent processes in the atmosphere, maps of the vertical mean poleward fluxes of sensible heat by transient eddies are shown in Fig. 28. By and large, the eddy fluxes are predominantly poleward in both hemispheres with a maximum localized in midlatitudes. The yearly map of $\widehat{\overline{v'T'}}$ shows midlatitude zones of strong poleward heat flux fairly uniform in the Southern Hemisphere, but with distinct maxima in the Northern Hemisphere over North America and eastern Asia. These are clearly associated with baroclinic disturbances along the polar front. Over the equator the meridional eddy fluxes are very small, as they are over the poles. However, near 70° latitude they can attain sizeable values, at some longitudes directed away from the poles.

The fluxes for the seasons are most intense in the winter hemisphere. In the Northern Hemisphere the influence of the land–sea distribution is very pronounced. In the Southern Hemisphere we find a rather uniform belt of maximum poleward heat flow around 45° S.

Peixóto's (1960) map of $\widehat{\overline{v'T'}}$ for the year 1950 and Peixóto's (1974) map for the International Geophysical Year (IGY) are very similar to our Fig.

28a. This suggests that the annual transient eddy heat flux does not vary much from year to year and that it is a relatively stable measure of the general circulation, at least on an annual basis.

The vertical distribution of the zonally averaged values for the various modes of sensible heat transport are given in Fig. 29 and in Tables A20 and A21. The patterns for the yearly and seasonal conditions are similar, except for the mean meridional fluxes in the tropics. This fact is also shown in the vertical-mean profiles in Fig. 30. Near 50° latitude the eddy transports exhibit two maxima in the vertical at about the 850- and 200-mbar levels, which are associated with the alternation of air masses and the fluctuations of the tropopause, respectively. The $[\overline{v'T'}]$ pattern between 20° S and 20° N in Fig. 29a shows that the transient eddies in the tropics transport heat toward the equator. Thereby, they act in an abnormal manner since they tend to heat rather than cool the inner tropics. This behavior resembles that of the eddies in the lower stratosphere, as discussed previously by Starr and Wallace (1964) based on Peixóto's (1960) analyses.

The standing eddy pattern ($[\bar{v}^*\bar{T}^*]$) has a different, usually weaker structure with much higher values in the Northern Hemisphere poleward of 40° N as shown also in the profiles (Fig. 30c). The standing eddies have a large seasonal variation in the Northern Hemisphere. In the winter their transport can exceed the transport by the transient eddies as inferred from the profiles. The eddy fluxes of sensible heat in Tables A20 and A21 confirm well the patterns shown by Newell et al. (1974, p. 40) even in the Southern Hemisphere.

The mean meridional circulation transports, $[\bar{v}]_{\text{ID}}[\bar{T}]$, were evaluated using the indirectly calculated values of $[\bar{v}]$. The vertical distribution in Fig. 29c shows that the mean meridional circulations are most active in the upper and lower levels. Again, the vertical cross section reveals the three-cell regime in both hemispheres. Large seasonal variations occur in the tropics since the mean meridional flow of sensible heat tends to be from the winter into the summer hemisphere, associated with the shifts in the Hadley cells. This is clearly shown in the corresponding vertical mean profiles in Fig. 30d.

The profiles of the various modes of meridional transport of sensible heat in Fig. 30 summarize the characteristics discussed above.

4.3.3. Meridional Transport of Potential Energy. The fluxes of potential energy are presented in Figs. 31 and 32 and in Tables A22 and A23. The eddy fluxes are very small compared with those of sensible heat. This is in agreement with the quasi-geostrophic character of the eddies, since for geostrophic flow

$$[v_g z] = \frac{g}{f}\left[\frac{\partial z}{\partial x} z\right] = 0$$

It is of interest to note that the transient eddy transport of potential energy across about 25° latitude is directed into the tropics in both hemispheres. This eddy transport is equivalent with a pressure forcing in the (x, y, z) system. Such a forcing was suggested by Mak (1969) as a possible energy source for tropical disturbances.

By far the most important mode is the mean meridional cell flux. As expected, the patterns in Figs. 31c and 32d are similar to those of the sensible heat flow in Figs. 29c and 30d, but with the sign reversed. The net cross-equatorial flux of potential energy is directed from the summer to the winter hemisphere. It, thereby, overcompensates for the flow of sensible heat, leaving a net residual transport of energy into the winter hemisphere.

4.3.4. Meridional Transport of Latent Heat. Maps showing the distribution of the yearly and seasonal fields of the meridional eddy transports of latent heat are presented in Fig. 33. The transports are poleward in each hemisphere, with a more uniform distribution in the Southern Hemi-

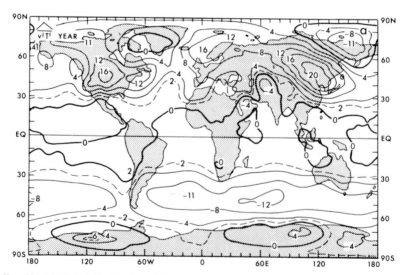

FIG. 28. (a) Global distribution of the vertical mean northward transport of sensible heat by transient eddies in °C m sec^{-1} for the 10-yr mean. (b) Same as (a) except for DJF season. (c) Same as (a) except for JJA season.

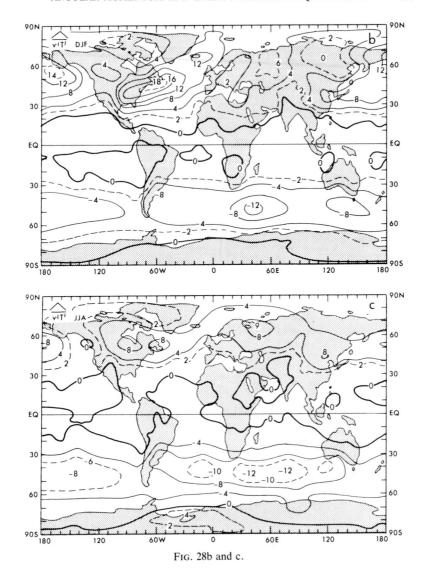

FIG. 28b and c.

sphere as expected. Two large centers dominate the flow over the Northern Hemisphere associated with the preferred locations of the baroclinic disturbances and the land–sea contrast. The maps of $\widehat{v'q'}$ resemble the corresponding maps of $\widehat{v'T'}$. However, the Northern Hemisphere maxima are shifted eastward toward the east coasts of North America and

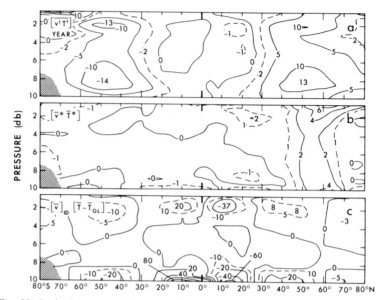

FIG. 29. Latitude–pressure diagrams of the zonal mean northward transport of sensible heat by (a) transient eddies, (b) stationary eddies, and (c) mean meridional circulations in °C m sec⁻¹ for the 10-yr mean.

Asia, and are located closer to the equator. The eastward shift is related to the greater availability of water over the oceans, and the equatorward shift to the greater humidity in the warm, tropical air.

The seasonal migration of the maximum of $\widehat{v'q'}$ over North America is large from about 30° N in DJF to 50° N in JJA. In terms of equivalent energy, the sensible heat (as shown in Fig. 28) dominates poleward of about 30° and the latent heat equatorward of that latitude. Comparing Fig. 33a with the maps of $\widehat{v'q'}$ for the year 1950 and the IGY of Peixóto (1958) and Peixóto et al. (1976), respectively, we find good agreement. However, a systematic difference is found over the North Pacific Ocean, where the earlier hand analyses show a broad belt of maximum northward flux between about 40° and 50° latitude, whereas the present objective analysis shows one clear center to the east of Asia. This difference may be considered a measure of the uncertainty due to oceanic data gaps (see Fig. 2), since it depends largely on the analysis scheme used.

The vertical structure of the transports of latent heat as well as their seasonal variability are displayed in Figs. 34 and 35 and in Tables A24 and A25. The transport of water vapor in all modes occurs mainly in the lower

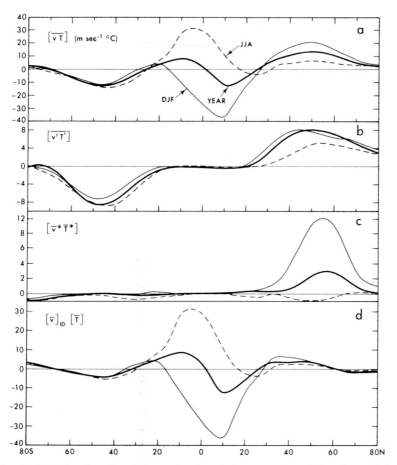

Fig. 30. Meridional profiles of the zonal and vertical mean northward flux of sensible heat by (a) all motions, (b) transient eddies, (c) stationary eddies, and (d) mean meridional circulations for the 10-yr period.

troposphere. The transient eddy contributions spread to upper levels, whereas the other modes are confined to the boundary layer. They undergo small seasonal variations, which are most pronounced in the Northern Hemisphere, as expected. The standing eddy contributions are mainly important in the Northern Hemisphere, where they show a considerable seasonal variation in intensity accompanied by a substantial meridional shift, in contrast with the situation in the Southern Hemisphere. The mean meridional circulation contributions are mainly important in the tropics, where they show a strong seasonal variation.

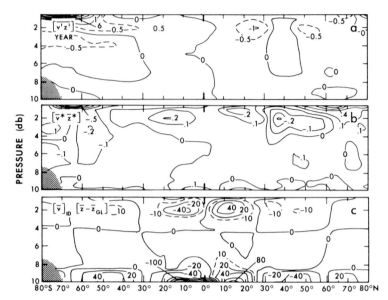

FIG. 31. Latitude–pressure diagrams of the zonal mean northward flux of potential energy by (a) transient eddies, (b) stationary eddies, and (c) mean meridional circulations in 10^2 gpm m sec^{-1} for the 10-yr period.

The profiles of the total flux of latent heat are also presented in Fig. 35a. They show a cross-equatorial transport which almost compensates for the residual sensible heat plus potential energy, as previously discussed.

The latent heat fluxes in Tables A24 and A25 can also be compared with the global cross sections by Peixóto *et al.* (1978) for the IGY. The present transient eddy fluxes are generally weaker at 850 and 700 mbar than those for the IGY data set. Comparisons with Oort and Rasmusson's (1971) vertical mean values for the 1958–1963 period show excellent agreement, except for the transient eddy fluxes of latent heat, which are about 30% weaker in the present sample. Some of the systematic differences, although small, are caused either by differences in data reduction or by real decade-to-decade differences in climate. These questions require further study of individual yearly analyses (see, e.g., Oort, 1977, 1983).

4.3.5. Meridional Transport of Kinetic Energy. To complete the analysis of the transports of various forms of energy, a discussion of the kinetic energy will be given now, in spite of the smallness of its contribution to the total energy balance. The results are presented in Figs. 36 and 37 and in Tables A26 and A27. The transports in the various modes occur mainly at the jet stream level in the upper troposphere. The cross sections

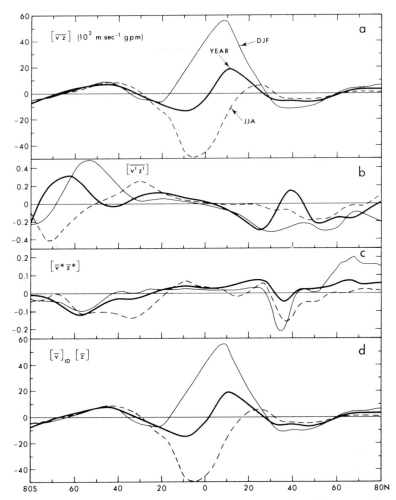

FIG. 32. Meridional profiles of the zonal and vertical mean northward transport of potential energy by (a) all motions, (b) transient eddies, (c) stationary eddies, and (d) mean meridional circulations.

show alternating positive and negative centers, leading to a well-defined convergence of total kinetic energy in middle latitudes. Among the various modes, the transient eddies attain the largest values. The stationary eddies are almost nonexistent in the Southern Hemisphere, but important in the Northern Hemisphere. They are even more important than the transient eddies during the DJF season at 30° N. The vertically integrated values in Fig. 37 show through the profiles some seasonal variability. It is worth noting that the contribution by the mean meridional cells to the total transport of kinetic energy is quite small, even in the tropical Hadley cells.

4.3.6. Meridional Transport of Total Energy. Before giving an integrated picture of the total energy transport in the atmosphere, some points are in order:

(1)   Only indirectly calculated $[\bar{v}]_{ID}$ have been used to evaluate the mean meridional circulation transports because, as we have seen before, the direct values are unrealistic in the Southern Hemisphere. Further, in the presentation of the mean meridional circulation fluxes of sensible heat, potential energy, and total energy their mean global values, averaged both horizontally and vertically, have been subtracted since they do not lead to a net transport. As pointed out by, e.g., Oort (1971), the mean meridional circulations transport large amounts of energy near the surface and the 200-mbar level. However, strong compensations occur between the flux of sensible plus latent heat and the flux of potential energy, especially when integrated in the vertical. Thus, the net effects of the mean meridional circulations are greatly reduced.

(2)   To convert the various fluxes to the same energy flux units, for example, to J m sec^{-1}, one should multiply the sensible heat flux in °C m

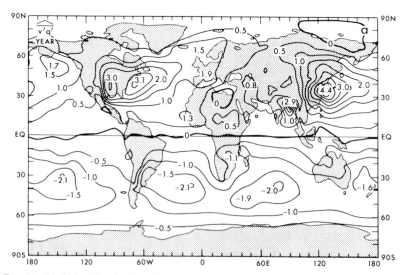

FIG. 33. (a) Global distribution of the vertical mean northward transport of humidity by transient eddies in g kg^{-1} m sec^{-1} for the 10-yr mean. (b) Same as (a) except for DJF season. (c) Same as (a) except for JJA season.

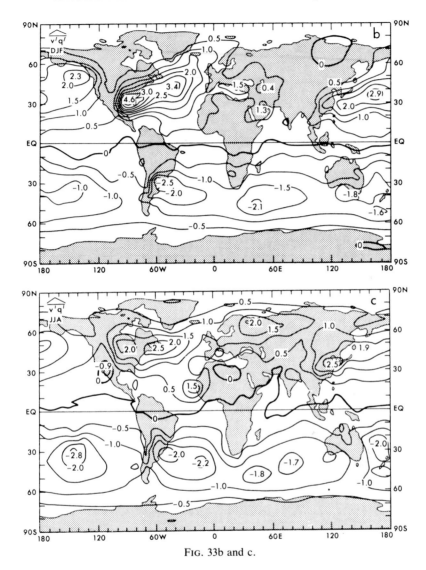

Fig. 33b and c.

sec^{-1} by 1.0, the geopotential flux values in gpm m sec^{-1} by 0.01, the water vapor flux values in g kg^{-1} m sec^{-1} by 2.5, and the kinetic energy flux values in m^3 sec^{-3} by 0.001.

(3) Similar latitude–height diagrams and profiles, but based on data sets which are generally less complete and less homogeneous both in space and time, have been published before by, e.g., Oort and Rasmusson

(1971) for the Northern Hemisphere, and Newell *et al.* (1972, 1974) for the tropics (and in some cases for the globe).

The combination of the previous zonal-mean cross sections and profiles, expressed in units of J m sec^{-1}, leads to the cross sections and profiles of the transport of total energy as shown in Figs. 38 and 39 and in Tables A28–A30. The vertical structure of each mode of total energy flux represents, of course, the net result of its various components.

The most important modes of total energy transport are the transient eddies and the mean meridional circulations. In fact, we find that between 20° S and 20° N, the mean meridional cell circulations are the important mechanism to transport energy meridionally in its various forms, whereas poleward of 30° N, the eddies play a dominant role for all energy forms except potential energy. Furthermore, for the total energy flux by the

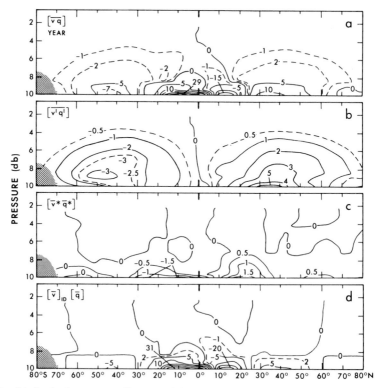

FIG. 34. Latitude–pressure diagrams of the zonal mean northward flux of specific humidity by (a) all motions, (b) transient eddies, (c) standing eddies, and (d) mean meridional circulations in g kg^{-1} m sec^{-1} for the 10-yr period.

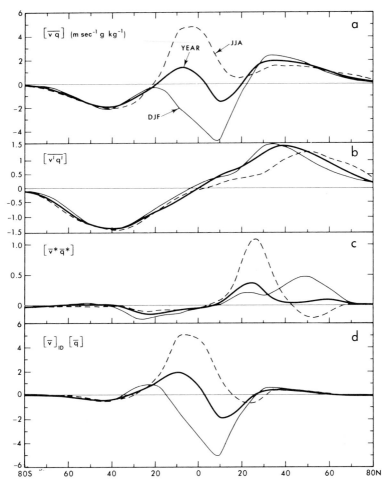

FIG. 35. Meridional profiles of the zonal and vertical mean northward transport of specific humidity by (a) all motions, (b) transient eddies, (c) stationary eddies, and (d) mean meridional circulations.

eddies the seasonal changes are large in the Northern and small in the Southern Hemisphere.

The net effect of the transient eddies on the atmospheric heat balance can be seen by studying the divergence of the energy transport. Thus, we present in Fig. 40 global maps of the vertical mean divergence of the transient eddy flux of energy. The centers of energy divergence, i.e., the sources of energy, are located around 30° N and 30° S. In the Southern

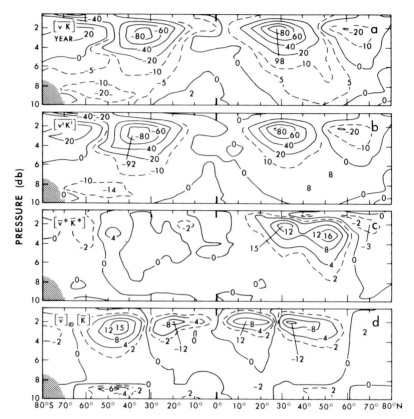

FIG. 36. Latitude–pressure diagrams of the zonal mean northward flux of kinetic energy by (a) all motions, (b) transient eddies, (c) stationary eddies, and (d) mean meridional circulations in 10 m³ sec⁻³ for the 10-yr mean.

Hemisphere a zonally uniform value of nearly 50 W m⁻² is found, whereas in the Northern Hemisphere we find two strong sources of energy over the Gulf Stream and Kuroshio regions with maximum divergence up to 150 W m⁻². Regions of convergence (sinks) of atmospheric energy are found poleward of about 45° latitude with centers of −100 W m⁻² (and less) over eastern Canada, northeast of Iceland, and over eastern Siberia. In the Southern Hemisphere, an extensive belt of convergence is found over the oceans around the rim of the Antarctic continent.

One should note that the field of the energy convergence by the time-mean flow is much more difficult to determine than the transient eddy field. When the time-mean flow is included, the patterns tend to be noisy with several maxima and minima around each latitude circle and a pole-ward shift of the oceanic divergence maxima of roughly 10° seems to take

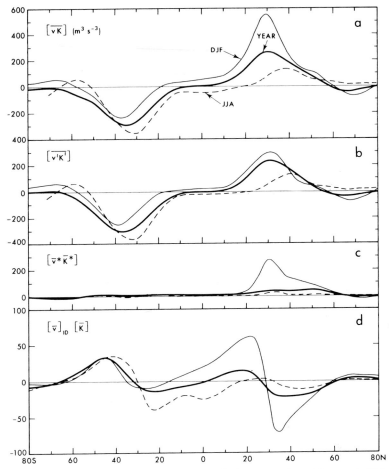

FIG. 37. Meridional profiles of the zonal and vertical mean northward transport of kinetic energy by (a) all motions, (b) transient eddies, (c) stationary eddies, and (d) mean meridional circulations.

place. In our opinion, the essence of the divergence of energy in the atmosphere is contained in the divergence of the transient eddy transports.

It is of interest to compare the yearly divergence map by transient eddies (Fig. 40a) with the earlier presented map of net incoming radiation (Fig. 17a). In the comparison, two other effects should be taken into account, namely the divergence by the time-mean atmospheric flow and the divergence by ocean currents. Since both terms are probably important, but cannot be determined well enough from the present data, we will speculate as to their relative significance. Obviously over the continents

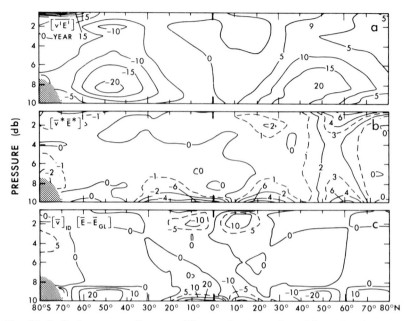

FIG. 38. Latitude–pressure diagrams of the zonal mean northward flow of total energy by (a) transient eddies, (b) stationary eddies, and (c) mean meridional circulations in °C m sec^{-1} for the 10-yr mean.

the differences between Figs. 17a and 40a must be entirely due to the time-mean atmospheric flow. In fact, from the comparison of the maps we would expect convergence of energy by the mean flow over western Canada and western Europe, and divergence over eastern Canada and Siberia. Over the oceans, the differences between the two maps are so large that the imbalance must be mainly due to oceanic heat transports. These would give divergence equatorward of 20° latitude, convergence in a zonal belt around 30° S, and strong convergence near 30° N localized over the Gulf Stream and Kuroshio, as well as general convergence over the eastern North Atlantic Ocean.

To compare the seasonal maps of net incoming radiation (Fig. 17b and 17c) with those of the transient eddy atmospheric divergence (Fig. 40b and 40c), the storage of energy in the atmosphere and oceans also must be taken into account. In a qualitative sense, over the continents we find the tendency for strong transient eddy convergence of energy in winter and for some divergence in summer, in general agreement with the conclusions reached by Campbell and Vonder Haar (1982), as discussed in Section 4.1.

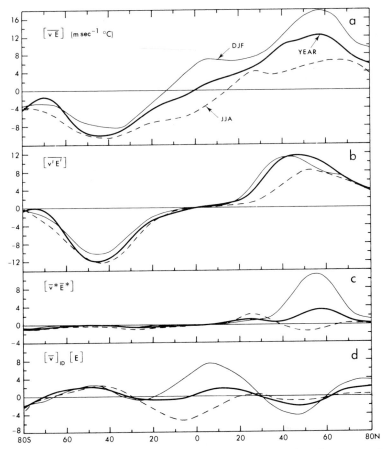

FIG. 39. Meridional profiles of the zonal and vertical mean northward transport of total energy by (a) all motions, (b) transient eddies, (c) stationary eddies, and (d) mean meridional circulations.

4.4. Vertical Transport of Energy

Going back to Eq. (4.6), and taking both the time and zonal averages, we obtain the following expression:

$$(\partial/\partial t)/[\overline{\rho E}] = -\partial([\overline{\rho v E}] + [\overline{pv}]) \cos \phi / R \cos \phi \; \partial \phi$$
$$-\partial([\overline{\rho w E}] + [\overline{pw}])/\partial z$$
$$-\partial([\bar{F}_{\mathrm{rad}}] + [\bar{F}_{\mathrm{con}}] + [\tau_{zz} w])/\partial z \qquad (4.12)$$

which we use to assess, in an indirect form, the vertical transport of energy. For a steady state the left-hand side of Eq. (4.12) vanishes. Thus, we can define a streamfunction for the total energy, ψ_E, such that

$$2\pi R \cos \phi \, [\bar{\tau}_{E\phi}] = -\partial\psi_E/\partial z \qquad (4.13a)$$

$$2\pi R \cos \phi \, [\bar{\tau}_{Ez}] = \partial\psi_E/R \, \partial\phi \qquad (4.13b)$$

where

$$[\overline{\tau_{E\phi}}] = [\overline{\rho v(E + p/\rho)}] \qquad (4.14a)$$

$$[\overline{\tau_{Ez}}] = [\overline{\rho w(E + p/\rho)}] + [\bar{F}_{\text{rad}}] + [\bar{F}_{\text{con}}] + [\overline{\tau_{zz} w}] \qquad (4.14b)$$

As was done in the case of angular momentum, we could separate each term in the various modes of transport, but this is not necessary for the present discussion. The main mechanisms of transporting energy upward are organized convection, associated with unstable conditions in the lower layers of the atmosphere, all types of cumulus activity, and smaller-scale turbulence. In contrast with the horizontal transport, the large-scale eddies play a less important role than the cumulus-scale eddies in the vertical transport (see Starr and Peixóto, 1971), especially in the tropics.

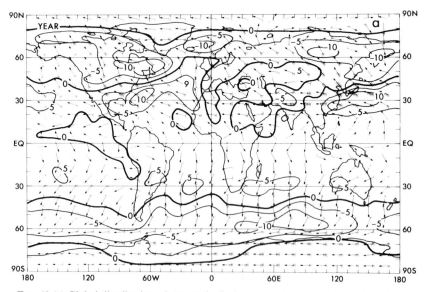

FIG. 40 (a) Global distribution of the vertically integrated horizontal flow of total energy by transient eddies for the 10-yr mean as shown by arrows. Each barb on the tail of an arrow represents a transport of 5×10^7 W m^{-1} (or in terms of a vertical mean value 5°C m sec^{-1}). Also shown are some isolines of its divergence, $\int_{25}^{p0} \text{div } \overline{\mathbf{v}'E'} \, dp/g$, labeled in units of 10 W m^{-2}. (b) Same as (a) except for DJF season. (c) Same as (a) except for JJA season.

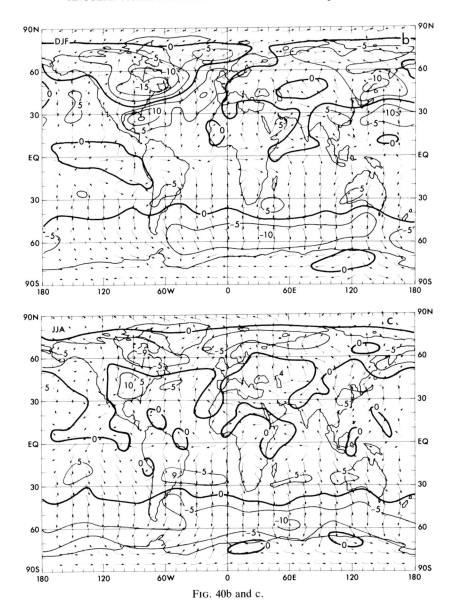

FIG. 40b and c.

At the top of the atmosphere we can neglect all terms on the right-hand side of Eq. (4.14b) except the radiation term. Thus, using the observed net radiation fluxes from Campbell and Vonder Haar (1980) (see Fig. 19d) as top boundary condition, we evaluated the ψ_E distribution from the horizontal fluxes alone by starting the integration of expression (4.13a) at one of the poles, where we assume $\psi_E = 0$. Figure 41 shows the resulting vertical distribution of ψ_E as a function of latitude for the year and for the DJF and JJA seasons. The streamlines of ψ_E represent the density of the flux of energy in units of 10^{15} W. It is of great interest to analyze the vertical energy fluxes inferred from these figures using Eq. (4.13b), which cannot be obtained directly from the present observations. To avoid the appearance of some closed streamlines induced by the tropical Hadley cells, the global mean value of the energy was subtracted in the construction of the streamlines. In the computations of the streamlines for DJF and JJA the (small) energy storage in the atmosphere was neglected.

In the yearly diagram (Fig. 41a), a large inflow of radiational energy occurs in the tropics, between about 30° S and 30° N. Roughly half of this energy passes unattenuated through the atmosphere, and is absorbed mainly in the tropical oceans, where it will be transported poleward by oceanic circulations. However, in the subtropics and middle latitudes most of the action seems to be due to the atmospheric circulations which transport the energy poleward until it is radiated back out to space. The

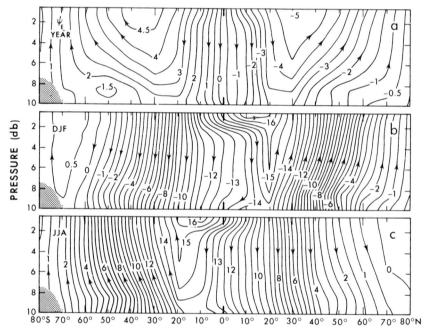

FIG. 41. Streamlines of the zonal mean transport of total energy in the atmosphere for the 10-yr period (a) and the DJF (b) and JJA (c) seasons. Units are 10^{15} W.

seasonal diagrams (Fig. 41b and c) show the enormous direct inflow of energy into the summer hemisphere and the corresponding outflow from the winter hemisphere. The tilting of the streamlines in the vertical is due to atmospheric motions, i.e., mean meridional circulations in the tropics and horizontal eddies in middle and high latitudes. The streamlines that intersect the 1000-mbar surface represent, approximately, the amount of energy which is given to or taken away from the earth's surface. Because of the small heat capacity of the land, and the small amounts of energy stored in snow and ice, the important energy exchange occurs with the oceans. Large amounts of heat are stored in the oceans of the summer hemisphere and lost by the oceans in the winter hemisphere. Only in the annual mean case is the rate of storage equal to zero, and are the values at the surface equal to the required oceanic heat transport.

4.5. Fulfillment of the Energy Balance

In this section we will summarize what happens to the solar energy once it has entered the atmosphere. We will study first the storage and transport of energy in the atmosphere, then the amount of energy that flows into the underlying surface, and finally we will analyze in some detail the storage and transport of energy in the oceans.

4.5.1. Atmosphere. The atmosphere has a limited heat capacity as shown by the curves of the rate of energy storage (S_A) obtained from our data and illustrated in Fig. 42a. For the DJF and JJA seasons the heat storage in the atmosphere amounts to less than 10 W m^{-2}, whereas in the transition seasons, March–May (MAM) and September–November (SON), larger values on the order of about 20 W m^{-2} occur at high latitudes. Let us consider, for our present discussion, Eq. (4.6) or (4.8), where the convergence due to horizontal atmospheric exchange seems to be one of the dominant atmospheric terms. Thus, the mean zonal divergence of energy (div T_A) was evaluated, and the resulting profiles are shown in Fig. 42b. On an annual basis about 20 W m^{-2} is lost between 40° S and 40° N, and about 50–70 W m^{-2} is gained poleward of 60°. The seasonal curves show a maximum divergence of 30–40 W m^{-2} in the tropics of the summer hemisphere, and a maximum convergence near the winter pole.

It is of some interest to analyze the contributions of the various modes of transport to the total convergence of energy in the atmosphere. Results are shown in Fig. 43. As expected from the earlier discussions, the contribution of the transient eddies to the total energy balance is most important

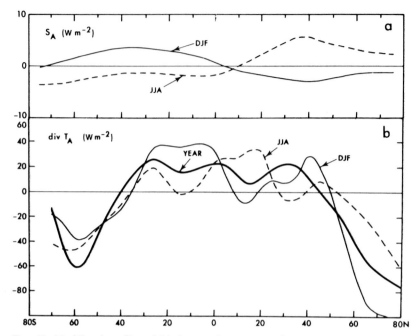

FIG. 42. Meridional profiles of (a) the zonal mean rate of energy storage in the atmosphere, and (b) the divergence of the vertically integrated atmospheric transport of energy.

in middle and high latitudes, whereas the mean meridional circulation contribution is dominant in the tropics and of some importance in middle and high latitudes. The stationary eddies only contribute significantly in the Northern Hemisphere during winter.

4.5.2. Atmosphere–Earth Exchange. The energy that is exchanged with the underlying earth's surface (F_{BA}) can, as Eq. (4.9) shows, be obtained as a residual. The estimated values are presented in Fig. 44. Between 30° S and 30° N the surface gains annually up to 40 W m^{-2}, and poleward of 30° it loses a similar amount. The seasonal variations are very large. In fact, a gain of about 90 W m^{-2} is found near 40° S during DJF, and a gain of about 60 W m^{-2} near 40° N during JJA. Losses of more than 100 W m^{-2} occur near 40° N during DJF and south of 40° S during JJA. The earlier estimates of F_{BA} by Oort and Vonder Haar (1976) for the Northern Hemisphere are also plotted in Fig. 44. The present curves are more smooth, but there is overall reasonable agreement between the two data sets.

The classical method to calculate the exchange of energy between the atmosphere and the earth's surface is based on a combination of radiation calculations and empirical drag-law formulae (see, e.g., Budyko, 1963). The total exchange is then evaluated as the sum of the separate contributions from radiation, sensible and latent heat exchange using as basic data the surface observations of cloudiness, surface wind, temperature, and humidity. Obviously, such calculations are tentative and can contain large systematic errors. Our present residual calculation is an attractive alternative method, since it is based on sound physical principles and does not involve empirical relations. A serious problem in our approach may be the possible unrepresentativeness of the radiation and atmospheric data sets

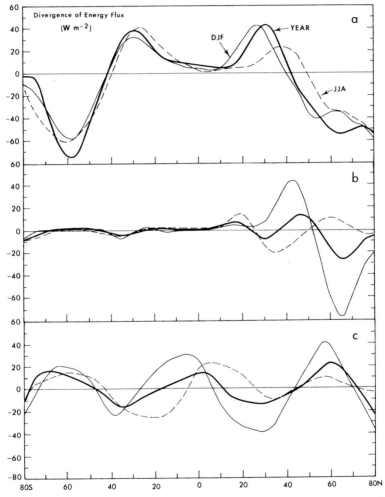

FIG. 43. Meridional profiles of the zonal mean divergence of atmospheric energy by (a) transient eddies, (b) stationary eddies, and (c) mean meridional circulations.

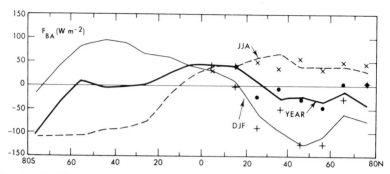

FIG. 44. Meridional profiles of the zonal mean flux of energy from the atmosphere to the earth's surface (calculated as a residual). Plotted symbols are from Oort and Vonder Haar (1976): (●) the year; (+) DJF; (×) JJA.

for long-term mean conditions. In other words, the two data sets may be incompatible in that they cover different time periods.

Because the heat capacity of the continental surface is very small, its effects in the annual variation of the global heat balance can be neglected. The same holds true for snow and ice, except at high latitudes. Therefore, on a large scale, the principal exchange must occur between the atmosphere and the oceans, and the curves in Fig. 44 should be regarded as a measure of the energy available for the world ocean.

4.5.3. Oceans. The oceanic heat storage plays a very important role in the annual variation of the earth's heat balance. Curves of the rate of heat storage (S_O) based on global monthly mean temperature analyses in the first 275 m of the world ocean, kindly provided by Levitus (1982), are presented in Fig. 45a. The choice of a depth of integration of 275 m was somewhat arbitrary. It was a compromise between the desire to include even the small annual variations occurring below the seasonal thermocline, perhaps down to a depth of 1000 m, and the problem of a rapid decrease of reliable mechanical and expendable bathythermograph data below 250 m. The storage curves in Fig. 45a depict an annual variation of nearly 100 W m^{-2} amplitude in middle latitudes of each hemisphere. Striking is the large DJF–JJA difference in storage near 60° S, whereas near 60° N the oceanic heat storage is negligible, due to the near nonexistence of oceans. The amount of energy left to be transported by ocean currents (div T_O) is given in Fig. 45b. Our residual estimates of the divergence of energy by ocean currents for annual mean conditions show that the oceans have to export about 40 W m^{-2} from the tropics and that poleward of 40° latitude the oceans import about 25 W m^{-2}. The energy conver-

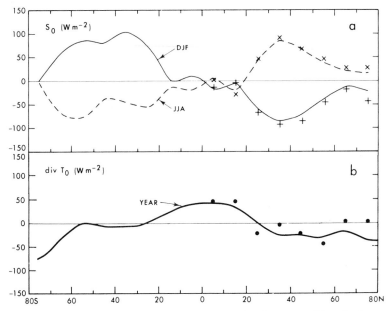

FIG. 45. Meridional profiles of (a) the zonal mean rate of storage of heat in the oceans from Levitus (1982), and (b) the divergence of the oceanic transport of energy (calculated as a residual). Plotted symbols are from Oort and Vonder Haar (1976): (●) the year; (+) DJF; (×) JJA.

gence by the oceans should vanish south of about 70° S, where no ocean is left. Thus, the calculated residual convergence at those latitudes must be spurious and is probably due to an underestimate of the atmospheric convergence south of 70° S (see Fig. 42b), leading to an overestimate of the oceanic convergence.

Seasonal curves of div T_O are not shown because no corrections for seasonal global balance were made in the calculations of F_{TA}, F_{BA}, and S_O on which the values of div T_O were based. This would be necessary before reasonable seasonable estimates of div T_O can be provided.

For the oceans it is difficult to assess the transport mechanisms responsible for the energy flux divergence curve as given in Fig. 45b. Numerical general circulation models may help to clarify this issue (for some early results see, e.g., Bryan and Lewis, 1979).

From the curve for $T_A + T_O$ in Fig. 20 and the curve for T_A in Fig. 39a one can obtain by subtraction the required ocean heat transport T_O. The resulting curve for T_O as well as the curves for T_A and $T_A + T_O$ are presented in Fig. 46. For comparison, earlier estimates of T_O by Oort and

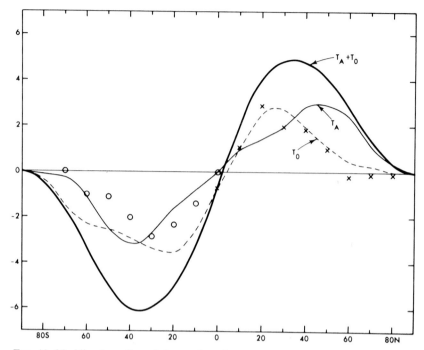

FIG. 46. Meridional profiles of the northward transports of energy in the atmosphere, oceans, and atmosphere plus oceans for annual mean conditions. Added are earlier estimates by Oort and Vonder Haar (1976) (×) and Trenberth (1979) (○).

Vonder Haar (1976) in the Northern Hemisphere and by Trenberth (1979) in the Southern Hemisphere are also shown.

The inspection of these profiles leads to the conclusion that for the mean annual heat balance of the climate system, the oceans dominate in low latitudes with maximum poleward heat transports of almost 3×10^{15} W near 25° N and of -3.5×10^{15} W near 20° S, whereas the atmosphere is more important in high latitudes of the Northern Hemisphere. In middle latitudes of the Southern Hemisphere the atmosphere and oceans appear to be of about equal importance with values of about -3×10^{15} W near 40° S, whereas the oceans dominate again at high southern latitudes. The values of T_O should vanish over land, i.e., south of 70° S; the nonzero value obtained can be considered a measure of the error in T_O. The agreement with the older calculations of Oort and Vonder Haar (1976) in the Northern Hemisphere is fairly good, except at 30° N where the recent T_A estimates are lower and the T_O estimates consequently higher than before. The comparison with Trenberth's (1979) estimates in the Southern

Hemisphere shows oceanic heat transports stronger overall by about 10^{15} W in our data. However, we find the same qualitative difference Trenberth found between the two hemispheres at 60° latitude, namely a stronger poleward heat flux near 60° S in the vicinity of the Antarctic circumpolar current than that at 60° N where there are mainly continents.

The uncertainty in the estimates of T_O is still quite large, especially in the Southern Hemisphere where the radiosonde network is inadequate to measure T_A reliably. This deficiency in the upper air data would probably tend to underestimate the atmospheric contribution and thereby overestimate the oceanic role in the total heat transport in middle latitudes of the Southern Hemisphere. As far as the seasonal heat budgets are concerned, the oceanic heat storage is probably the least known component. Another problem may relate to the fact that the atmospheric, oceanic, and radiation samples were not taken during the same years. Thus, some differences may be expected only because of year-to-year variations. In spite of all these problems, our results should be taken seriously since they are based on less assumptions than any of the other methods presently used to derive oceanic heat transports. The apparent controversy with conventional oceanographic estimates, such as those by Hall and Bryden (1982) and by Hastenrath (1982), which suggest much smaller oceanic heat transports, still remains. For a more thorough examination of these questions, see Oort *et al.* (1983).

5. Some Implications for the Global Energy Cycle of the Climatic System

For a global view of the atmospheric energetics we will use the formulation of the energy cycle as devised by Lorenz (1955). In this scheme, the inert part of the potential plus internal energy is subtracted to obtain the so-called available potential energy. The inert part is obtained through a thought experiment by redistributing the mass adiabatically so that surfaces of equal potential temperature are horizontal. The interested reader is referred to further discussions of available potential energy by Peixóto (1965) and Pearce (1978).

5.1. Equations

Following Lorenz' (1955) formulation and the notations used by Peixóto and Oort (1974), one can write the zonal mean and eddy parts of the available potential energy as

$$P_M = \frac{c_p}{2} \int \gamma ([\bar{T}] - \bar{\tilde{T}})^2 \, dm \tag{5.1}$$

$$P_E = \frac{c_p}{2} \int \gamma [\overline{T'^2} + \bar{T}^{*2}] \, dm \tag{5.2}$$

or

$$P_E = P_{TE} + P_{SE} \tag{5.2'}$$

and the zonal mean and eddy kinetic energy as

$$K_M = \tfrac{1}{2} \int ([\bar{u}]^2 + [\bar{v}]^2) \, dm \tag{5.3}$$

$$K_E = \tfrac{1}{2} \int [\overline{u'^2} + \overline{v'^2}] \, dm + \tfrac{1}{2} \int [\bar{u}^{*2} + \bar{v}^{*2}] \, dm \tag{5.4}$$

or

$$K_E = K_{TE} + K_{SE} \tag{5.4'}$$

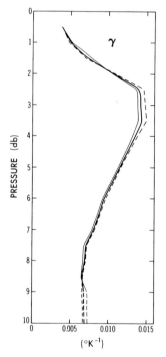

Fig. 47. Vertical profiles of the global mean static stability factor γ for the 10-yr mean (thick solid line), the DJF (thin solid line), and the JJA seasons (dashed line).

The static stability parameter in the definitions of P_M and P_E is given by

$$\gamma = -(\theta/T)(R/c_p p)(\partial\bar{\bar{\theta}}/\partial p)^{-1} \tag{5.5}$$

where $\bar{\bar{\theta}}$ indicates the global average of θ. Figure 47 shows the vertical dependence of γ for DJF, JJA, and the year computed from our data. The curves are quite similar to our previous hemispheric curves (Peixóto and Oort, 1974) for the Northern Hemisphere only. Below 900 mbar the curves were extrapolated to avoid unrealistic features due to topographic effects.

The balance equations for these basic forms of energy can be written symbolically as

$$\partial P_M/\partial t = G(P_M) - C(P_M, P_E) - C(P_M, K_M) + B(P_M) \tag{5.6}$$

$$\partial P_E/\partial t = G(P_E) + C(P_M, P_E) - C(P_E, K_E) + B(P_E) \tag{5.7}$$

$$\partial K_M/\partial t = C(P_M, K_M) + C(K_E, K_M) - D(K_M) + B(K_M) \tag{5.8}$$

$$\partial K_E/\partial t = C(P_E, K_E) - C(K_E, K_M) - D(K_E) + B(K_E) \tag{5.9}$$

The connecting links between the various energy forms are illustrated in a Lorenz' box diagram in Fig. 48. Of the various generation (G), conversion (C), dissipation (D), and boundary (B) terms in Eqs. (5.6)–(5.9) only the following terms were evaluated:

$$C(P_M, P_E) = -c_p \int \gamma[\overline{v'T'} + \bar{v}^*\bar{T}^*] \frac{\partial[\bar{T}]}{R\,\partial\phi}\,dm \tag{5.10}$$

$$C(P_M, K_M) = -\int [\bar{v}]g\,\frac{\partial[\bar{z}]}{R\,\partial\phi}\,dm \tag{5.11}$$

$$B(P_M) = c_p \iint_{Eq} \gamma[\bar{v}]([\bar{T}] - \bar{\bar{T}})^2\,dx\,dp/2g$$

$$+ \iint_{Eq} [\bar{v}]([\bar{z}] - \bar{\bar{z}})\,dx\,dp \tag{5.12}$$

$$C(K_E, K_M) = \int [\overline{v'u'} + \bar{v}^*\bar{u}^*] \cos\phi\left\{\frac{\partial([\bar{u}]/\cos\phi)}{R\,\partial\phi}\right\}\,dm \tag{5.13}$$

Of the remaining terms the "$\partial/\partial t$" and boundary terms [except $B(P_M)$] can be neglected for hemispheric integrals. The important unknown terms are $G(P_M)$, $G(P_E)$, $C(P_E, K_E)$, $D(K_M)$, and $D(K_E)$. As in our earlier article (Oort and Peixóto, 1974), we will assume a value for the generation of eddy available potential energy $G(P_E)$ to close the cycle. Then the

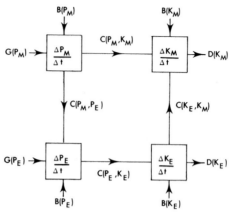

FIG. 48. Schematic box diagram of the hemispheric (or global when boundary terms are left out) energy cycle.

other unknown terms $G(P_M)$, $C(P_E, K_E)$, $D(K_M)$, and $D(K_E)$ can be computed as the residual in Eqs. (5.6)–(5.9), respectively.

5.2. Spatial Distribution of Energy

Meridional and vertical profiles of the mean and eddy energy components are shown in Figs. 49 and 50. The curves represent the contributions to the global integrals to be presented in a later subsection.

The meridional profiles in Fig. 49 show the following important differences between the Northern and Southern Hemisphere.

(1) Poleward of 50° latitude the contributions to P_M are greater in the Southern than in the Northern Hemisphere for all seasons. This is because the temperatures are lower over the Antarctic continent than over the Arctic Ocean, leading to a larger meridional temperature gradient in the Southern Hemisphere.

(2) On the other hand, the eddy activity shown in P_{TE} and P_{SE} is higher in the Northern than in the Southern Hemisphere. The very large values of P_{TE} in the northern midlatitudes for the year are not due to synoptic-scale eddies but to the large seasonal cycle of the temperature over the northern continents.

(3) The K_M values are in general smaller in the Northern than in the Southern Hemisphere, except during winter.

(4) The K_{TE} values are of comparable magnitude in both hemispheres.

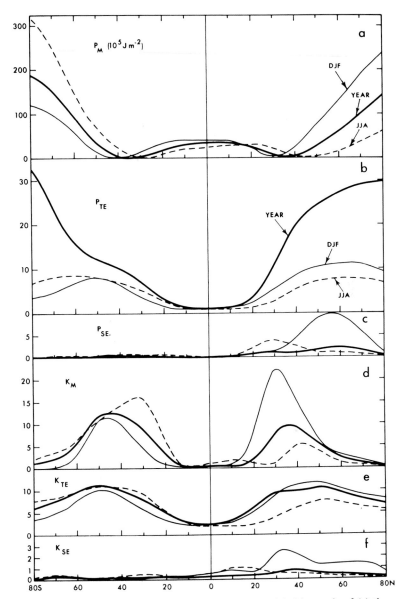

FIG. 49. Meridional profiles of the contributions to the global integrals of (a) the mean available potential energy, (b) the transient eddy available potential energy, (c) the standing eddy available potential energy, (d) the mean kinetic energy, (e) the transient eddy kinetic energy, and (f) the standing eddy kinetic energy.

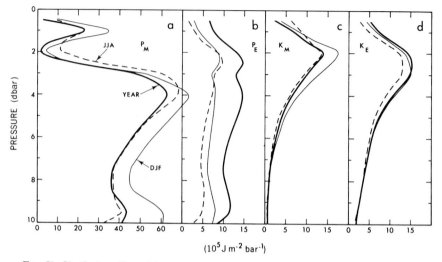

FIG. 50. Vertical profiles of the contributions to the global integrals of (a) mean available potential energy, (b) transient plus standing eddy available potential energy, (c) mean kinetic energy, and (d) transient plus standing eddy kinetic energy.

(5) As we have seen in earlier sections for other parameters, the seasonal variations are in general larger in the Northern than in the Southern Hemmisphere.

(6) The stationary eddies are only important in the Northern Hemisphere.

The vertical profiles of the global averages of the various forms of energy (as illustrated in Fig. 50) show that there is a seasonal variation even when the global energy content of the atmosphere is considered. It seems, from the energetics point of view, that the troposphere is more active during the northern winter (DJF) than during the northern summer (JJA) for all energy components.

The results in both Figs. 49 and 50 compare well with the Northern Hemisphere results presented by Peixóto and Oort (1974), allowing for the differences in defining the reference state in case of P_M.

5.3 Spatial Distribution of Energy Conversions

The contributions to the global integrals of the energy conversion terms are shown in Figs. 51 and 52. The most striking differences between the two hemispheres in Fig. 51 and between the seasonal curves for the globe in Fig. 52 are again due to the differences in the amplitude of the seasonal cycle. Aside from this fact, the two hemispheres appear to behave much alike.

All curves compare well with the curves presented and discussed by Oort and Peixóto (1974) for the Northern and by Peixóto and Corte-Real (1982) for the Southern Hemisphere.

5.4. Energy Cycle

The energetics of the atmospheric general circulation are summarized in the form of box diagrams in Figs. 53–55 for the year, DJF, and JJA, respectively. Except for the numbers between parentheses, all values are directly computed from our data using Eqs. (5.1)–(5.4) and (5.10)–(5.13).

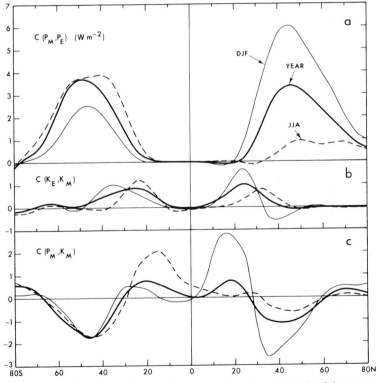

FIG. 51. Meridional profiles of the contributions to the global integrals of the conversion rates (a) from mean to eddy available potential energy, (b) from eddy to mean kinetic energy, and (c) from mean available potential to mean kinetic energy.

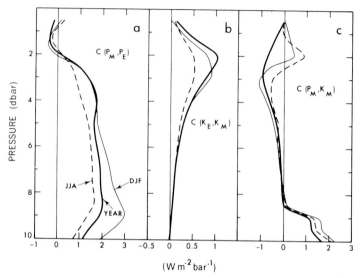

FIG. 52. Same as Fig. 51 except for vertical profiles.

The values of $G(P_E)$ are assumed guesses, whereas those for $G(P_M)$, $D(K_M)$, and $D(K_E)$ are calculated as a residual using Eqs. (5.6)–(5.9).

The values for the generation of eddy available potential energy, the basic unknown in this study, were chosen based on tentative evidence from earlier diagnostic and numerical model studies referred to by Oort and Peixóto (1974). For instance, in summer the important driving of the energy cycle is thought to occur through $G(P_E)$, whereas the baroclinic energy conversion $C(P_M, P_E)$ seems to play a less important role. The final values shown for $G(P_E)$ are speculative, but most likely of the correct order of magnitude.

Considering each hemisphere separately, the figures show, as most important features, maximum generation in the summer hemisphere, a strong flux of P_M across the equator from summer to winter hemisphere, and finally, the strongest dissipation in the winter hemisphere.

To get a better overview of the atmospheric energetics, some of the more important parameters, such as the mean plus eddy energy amounts, P and K, the generation and dissipation rates, $G(P)$ and $D(K)$, and the energy ratios, P_E/P and K_E/K, are tabulated in Table V. The inspection of this table shows that, on a global basis, the generation and dissipation values clearly indicate that the greatest atmospheric activity occurs during northern winter. Similarly the total energy amounts during DJF are substantially larger than those during JJA.

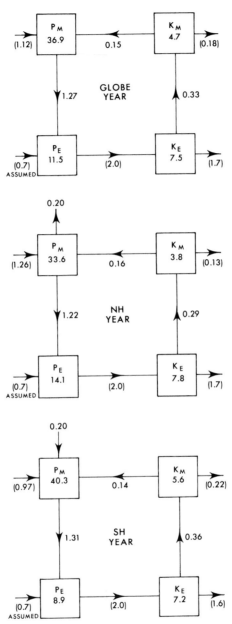

FIG. 53. Box diagrams of the energy cycle for the 10-yr mean. Energy amounts are in units of 10^5 J m^{-2}; generation, conversion, and dissipation rates are in units of W m^{-2}. Numbers in parentheses were calculated as residuals, except for the $G(P_E)$ values, which were assumed. NH, Northern Hemisphere; SH, Southern Hemisphere.

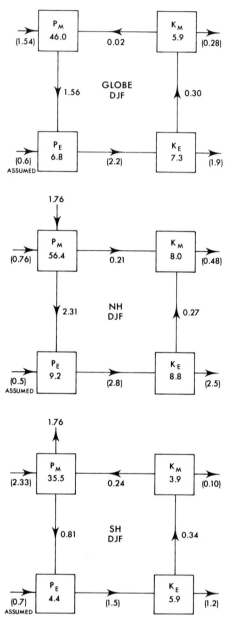

FIG. 54. Same as Fig. 53 except for DJF season.

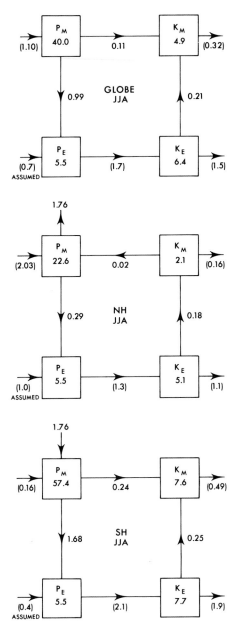

FIG. 55. Same as Fig. 53 except for JJA season.

TABLE V. SEASONAL VALUES OF AVAILABLE POTENTIAL AND KINETIC ENERGY, THEIR GENERATION AND DISSIPATION RATES, AND THE EDDY TO MEAN ENERGY RATIOS

Time[a]	Space[b]	P (10^5 J m^{-2})	K (10^5 J m^{-2})	$G(P)$ (W m^{-2})	$D(K)$ (W m^{-2})	P_E/P (%)	K_E/K (%)
Year	GL	48.4	12.2	1.9	1.9	24	61
	NH	47.7	11.6	2.0	1.8	30	67
	SH	49.2	12.8	1.7	1.9	18	56
DJF	GL	52.8	13.2	2.2	2.2	13	55
	NH	65.6	16.8	1.3	3.0	14	52
	SH	39.9	9.8	3.0	1.3	11	60
JJA	GL	45.5	11.3	1.8	1.8	12	57
	NH	28.1	7.2	3.0	1.3	20	71
	SH	62.9	15.3	0.6	2.4	9	50
Year[c]	NH	55	15	2.3	2.3	36	67
Year[d]	NH	49.1	12.4	2.2	2.0	32	71
DJF[e]	GL	47.1	12.3			15	59
JJA[e]	GL	38.9	10.4			15	52

[a] DJF, December–February; JJA, June–August.
[b] GL, Globe; NH, Northern Hemisphere; SH, Southern Hemisphere.
[c] From Oort (1964, mixed space–time domain).
[d] From Oort and Peixóto (1974).
[e] From Newell et al. (1974; p. 100, Lorenz approximation).

In order to assess the relative importance of the eddy circulations for the achievement of the heat balance, the ratios of the eddy and mean total values are also presented in the last two columns of Table V. They show that for comparable seasons the northern eddies contain a larger fraction of the total energy than the southern eddies. This can be explained by the greater inhomogeneity of the earth's surface in the Northern Hemisphere, leading to a stronger generation of eddies in that hemisphere. The, perhaps unexpected, large values for the yearly ratios P_E/P and K_E/K compared with those of the DJF and JJA ratios is due to the inclusion of the winter–summer variations in the transient eddy contributions to P_E and K_E.

Some earlier results by Oort (1964), Oort and Peixóto (1974), and Newell et al. (1974), using the same approximate formulation for the mixed space–time domain from Lorenz (1955), are added at the bottom of Table V. They compare well with the present results. We should mention that the use of a different formulation, such as the "exact" expression for available potential energy, would lead to different values for P_M, P_E, $G(P_M)$, $G(P_E)$, and $C(P_M, P_E)$, whereas the remaining terms involving kinetic energy would remain the same. A thorough comparison between the approximate and exact values of available potential energy was recently published by Min and Horn (1982) for the Northern Hemisphere during the FGGE year.

A final comment concerns the dissipation of kinetic energy. Kung (1969) has presented multiannual-mean dissipation rates over North America of 4.3, 5.2, and 3.3 W m^{-2} for the year, DJF and JJA seasons, respectively. These values are about a factor 2 larger than those for the Northern Hemisphere presented in Table V. The differences are probably due to the exclusion of the tropics in Kung's results, where the dissipation is less than 1 W m^{-2} (Holopainen, 1969). Combining these tropical and midlatitude estimates into a hemispheric average would yield a value on the order of 2 or 3 W m^{-2}, in better agreement with our tabulated values.

6. Concluding Remarks

In this section we will present a summary of the most important results obtained regarding the global angular momentum and energy cycles in the atmosphere. The results are based on a homogeneous set of daily surface ship and rawinsonde data from the 10-yr period May 1963 through April 1973. Taken as a whole, the results constitute one of the most complete summaries of the global balances made to date. Most of the data used were obtained directly from the original observations without the use of simplifying assumptions or of theoretically derived relationships in order to derive a picture of the general circulation of the atmosphere that is as objective as possible. However, there is one notable exception where an indirectly computed parameter was used, namely the mean meridional circulation outside the tropics. Especially in the Southern Hemisphere, the location and strength of the mean meridional circulations are difficult to measure directly, and their middle- and high-latitude values were, therefore, inferred from momentum balance considerations (see Appendix B).

Since one of our important objectives was to give a comprehensive picture of the global general circulation, much space in this article was devoted to a documentation of the three-dimensional structure of the various meteorological parameters of interest. Thus, the structure of the basic circulation, the temperature, humidity, and kinetic energy fields, both regarding their long-term mean values and their day-to-day and seasonal variability characteristics, were discussed in considerable detail. Following a theoretical formulation of the horizontal and vertical exchange mechanisms, the three-dimensional structure of the transports by the transient eddy, stationary eddy, and mean meridional circulations was shown in detail. These transports play a dominant role in preserving the global balances of angular momentum, water vapor, and energy in its various forms.

A cautionary note should be made here concerning the statistics in middle and high latitudes of the Southern Hemisphere. The large data gaps and the tendency at some stations to lose track of the rawinsonde balloons during strong wind conditions have probably led to an underestimate of the zonal wind speed, the kinetic energy, and possibly the fluxes of angular momentum and heat in the upper troposphere of the 40°–60° S belt (see, e.g., Van Loon *et al.*, 1971; Swanson and Trenberth, 1981). However, in view of the large year-to-year variations in the circulation of the Southern Hemisphere (Trenberth and Van Loon, 1981), several sample years of good data coverage, such as the one taken during the FGGE period, will be needed to establish what the actual biases in the present results are.

Much of the importance of the information, presented graphically in the form of maps, cross sections, and profiles in the main text and in the form of tables in Appendix A, stems from its coherent and comprehensive scale, but this cannot be easily summarized in this short section. Therefore, the reader should consult the specific sections for questions concerning the general climatology, and we will restrict ourselves here to the discussion of a few highlights selected from the previous sections.

Some of the main conclusions from our study of the angular momentum balance are the following:

(1) The largest seasonal variations in the angular momentum and its transports occur in the Northern Hemisphere.

(2) By and large, the stationary eddies are only important in the Northern Hemisphere.

(3) The results of the present study agree very well with those from earlier studies by, e.g., Newell *et al.* (1972, 1974).

(4) The detailed sections depicting the flow of angular momentum show that the region of subtropical easterlies in the winter hemisphere, where the surface winds are strongest, is the main source of angular momentum for the atmosphere. Most of this angular momentum is transported toward the winter pole at high levels in the troposphere, and then transported downward into the sink region of midlatitude westerlies (maximum transports of $+36 \times 10^{18}$ and -32×10^{18} kg m^2 sec^{-2} for the Northern and Southern Hemisphere, respectively). A similar transport pattern occurs in summer, but it is considerably weaker than in winter, especially during the northern summer (maximum transports of 14×10^{18} and -24×10^{18} kg

$m^2 \; sec^{-2}$ for the Northern and Southern Hemisphere, respectively). There is some cross-equatorial flow of angular momentum into the summer hemisphere.

(5) The zonal-mean (land plus ocean) values of the surface stress determined from the budget of angular momentum in the atmosphere agree well with the zonally averaged oceanic wind stress values derived from surface ship observations by Priestley (1951) and Hellerman (1967, 1982) using surface drag formulations.

(6) Mountain torques appear to give minor, but still important, contributions to the total torques around the latitude. At most latitudes the mountain effects seem to reinforce the surface friction effects, except possibly in the Southern Hemisphere.

As far as energetics are concerned we may summarize our main results as follows:

(1) Updated radiation estimates by Campbell and Vonder Haar (1980) were used to determine the required global energy transports by the ocean–atmosphere (Fig. 20). In the annual mean, a maximum poleward transport of 5.0×10^{15} W should occur near 35° N and of -6.1×10^{15} W near 35° S.

(2) A breakdown of the annual mean transports into their atmospheric and oceanic components (Fig. 46) shows that the atmospheric contributions are approximately antisymmetric with respect to the equator with extrema of about 3×10^{15} W near 45° N and -3×10^{15} W near 40° S. The extrema in oceanic transport are found near 25° latitude in both hemispheres and are on the order of 3×10^{15} W. The Southern Hemisphere maximum appears to be somewhat stronger than the one in the Northern Hemisphere. Poleward of 30° latitude the calculated oceanic transports are much stronger in the Southern than in the Northern Hemisphere. This conclusion is different from the one reached by Trenberth (1979), who in a similar calculation found oceanic poleward heat transports in the Southern Hemisphere which were comparable to those in the Northern Hemisphere (except for a secondary maximum near 60° S, which did not occur at 60° N). This discrepancy points out the degree of uncertainty involved in the indirect calculations of the oceanic heat transports. Nevertheless, all planetary balance calculations indicate that the oceans play a very important role in transporting heat poleward. They also emphasize the existing discrepancy with classical oceanographic methods (e.g., Hall and Bryden, 1982; Hastenrath, 1982), which suggest considerably smaller tropical oceanic heat fluxes.

(3) A thorough quantitative description of the global fluxes of sensible heat, geopotential energy, latent heat, and kinetic energy by the transient eddies, stationary eddies, and mean meridional circulations (Figs. 28–40), as well as their role in the global energetics, were presented with the aid of many tables, maps, cross sections, and profiles. Earlier results indicating the importance of the mean meridional circulation fluxes of potential energy, sensible and latent heat in the tropics and the importance of the eddy fluxes of sensible and latent heat in middle and high latitudes were confirmed in the present study. However, a new feature of this study is that for the first time a comprehensive and fairly uniform global data set is used in all evaluations, leading to probably more reliable results.

(4) Zonal cross sections of the streamlines of total energy in the latitude–pressure plane (Fig. 41) illustrate vividly the enormous direct inflow of energy into the summer hemisphere and the corresponding outflow from the winter hemisphere. The role of the atmosphere is evident in the vertical tilt of the annual streamlines, whereas their intersection with the earth's surface indicates the role of oceanic heat storage and transports.

(5) The storage of heat in the atmosphere is rather insignificant during the DJF and JJA seasons, reaching a maximum of only 5 W m^{-2} during summer near 40° latitude (Fig. 42a). On the other hand, the storage in the oceans (Fig. 45a) is a dominant factor in the overall heat balance, reaching values of 50–100 W m^{-2}. In the Northern Hemisphere the extrema are fairly localized near 35° N, whereas in the Southern Hemisphere they tend to be flatter, covering a much wider latitude zone.

(6) A comparison of the patterns of the divergence of energy in the atmosphere and oceans (Figs. 42b and 45b) show the relative contributions of the two media in transporting the heat from the tropics to high latitudes.

Finally, our evaluation of the integrated energy cycle according to Lorenz' (1955) formulation has led to the following conclusions:

(1) Poleward of 50° latitude the contributions to the hemispheric mean available potential energy P_M are greater in the Southern than in the Northern Hemisphere for all seasons due to the presence of the Antarctic continent. However, the activity in the eddy components P_{TE} and P_{SE} is larger in the Northern Hemisphere.

(2) The values of the mean kinetic energy K_M are larger in the Southern than in the Northern Hemisphere, whereas the transient eddy kinetic energy is of the same magnitude in both hemispheres. The stationary eddies are only important in the Northern Hemisphere.

(3) Assuming certain reasonable values for the generation of eddy available potential energy $G(P_E)$, estimates for the entire four-box energy cycle could be derived. For the globe as a whole, the intensity of the energy cycle, i.e., the total throughput of the system from generation to dissipation, was determined to be 1.9 W m^{-2} for the year, 2.2 W m^{-2} for DJF, and 1.8 W m^{-2} for JJA. Other evidence that the global energetics are more intense during the northern winter is that the values of P and K for DJF are larger than those for JJA.

(4) For comparable seasons the northern eddies contain a larger fraction of the total energy than the southern eddies. Apparently this is due to the presence of the major continents and mountain barriers in the Northern Hemisphere.

(5) Comparing the two hemispheres, we compute a maximum eddy plus mean potential energy generation of 3.0 W m^{-2} in the summer hemisphere, a strong cross-equatorial flow of P_M of 1.8 W m^{-2} from the summer to the winter hemisphere, and a dissipation between 2.5 and 3.0 W m^{-2} in the winter hemisphere.

APPENDIX A. TABLES

In this appendix zonally averaged values of some selected parameters are tabulated for the year and for the DJF and JJA seasons. The data are given for each 5° latitude circle between 80° S and 80° N (except for 75° N and 75° S) at eight pressure levels: 50, 100, 200, 300, 500, 700, 850, and 1000 mbar. In the last three columns of each table also areal averages are shown for the Northern Hemisphere (NH), the Southern Hemisphere (SH), and the globe. In the last row, entitled "MEAN," mass-weighted vertically averaged values are presented, which were computed using the following formula for the "MEAN" of an arbitrary parameter A:

$$[\hat{A}] = \left[\int_{25\ \text{mbar}}^{p_0} A\ dp/(p_0 - 25\ \text{mbar}) \right]$$

Here p_0 is the surface pressure which equals a global-mean value of 1012.5 mbar at sea level, but is reduced over mountains according to an assumed standard atmosphere profile. One should note that in the zonal averages only the grid point values above the earth's surface were included. To obtain a vertically integrated, rather than an averaged, value, one should multiply the value under "MEAN" by the appropriate factor at the corresponding latitude as presented in Table A1.

The quantities tabulated are as follows:

Table A.1.   Multiplication factor to convert the vertically averaged values (row "MEAN") in Tables A2–A30 to vertically integrated values, while taking into account the mean mass distribution over the globe

Table A2.   Zonal wind $[\bar{u}]$

Table A3.   Meridional wind $[\bar{v}]$

Table A4.   Meridional wind computed indirectly from momentum balance poleward of 20° latitude $[\bar{v}]_{ID}$

Table A5.   "Vertical velocity" $[\bar{\omega}]_{ID}$

Table A6.   Temperature $[\bar{T}]$

Table A7.   Geopotential height departure from NMC standard atmosphere $[\bar{z}] - z_{SA}$

Table A8.   Specific humidity $[\bar{q}]$

Table A9.   Transient eddy kinetic energy $[\bar{K}_{TE}]$

Table A10.   Standing eddy kinetic energy $[\bar{K}_{SE}]$

Table A11.   Zonal mean kinetic energy $[\bar{K}_M]$

Table A12.   Temporal standard deviation of zonal wind $[\sigma(u)]$

Table A13.   Temporal standard deviation of meridional wind $[\sigma(v)]$

Table A14.   Temporal standard deviation of temperature $[\sigma(T)]$

Table A15.   Temporal standard deviation of geopotential height $[\sigma(z)]$

Table A16.   Temporal standard deviation of specific humidity $[\sigma(q)]$

Table A17.   Northward transport of momentum by transient eddies $[\overline{v'u'}]$

Table A18.   Northward transport of momentum by stationary eddies $[\bar{v}^*\bar{u}^*]$

Table A19.   Northward transport of momentum by mean meridional circulations $[\bar{v}]_{ID}[\bar{u}]$

Table A20.   Northward transport of sensible heat by transient eddies $[\overline{v'T'}]$

Table A21.   Northward transport of sensible heat by stationary eddies $[\bar{v}^*\bar{T}^*]$

Table A22.   Northward transport of geopotential energy by transient eddies $[\overline{v'z'}]$

Table A23.   Northward transport of geopotential energy by stationary eddies $[\bar{v}^*\bar{z}^*]$

Table A24.   Northward transport of humidity by transient eddies $[\overline{v'q'}]$

Table A25.   Northward transport of humidity by stationary eddies $[\bar{v}^*\bar{q}^*]$

TABLE A1. MULTIPLICATION FACTOR TO CONVERT THE VERTICALLY AVERAGED VALUES (ROW "MEAN") IN TABLES A2–A30 TO VERTICALLY INTEGRATED VALUES, WHILE TAKING INTO ACCOUNT THE MEAN MASS DISTRIBUTION OVER THE GLOBE (i.e., LESS MASS OVER THE MOUNTAIN)[a]

Latitude	Multiplication factor	Latitude	Multiplication factor
80S	0.748	5N	0.982
75S	0.777	10N	0.985
70S	0.903	15N	0.982
65S	0.991	20N	0.976
60S	1.000	25N	0.968
55S	1.000	30N	0.944
50S	0.999	35N	0.930
45S	0.998	40N	0.937
40S	0.997	45N	0.952
35S	0.992	50N	0.956
30S	0.985	55N	0.966
25S	0.980	60N	0.962
20S	0.981	65N	0.953
15S	0.977	70N	0.961
10S	0.985	75N	0.971
5S	0.980	80N	0.974
EQ	0.982		

[a] For simplicity, the sea level pressure is assumed to be uniform over the globe.

TABLE A2. ZONAL WIND [ū] (IN m sec⁻¹)

UNITS = 1.E 00 M/S

YEAR 63-73

p (MB)	80S	70S	65S	60S	55S	50S	45S	40S	35S	30S	25S	20S	15S	10S	5S	EQ	5N	10N	15N	20N	25N	30N	35N	40N	45N	50N	55N	60N	65N	70N	80N	SH	NH	GLOBE
50	10.5	13.3	14.1	14.2	13.4	11.6	8.8	5.5	2.0	-1.3	-3.9	-5.7	-5.7	-4.5	-3.0	-2.5	-4.0	-6.4	-7.4	-6.8	-4.1	-0.6	2.4	4.6	6.0	6.7	7.0	7.2	6.8	3.9		2.31	0.01	1.16
100	9.3	13.3	15.3	17.1	18.7	19.5	19.5	18.4	17.5	15.5	12.0	7.4	2.7	-1.0	-2.6	-3.7	-4.0	-3.8	-0.7	4.1	9.7	14.2	15.9	15.6	13.9	11.8	10.0	8.8	7.9	7.0	4.3	10.41	6.59	8.50
200	6.3	9.3	11.7	14.8	18.4	21.5	23.8	25.6	26.8	26.7	23.2	16.9	8.8	1.9	-1.9	-3.8	-2.5	0.7	6.7	13.6	20.0	24.8	25.7	24.0	20.5	16.7	13.3	10.5	8.7	7.3	4.2	14.72	12.33	13.52
300	5.3	7.6	10.3	14.1	18.2	22.1	22.3	22.0	21.0	17.9	12.8	6.3	0.7	-2.4	-4.0	-2.9	-0.3	4.5	10.2	15.7	19.5	20.7	20.0	17.9	15.3	12.7	10.2	8.6	6.9	4.8	2.9	12.26	10.11	11.19
500	2.4	4.0	6.7	10.3	13.9	16.1	16.3	15.4	13.6	11.3	8.4	4.9	1.3	-1.6	-3.4	-4.1	-3.4	-0.6	3.5	7.6	10.2	11.6	12.2	11.6	10.3	8.6	6.9	5.8	4.8	2.9	1.5	6.97	4.94	5.95
700	1.1	0.3	2.5	6.1	9.6	11.6	11.5	10.1	7.9	5.4	2.9	0.6	-1.1	-2.3	-3.0	-3.4	-3.8	-2.5	0.0	2.7	4.8	6.1	6.6	6.9	6.5	5.5	4.3	3.4	2.7	1.5		3.55	1.94	2.74
850	-2.3	-1.2	2.5	6.3	8.6	8.4	6.8	4.4	1.8	-0.3	-1.9	-2.8	-3.1	-2.8	-2.7	-3.1	-2.9	-1.9	-0.1	1.5	3.0	3.8	3.9	3.9	3.9	3.4	2.5	1.9	1.4	0.8		1.26	0.37	0.82
1000	.....	1.3	3.5	5.0	5.9	6.0	4.9	3.2	1.2	-1.0	-3.0	-4.2	-4.5	-3.8	-2.9	-2.0	-1.8	-2.8	-3.9	-4.0	-2.7	-0.9	1.0	1.9	2.2	2.5	1.5	0.8	0.5	-0.0	-0.5	0.26	-1.16	-0.35
MEAN	4.2	5.0	6.8	9.8	12.8	14.7	14.8	13.9	12.4	10.5	7.9	4.6	1.2	-1.5	-2.8	-3.3	-2.7	-0.3	3.1	6.9	10.0	11.5	11.7	10.7	9.4	7.9	6.4	5.4	4.5	2.6		6.66	4.72	5.70

DJF 63-73

p (MB)	80S	70S	65S	60S	55S	50S	45S	40S	35S	30S	25S	20S	15S	10S	5S	EQ	5N	10N	15N	20N	25N	30N	35N	40N	45N	50N	55N	60N	65N	70N	80N	SH	NH	GLOBE
50	0.3	0.3	0.5	0.7	0.7	0.3	-0.9	-2.9	-5.7	-8.7	-11.4	-13.1	-12.6	-10.4	-6.9	-4.2	-3.1	-2.7	-1.5	0.7	4.2	7.5	9.7	11.3	12.2	13.1	13.7	14.4	15.0	14.4	8.4	-5.92	5.84	-0.04
100	2.6	3.6	5.3	7.8	10.8	13.5	15.1	15.4	13.7	10.6	5.8	0.9	-2.7	-4.6	-4.1	-3.1	-1.5	1.3	7.4	15.1	22.0	25.9	24.8	21.5	18.0	14.9	13.3	12.5	12.1	11.2	6.9	5.22	12.91	9.07
200	4.1	5.2	7.6	11.5	16.2	20.3	22.8	23.6	22.2	19.3	14.3	8.3	2.4	-1.8	-2.8	-2.2	1.4	6.6	15.7	25.9	34.8	39.4	35.7	28.6	22.0	16.5	12.8	10.4	9.0	7.8	4.7	10.62	18.43	14.52
300	3.7	4.8	8.1	13.2	18.5	22.2	22.6	21.1	18.1	14.4	10.0	5.4	1.4	-1.5	-2.6	-2.7	-0.1	4.0	11.5	20.7	28.6	31.7	29.5	25.2	20.3	15.5	11.9	9.3	7.8	6.6	4.2	9.43	15.18	12.30
500	1.3	1.7	4.7	9.4	14.0	17.0	16.8	14.7	11.1	7.3	3.7	0.9	-0.8	-1.9	-2.3	-2.8	-2.8	-0.8	2.8	10.6	13.5	17.0	16.0	13.5	10.6	8.2	6.4	5.1	4.2	3.0		5.39	7.69	6.54
700	0.6	-1.2	1.2	5.3	9.3	11.7	11.4	9.3	6.3	3.1	0.6	-1.0	-1.4	-1.4	-1.3	-2.1	-3.1	-2.3	-1.3	2.6	6.2	8.4	9.4	9.1	8.4	5.1	5.5	4.8	4.2	3.3		2.87	3.23	3.05
850	-2.0	-1.1	2.4	6.1	8.1	7.6	5.5	4.5	2.4	0.1	-2.2	-3.8	-4.1	-3.3	-1.9	-1.7	-2.0	-3.3	-5.1	-5.3	-4.4	-2.0	0.7	2.5	3.2	2.8	2.6	1.4				1.24	0.52	0.88
1000	.....	0.6	2.1	3.8	5.2	5.5	4.5	2.4	0.1	-2.2	-3.8	-4.1	-3.3	-1.9	-1.7	-2.0	-3.3	-5.1	-5.3	-4.4	-2.0	0.7	2.5	3.2	2.8	2.6	1.4					0.10	-1.35	-0.53
MEAN	1.9	1.9	4.0	7.6	11.3	13.7	13.8	12.2	9.5	6.4	3.3	0.6	-1.2	-2.2	-2.8	-3.1	-2.6	-1.1	3.2	9.0	14.3	17.3	17.3	15.4	12.8	10.2	8.2	6.7	5.9	5.0	3.1	4.51	7.76	6.13

JJA 63-73

p (MB)	80S	70S	65S	60S	55S	50S	45S	40S	35S	30S	25S	20S	15S	10S	5S	EQ	5N	10N	15N	20N	25N	30N	35N	40N	45N	50N	55N	60N	65N	70N	80N	SH	NH	GLOBE
50	14.6	18.4	20.1	21.2	21.1	19.8	17.2	13.7	9.8	6.2	3.5	1.4	0.4	-0.1	-1.4	-3.6	-8.2	-13.4	-16.2	-17.0	-14.9	-11.1	-7.2	-4.1	-2.3	-1.5	-1.4	-1.7	-2.0	-1.6		8.39	-8.19	0.10
100	11.7	18.4	20.9	22.7	23.7	24.0	23.4	21.8	20.4	17.1	12.1	6.0	0.9	-2.5	-5.5	-8.5	-11.7	-12.1	-10.2	-5.5	0.5	5.8	9.2	9.4	7.7	5.4	3.5	2.5	1.9	0.9		14.01	-1.59	6.21
200	8.0	10.9	12.5	14.4	17.0	20.0	24.0	28.4	32.5	34.6	30.9	22.8	12.3	2.7	-4.7	-7.7	-5.5	-6.1	-5.6	-3.6	-1.5	1.8	6.7	12.2	15.6	15.4	13.4	10.6	8.2	6.5		17.12	4.55	10.83
300	6.5	8.7	10.2	12.2	14.7	17.3	20.0	23.0	26.0	27.7	25.0	18.8	9.9	1.7	-3.2	-6.2	-6.5	-4.6	-4.7	-4.1	-2.8	-0.2	3.1	6.3	8.7	9.3	8.6	6.9	5.3	4.6	3.5	13.88	4.10	8.99
500	4.5	6.4	9.0	11.6	13.5	14.5	15.2	15.4	14.8	12.7	8.8	3.4	-1.3	-3.6	-4.5	-4.6	-4.7	-4.1	-2.8	-0.7	1.3	3.2	4.2	5.2	5.3	4.3	3.2	2.8	2.9	2.2		7.78	1.79	4.79
700	1.7	0.4	2.3	5.6	8.7	10.7	11.1	10.6	9.2	7.2	4.9	2.2	-0.5	-2.7	-3.4	-3.3	-3.0	-2.4	-0.7	1.3	3.2	4.3	5.2	5.3	4.3	3.2	2.8	2.9	2.2	1.5		3.96	0.67	2.30
850	-2.6	-1.6	1.9	5.7	8.1	8.5	7.9	5.9	3.4	0.7	-1.9	-3.7	-4.5	-3.7	-1.1	-0.2	-1.5	-3.0	-3.1	-1.1	0.2	1.5	2.3	2.8	3.1	2.6	1.7	1.5	1.5	1.5		1.21	0.33	0.77
1000	.....	1.6	3.7	5.1	5.9	6.1	5.5	4.3	2.6	0.2	-2.3	-4.4	-5.6	-5.3	-4.0	-2.2	-0.5	-0.6	-2.4	-3.5	-3.0	-1.6	0.2	1.2	1.6	2.0	1.3	0.7	0.1	-0.3	0.2	0.34	-0.81	-0.16
MEAN	5.4	6.2	7.5	9.9	12.3	14.1	14.9	15.3	14.5	11.9	7.9	3.0	-1.2	-3.3	-4.3	-4.3	-4.7	-4.9	-4.6	-3.5	-1.2	2.3	5.8	8.1	8.4	7.6	5.8	4.2	3.6	2.4		8.03	1.05	4.56

TABLE A3. MERIDIONAL WIND $[\bar{v}]$ (IN m sec^{-1})

V UNITS= 1.E 00 M/S

YEAR 63-73

P (MB)	80S	70S	65S	60S	55S	50S	45S	40S	35S	30S	25S	20S	15S	10S	5S	EQ	5N	10N	15N	20N	25N	30N	35N	40N	45N	50N	55N	60N	65N	70N	80N	SH	NH	GLOBE
50	-1.5	-2.2	-2.1	-1.7	-1.1	-0.4	-0.0	0.3	0.4	0.4	0.2	0.1	0.1	0.1	0.1	0.1	0.1	0.1	0.1	0.1	-0.2	-0.1	0.0	0.1	0.1	0.1	0.0	0.0	0.1	0.1	-0.6	-0.22	-0.00	-0.11
100	-0.4	-1.0	-1.2	-1.2	-0.9	-0.4	-0.4	0.3	0.5	0.6	0.6	0.4	0.2	0.4	0.2	0.4	0.4	0.5	0.3	0.5	0.1	-0.1	-0.2	-0.1	-0.1	0.0	0.1	0.1	0.1	0.2	-0.4	-0.04	0.16	0.06
200	-0.8	-0.3	0.2	0.7	1.0	1.2	1.5	1.5	0.9	0.7	0.6	0.4	-0.2	-0.7	-0.4	0.1	0.7	1.2	1.1	0.8	0.4	0.0	-0.4	-0.7	-0.6	-0.3	-0.1	0.1	0.2	0.3	0.51	0.20	0.35
300	-0.5	-0.7	-0.3	0.3	0.2	0.7	1.0	1.0	0.9	0.7	0.6	0.5	0.3	0.3	0.3	0.2	0.3	0.3	0.1	-0.2	-0.5	-0.6	-0.4	-0.1	0.2	0.3	0.42	0.04	0.23
500	0.2	-0.5	-0.5	-0.3	-0.3	-0.1	0.0	0.2	0.2	0.1	-0.0	-0.1	0.1	0.1	0.1	-0.0	-0.1	-0.1	-0.3	-0.4	-0.4	-0.2	-0.1	-0.3	-0.4	-0.4	-0.2	-0.1	-0.0	0.1	0.1	-0.08	-0.04	-0.06
700	-1.0	-0.4	-0.2	-0.2	-0.1	0.0	0.2	0.1	-0.0	-0.2	-0.2	-0.1	-0.2	-0.1	-0.0	0.0	0.1	0.0	0.0	-0.1	0.0	0.1	0.2	0.2	0.2	0.1	-0.0	-0.1	-0.1	-0.3	-0.21	-0.05	-0.13
850	.....	-0.2	-0.4	-0.6	-0.7	-0.5	-0.5	-0.5	-0.3	-0.5	-0.5	-0.3	0.0	0.0	0.4	0.5	0.3	-0.1	0.0	0.1	-0.0	-0.0	-0.1	-0.3	-0.21	-0.03	-0.12
1000	.....	-0.9	-1.2	-1.3	-1.1	-0.8	-0.3	0.4	1.0	1.5	2.0	2.2	2.0	1.6	1.4	0.6	-1.0	-1.6	-1.5	-1.1	-1.1	-0.7	-0.5	-0.3	0.0	0.3	-0.0	-0.6	-0.5	0.1	0.53	-0.49	0.09
MEAN	-0.4	-0.5	-0.4	-0.3	-0.1	0.0	0.2	0.2	0.2	0.2	0.2	0.1	0.1	0.1	0.1	0.1	0.1	0.0	0.0	0.1	-0.0	-0.2	-0.3	-0.2	-0.2	-0.0	-0.0	0.0	0.0	0.0	0.0	0.05	-0.02	0.02

DJF 63-73

P (MB)	80S	70S	65S	60S	55S	50S	45S	40S	35S	30S	25S	20S	15S	10S	5S	EQ	5N	10N	15N	20N	25N	30N	35N	40N	45N	50N	55N	60N	65N	70N	80N	SH	NH	GLOBE
50	-0.3	-0.6	-0.6	-0.5	-0.3	-0.1	0.0	0.1	0.0	-0.0	-0.0	-0.1	-0.1	0.0	0.2	0.4	0.4	0.2	0.3	0.3	0.2	-0.0	-0.1	-0.1	-0.1	-0.0	-0.0	-0.1	-0.1	0.1	-1.1	-0.07	0.07	-0.00
100	-0.7	-0.8	-0.7	-0.5	-0.4	-0.2	0.0	0.3	0.5	0.5	0.7	0.8	0.6	0.3	0.6	0.5	0.9	1.2	1.4	1.3	0.8	0.1	-0.5	-0.6	-0.4	0.0	0.0	0.1	0.2	0.2	-1.1	0.22	0.41	0.31
200	-0.7	-0.1	0.2	0.5	0.7	0.8	0.9	0.9	0.8	0.7	0.8	1.1	1.3	1.4	2.3	3.6	3.4	2.7	2.0	1.1	-0.1	-0.1	-1.3	-1.6	-1.2	-0.6	-0.3	-0.1	-0.1	0.1	0.5	1.02	0.92	0.97
300	-1.3	-0.9	-0.3	0.3	0.8	1.0	1.0	0.8	0.6	0.8	0.9	1.0	0.9	1.0	1.0	1.0	0.7	0.9	0.7	-0.2	-0.9	-1.3	-1.0	-0.4	-0.0	0.0	0.1	0.0	0.0	0.7	0.63	0.20	0.42
500	0.0	-0.3	-0.2	-0.1	0.1	0.1	0.2	0.3	0.3	0.4	0.2	0.2	0.1	0.0	-0.0	-0.2	-0.1	0.0	0.3	0.4	-0.4	-0.6	-0.5	-0.3	-0.2	-0.1	-0.1	-0.1	-0.1	0.0	0.5	0.06	-0.05	0.01
700	-1.2	-0.1	0.0	-0.0	-0.0	-0.1	-0.1	-0.1	-0.3	-0.3	-0.4	-0.3	-0.3	-0.4	-0.5	-0.7	-0.5	-0.2	-0.1	0.1	0.1	0.3	0.1	0.1	-0.2	-0.1	-0.1	0.3	0.2	0.1	0.2	-0.30	-0.11	-0.21
850	.....	-0.1	-0.2	-0.4	-0.5	-0.3	-0.2	-0.1	-0.3	-0.4	-0.6	-0.6	-0.7	-0.6	-0.7	-0.9	-1.3	-1.5	-1.0	-0.2	0.0	0.1	0.2	0.3	0.4	0.3	-0.0	-0.1	-0.1	-0.6	-0.46	-0.31	-0.39
1000	.....	0.4	-0.4	-0.4	-0.9	-1.3	-1.4	-1.0	-0.2	0.6	1.0	1.3	1.5	1.2	0.5	0.2	0.0	-1.3	-3.0	-2.5	-1.7	-1.0	-0.7	-0.4	-0.1	-0.1	-0.4	-0.6	-0.3	0.0	0.5	0.17	-1.27	-0.46
MEAN	-0.6	-0.2	-0.1	-0.1	-0.0	0.0	0.1	0.2	0.2	0.2	0.2	0.3	0.2	0.2	0.2	0.3	0.3	0.3	0.3	0.3	-0.0	-0.4	-0.4	-0.6	-0.4	-0.2	-0.1	-0.1	-0.0	-0.1	0.3	0.11	-0.01	0.05

JJA 63-73

P (MB)	80S	70S	65S	60S	55S	50S	45S	40S	35S	30S	25S	20S	15S	10S	5S	EQ	5N	10N	15N	20N	25N	30N	35N	40N	45N	50N	55N	60N	65N	70N	80N	SH	NH	GLOBE
50	-2.0	-4.9	-4.9	-3.9	-2.3	-0.9	0.5	0.8	0.8	0.7	0.4	0.2	0.1	0.0	-0.1	-0.2	-0.1	-0.1	-0.1	-0.1	0.0	0.0	0.0	-0.0	-0.0	-0.0	0.0	0.1	0.0	0.0	-0.0	-0.52	-0.05	-0.29
100	0.2	-0.9	-1.2	-1.1	-0.8	-0.3	0.0	0.4	0.5	0.3	-0.1	-0.3	-0.4	-0.3	-0.2	-0.2	-0.2	-0.3	-0.4	-0.3	0.0	0.0	0.0	0.0	0.1	0.1	0.1	0.0	0.0	0.0	0.3	-0.15	-0.07	-0.11
200	-0.4	-0.1	0.5	1.1	1.2	1.2	1.4	1.6	1.7	1.7	0.9	-0.3	-1.6	-2.7	-3.1	-3.0	-2.2	-1.1	-0.4	-0.1	-0.1	-0.1	0.0	0.0	-0.4	-0.6	-0.2	-0.3	-0.0	0.1	0.3	-0.09	-0.50	-0.30
300	-0.3	-0.6	-0.1	0.5	0.8	0.9	0.9	0.8	0.7	0.5	0.1	-0.3	-0.5	-0.7	-0.7	-0.5	-0.2	-0.0	0.0	0.2	0.1	0.1	0.1	0.1	-0.1	-0.2	-0.3	-0.2	0.1	0.1	0.1	0.19	-0.09	0.05
500	0.8	-0.6	-0.5	-0.4	-0.3	-0.2	-0.0	-0.0	-0.0	-0.2	-0.4	-0.4	-0.3	-0.4	-0.3	-0.2	-0.0	0.1	0.2	0.1	0.0	0.0	0.0	-0.0	-0.0	0.1	0.0	-0.0	-0.1	-0.0	-0.27	0.02	-0.13
700	0.0	-0.2	-0.3	-0.4	-0.3	-0.2	-0.0	-0.0	-0.0	-0.3	-0.7	-0.8	-0.6	-0.6	-0.5	-0.2	0.0	0.1	0.1	0.2	0.1	0.1	0.0	0.0	-0.1	-0.1	0.1	0.2	0.1	0.0	-0.1	-0.17	0.07	-0.05
850	.....	-0.2	-0.6	-1.1	-1.2	-1.0	-0.8	-0.6	-0.6	-0.7	-0.5	-0.1	0.6	1.3	1.6	1.4	1.0	0.6	0.3	0.2	0.1	0.2	0.1	0.0	0.0	0.1	0.2	0.1	-0.0	-0.2	-0.4	-0.05	0.25	0.10
1000	.....	-2.2	-2.1	-2.2	-2.1	-1.7	-1.1	-0.5	0.3	1.0	1.7	2.6	3.2	3.2	2.9	2.8	2.5	0.9	-0.2	-0.4	-0.0	-0.1	-0.1	0.0	0.0	0.3	0.8	0.5	-0.2	-0.7	-0.3	0.67	0.43	0.56
MEAN	0.1	-0.6	-0.6	-0.5	-0.3	-0.2	-0.0	0.1	0.1	0.1	0.0	0.0	0.0	0.0	0.0	0.1	0.1	0.1	0.0	0.1	0.0	-0.0	-0.1	0.0	-0.1	-0.1	0.0	0.1	-0.0	-0.1	-0.1	-0.06	0.00	-0.03

TABLE A4. MERIDIONAL WIND COMPUTED INDIRECTLY FROM MOMENTUM BALANCE POLEWARD OF 20° LATITUDE [v̄]$_{ID}$ (IN m sec^{-1})

V-ID   YEAR 63-73   UNITS= 1.E 00 M/S

YEAR 63-73

p (MB)	80S	70S	65S	60S	55S	50S	45S	40S	35S	30S	25S	20S	15S	10S	5S	EQ	5N	10N	15N	20N	25N	30N	35N	40N	45N	50N	55N	60N	65N	70N	80N	SH	NH	GLOBE
50	0.1	0.1	0.0	-0.0	-0.0	-0.0	-0.0	-0.0	-0.0	-0.0	-0.0	-0.0	-0.0	-0.0	-0.0	-0.0	0.1	0.4	0.4	0.2	0.0	-0.1	-0.1	-0.0	-0.0	-0.0	-0.0	0.0	-0.0	-0.0	0.2	0.00	-0.01	-0.00
100	-0.0	0.1	0.1	0.1	0.1	0.2	0.3	0.1	0.0	-0.0	-0.1	-0.4	-0.6	-0.8	-0.6	-0.1	0.4	1.1	0.7	0.4	0.1	-0.1	-0.1	-0.2	-0.0	-0.0	-0.0	0.0	0.0	0.1	0.1	-0.04	0.06	0.01
200	-0.1	-0.1	-0.0	0.1	0.2	0.3	0.3	0.2	0.1	-0.0	-0.1	-0.4	-0.6	-0.8	-0.6	-0.1	0.6	1.1	1.0	0.4	0.1	-0.1	-0.2	-0.2	-0.2	-0.1	-0.0	0.0	0.1	0.0	0.0	-0.17	0.19	0.01
300	-0.3	-0.1	0.0	0.1	0.3	0.3	0.2	0.1	0.0	-0.0	-0.1	-0.2	-0.2	-0.1	0.0	0.3	0.3	0.3	0.2	0.0	-0.0	-0.1	-0.1	-0.2	-0.2	-0.1	-0.1	0.0	0.1	0.0	0.1	0.01	0.04	0.02
500	-0.1	-0.1	0.0	0.1	0.1	0.1	0.1	0.0	0.0	-0.0	-0.2	-0.0	-0.0	-0.2	-0.3	-0.3	-0.1	-0.0	-0.0	-0.1	-0.0	-0.1	-0.1	-0.1	-0.1	-0.0	-0.0	0.0	-0.0	0.0	-0.0	-0.06	-0.02	-0.04
700	-0.0	-0.0	-0.0	0.0	0.0	0.0	0.1	0.0	0.0	0.0	0.0	0.0	-0.2	-0.1	-0.0	-0.2	-0.3	-0.3	-0.1	-0.0	-0.0	-0.0	-0.0	-0.0	-0.0	-0.0	-0.0	0.0	0.0	0.0	0.0	-0.02	-0.06	-0.04
850	-0.0	-0.0	0.0	0.0	0.0	0.0	0.1	0.0	0.0	0.0	0.0	0.0	0.0	0.3	0.4	0.2	-0.0	-0.2	-0.1	-0.1	-0.0	-0.0	-0.0	-0.0	-0.0	-0.0	0.0	0.0	0.0	0.0	0.0	0.10	-0.06	0.01
1000	0.5	-0.0	-0.0	-0.3	-0.5	-0.7	-0.8	-0.6	-0.3	0.1	0.4	1.7	1.9	1.7	1.5	1.3	0.5	-1.1	-1.3	-0.8	-0.1	0.4	0.6	0.7	0.7	0.4	0.0	-0.3	-0.4	-0.2	-0.2	0.36	-0.08	0.19
MEAN	-0.0	0.0	-0.0	0.0	0.0	0.0	0.0	0.0	-0.0	-0.0	-0.0	0.0	0.0	-0.0	-0.0	-0.0	0.0	0.0	0.0	0.0	0.0	-0.0	-0.0	-0.0	-0.0	-0.0	-0.0	0.0	0.0	0.0	0.0	-0.00	0.00	0.00

DJF 63-73

p (MB)	80S	70S	65S	60S	55S	50S	45S	40S	35S	30S	25S	20S	15S	10S	5S	EQ	5N	10N	15N	20N	25N	30N	35N	40N	45N	50N	55N	60N	65N	70N	80N	SH	NH	GLOBE
50	0.0	0.0	0.0	0.0	0.1	0.1	0.1	0.1	-0.0	-0.0	-0.1	-0.1	-0.1	-0.1	0.1	0.2	0.2	0.1	0.1	0.3	0.2	0.0	-0.0	-0.1	-0.1	-0.1	-0.1	-0.0	-0.0	0.0	0.4	0.00	0.04	0.02
100	0.0	0.0	0.0	0.1	0.1	0.1	0.1	0.1	0.0	0.0	-0.1	-0.1	-0.1	-0.1	0.2	0.3	0.7	1.1	0.8	0.3	0.2	-0.0	-0.1	-0.1	-0.1	-0.1	-0.0	-0.0	0.0	0.1	0.1	0.07	0.26	0.17
200	-0.1	-0.0	-0.0	0.0	0.1	0.3	0.3	0.2	0.0	0.0	-0.1	-0.4	-0.4	-0.4	0.3	1.9	3.4	3.3	1.0	0.6	0.2	-0.0	-0.4	-0.3	-0.3	-0.2	-0.1	-0.0	0.1	0.1	0.0	0.45	0.93	0.69
300	-0.2	-0.1	-0.0	0.1	0.2	0.2	0.3	0.2	0.0	0.0	-0.1	-0.2	-0.2	-0.2	0.7	0.8	0.9	1.0	0.6	0.3	0.1	-0.0	-0.3	-0.3	-0.2	-0.1	-0.0	0.0	0.1	0.1	0.0	0.18	0.19	0.19
500	-0.1	-0.1	0.0	0.1	0.1	0.1	0.1	0.1	-0.0	-0.1	-0.1	-0.1	-0.1	-0.1	-0.2	-0.3	-0.2	-0.1	-0.0	0.1	0.1	-0.1	-0.1	-0.1	-0.1	-0.0	0.0	0.1	0.1	0.1	-0.0	-0.07	-0.07	-0.07
700	-0.1	-0.1	0.0	0.0	0.1	0.1	0.1	0.1	-0.0	-0.0	-0.1	-0.0	-0.3	-0.5	-0.3	-0.7	-0.7	-0.4	-0.2	-0.0	-0.0	-0.1	-0.0	-0.0	-0.0	-0.0	0.0	0.0	0.0	0.0	0.0	-0.15	-0.17	-0.16
850	-0.0	-0.0	0.0	0.0	0.0	0.0	0.0	0.0	0.0	0.0	0.0	0.0	-0.0	-0.4	-0.4	-0.8	-1.1	-1.4	-0.6	-0.0	-0.1	-0.0	-0.0	-0.0	0.0	0.0	0.0	0.0	0.0	0.0	0.0	-0.20	-0.38	-0.29
1000	0.4	-0.0	-0.0	-0.3	-0.5	-0.7	-0.8	-0.6	-0.1	0.4	0.5	0.6	0.9	0.4	0.1	-0.1	-3.1	-2.4	-1.3	-0.4	0.7	1.1	1.1	1.1	0.9	0.4	0.0	-0.4	-0.5	-0.3	0.05	-0.63	-0.24
MEAN	0.4	0.0	0.0	0.0	0.0	0.0	0.0	0.0	0.0	0.0	0.0	0.0	0.0	0.0	0.0	0.0	0.0	0.0	0.0	0.0	0.0	0.0	0.0	0.0	0.0	0.0	0.0	0.0	0.0	0.0	0.0	0.00	0.00	0.00

JJA 63-73

p (MB)	80S	70S	65S	60S	55S	50S	45S	40S	35S	30S	25S	20S	15S	10S	5S	EQ	5N	10N	15N	20N	25N	30N	35N	40N	45N	50N	55N	60N	65N	70N	80N	SH	NH	GLOBE
50	0.0	0.0	0.0	0.1	0.1	0.2	0.2	0.2	0.0	0.0	-0.0	-0.1	-0.1	0.0	-0.0	-0.2	-0.2	-0.2	0.1	0.0	0.0	0.0	-0.0	-0.0	-0.0	-0.0	-0.0	0.0	0.0	0.0	0.0	-0.01	-0.05	-0.03
100	0.1	0.0	0.0	0.1	0.1	0.1	0.2	0.2	0.0	0.0	-0.1	-0.1	-0.2	-0.4	-0.2	-0.4	-0.3	-0.2	0.0	0.1	0.1	0.0	-0.0	-0.1	-0.0	-0.0	-0.0	0.0	0.0	0.0	0.0	-0.10	-0.07	-0.09
200	-0.2	0.0	0.1	0.1	0.2	0.2	0.2	0.2	0.0	0.0	-0.4	-0.8	-1.5	-2.8	-3.2	-3.1	-2.2	-1.2	0.0	0.2	0.2	0.1	0.0	-0.2	-0.1	-0.1	-0.1	0.0	0.0	0.0	0.0	-0.81	-0.44	-0.63
300	-0.3	-0.1	-0.0	0.1	0.1	0.2	0.3	0.3	0.1	0.0	-0.2	-0.3	-0.4	-0.4	-0.6	-0.8	-0.6	-0.3	0.0	0.1	0.1	0.1	-0.0	-0.1	-0.1	-0.1	-0.1	0.0	0.0	0.0	0.0	-0.17	-0.11	-0.14
500	-0.1	-0.1	-0.0	0.1	0.1	0.1	0.1	0.1	0.1	0.0	-0.0	0.0	-0.0	-0.3	-0.4	-0.3	-0.4	-0.3	-0.0	0.1	0.1	0.1	0.0	-0.0	-0.0	-0.0	-0.0	0.0	0.0	0.0	0.0	-0.07	-0.00	-0.04
700	-0.1	-0.0	-0.0	0.0	0.0	0.0	0.1	0.1	0.1	0.1	0.0	0.1	0.1	0.3	0.5	0.5	0.5	0.3	0.0	0.1	0.1	0.1	0.0	-0.0	-0.0	-0.0	0.0	0.0	0.0	0.0	0.0	0.11	0.05	0.08
850	-0.0	-0.0	0.0	0.0	0.0	0.0	0.0	0.0	0.1	0.1	0.1	0.3	1.3	1.5	1.3	1.3	0.9	0.5	0.1	-0.1	-0.1	-0.0	-0.0	0.0	0.0	0.0	0.0	0.0	0.0	0.0	0.0	0.37	0.19	0.28
1000	0.3	0.0	0.0	-0.2	-0.4	-0.7	-0.9	-0.7	-0.5	-0.2	0.4	1.0	2.4	3.2	2.8	2.7	2.5	0.8	-0.1	-0.4	-0.5	-0.2	0.3	0.4	0.5	0.3	0.0	-0.1	-0.0	-0.0	0.0	0.68	0.51	0.60
MEAN	-0.0	0.0	0.0	0.0	0.0	0.0	0.0	0.0	0.0	0.0	0.0	0.0	0.0	0.0	0.0	0.0	0.0	0.0	0.0	0.0	0.0	0.0	0.0	0.0	0.0	0.0	0.0	0.0	0.0	0.0	0.0	-0.00	-0.00	-0.00

TABLE A5. "VERTICAL VELOCITY" $[\bar{\omega}]_{1D}$ (IN 10^{-4} mbar sec^{-1})

W-1D

YEAR 63-73

P (MB)	80S	70S	65S	60S	55S	50S	45S	40S	35S	30S	25S	20S	15S	10S	5S	EQ	5N	10N	15N	20N	25N	30N	35N	40N	45N	50N	55N	60N	65N	70N	80N	SH	NH	GLOBE
50	-0.0	-0.0	0.0	0.0	0.0	0.0	0.0	0.0	0.0	0.0	0.0	-0.0	-0.0	-0.0	-0.0	-0.0	-0.0	-0.0	-0.0	-0.0	-0.0	0.0	0.0	0.0	0.0	0.0	-0.0	-0.0	-0.0	-0.0	0.0	0.00	-0.00	-0.00
100	-0.1	-0.0	0.0	0.1	0.0	-0.0	-0.0	0.0	0.1	0.2	0.1	0.1	0.1	0.0	0.1	0.1	0.1	0.1	0.1	0.1	0.1	0.0	0.0	0.0	0.0	0.0	-0.0	-0.0	-0.0	-0.0	0.1	0.01	-0.01	-0.00
200	-0.1	-0.1	-0.1	-0.0	-0.0	-0.0	0.1	0.2	0.2	0.2	0.3	0.3	0.3	0.3	-0.5	-0.8	-1.0	-0.2	0.0	0.1	-0.1	-0.1	-0.0	-0.1	0.0	-0.1	-0.1	-0.1	-0.1	-0.0	0.1	0.01	-0.02	-0.00
300	-0.1	-0.2	-0.2	-0.2	-0.3	-0.1	0.3	0.4	0.4	0.5	0.3	0.4	0.4	0.5	-0.9	-1.4	-1.7	-0.2	0.0	0.2	0.0	0.2	-0.0	-0.0	-0.1	-0.2	-0.3	-0.3	-0.2	0.0	0.2	0.01	-0.02	-0.01
500	-0.0	-0.5	-0.6	-0.5	-0.4	-0.5	-0.1	0.3	0.4	0.7	0.5	0.3	0.4	0.4	-1.0	-1.6	-2.1	-0.3	0.0	0.1	0.6	-0.0	-0.0	0.1	-0.3	-0.6	-0.6	-0.3	0.1	0.2	0.4	0.02	-0.03	-0.01
700	0.1	-0.5	-0.7	-0.6	-0.5	-0.1	0.6	0.6	0.7	0.8	0.5	0.6	0.8	0.3	-0.9	-1.5	-2.3	-0.8	0.7	1.3	1.3	0.6	-0.1	0.0	0.1	-0.4	-0.7	-0.7	-0.3	0.1	0.6	0.06	-0.10	-0.01
850	•••••	-0.5	-0.8	-0.7	-0.6	-0.1	0.6	0.8	0.8	0.8	0.6	0.8	0.7	-0.3	-0.8	-1.7	-0.9	0.1	1.3	1.2	1.3	0.6	-0.0	0.1	0.1	-0.4	-0.7	-0.8	-0.3	0.1	0.6	0.08	-0.10	-0.01
1000	•••••	-0.0	-0.1	-0.1	-0.1	-0.0	0.1	0.1	0.1	0.1	0.0	0.1	0.1	-0.0	-0.0	-0.2	-0.2	-0.2	0.1	0.1	0.1	0.0	0.0	-0.0	-0.0	-0.0	-0.0	-0.0	-0.0	0.1	0.1	0.02	-0.02	-0.00
MEAN	0.0	-0.3	-0.4	-0.3	-0.3	-0.3	-0.1	0.4	0.5	0.5	0.4	0.4	0.4	0.4	0.2	-0.7	-1.0	-1.5	-0.4	0.6	0.9	0.9	0.4	-0.0	0.0	0.1	-0.3	-0.4	-0.5	-0.2	0.0	0.3	0.04	-0.05	-0.01

DJF 63-73

P (MB)	80S	70S	65S	60S	55S	50S	45S	40S	35S	30S	25S	20S	15S	10S	5S	EQ	5N	10N	15N	20N	25N	30N	35N	40N	45N	50N	55N	60N	65N	70N	80N	SH	NH	GLOBE
50	0.0	-0.0	-0.0	0.0	0.0	0.0	0.0	0.0	0.0	0.0	0.0	0.0	0.0	0.0	-0.1	-0.0	0.1	0.1	0.0	0.0	0.0	0.0	0.0	0.0	0.0	-0.0	-0.0	-0.0	0.0	0.0	-0.0	0.01	0.01	-0.00
100	-0.0	-0.0	0.0	0.0	-0.0	-0.0	0.0	0.2	0.0	0.0	0.1	-0.1	-1.0	-0.8	-1.2	-0.2	-0.1	-0.1	-0.1	0.1	0.1	0.1	0.0	0.1	0.0	-0.1	-0.1	-0.1	-0.0	-0.0	-0.0	-0.03	0.03	-0.00
200	-0.0	-0.0	-0.1	-0.1	-0.1	-0.1	0.0	0.2	0.0	0.2	0.0	-0.1	-1.0	-0.8	-1.2	-0.5	-0.1	-0.8	-1.2	-0.2	0.2	0.7	0.9	0.1	-0.1	0.0	-0.2	-0.2	-0.1	0.0	0.2	-0.30	0.30	-0.00
300	-0.1	-0.1	-0.2	-0.2	-0.3	-0.2	-0.0	0.5	0.6	0.2	0.2	-0.4	-2.1	-1.8	-2.3	-1.7	-0.6	1.9	3.1	1.6	0.7	1.6	0.2	-0.1	-0.2	-0.4	-0.5	-0.4	-0.2	0.1	0.4	-0.61	0.61	-0.01
500	-0.2	-0.2	-0.4	-0.5	-0.6	-0.6	-0.0	0.8	1.0	0.5	0.1	-0.7	-2.5	-1.9	-2.0	-1.6	-0.9	1.6	3.1	2.0	2.3	1.9	0.1	-0.2	-0.7	-0.8	-0.8	-0.6	-0.3	0.1	0.8	-0.61	0.60	-0.01
700	-0.2	-0.2	-0.6	-0.6	-0.6	-0.6	-0.0	0.9	1.1	0.7	0.1	-0.3	-1.9	-1.4	-1.4	-1.6	-1.6	0.8	2.7	2.3	1.8	1.7	-0.0	-0.2	-0.8	-1.0	-1.0	-0.8	-0.3	0.1	1.1	-0.44	0.42	-0.01
850	•••••	-0.3	-0.7	-0.7	-0.6	-0.6	-0.0	0.9	1.2	0.7	0.1	0.2	-1.1	-1.0	-0.9	-1.1	-1.6	-0.2	1.4	2.4	1.6	1.5	-0.1	-0.2	-0.8	-1.1	-1.1	-0.8	-0.3	0.2	1.1	-0.22	0.20	-0.01
1000	•••••	-0.0	-0.1	-0.1	-0.1	-0.1	-0.0	0.1	0.1	0.1	0.1	0.2	-0.1	-0.0	-0.0	-0.3	-0.2	0.2	0.2	0.2	0.2	0.1	-0.0	-0.0	-0.1	-0.1	-0.1	-0.0	-0.0	0.1	0.1	0.00	-0.01	-0.00
MEAN	-0.1	-0.2	-0.4	-0.4	-0.4	-0.4	-0.0	0.6	0.7	0.4	0.1	-0.2	-1.5	-1.2	-1.3	-1.2	-0.9	0.8	2.1	1.3	1.5	1.3	0.1	-0.1	-0.5	-0.6	-0.6	-0.5	-0.2	0.1	0.6	-0.37	0.37	-0.00

JJA 63-73

P (MB)	80S	70S	65S	60S	55S	50S	45S	40S	35S	30S	25S	20S	15S	10S	5S	EQ	5N	10N	15N	20N	25N	30N	35N	40N	45N	50N	55N	60N	65N	70N	80N	SH	NH	GLOBE
50	-0.0	0.0	-0.0	-0.0	-0.0	-0.0	0.0	0.0	0.0	0.0	0.0	0.1	0.1	0.1	-0.0	-0.0	-0.0	0.1	-0.0	-0.0	-0.0	0.0	0.2	0.0	0.0	0.0	0.0	0.0	0.0	0.0	0.0	0.01	-0.01	-0.00
100	-0.1	0.0	-0.0	-0.0	-0.0	-0.1	-0.1	0.0	0.0	0.1	0.1	0.2	0.1	0.1	0.0	-0.1	-0.1	-0.1	-0.1	-0.0	-0.0	0.1	0.4	-0.0	-0.0	-0.0	-0.0	0.0	0.0	0.0	0.0	0.03	-0.03	-0.00
200	-0.1	-0.0	-0.0	-0.1	-0.1	-0.1	-0.1	0.0	0.1	0.5	0.4	0.3	1.2	1.3	0.1	-0.5	-1.0	-1.2	-0.8	-0.4	-0.0	0.4	0.2	-0.0	-0.1	-0.1	-0.1	-0.0	-0.0	0.0	0.2	0.28	-0.29	-0.00
300	-0.3	-0.1	-0.1	-0.1	-0.2	-0.3	-0.2	0.0	0.3	0.9	0.9	0.8	2.4	2.4	0.3	-1.0	-2.3	-2.4	-1.5	-0.6	-0.0	0.7	0.4	-0.0	-0.1	-0.2	-0.2	-0.2	-0.0	0.0	0.4	0.58	-0.59	-0.00
500	-0.3	-0.3	-0.3	-0.4	-0.5	-0.6	-0.2	0.3	0.6	1.2	1.2	1.3	3.0	2.4	-0.1	-2.8	-2.3	-0.7	0.0	1.0	1.0	0.4	-0.0	-0.3	-0.4	-0.2	-0.3	-0.4	-0.0	0.1	0.0	0.60	-0.62	-0.01
700	-0.2	-0.3	-0.4	-0.5	-0.7	-0.8	-0.2	0.4	0.6	1.1	1.4	1.3	3.0	2.0	-0.5	-1.4	-2.8	-2.1	-1.3	-0.7	0.2	1.1	0.3	0.1	0.1	-0.2	-0.4	-0.5	-0.2	0.1	0.1	0.57	-0.59	-0.01
850	•••••	-0.4	-0.4	-0.6	-0.8	-0.8	-0.2	0.4	0.6	1.0	1.5	1.4	2.0	0.5	-0.8	-0.7	-1.8	-1.5	-0.9	-0.7	0.2	1.2	0.3	0.1	0.1	-0.3	-0.4	-0.5	-0.3	0.1	0.1	0.37	-0.40	-0.01
1000	•••••	-0.0	-0.0	-0.1	-0.1	-0.0	-0.0	0.1	0.1	0.1	0.1	0.2	0.2	0.0	-0.1	0.0	-0.2	-0.3	-0.1	-0.0	0.0	0.1	0.0	0.0	0.0	-0.0	-0.0	-0.0	-0.0	0.0	0.0	0.04	-0.05	-0.00
MEAN	-0.2	-0.2	-0.2	-0.3	-0.4	-0.5	-0.2	0.2	0.4	0.8	0.9	0.8	2.0	1.5	-0.2	-0.8	-1.9	-1.6	-1.0	-0.5	0.1	0.8	0.3	0.0	0.0	-0.2	-0.2	-0.3	-0.1	0.0	0.1	0.41	-0.43	-0.01

453

TABLE A6. TEMPERATURE [$\bar{T}$] (IN °C)

YEAR 63-73   UNITS= 1.E 00 °C

P (MB)	80S	70S	65S	60S	55S	50S	45S	40S	35S	30S	25S	20S	15S	10S	5S	EQ	5N	10N	15N	20N	25N	30N	35N	40N	45N	50N	55N	60N	65N	70N	80N	SH	NH	GLOBE
50	-61.1	-58.7	-57.4	-56.2	-55.3	-54.9	-55.3	-56.5	-57.9	-59.5	-60.8	-62.0	-62.9	-63.5	-63.9	-64.4	-64.0	-63.3	-62.4	-61.4	-60.3	-58.9	-57.2	-55.6	-54.3	-53.4	-52.8	-52.7	-52.8	-53.7		-59.80	-59.23	-59.51
100	-60.4	-58.3	-57.1	-55.9	-55.2	-55.3	-56.7	-59.3	-62.9	-67.0	-70.9	-74.2	-76.6	-78.0	-78.1	-78.5	-78.4	-78.0	-76.7	-74.7	-71.7	-68.3	-64.4	-60.4	-57.2	-54.7	-53.2	-52.0	-52.0	-52.4		-67.03	-66.42	-66.72
200	-59.3	-58.0	-57.3	-56.6	-56.1	-55.7	-55.5	-55.0	-54.5	-54.1	-53.9	-53.8	-53.8	-53.6	-53.4	-53.5	-53.7	-54.3	-54.9	-55.3	-55.0	-54.4	-53.6	-52.9	-52.0	-51.4	-54.9	-53.72	-54.35
300	-58.4	-56.7	-55.5	-53.0	-51.5	-48.7	-46.4	-43.9	-41.1	-38.2	-35.5	-33.5	-32.3	-31.7	-31.4	-31.2	-31.3	-31.9	-32.8	-34.5	-36.9	-39.9	-42.8	-45.2	-47.2	-48.7	-49.9	-51.0	-51.9	-53.0	-40.7	-39.2	-40.00
500	-38.0	-35.4	-33.3	-30.6	-27.5	-24.3	-21.1	-18.0	-15.0	-12.2	-9.6	-7.5	-6.4	-5.8	-5.7	-5.6	-5.6	-6.0	-6.8	-8.6	-10.9	-13.9	-17.1	-20.0	-22.8	-25.1	-27.2	-29.0	-30.7	-33.4	-15.74	-14.39	-15.06
700	-23.9	-21.2	-18.7	-15.5	-12.1	-8.6	-5.3	-2.0	1.1	3.8	6.1	7.7	8.6	9.0	9.1	9.3	9.6	9.9	9.7	9.1	7.5	5.2	2.2	-0.9	-4.3	-7.5	-10.1	-12.2	-14.2	-16.1	-19.3	0.03	1.14	0.59
850	*****	-14.6	-12.1	-8.8	-5.1	-1.5	1.9	5.2	8.2	11.2	13.6	15.4	16.8	17.2	17.6	18.0	18.5	18.3	17.5	15.6	13.0	9.8	6.6	3.3	-0.1	-2.8	-5.2	-7.5	-9.9	-13.9	7.80	9.02	8.41
1000	*****	-2.2	-0.9	1.1	3.6	6.5	10.0	13.5	16.7	19.5	21.7	23.6	25.1	26.1	26.5	26.7	27.0	26.4	25.3	23.5	21.2	18.2	14.7	11.0	8.1	4.8	1.4	-2.1	-7.2	-13.3	16.11	17.61	16.76
MEAN	-44.5	-37.9	-33.8	-31.3	-28.7	-26.1	-23.8	-21.7	-19.8	-18.2	-16.8	-15.1	-14.5	-14.1	-13.8	-14.1	-14.7	-16.1	-18.4	-20.8	-22.8	-24.3	-26.0	-27.3	-28.8	-30.5	-31.7	-34.0	-21.12	-20.24	-20.68

DJF 63-73

P (MB)	80S	70S	65S	60S	55S	50S	45S	40S	35S	30S	25S	20S	15S	10S	5S	EQ	5N	10N	15N	20N	25N	30N	35N	40N	45N	50N	55N	60N	65N	70N	80N	SH	NH	GLOBE
50	-38.6	-40.3	-41.8	-44.0	-47.0	-50.2	-53.5	-56.5	-58.9	-60.9	-62.3	-63.6	-64.5	-65.4	-66.1	-66.7	-66.5	-66.1	-65.3	-64.1	-65.3	-64.3	-63.3	-62.1	-60.4	-58.4	-57.0	-56.3	-56.3	-57.1	-61.0	-65.5	-57.35	-62.08	-59.71
100	-40.8	-42.3	-43.8	-45.4	-49.4	-49.1	-52.9	-57.2	-61.8	-66.1	-70.1	-73.3	-76.0	-78.0	-79.1	-79.2	-79.3	-77.8	-75.4	-71.8	-67.5	-63.0	-59.1	-56.5	-55.2	-55.6	-56.5	-57.5	-59.2	-62.7	-65.7	-67.7	-74.66	-71.71	-74.66
200	-44.9	-46.0	-47.0	-48.4	-50.2	-52.1	-53.6	-54.4	-54.4	-53.9	-53.4	-53.7	-53.7	-53.9	-54.3	-53.5	-55.0	-55.8	-56.2	-55.9	-55.6	-55.8	-56.3	-57.1	-58.0	-59.2	-55.27	-55.08	-53.67
300	-52.4	-51.7	-50.6	-49.2	-47.2	-45.0	-42.7	-40.2	-37.7	-35.4	-33.4	-32.0	-31.5	-31.3	-31.2	-31.5	-32.5	-34.1	-36.9	-40.6	-44.8	-48.1	-50.4	-52.2	-53.5	-54.5	-55.6	-56.1	-58.1	-38.2	-42.40	-40.23
500	-33.1	-31.1	-29.3	-27.1	-24.1	-20.8	-17.3	-14.1	-11.3	-9.1	-7.4	-6.3	-5.8	-5.7	-5.6	-5.5	-5.5	-6.4	-8.1	-11.3	-15.2	-19.5	-23.6	-27.1	-30.0	-32.3	-34.1	-35.7	-37.3	-39.8	-13.39	-18.05	-15.74
700	-19.1	-17.1	-15.1	-12.6	-9.4	-5.8	-2.1	1.5	4.5	6.3	8.3	9.1	9.3	9.1	9.4	9.7	10.0	9.2	7.5	4.4	0.5	-3.8	-8.0	-11.8	-15.0	-17.5	-19.6	-21.5	-23.4	-26.3	2.19	-2.87	-0.35
850	*****	-9.3	-7.6	-5.4	-2.4	1.1	4.8	8.5	11.7	14.6	16.4	17.5	17.6	17.4	17.6	17.7	17.8	16.7	14.7	11.3	7.5	3.2	-1.1	-5.2	-8.9	-11.1	-14.2	-16.5	-18.5	-22.0	10.21	4.09	7.14
1000	*****	-1.3	0.1	2.3	5.0	7.9	11.6	15.5	19.0	21.9	23.8	25.2	26.1	26.7	26.8	26.8	26.4	25.4	23.6	21.0	17.8	13.9	10.2	6.2	4.3	-1.4	-6.7	-11.6	-18.2	-23.7	17.48	14.22	16.06
MEAN	-35.4	-30.8	-27.9	-26.4	-24.8	-23.0	-21.0	-19.0	-17.4	-16.1	-15.1	-14.5	-14.3	-14.5	-14.3	-14.3	-14.2	-14.9	-16.3	-18.6	-21.9	-25.2	-27.8	-29.8	-31.9	-33.5	-35.4	-37.4	-39.0	-41.6	-18.81	-23.68	-21.24

JJA 63-73

P (MB)	80S	70S	65S	60S	55S	50S	45S	40S	35S	30S	25S	20S	15S	10S	5S	EQ	5N	10N	15N	20N	25N	30N	35N	40N	45N	50N	55N	60N	65N	70N	80N	SH	NH	GLOBE
50	-83.5	-76.8	-72.7	-68.5	-64.2	-60.4	-58.1	-57.1	-57.4	-58.3	-59.4	-60.4	-61.1	-61.6	-62.0	-62.5	-62.4	-62.0	-61.4	-60.6	-59.6	-58.6	-57.2	-55.4	-53.3	-51.2	-49.1	-47.4	-45.4	-43.3	-41.5	-62.47	-56.16	-59.31
100	-78.6	-73.6	-69.9	-65.6	-61.5	-58.3	-57.0	-57.6	-60.6	-64.6	-68.9	-72.5	-74.9	-76.1	-76.2	-76.1	-75.1	-73.7	-71.7	-69.9	-75.1	-73.9	-71.7	-69.4	-66.1	-61.9	-57.4	-53.5	-50.7	-48.4	-46.7	-45.1	-42.7	-68.34	-64.87	-66.60
200	-71.1	-68.3	-66.4	-64.1	-61.6	-59.2	-57.3	-56.0	-55.1	-54.8	-54.3	-54.3	-54.1	-53.9	-53.7	-53.6	-53.2	-52.7	-52.2	-52.4	-52.4	-52.6	-52.5	-51.5	-51.1	-49.7	-48.1	-46.4	-44.3	-43.4	-43.7	-48.7	-57.32	-51.84	-54.58
300	-63.2	-60.9	-59.2	-57.1	-54.8	-52.4	-50.0	-47.3	-44.0	-40.4	-37.1	-34.5	-33.1	-32.3	-32.0	-31.8	-31.5	-31.6	-31.4	-31.1	-31.0	-31.0	-32.7	-36.2	-38.8	-41.1	-42.8	-44.1	-45.1	-45.8	-46.6	-63.02	-35.62	-39.32
500	-41.4	-38.5	-36.2	-33.5	-30.6	-27.7	-24.9	-22.0	-18.7	-15.2	-11.6	-8.6	-6.9	-6.0	-6.0	-5.9	-5.9	-6.4	-7.9	-10.1	-12.4	-14.7	-16.7	-18.4	-19.9	-21.4	-23.8	-17.91	-10.12	-14.01
700	-27.3	-24.2	-21.2	-17.9	-14.4	-11.2	-8.3	-5.4	-2.4	0.7	3.8	6.3	7.7	8.6	8.7	8.9	9.2	9.7	10.2	10.6	10.7	9.9	8.4	6.5	3.7	0.8	-1.3	-3.1	-4.6	-6.0	-8.9	-2.00	5.70	1.87
850	*****	-18.8	-15.3	-11.2	-7.1	-3.6	-0.6	2.1	4.9	7.7	10.5	12.9	14.6	15.7	16.5	17.2	17.9	18.5	19.2	19.7	19.5	18.6	16.6	14.6	12.2	9.4	7.1	5.4	3.7	1.6	-2.3	5.47	14.24	9.87
1000	*****	-4.6	-2.6	-0.3	2.3	5.2	8.4	11.6	14.4	17.2	19.8	22.0	23.9	25.3	26.0	26.4	26.9	27.2	27.1	26.6	25.8	24.5	22.6	19.3	15.7	11.8	10.8	9.8	7.9	5.3	-0.2	14.60	21.05	17.40
MEAN	-52.3	-44.0	-38.9	-35.5	-32.2	-29.2	-26.6	-24.3	-22.2	-20.2	-18.3	-16.5	-15.7	-14.7	-14.3	-14.0	-13.7	-14.3	-13.4	-13.2	-13.5	-14.5	-15.9	-17.1	-18.0	-19.4	-20.2	-21.2	-22.7	-24.0	-23.2	-16.34	-19.83

TABLE A7. GEOPOTENTIAL HEIGHT DEPARTURE FROM NMC STANDARD ATMOSPHERE $[\bar{z}] - z_{SA}$ (IN gpm $\approx$ m)

Z   YEAR 63-73   UNITS= 1.E 00 GPM

P (MB)	80S	70S	65S	60S	55S	50S	45S	40S	35S	30S	25S	20S	15S	10S	5S	EQ	5N	10N	15N	20N	25N	30N	35N	40N	45N	50N	55N	60N	65N	70N	80N	SH	NH	GLOBE
50	-781	-704	-619	-499	-350	-203	-87	-3	42	66	76	72	54	41	35	36	38	50	65	70	64	46	23	-11	-58	-107	-156	-212	-272	-334	-374	-102	-15	-59
100	-1006	-847	-715	-556	-379	-206	-54	73	175	258	320	359	372	378	390	387	390	391	383	383	348	285	200	109	22	-63	-140	-208	-273	-346	-431	51	162	106
200	-923	-811	-703	-563	-395	-219	-48	113	259	389	496	573	611	630	647	642	646	647	634	636	603	435	307	167	34	-87	-188	-273	-346	-410	-506	179	279	229
300	-836	-743	-647	-520	-365	-202	-45	98	222	328	413	473	503	517	521	525	521	517	521	532	522	498	443	365	261	144	27	-34	-182	-335	-398	135	122	177
500	-600	-543	-477	-386	-265	-154	-39	62	142	205	248	275	285	288	289	293	299	299	299	290	263	222	165	96	19	-57	-122	-178	-225	-267	-331	56	122	89
700	-418	-392	-351	-290	-211	-125	-44	25	75	109	128	137	138	134	137	141	144	144	144	144	134	117	88	51	5	-41	-82	-117	-146	-170	-204	9	52	31
850	.....	-282	-261	-220	-162	-98	-37	13	44	62	67	66	60	54	50	49	52	58	63	62	51	38	35	5	-22	-47	-69	-95	-96	-111	-4		18	4
1000	.....	-147	-142	-137	-111	-70	-23	14	33	39	31	14	-3	-17	-27	-24	-24	-11	5	24	38	35	27	9	-14	-25	-26	-25	-13	-4		-2	-15	
MEAN	-681	-558	-469	-378	-268	-152	-43	52	128	187	229	255	265	265	267	269	272	272	272	267	245	210	154	87	16	-53	-113	-164	-209	-247	-304	48	113	80

DJF 63-73

P (MB)	80S	70S	65S	60S	55S	50S	45S	40S	35S	30S	25S	20S	15S	10S	5S	EQ	5N	10N	15N	20N	25N	30N	35N	40N	45N	50N	55N	60N	65N	70N	80N	SH	NH	GLOBE
50	17	1	1	11	29	51	71	84	88	85	77	58	26	0	-6	-17	-21	-24	-27	-36	-64	-106	-161	-221	-296	-389	-491	-598	-717	-836	-1061	40	-238	-99
100	-305	-292	-255	-194	-106	-4	97	192	271	333	372	389	386	379	381	383	385	385	367	331	249	131	-4	-137	-256	-369	-471	-569	-667	-761	-921	193	-11	91
200	-560	-521	-458	-360	-221	-59	108	266	397	502	574	617	631	633	640	642	642	610	547	424	260	75	-102	-252	-380	-481	-565	-641	-707	-814	289	120	205
300	-602	-560	-495	-394	-251	-88	75	223	336	420	474	505	515	516	522	525	528	503	451	349	217	67	-84	-223	-344	-438	-514	-579	-635	-726	212	86	149
500	-444	-423	-381	-313	-213	-97	19	124	198	248	274	285	286	284	290	290	266	212	140	54	-39	-133	-218	-336	-380	-418	-477	96	41	68
700	-310	-316	-295	-251	-183	-102	-20	49	124	198	133	135	131	129	132	139	140	133	110	77	32	-18	-76	-129	-205	-231	-253	-265	25	8	17
850	.....	-235	-227	-198	-149	-90	-31	16	44	56	57	53	48	45	44	47	53	60	66	62	52	38	29	7	-32	-65	-91	-111	-125	-135	-150	-9	1	-3
1000	.....	-192	-183	-164	-127	-79	-31	7	22	22	11	-6	-21	-29	-31	-33	-31	-21	-3	17	38	29	29	7	-16	-48	-35	-26	-7	1		-41	-4	-25
MEAN	-402	-364	-320	-262	-175	-75	26	116	183	228	254	264	264	259	260	260	263	267	260	240	190	121	34	-57	-144	-224	-286	-340	-390	-429	-495	100	23	62

JJA 63-73

P (MB)	80S	70S	65S	60S	55S	50S	45S	40S	35S	30S	25S	20S	15S	10S	5S	EQ	5N	10N	15N	20N	25N	30N	35N	40N	45N	50N	55N	60N	65N	70N	80N	SH	NH	GLOBE
50	-1367	-1266	-1137	-952	-720	-482	-278	-121	-26	32	68	77	84	87	92	96	103	135	177	223	261	290	314	327	333	345	362	379	394	425		-232	243	5
100	-1541	-1258	-1060	-841	-612	-396	-210	-51	77	185	273	333	358	371	374	379	387	393	416	446	469	472	401	346	288	246	217	197	179	162		-71	367	147
200	-1174	-1017	-881	-717	-538	-360	-192	-28	112	240	355	428	534	589	618	636	642	655	665	662	637	521	468	349	235	146	79	25	-22	-91		85	120	274
300	-987	-856	-742	-604	-451	-300	-157	-19	112	240	355	443	489	512	515	517	522	527	537	544	541	521	466	384	288	191	109	42	-15	-71	-162	69	371	220
500	-692	-606	-525	-423	-309	-196	-92	1	84	157	217	261	281	289	288	292	296	303	310	310	301	274	230	170	107	55	12	-24	-60	-120		23	209	116
700	-459	-419	-369	-301	-220	-138	-63	2	53	92	119	136	140	140	138	138	140	145	150	153	154	140	114	80	42	10	-15	-37	-58	-93		0	95	48
850	.....	-293	-263	-217	-160	-98	-41	-7	29	53	75	71	63	58	53	54	51	54	57	61	69	68	54	33	11	-7	-23	-35	-45	-63		34	13	13
1000	.....	-74	-81	-94	-85	-57	-19	18	42	55	51	36	13	-5	-20	-24	-26	-19	-4	16	37	34	35	22	9	-8	-21	-24	-21	-16		-6	-2	-4
MEAN	-868	-684	-564	-458	-339	-220	-110	-11	72	145	204	245	263	268	271	268	271	273	283	295	302	305	283	240	184	127	81	45	17	-9	-52	5	213	108

455

TABLE A8. SPECIFIC HUMIDITY $[\bar{q}]$ (IN g kg^{-1})

YEAR 63-73   UNITS= 1.E 00 G/KG

Q

P (MB)	80S	70S	65S	60S	55S	50S	45S	40S	35S	30S	25S	20S	15S	10S	5S	EQ	5N	10N	15N	20N	25N	30N	35N	40N	45N	50N	55N	60N	65N	70N	80N	SH	NH	GLOBE
300	0.0	0.0	0.1	0.0	0.1	0.1	0.1	0.1	0.2	0.2	0.2	0.2	0.3	0.3	0.3	0.3	0.3	0.3	0.3	0.3	0.3	0.2	0.2	0.1	0.1	0.1	0.1	0.1	0.1	0.1	0.1	0.18	0.19	0.18
500	0.2	0.2	0.3	0.4	0.4	0.5	0.6	0.8	0.9	1.1	1.3	1.5	1.8	2.0	2.1	2.3	2.2	1.9	1.7	1.5	1.4	1.2	1.0	0.9	0.8	0.7	0.6	0.5	0.5	0.4	0.3	1.16	1.24	1.20
700	0.5	0.7	0.8	1.0	1.2	1.5	1.8	2.1	2.5	3.0	3.6	4.2	4.8	5.3	5.8	6.0	5.9	5.4	4.8	4.3	3.8	3.2	2.8	2.5	2.2	2.0	1.8	1.5	1.4	1.2	0.9	3.22	3.44	3.33
850	.....	1.1	1.4	1.7	1.9	2.4	3.0	3.6	4.2	5.0	5.9	6.9	8.0	9.9	10.4	10.7	10.5	9.9	9.0	8.1	7.0	6.1	5.2	4.4	3.9	3.4	3.0	2.6	2.2	1.9	1.3	6.16	6.28	6.22
900	.....	1.3	1.7	2.0	2.2	2.9	3.7	4.4	5.2	6.3	7.3	8.5	9.8	11.0	11.8	12.2	12.5	12.1	11.0	9.4	8.1	7.0	6.0	5.2	4.6	3.9	3.4	2.9	2.5	2.0	1.4	7.31	7.55	7.43
950	.....	1.5	1.9	2.5	3.4	4.1	4.9	5.9	6.8	8.6	9.5	11.5	13.1	13.1	14.2	14.5	14.6	14.2	13.1	11.8	10.3	8.7	7.5	6.3	5.2	4.4	3.8	3.2	2.7	2.2	1.6	8.39	8.96	8.67
1000	.....	2.2	2.5	3.4	4.1	5.1	5.7	6.4	7.7	8.8	10.8	12.4	13.9	15.4	16.6	17.6	17.9	17.8	17.6	16.9	15.6	13.9	13.2	12.2	10.4	8.6	6.8	5.7	4.8	4.0	2.5	10.34	12.09	11.10
MEAN	0.2	0.4	0.6	0.8	1.0	1.3	1.5	1.8	2.1	2.4	2.8	3.3	3.7	4.1	4.3	4.5	4.4	4.2	3.8	3.4	2.9	2.4	2.0	1.7	1.5	1.3	1.2	1.0	0.9	0.7	0.5	2.51	2.59	2.55

DJF 63-73

P (MB)	80S	70S	65S	60S	55S	50S	45S	40S	35S	30S	25S	20S	15S	10S	5S	EQ	5N	10N	15N	20N	25N	30N	35N	40N	45N	50N	55N	60N	65N	70N	80N	SH	NH	GLOBE
300	0.0	0.0	0.1	0.1	0.1	0.1	0.1	0.2	0.2	0.2	0.3	0.3	0.3	0.3	0.3	0.3	0.3	0.3	0.2	0.2	0.2	0.2	0.1	0.1	0.1	0.0	0.0	0.0	0.0	0.0	0.0	0.21	0.15	0.18
500	0.3	0.4	0.4	0.5	0.7	0.8	1.0	1.2	1.4	1.7	2.0	2.2	2.3	2.3	2.3	2.2	1.9	1.5	1.2	1.0	0.9	0.8	0.7	0.5	0.5	0.4	0.3	0.3	0.2	0.2	0.2	1.41	0.89	1.15
700	0.7	0.9	1.0	1.2	1.5	1.8	2.2	2.6	3.2	3.8	4.6	5.3	5.7	5.9	5.9	5.3	4.5	3.7	3.1	2.6	2.2	1.8	1.5	1.2	1.1	0.9	0.8	0.7	0.6	0.5	0.4	3.78	2.56	3.17
850	.....	1.5	1.8	2.2	2.8	3.4	4.2	5.1	6.1	7.2	8.3	9.4	10.2	10.6	10.7	10.6	10.0	8.9	7.7	6.6	5.3	4.3	3.5	2.9	2.3	1.9	1.5	1.3	1.1	0.9	0.6	6.97	5.03	5.99
900	.....	1.8	2.1	2.7	3.4	4.3	5.1	6.2	7.3	8.3	9.5	10.8	11.8	12.4	12.5	12.1	11.1	9.8	8.3	6.7	5.4	4.4	3.6	2.8	2.2	1.7	1.4	1.1	0.9	0.6	0.5	8.16	6.22	7.19
950	.....	2.1	2.5	3.1	3.9	4.9	5.9	7.1	8.3	9.6	11.0	12.5	13.6	14.3	14.5	14.1	13.1	11.5	9.8	8.0	6.7	5.4	4.1	3.2	2.5	2.0	1.7	1.4	1.2	0.5	0.2	9.36	7.38	8.40
1000	.....	2.7	3.1	3.8	4.6	5.6	7.1	8.3	9.6	10.8	12.7	14.3	15.5	16.5	17.2	17.7	17.0	15.5	13.6	11.5	9.5	7.6	6.2	4.9	4.3	3.3	2.5	1.9	1.1	0.5	0.4	11.27	10.57	10.97
MEAN	0.3	0.6	0.8	1.0	1.2	1.5	1.7	2.0	2.4	2.8	3.3	3.7	4.1	4.2	4.5	4.4	4.2	3.8	3.2	2.7	2.2	1.7	1.4	1.1	0.9	0.8	0.6	0.5	0.4	0.4	0.2	2.86	2.06	2.46

JJA 63-73

P (MB)	80S	70S	65S	60S	55S	50S	45S	40S	35S	30S	25S	20S	15S	10S	5S	EQ	5N	10N	15N	20N	25N	30N	35N	40N	45N	50N	55N	60N	65N	70N	80N	SH	NH	GLOBE
300	0.0	0.0	0.1	0.1	0.1	0.1	0.1	0.1	0.1	0.1	0.2	0.2	0.2	0.3	0.3	0.3	0.3	0.3	0.3	0.3	0.3	0.3	0.3	0.3	0.2	0.2	0.2	0.1	0.1	0.1	0.1	0.14	0.26	0.20
500	0.2	0.2	0.2	0.3	0.3	0.3	0.4	0.5	0.6	0.7	0.9	1.0	1.3	1.6	1.9	2.2	2.4	2.3	2.2	2.2	2.0	1.8	1.5	1.3	1.2	1.1	1.0	0.9	0.8	0.7	0.6	0.90	1.67	1.28
700	0.4	0.5	0.7	0.8	1.0	1.2	1.4	1.6	1.9	2.2	2.6	3.1	3.8	4.5	5.3	6.0	6.2	5.9	5.5	5.0	4.4	4.0	3.7	3.5	3.3	3.0	2.7	2.4	2.1	1.8	1.6	2.64	4.45	3.55
850	.....	1.1	1.4	1.9	2.6	3.5	4.0	4.7	5.5	6.6	7.8	8.9	9.9	10.5	10.9	11.0	11.7	12.2	12.7	12.8	11.7	11.0	10.3	9.3	8.5	7.1	6.0	5.0	4.0	3.5	2.6	5.36	7.84	6.60
900	.....	1.4	1.9	2.6	3.2	4.3	5.0	5.7	6.8	8.1	9.6	10.8	11.7	12.4	12.7	12.8	13.5	14.8	15.0	15.1	14.5	13.5	12.7	11.3	10.3	9.3	8.5	7.0	6.3	5.2	2.8	6.49	9.16	7.82
950	.....	2.1	2.9	3.6	4.3	5.1	5.8	6.8	7.9	9.0	10.2	11.3	12.7	13.5	14.3	14.8	15.0	15.1	16.2	15.1	13.7	11.5	10.2	9.0	7.9	6.8	6.3	5.8	5.1	4.3	3.0	7.42	10.79	9.06
1000	.....	2.8	3.7	4.6	5.6	6.1	6.7	7.9	9.2	10.7	12.3	14.2	15.7	16.6	17.2	17.8	18.3	18.0	17.3	16.2	15.1	13.7	12.7	11.5	9.1	7.5	6.7	6.1	5.4	4.6	3.3	9.34	13.70	11.24
MEAN	0.1	0.3	0.5	0.7	0.9	1.1	1.3	1.5	1.7	1.9	2.2	2.7	3.1	3.7	4.1	4.4	4.7	4.6	4.4	4.1	3.7	3.2	2.9	2.6	2.3	2.1	2.0	1.8	1.6	1.4	1.1	2.15	3.23	2.69

TABLE A9. TRANSIENT EDDY KINETIC ENERGY [$\bar{K}_{TE}$] (IN m² sec⁻²)

KTE   UNITS= 1.E 00 M2/S2

YEAR 63-73

p (MB)	80S	70S	65S	60S	55S	50S	45S	40S	35S	30S	25S	20S	15S	10S	5S	EQ	5N	10N	15N	20N	25N	30N	35N	40N	45N	50N	55N	60N	65N	70N	80N	SH	NH	GLOBE
50	59	79	95	84	78	69	59	49	42	39	35	37	34	32	29	29	31	33	38	44	49	49	47	45	47	53	60	70	79	86	77	48	49	48
100	45	68	76	77	75	71	66	63	62	63	56	56	48	43	42	44	49	60	81	106	119	113	87	66	56	53	54	55	56	47	47	58	73	66
200	60	85	104	127	151	172	186	195	201	200	175	135	94	65	55	55	64	83	121	175	229	253	224	195	181	166	146	119	97	80	56	132	148	140
300	112	148	179	214	244	259	251	231	207	181	146	105	67	41	34	34	45	72	116	166	195	202	230	213	240	244	215	189	168	134	56	145	148	147
500	58	82	99	115	126	130	122	107	91	74	59	43	29	21	18	16	20	30	47	78	99	108	116	116	116	117	108	99	93	86	68	69	68	69
700	38	46	53	62	68	70	66	59	50	40	31	24	18	14	14	12	14	17	20	25	33	41	47	48	55	61	62	56	50	47	46	39	37	38
850		35	41	48	54	57	54	48	40	32	25	18	15	16	16	14	12	16	18	23	29	35	35	40	44	49	48	44	39	37	35	31	29	30
1000		17	20	22	21	21	21	20	19	16	14	11	10	6	8	5	5	7	11	15	20	22	29	31	32	34	31	30	27	24	20	14	18	16
MEAN	62	77	86	99	110	115	111	103	93	82	69	52	37	27	22	22	25	31	44	63	84	97	97	97	102	107	104	94	86	79	68	70	72	71

DJF 63-73

p (MB)	80S	70S	65S	60S	55S	50S	45S	40S	35S	30S	25S	20S	15S	10S	5S	EQ	5N	10N	15N	20N	25N	30N	35N	40N	45N	50N	55N	60N	65N	70N	80N	SH	NH	GLOBE
50	9	12	15	17	18	18	18	17	18	19	20	21	20	23	27	29	27	25	22	20	19	16	18	17	18	18	18	17	15	12	10	20	44	32
100	10	18	26	34	43	51	56	59	58	55	49	42	40	43	47	50	49	48	51	59	65	68	67	64	60	61	64	67	71	75	73	45	61	53
200	24	46	69	101	138	171	187	187	169	147	120	90	66	54	56	56	61	72	94	128	168	201	207	188	166	147	130	112	100	90	74	108	128	118
300	84	123	162	205	241	257	240	205	161	120	88	61	43	33	31	33	39	50	72	104	148	198	233	251	259	256	238	209	184	166	140	121	154	137
500	37	55	74	95	110	115	107	90	68	48	35	26	20	18	21	21	25	34	48	66	85	103	120	132	139	136	124	114	108	100	79	54	79	66
700	25	31	38	48	56	60	57	47	41	31	25	20	16	15	13	13	14	15	19	26	35	46	55	58	66	72	72	67	61	58	55	32	42	37
850	23	23	31	38	43	47	44	38	31	25	19	15	14	13	10	10	11	13	14	20	26	34	43	49	54	58	57	52	47	44	42	25	32	29
1000	12	16	18	18	18	18	17	17	15	11	8	7	7	5	4	4	5	5	7	11	14	19	24	29	30	35	38	35	30	26	21	11	20	15
MEAN	37	50	63	80	96	104	100	89	73	57	45	34	27	24	23	24	26	29	38	52	71	91	105	112	116	117	112	102	95	89	81	56	74	65

JJA 63-73

p (MB)	80S	70S	65S	60S	55S	50S	45S	40S	35S	30S	25S	20S	15S	10S	5S	EQ	5N	10N	15N	20N	25N	30N	35N	40N	45N	50N	55N	60N	65N	70N	80N	SH	NH	GLOBE
50	45	55	60	60	57	53	46	38	31	26	21	17	18	19	22	24	22	19	17	15	14	13	13	13	12	12	12	11	11	10	8	33	15	24
100	50	64	67	60	67	64	59	55	51	57	55	48	41	38	37	39	37	33	32	35	41	45	42	43	35	31	26	26	19	17	13	52	33	43
200	97	119	131	131	144	158	193	208	208	208	173	116	75	61	52	56	61	72	88	105	158	138	101	88	62	62	43	42	42	42	42	132	92	112
300	145	160	179	202	226	243	250	249	239	214	167	116	75	51	46	46	45	42	50	64	76	76	73	76	76	80	80	79	76	73	73	154	93	123
500	79	97	108	118	126	128	124	117	105	90	71	51	35	27	24	26	24	20	18	20	27	35	41	50	64	76	80	79	76	76	73	76	42	59
700	53	56	63	71	75	75	71	65	57	47	37	27	20	16	14	13	14	16	18	20	22	25	28	37	44	46	46	43	43	38	38	34	27	30
850		38	43	50	58	62	60	55	48	31	25	17	14	10	7	7	5	7	8	11	13	18	20	25	30	35	32	30	29	18	13	34	21	27
1000		19	22	24	23	22	22	22	18	13	8	7	7	4	5	4	5	7	7	8	8	11	15	18	20	21	21	21	19	16	16	15	11	13
MEAN	80	85	90	99	107	111	111	108	103	93	74	54	38	27	22	20	19	21	21	25	32	42	51	60	71	77	77	72	66	62	55	73	44	58

TABLE A10. STANDING EDDY KINETIC ENERGY $[\bar{K}_{SE}]$ (IN m^2 sec^{-2})

KSE YEAR 63-73 UNITS= 1.E 00 M2/S2

P (MB)	80S	70S	65S	60S	55S	50S	45S	40S	35S	30S	25S	20S	15S	10S	5S	EQ	5N	10N	15N	20N	25N	30N	35N	40N	45N	50N	55N	60N	65N	70N	80N	SH	NH	GLOBE
50	3	4	4	1	2	1	1	1	1	1	0	0	0	0	0	1	0	0	0	0	0	1	3	5	7	8	8	8	10	13	11	1	4	2
100	1	2	2	3	3	3	3	3	1	8	4	10	10	8	11	4	6	11	10	10	6	7	13	15	10	6	6	7	7	8	5	3	8	5
200	1	3	3	3	3	2	3	3	3	4	8	12	10	6	14	10	11	14	13	10	10	20	20	30	16	10	6	7	7	8	3	3	14	10
300	2	4	4	4	3	3	3	4	4	4	7	7	6	2	7	6	1	7	1	4	7	15	8	19	15	13	12	11	9	8	3	3	9	6
500	2	4	3	3	3	2	2	3	3	2	2	2	1	1	0	0	1	1	2	3	3	6	8	8	8	7	7	7	6	5	2	2	4	3
700	3	5	5	7	2	1	2	2	2	1	0	1	1	2	3	3	3	3	3	3	3	3	4	5	8	4	4	4	6	2	1	2	3	2
850	*****	8	7	4	2	1	1	1	1	0	1	1	2	3	3	3	3	3	3	3	2	1	1	2	3	3	3	3	2	1	0	2	3	2
1000	*****	0	0	1	0	0	1	1	1	2	2	2	2	4	4	4	4	3	3	5	3	2	2	1	1	0	1	0	0	0	0	1	2	2
MEAN	2	4	3	3	2	2	2	2	2	2	3	2	2	3	3	3	3	4	4	3	4	7	10	10	8	6	6	6	5	5	2	2	6	4

DJF 63-73

P (MB)	80S	70S	65S	60S	55S	50S	45S	40S	35S	30S	25S	20S	15S	10S	5S	EQ	5N	10N	15N	20N	25N	30N	35N	40N	45N	50N	55N	60N	65N	70N	80N	SH	NH	GLOBE
50	0	1	1	0	0	0	1	2	4	1	0	3	2	2	5	5	5	3	2	3	6	10	15	21	28	35	40	45	55	71	60	1	20	10
100	1	1	1	1	2	2	5	7	7	6	6	6	17	9	18	14	18	21	17	12	18	32	34	29	22	20	25	30	35	40	21	6	23	15
200	2	3	3	3	3	4	4	7	8	13	13	15	30	16	23	19	29	30	18	29	45	87	86	55	31	26	31	32	30	28	11	10	39	24
300	4	8	7	5	4	4	6	6	4	5	6	4	4	6	11	6	11	17	18	21	42	69	68	54	37	35	38	35	31	27	10	5	33	19
500	3	6	5	5	4	3	3	4	2	2	2	1	1	1	2	4	2	5	6	4	17	23	28	26	21	20	21	19	17	15	6	3	14	8
700	2	7	6	6	3	2	2	2	2	1	2	2	3	2	3	4	3	5	6	5	9	9	13	13	11	10	11	10	8	7	6	3	7	5
850	*****	6	6	3	2	2	2	2	2	1	4	2	2	4	5	4	2	3	2	4	5	4	6	8	7	7	7	8	6	4	2	2	4	4
1000	*****	1	2	3	2	2	2	2	3	3	6	6	8	6	8	7	5	2	3	2	2	3	3	3	4	3	4	4	4	3	1	2	3	3
MEAN	2	4	3	3	2	2	2	3	4	3	4	4	9	4	7	6	7	9	9	10	18	30	33	27	20	19	21	20	20	19	10	4	17	10

JJA 63-73

P (MB)	80S	70S	65S	60S	55S	50S	45S	40S	35S	30S	25S	20S	15S	10S	5S	EQ	5N	10N	15N	20N	25N	30N	35N	40N	45N	50N	55N	60N	65N	70N	80N	SH	NH	GLOBE
50	2	7	8	6	4	3	2	2	4	1	0	1	1	1	2	3	2	2	2	2	2	2	1	1	1	1	0	0	0	0	0	2	1	2
100	2	5	5	4	3	2	2	3	4	6	6	6	1	6	6	4	4	11	30	37	18	12	12	14	9	3	3	2	2	2	0	5	15	10
200	3	5	6	4	4	3	7	9	14	25	22	16	17	22	13	18	18	30	31	31	21	18	12	18	13	13	11	8	6	5	0	14	21	17
300	5	6	6	4	3	3	7	7	8	17	19	10	6	10	6	6	2	17	34	31	19	8	8	13	13	15	13	11	10	9	3	7	21	8
500	4	7	6	4	3	2	4	5	4	4	7	7	5	3	3	3	3	5	11	8	9	4	5	6	6	8	7	6	6	5	2	4	6	4
700	5	8	6	3	3	2	3	3	3	4	7	7	3	1	3	5	7	11	11	8	5	5	4	5	4	4	4	4	2	2	1	4	5	4
850	*****	9	8	4	2	2	2	2	2	2	2	3	3	3	3	4	8	13	16	11	7	5	3	4	3	1	2	1	1	1	0	3	6	4
1000	*****	0	1	1	2	2	2	3	3	3	2	2	2	2	2	4	8	13	16	10	6	4	3	2	2	1	1	0	1	0	1	2	6	4
MEAN	3	6	5	3	3	2	2	4	5	7	6	6	4	4	5	5	6	11	13	10	7	6	8	8	7	7	6	5	5	4	1	5	8	6

TABLE A11. ZONAL MEAN KINETIC ENERGY $[\bar{K}_M]$ (IN m² sec⁻²)

UNITS = 1.E 00 M2/S2

KM — YEAR 63-73

p (MB)	80S	70S	65S	60S	55S	50S	45S	40S	35S	30S	25S	20S	15S	10S	5S	EQ	5N	10N	15N	20N	25N	30N	35N	40N	45N	50N	55N	60N	65N	70N	80N	SH	NH	GLOBE
50	55	87	99	101	89	66	39	15	2	0	7	16	16	10	4	3	8	20	27	23	8	0	3	10	17	22	24	25	25	22	7	30	15	22
100	43	88	116	146	190	174	178	328	153	119	72	27	3	0	0	6	8	7	20	23	47	101	126	331	210	138	69	50	38	31	24	9	87	47	67
200	20	43	68	109	168	231	283	359	357	270	142	40	0	0	7	7	3	0	23	92	200	307	331	287	210	117	88	54	52	37	26	8	160	122	141
300	14	29	53	99	164	222	245	247	241	220	160	81	20	0	1	8	3	5	0	51	123	190	213	200	160	117	80	52	37	26	11	80	115	84	100
500	2	8	22	53	96	130	133	118	92	63	35	12	1	1	5	9	8	5	0	6	29	52	67	73	66	52	37	23	16	9	4	48	29	38
700	0	0	3	18	46	67	66	51	31	14	4	0	0	2	7	5	7	7	4	1	3	11	18	21	23	20	15	9	5	3	1	19	9	14
850	2	2	0	0	20	36	35	23	9	1	0	9	0	9	4	3	7	7	4	0	1	0	0	1	7	7	1	1	0	0	0	9	3	6
1000	.....	0	6	12	17	18	12	5	0	0	7	4	7	3	4	2	1	3	7	8	3	0	0	1	2	3	1	0	0	0	0	7	2	5
MEAN	14	23	34	58	92	121	127	120	107	91	62	30	8	2	4	6	5	4	6	19	48	80	95	91	73	54	37	24	18	13	4	57	37	47

DJF 63-73

p (MB)	80S	70S	65S	60S	55S	50S	45S	40S	35S	30S	25S	20S	15S	10S	5S	EQ	5N	10N	15N	20N	25N	30N	35N	40N	45N	50N	55N	60N	65N	70N	80N	SH	NH	GLOBE
50	0	3	8	6	0	0	0	4	16	37	64	85	79	54	24	8	4	3	1	0	1	9	27	47	63	74	85	94	103	111	104	35	30	39	35
100	3	6	14	30	58	114	114	119	94	55	17	3	3	1	1	4	1	0	28	113	241	335	307	231	161	111	87	78	73	63	30	87	123	81
200	8	13	28	65	131	206	259	277	247	186	103	34	1	0	1	2	1	21	124	335	606	776	638	408	241	136	81	71	54	40	30	210	251	176
300	6	11	33	86	172	245	256	222	164	104	50	14	1	0	0	3	0	7	68	214	410	503	434	317	206	119	71	43	33	21	8	176	174	129
500	0	0	11	43	98	144	141	107	62	26	4	0	0	1	0	3	0	2	47	116	145	127	91	56	34	14	8	2	0	0	0	48	57	48
700	0	0	0	14	43	68	64	43	19	4	0	0	1	1	1	4	7	7	7	19	47	35	43	41	35	23	14	8	5	3	0	16	16	16
850	2	2	0	0	18	33	28	15	4	0	0	2	5	8	2	2	7	2	1	3	19	35	13	15	11	8	5	0	2	0	0	6	6	6
1000	.....	0	2	7	13	15	10	2	0	2	0	7	14	12	5	1	5	5	5	0	0	0	3	5	11	3	3	0	0	0	0	5	5	5
MEAN	3	12	35	75	111	117	100	72	45	23	9	5	4	5	4	5	4	7	26	83	167	220	200	150	100	63	41	29	24	19	7	59	78	59

JJA 63-73

p (MB)	80S	70S	65S	60S	55S	50S	45S	40S	35S	30S	25S	20S	15S	10S	5S	EQ	5N	10N	15N	20N	25N	30N	35N	40N	45N	50N	55N	60N	65N	70N	80N	SH	NH	GLOBE
50	105	168	202	224	221	196	148	94	48	19	6	6	0	0	0	6	34	89	131	145	111	62	26	8	2	2	1	1	3	2	2	71	50	60
100	68	169	219	258	281	287	274	257	238	208	147	72	18	0	0	15	36	68	73	52	15	3	1	17	42	44	29	14	3	1	0	145	31	88
200	31	59	77	103	144	202	287	402	528	598	478	260	78	12	0	26	35	29	12	1	1	0	35	118	198	188	124	66	33	15	6	5	224	57	141
300	21	38	52	77	108	150	202	265	338	383	314	176	50	15	3	19	21	15	10	3	1	0	0	22	74	120	117	89	55	33	23	10	149	38	94
500	5	10	20	40	67	90	104	115	118	109	80	38	9	10	10	10	10	10	10	5	0	0	3	4	20	43	37	37	23	13	6	53	15	34
700	1	2	1	2	15	37	57	61	42	26	11	2	1	0	0	0	0	0	0	0	0	1	4	0	13	13	4	2	3	1	0	21	6	13
850	.....	3	1	16	32	36	29	17	5	0	0	2	7	5	0	0	5	0	0	5	0	0	1	1	2	3	4	1	1	0	0	11	2	6
1000	.....	1	6	12	17	18	14	9	3	0	2	9	15	14	0	0	0	0	0	2	4	0	0	1	0	1	1	0	0	0	0	9	1	6
MEAN	23	37	47	65	89	112	128	142	156	158	122	64	18	3	5	10	14	18	17	13	7	11	13	31	51	39	23	12	9	8	4	76	20	48

459

TABLE A12. TEMPORAL STANDARD DEVIATION OF ZONAL WIND [$\sigma(u)$] (IN m sec^{-1})

SD-U   YEAR 63-73   UNITS= 1.E 00 M/S

YEAR 63-73

P (HB)	80S	70S	65S	60S	55S	50S	45S	40S	35S	30S	25S	20S	15S	10S	5S	EQ	5N	10N	15N	20N	25N	30N	35N	40N	45N	50N	55N	60N	65N	70N	80N	SH	NH	GLOBE
50	8.9	10.9	11.4	11.2	10.6	9.7	8.9	8.3	8.0	7.9	7.7	7.7	7.5	7.1	6.6	6.5	6.8	7.3	8.0	8.8	9.2	9.2	8.7	8.3	8.6	9.0	9.5	9.9	10.0	9.0		8.58	8.44	8.51
100	7.2	9.3	9.8	9.6	9.2	8.8	8.6	8.7	9.0	9.2	8.9	8.4	8.1	7.7	7.6	7.8	8.4	9.5	11.2	13.0	13.8	13.1	10.9	9.0	8.0	7.6	7.6	7.6	7.0	7.0		8.64	9.81	9.22
200	7.9	9.2	9.8	10.6	11.5	12.5	13.3	14.1	14.8	15.2	14.5	12.8	10.6	8.8	8.2	8.1	9.0	10.5	12.9	15.9	18.1	18.5	16.2	14.1	13.2	12.6	11.7	10.5	9.5	8.7		11.63	12.99	12.26
300	10.8	12.2	13.8	14.7	15.2	15.4	14.9	14.7	14.4	14.3	11.3	9.0	7.0	6.2	6.0	6.6	7.7	9.9	13.0	15.4	16.1	15.2	14.6	14.8	15.1	15.0	14.2	13.3	12.6	11.5		11.77	12.43	12.10
500	7.7	9.4	10.0	10.1	10.6	10.1	9.6	9.2	8.5	7.4	6.1	5.1	4.6	4.6	4.6	4.6	4.8	5.3	6.4	8.2	9.8	10.0	9.7	9.7	10.1	10.6	10.1	9.7	9.5	9.3		8.17	8.44	8.30
700	5.7	6.9	7.3	7.7	7.9	8.1	8.0	7.7	7.2	6.7	6.1	5.4	4.8	4.6	4.4	4.2	4.4	4.8	5.1	5.6	6.5	6.9	7.0	7.0	7.4	7.8	7.8	7.4	7.0	6.7		6.25	6.24	6.24
850	5.9	6.5	6.9	7.3	7.5	7.5	7.4	7.0	6.5	6.1	5.4	4.5	3.9	3.6	3.6	3.6	3.9	4.5	5.0	5.6	6.5	6.9	6.3	6.6	6.9	6.8	6.5	6.1	5.9	5.7	••••••	5.57	5.37	5.47
1000	4.5	5.0	5.3	5.5	5.4	5.2	4.9	4.6	4.2	3.8	3.4	3.0	2.7	2.3	2.6	2.1	2.3	2.6	3.1	3.7	4.3	4.5	4.5	4.9	5.4	5.7	5.4	5.1	4.7	4.5	••••••	4.05	3.75	3.92
MEAN	7.8	8.8	9.1	9.5	9.8	10.0	9.9	9.7	9.4	9.1	8.5	7.5	6.5	5.6	5.2	5.1	5.5	6.1	7.2	8.7	10.0	10.4	9.9	9.6	9.8	9.8	9.6	9.1	8.5	8.0		8.06	8.40	8.23

DJF 63-73

P (HB)	80S	70S	65S	60S	55S	50S	45S	40S	35S	30S	25S	20S	15S	10S	5S	EQ	5N	10N	15N	20N	25N	30N	35N	40N	45N	50N	55N	60N	65N	70N	80N	SH	NH	GLOBE
50	3.5	4.2	4.6	4.8	4.7	4.6	4.6	4.8	5.1	5.2	5.3	5.5	5.9	6.1	6.1	6.2	6.1	5.5	5.2	5.4	6.0	6.5	6.9	7.4	7.8	8.3	8.8	9.2	9.7	10.1	11.1	5.12	7.06	6.09
100	3.5	4.7	5.5	6.3	7.0	7.4	7.6	7.8	7.9	8.0	7.8	8.0	8.4	8.8	8.0	7.9	8.2	7.6	7.3	7.4	7.8	8.3	8.6	8.3	8.1	8.1	8.1	8.0	8.3	8.6	8.6	7.24	8.26	7.75
200	5.1	7.3	8.7	10.1	11.5	12.6	13.3	13.5	13.1	12.5	11.5	10.1	8.7	8.0	8.2	8.4	9.8	11.4	12.9	14.1	15.5	15.7	15.0	13.9	13.0	12.4	11.8	11.9	11.1	9.5	9.1	10.32	11.07	10.69
300	9.4	11.6	12.9	13.9	15.2	14.8	13.9	12.7	11.4	9.9	8.4	7.1	6.3	6.7	7.4	8.7	10.4	12.5	15.7	15.7	15.3	11.3	11.5	10.8	10.4	10.2	10.1	10.1	9.5	9.1		10.51	11.88	11.20
500	6.1	8.0	8.9	9.5	9.8	10.0	9.7	9.2	8.3	7.2	6.3	5.6	4.8	4.8	4.9	4.8	4.9	4.7	4.6	4.6	5.6	6.3	6.9	7.3	8.0	8.4	8.4	8.0	7.3	7.5	7.08	8.68	7.88
700	4.3	5.8	6.5	7.0	7.4	7.1	6.4	5.7	5.1	4.7	4.4	4.3	4.1	3.7	3.6	3.7	3.6	4.0	4.2	4.2	4.9	5.2	5.8	6.4	6.8	7.2	7.5	7.4	7.0	6.7	6.6	5.72	6.41	6.07
850	5.0	5.9	6.3	6.7	6.9	6.7	6.2	5.6	5.0	4.5	4.2	4.2	4.0	3.8	3.6	2.0	2.2	2.3	2.6	2.9	3.2	3.4	4.2	4.9	5.0	5.4	5.8	5.5	5.2	5.1	4.9	5.08	5.51	5.30
1000	3.4	3.9	4.4	4.8	5.1	5.0	4.7	4.4	3.9	3.6	3.2	2.9	2.6	2.7	2.2	2.0	2.2	2.3	2.5	2.6	2.9	3.3	3.3	3.8	4.3	4.7	4.6	4.5	4.2	4.3	4.5	3.77	3.93	3.84
MEAN	5.8	7.0	7.7	8.4	8.9	9.2	9.1	8.7	8.1	7.4	6.7	6.0	5.5	5.2	5.3	5.2	5.6	6.2	7.2	8.2	9.1	9.6	9.9	10.1	10.0	10.0	9.5	8.9	8.7	8.9	9.1	7.02	8.06	7.54

JJA 63-73

P (HB)	80S	70S	65S	60S	55S	50S	45S	40S	35S	30S	25S	20S	15S	10S	5S	EQ	5N	10N	15N	20N	25N	30N	35N	40N	45N	50N	55N	60N	65N	70N	80N	SH	NH	GLOBE
50	7.1	8.4	9.0	9.1	9.0	8.6	8.0	7.3	6.6	6.0	5.4	4.9	5.0	5.3	5.7	5.9	5.6	5.2	4.8	4.6	4.5	4.4	4.2	4.0	3.8	3.8	3.8	3.7	3.6	3.4	3.1	6.60	4.40	5.50
100	7.1	7.9	8.2	8.4	8.6	8.5	8.1	7.9	7.9	8.2	8.1	7.8	7.3	7.1	7.2	7.4	7.2	7.7	8.2	7.7	7.3	7.0	7.7	7.0	6.4	6.0	5.7	5.2	4.8	4.4	4.1	7.81	6.37	7.09
200	9.6	10.0	10.0	10.0	11.6	12.7	13.8	14.6	15.3	15.3	14.8	13.3	11.1	9.6	8.2	7.7	7.3	8.2	9.4	11.4	12.5	14.1	14.2	13.4	12.5	12.9	12.4	11.3	10.1	8.9	7.9	11.07	9.73	10.40
300	12.1	12.4	12.8	13.3	14.0	14.7	15.1	15.3	15.3	14.8	13.0	10.7	8.8	7.1	6.2	5.5	5.3	6.2	7.1	9.0	11.7	13.6	13.6	13.3	13.6	13.4	12.6	11.6	11.7	11.2		11.72	9.20	10.46
500	8.7	10.2	10.6	10.7	10.8	10.8	10.6	10.3	10.0	9.6	8.8	7.5	6.4	5.6	4.9	4.5	4.7	5.0	5.8	6.4	7.1	7.9	6.6	6.1	7.1	8.6	8.8	8.7	8.4	8.5	8.4	8.47	6.44	7.45
700	6.7	7.4	7.7	8.0	8.2	8.4	8.3	7.6	7.1	6.4	5.7	5.0	4.3	4.1	4.1	4.3	4.5	4.3	4.1	4.0	3.6	3.4	5.4	5.3	5.5	5.9	5.9	5.4	5.3	4.8		6.51	5.36	5.93
850	6.1	6.6	7.0	7.5	7.6	7.6	7.4	6.9	6.2	5.6	5.1	4.8	4.6	4.4	3.6	3.4	3.4	3.4	3.3	2.5	2.2	2.2	2.3	2.5	2.9	3.3	3.3	3.8	4.3	4.5	4.5	5.64	4.68	5.16
1000	4.8	5.2	5.6	5.5	5.5	5.1	4.8	4.6	4.4	4.1	3.6	3.1	2.5	2.6	2.3	2.2	2.3	2.5	2.9	3.3	3.3	3.8	4.3	4.5	4.2	4.5	4.3	4.5	4.2	4.3	4.5	4.12	3.17	3.71
MEAN	8.5	8.9	9.0	9.3	9.6	9.8	9.7	9.6	9.2	8.3	7.2	6.2	5.4	5.0	4.8	4.8	4.9	5.0	5.4	6.2	7.0	6.7	7.0	7.3	7.8	8.1	7.8	7.5	7.2	6.8	7.2	7.96	6.35	7.16

TABLE A13. TEMPORAL STANDARD DEVIATION OF MERIDIONAL WIND $[\sigma(v)]$ (IN m sec^{-1})

UNITS = 1.E 00 M/S

SD-V YEAR 63-73

P (MB)	80S	70S	65S	60S	55S	50S	45S	40S	35S	30S	25S	20S	15S	10S	5S	EQ	5N	10N	15N	20N	25N	30N	35N	40N	45N	50N	55N	60N	65N	70N	80N	SH	NH	GLOBE
50	6.2	6.3	6.2	6.0	5.6	5.2	4.8	4.4	4.0	3.7	3.5	3.3	3.4	3.6	3.8	3.9	3.7	3.5	3.3	3.2	3.3	3.5	3.9	4.3	4.7	5.3	6.1	6.9	7.7	8.3	8.5	4.33	4.53	4.43
100	6.2	7.1	7.4	7.6	7.6	7.6	7.3	7.0	6.7	6.3	5.7	5.3	5.1	5.2	5.2	5.2	5.2	5.5	5.9	6.4	6.8	6.8	7.0	7.0	6.9	7.0	7.1	7.2	7.3	7.3	6.7	6.40	6.35	6.37
200	7.6	9.2	10.6	11.9	13.0	13.7	14.0	13.9	13.5	12.8	11.8	10.3	8.6	7.3	6.8	6.6	6.8	7.3	8.2	9.6	11.0	12.4	13.2	13.6	13.5	13.1	12.3	11.2	10.1	9.2	7.5	10.69	10.31	10.50
300	10.3	12.0	13.7	15.4	16.5	16.9	16.5	15.4	14.0	12.4	10.7	9.2	7.2	5.8	5.1	4.7	4.9	5.4	6.2	7.4	8.4	9.4	11.2	12.8	14.3	15.5	15.8	15.1	14.1	13.1	11.2	11.12	10.48	10.80
500	7.6	8.7	9.9	11.1	11.8	12.0	11.5	10.7	9.4	8.0	6.8	5.6	4.7	3.9	3.5	3.4	3.8	4.4	5.2	6.2	7.4	8.4	9.4	10.2	10.8	10.9	10.6	10.1	9.8	9.3	7.6	7.62	7.20	7.41
700	6.7	6.6	7.3	8.1	8.6	8.7	8.3	7.8	6.9	6.0	5.1	4.3	3.8	3.3	3.2	3.1	3.3	3.7	4.3	4.9	5.9	6.7	7.3	7.9	7.8	7.6	7.2	6.9	6.7	6.1	6.0	5.78	5.48	5.63
950	5.9	6.1	6.9	7.4	7.6	7.3	6.9	6.3	5.5	4.7	4.0	3.5	3.2	3.1	3.1	3.0	3.1	3.5	3.9	4.5	5.1	5.8	6.2	6.6	6.6	6.9	6.7	6.7	6.3	6.1	6.0	5.20	4.96	5.08
1000	4.3	4.9	5.4	5.6	5.6	5.4	5.2	4.9	4.4	3.9	3.3	2.8	2.5	2.4	2.3	2.2	2.4	2.8	3.3	3.9	4.5	4.6	4.8	5.3	5.6	5.9	5.7	5.5	5.3	4.9	4.6	4.12	3.87	4.01
MEAN	7.6	8.1	8.8	9.6	10.2	10.4	10.2	9.6	8.8	7.8	6.8	5.8	4.9	4.3	4.0	3.8	3.9	4.2	4.7	5.4	6.3	7.4	8.3	8.9	9.4	9.7	9.7	9.4	9.0	8.6	8.0	7.24	6.90	7.07

DJF 63-73

P (MB)	80S	70S	65S	60S	55S	50S	45S	40S	35S	30S	25S	20S	15S	10S	5S	EQ	5N	10N	15N	20N	25N	30N	35N	40N	45N	50N	55N	60N	65N	70N	80N	SH	NH	GLOBE
50	2.6	2.8	3.1	3.5	3.7	3.8	3.6	3.6	3.6	3.7	3.4	3.4	3.4	3.6	3.8	4.1	4.2	4.0	3.7	3.4	3.7	4.1	4.7	5.3	5.9	6.8	7.8	8.9	10.0	11.0	11.2	3.59	5.44	4.52
100	3.0	3.8	4.5	5.4	6.2	6.9	7.3	7.4	7.2	6.7	6.1	5.5	5.3	5.4	5.7	5.9	6.0	6.2	6.7	7.2	7.7	7.7	7.8	7.9	8.3	8.8	9.0	8.5	8.8	9.0	8.5	5.92	7.21	6.56
200	4.7	6.2	7.9	10.0	12.0	13.5	14.0	13.8	12.9	11.7	10.3	8.8	7.4	6.5	6.4	6.5	7.0	7.8	8.8	9.2	10.9	12.8	14.2	14.4	13.6	12.8	12.1	11.4	10.8	10.3	9.7	9.64	10.84	10.24
300	8.9	10.4	12.5	14.6	16.1	16.8	16.1	14.7	12.6	10.5	8.7	7.1	5.8	5.1	4.9	5.1	5.6	6.4	7.8	9.6	11.7	13.6	15.0	15.8	16.1	15.5	15.4	14.0	13.3	11.9	10.0	10.09	11.57	10.83
500	6.0	6.8	8.3	9.9	11.1	11.5	11.0	9.8	8.3	6.7	5.6	4.7	4.1	3.7	3.5	3.6	3.8	4.2	5.1	5.9	6.7	7.4	7.5	8.0	8.4	8.5	8.2	7.9	7.6	7.5	10.6	6.80	8.09	7.45
700	5.6	5.2	5.9	6.9	7.6	7.9	7.7	7.1	6.3	5.4	4.7	4.1	3.7	3.4	3.2	3.0	3.3	3.6	3.9	4.6	5.5	6.7	7.4	7.5	8.0	8.2	7.9	7.6	7.5	7.5	10.9	5.29	5.94	5.62
950	4.5	5.0	5.9	6.6	6.9	6.6	6.2	5.6	5.0	4.3	3.8	3.1	2.9	3.0	2.9	2.8	2.9	3.5	4.1	4.6	5.5	6.7	7.2	7.4	7.5	7.2	6.9	6.7	6.6	4.8	6.4	4.69	5.27	4.98
1000	4.0	4.3	4.6	4.9	5.0	5.0	4.8	4.5	4.0	3.5	3.0	2.5	2.2	2.0	1.9	2.0	2.2	2.8	3.5	4.1	4.6	5.3	5.5	6.1	6.3	6.4	6.0	5.9	5.5	5.0	4.8	3.70	4.15	3.89
MEAN	5.8	6.2	7.2	8.4	9.4	9.9	9.7	9.0	8.0	6.9	5.9	5.1	4.4	4.0	3.9	3.9	4.1	4.5	5.2	6.3	7.4	8.7	9.5	9.9	10.1	10.2	9.9	9.6	9.3	8.7	9.0	6.53	7.59	7.06

JJA 63-73

P (MB)	80S	70S	65S	60S	55S	50S	45S	40S	35S	30S	25S	20S	15S	10S	5S	EQ	5N	10N	15N	20N	25N	30N	35N	40N	45N	50N	55N	60N	65N	70N	80N	SH	NH	GLOBE
50	6.3	6.3	6.3	6.1	5.9	5.7	5.3	4.9	4.4	4.0	3.6	3.4	3.1	3.4	3.7	3.8	3.6	3.4	3.1	3.0	2.8	2.8	2.8	3.0	3.1	3.1	3.0	2.9	2.8	2.8	2.8	4.50	3.07	3.78
100	7.1	8.2	8.2	8.1	7.8	7.8	7.4	7.2	7.0	6.9	6.5	5.9	5.3	5.0	4.9	4.8	4.7	4.7	4.3	4.4	4.4	4.7	5.2	5.6	5.9	5.8	5.5	5.1	4.7	4.1	3.7	6.48	4.86	5.67
200	10.0	10.8	11.7	12.6	13.1	13.5	13.7	14.0	14.2	14.1	13.0	11.7	9.1	7.4	6.6	6.1	5.8	5.8	6.1	7.6	8.6	9.2	11.1	12.7	13.1	12.7	11.7	10.4	9.2	8.2	6.6	11.23	8.72	9.98
300	12.0	12.9	13.9	15.1	16.0	16.5	16.5	16.2	15.4	12.8	10.6	9.1	8.4	6.4	5.3	4.6	4.2	4.1	4.4	4.9	5.9	7.5	9.5	11.3	12.8	13.9	14.2	14.0	13.6	13.3	12.2	11.95	8.51	10.23
500	9.1	9.5	10.2	11.0	11.6	11.7	11.3	10.5	9.4	8.0	6.7	5.4	4.3	3.7	3.4	3.4	3.3	3.4	3.7	4.0	4.6	5.3	6.1	7.0	8.0	8.8	9.1	9.0	8.9	8.7	8.9	8.21	5.86	7.03
700	7.7	7.5	8.1	8.8	9.1	9.0	8.6	8.2	7.5	6.7	5.7	4.7	4.0	3.0	3.1	3.1	3.1	3.1	3.6	3.9	4.2	4.7	5.0	5.7	6.0	6.6	6.8	6.6	6.3	6.3	6.3	6.20	4.78	5.49
950	6.2	6.4	7.1	7.8	8.0	7.9	7.5	7.0	6.2	5.2	4.2	3.5	2.9	2.9	2.9	2.9	3.0	3.3	3.8	4.1	4.5	4.9	5.4	5.8	5.6	5.4	5.4	5.4	4.4	5.4	4.4	5.49	4.29	4.89
1000	4.4	5.1	5.5	5.7	5.6	5.4	5.0	4.6	4.2	3.5	2.9	2.5	2.3	2.1	2.1	2.3	2.3	2.5	2.8	3.2	3.4	4.0	4.5	4.8	4.8	4.8	4.6	4.4	4.4	4.1	4.4	4.25	3.15	3.77
MEAN	8.9	9.2	9.9	10.3	10.5	10.3	10.0	9.5	8.8	7.7	6.5	5.4	4.4	4.0	3.7	3.6	3.6	3.8	4.1	4.6	5.4	6.3	7.1	7.8	8.2	8.3	8.0	7.7	7.4	7.1	7.1	7.68	5.67	6.68

461

TABLE A14. TEMPORAL STANDARD DEVIATION OF TEMPERATURE $[\sigma(T)]$ (IN °C)

SD-T YEAR 63–73 UNITS = 1.E 00 °C

P (MB)	80S	70S	65S	60S	55S	50S	45S	40S	35S	30S	25S	20S	15S	10S	5S	EQ	5N	10N	15N	20N	25N	30N	35N	40N	45N	50N	55N	60N	65N	70N	80N	SH	NH	GLOBE
50	18.8	15.5	13.2	10.5	7.7	5.2	3.7	3.0	2.8	2.6	2.6	2.7	2.8	3.0	3.1	3.2	3.2	3.0	2.9	2.9	2.9	2.8	2.8	2.9	3.3	3.9	4.7	5.7	6.9	8.2	10.7	5.07	3.87	4.47
100	16.1	13.7	11.6	9.0	6.4	4.5	4.0	4.1	4.2	4.0	3.6	3.1	2.8	3.0	2.9	3.1	3.0	3.0	2.9	2.9	3.1	3.5	3.8	3.7	3.6	3.7	4.1	4.8	5.7	6.7	8.8	5.06	3.81	4.43
200	11.8	10.6	9.6	8.4	7.2	6.3	5.6	4.9	4.2	3.5	2.7	2.1	1.7	1.6	1.6	1.6	1.6	1.8	1.8	2.3	3.0	3.7	4.5	5.0	5.3	5.5	5.7	5.9	6.1	6.5	7.6	4.40	3.81	4.11
300	5.3	5.0	4.9	4.8	4.8	4.7	4.7	4.5	4.2	3.8	2.5	1.9	1.5	1.4	1.4	1.5	1.5	1.5	1.9	2.5	3.6	4.8	5.7	6.0	5.9	5.7	5.6	5.5	5.4	5.5	5.8	3.39	3.95	3.67
500	4.8	5.0	5.0	5.1	5.3	5.4	5.3	5.0	4.6	3.9	3.2	2.3	1.8	1.4	1.3	1.3	1.4	1.4	1.6	2.1	3.2	4.6	5.8	6.7	7.3	7.7	7.8	7.9	7.9	7.7	8.5	3.50	4.56	4.03
700	5.3	4.9	4.9	4.9	5.0	5.0	5.0	4.8	4.5	3.9	3.2	2.5	2.0	1.7	1.5	1.4	1.4	1.7	2.3	3.5	4.8	5.8	6.1	7.2	7.7	7.9	8.0	8.2	8.3	8.5	9.4	3.48	4.81	4.15
850	5.7	5.4	5.1	4.9	4.9	5.0	4.9	4.9	4.6	4.0	3.2	2.4	1.8	1.7	1.6	1.4	1.6	2.1	3.1	4.4	5.7	6.7	7.7	8.4	8.8	9.0	9.3	9.6	9.7	9.4	11.6	3.78	5.40	4.59
1000	*****	7.9	6.5	5.5	4.7	4.2	4.0	3.9	3.7	3.4	3.0	2.7	2.3	1.9	1.7	1.5	1.4	1.3	1.6	2.3	3.0	3.9	4.8	5.1	5.6	5.1	5.1	5.6	5.1	8.3	11.4	3.51	4.14	3.79
MEAN	8.2	7.1	6.5	5.9	5.4	5.0	4.8	4.7	4.4	3.9	3.2	2.6	2.1	1.8	1.7	1.6	1.6	1.7	2.0	2.5	3.5	4.5	5.5	6.2	6.6	6.9	7.1	7.5	7.7	8.0	8.3	3.81	4.50	4.16

SD-T DJF 63–73

P (MB)	80S	70S	65S	60S	55S	50S	45S	40S	35S	30S	25S	20S	15S	10S	5S	EQ	5N	10N	15N	20N	25N	30N	35N	40N	45N	50N	55N	60N	65N	70N	80N	SH	NH	GLOBE
50	2.5	2.4	2.5	2.6	2.8	2.8	2.6	2.5	2.4	2.3	2.3	2.4	2.6	2.8	3.0	3.1	3.0	2.8	2.7	2.6	2.6	2.6	2.6	2.8	3.2	3.8	4.4	5.0	5.5	5.7	5.5	2.60	3.32	2.96
100	1.6	2.2	2.7	3.1	3.6	3.8	3.8	3.7	3.5	3.2	2.9	2.5	2.3	2.4	2.4	2.4	2.4	2.5	2.6	2.6	2.6	2.9	3.3	3.3	3.4	3.6	4.0	4.4	4.7	4.8	4.6	2.88	3.22	3.05
200	2.9	3.9	4.5	5.0	5.3	5.4	5.0	4.4	3.7	3.0	2.4	1.9	1.7	1.5	1.6	1.5	1.6	1.7	1.9	2.4	3.1	3.9	4.5	5.0	5.2	5.3	5.3	5.2	5.1	4.8	4.3	3.12	3.54	3.33
300	3.2	3.4	3.4	3.6	3.8	3.9	3.8	3.5	3.3	2.9	2.5	2.0	1.7	1.5	1.4	1.3	1.3	1.5	1.6	2.0	3.0	3.4	3.4	3.3	3.3	3.3	3.3	3.4	3.4	3.3	3.3	2.68	2.72	2.70
500	2.8	3.3	3.6	4.0	4.3	4.3	4.1	3.6	3.1	2.5	2.1	1.7	1.5	1.3	1.3	1.3	1.3	1.5	1.7	2.1	3.2	4.2	4.8	5.0	4.8	4.8	5.0	5.0	4.8	4.3	3.3	2.62	2.92	2.77
700	2.9	3.3	3.7	4.0	4.2	4.3	4.2	3.8	3.1	2.7	2.1	1.7	1.5	1.5	1.5	1.5	1.5	1.7	2.1	2.7	3.5	4.2	4.8	5.1	5.2	5.3	5.3	5.3	5.4	5.4	5.1	2.71	3.23	2.92
850	2.9	3.5	3.9	4.2	4.4	4.4	4.5	4.0	3.5	2.8	2.5	2.1	1.8	1.5	1.5	1.5	1.5	1.9	2.5	3.2	3.8	4.3	5.1	5.8	6.0	5.8	5.6	6.2	6.5	5.7	*****	2.90	3.60	3.25
1000	1.7	1.7	2.0	2.4	2.8	3.0	3.0	3.1	2.8	2.4	2.0	1.8	1.5	1.4	1.2	1.2	1.1	1.0	1.1	1.5	2.0	2.4	2.7	2.9	3.1	3.1	4.4	5.6	5.7	5.0	4.7	2.51	3.26	2.28
MEAN	2.7	3.1	3.4	3.7	4.0	4.1	4.0	3.8	3.3	2.9	2.4	1.9	1.7	1.5	1.5	1.5	1.5	1.6	1.9	2.3	2.9	3.4	3.8	4.2	4.5	4.6	4.8	5.0	5.1	4.7	5.0	2.71	3.26	2.98

SD-T JJA 63–73

P (MB)	80S	70S	65S	60S	55S	50S	45S	40S	35S	30S	25S	20S	15S	10S	5S	EQ	5N	10N	15N	20N	25N	30N	35N	40N	45N	50N	55N	60N	65N	70N	80N	SH	NH	GLOBE
50	3.3	3.6	3.5	3.5	3.4	3.4	3.2	2.9	2.7	2.4	2.3	2.2	2.3	2.2	2.3	2.4	2.3	2.2	2.2	2.1	2.0	1.9	1.9	1.8	1.8	1.7	1.7	1.7	1.8	2.0	2.0	2.71	2.00	2.35
100	3.7	4.6	4.6	4.5	4.2	3.8	3.6	3.5	3.4	3.3	3.0	2.7	2.5	2.3	2.2	2.4	2.8	2.8	2.6	2.6	2.6	2.8	3.1	3.3	3.1	2.8	2.5	2.2	2.1	2.0	1.9	3.30	2.63	2.99
200	4.3	5.0	5.3	5.6	5.8	5.8	5.2	4.6	4.0	3.4	3.0	2.3	1.8	1.5	1.6	1.6	1.6	1.5	1.5	1.6	2.0	2.2	2.8	3.5	4.3	4.9	5.2	5.0	4.7	3.8	3.8	3.62	2.89	3.25
300	2.5	3.1	3.3	3.4	3.4	3.4	3.4	3.4	3.5	3.4	2.9	2.3	1.8	1.3	1.3	1.3	1.3	1.3	1.4	1.6	2.0	2.5	3.2	3.7	3.8	3.7	3.6	3.5	3.3	3.3	3.4	2.64	2.50	2.57
500	3.8	4.0	4.2	4.3	4.2	4.3	4.2	4.0	3.7	3.4	2.9	2.4	1.9	1.5	1.3	1.3	1.3	1.4	1.5	1.8	2.1	2.6	3.1	3.4	3.7	3.7	3.9	3.9	4.0	4.1	4.1	3.02	2.46	2.74
700	4.5	4.4	4.3	4.3	4.2	4.1	4.0	3.8	3.6	3.3	2.9	2.5	2.1	1.7	1.4	1.4	1.3	1.3	1.5	1.7	2.1	2.6	2.7	3.2	3.9	3.7	4.2	4.6	4.8	4.2	4.2	3.05	2.47	2.76
850	5.0	4.8	4.8	4.7	4.5	4.0	3.9	3.7	3.7	2.6	2.3	2.1	2.1	1.9	1.3	1.3	1.3	1.5	1.6	2.1	2.6	3.2	3.2	3.2	2.8	3.2	3.9	4.3	4.6	4.4	4.4	3.18	2.75	2.96
1000	5.1	4.8	4.3	3.6	3.1	3.1	2.8	2.7	2.7	2.6	2.3	2.1	1.9	1.7	1.4	1.4	1.3	1.3	1.4	1.6	2.0	2.3	2.5	2.8	2.7	3.0	3.4	3.8	3.8	4.2	3.7	2.67	2.09	2.41
MEAN	3.8	4.3	4.4	4.3	4.2	4.1	4.0	3.8	3.6	3.3	2.9	2.4	2.0	1.6	1.5	1.5	1.5	1.5	1.6	1.9	2.3	2.8	3.3	3.6	3.7	3.7	3.8	3.9	4.0	3.7		3.04	2.55	2.80

462

TABLE A15. TEMPORAL STANDARD DEVIATION OF GEOPOTENTIAL HEIGHT [σ(z)] (IN gpm)

SD-Z   YEAR 63-73   UNITS= 1.E 00 GPM

P (MB)	80S	70S	65S	60S	55S	50S	45S	40S	35S	30S	25S	20S	15S	10S	5S	EQ	5N	10N	15N	20N	25N	30N	35N	40N	45N	50N	55N	60N	65N	70N	80N	SH	NH	GLOBE
50	632	564	503	424	333	247	178	128	98	78	70	68	69	73	77	80	84	87	95	109	133	162	195	228	265	307	355	407	463	519	620	179	217	198
100	531	435	373	309	252	206	171	144	122	101	82	67	58	55	57	59	60	62	66	80	115	159	202	236	261	294	308	338	369	399	454	154	185	169
200	281	244	226	212	204	199	192	180	159	132	100	71	52	42	41	40	41	43	52	77	124	182	232	266	280	287	291	297	277	308	316	131	173	152
300	195	182	180	182	183	189	183	169	146	116	85	58	39	30	28	19	18	17	22	34	58	86	114	139	157	170	178	183	195	274	267	112	149	130
500	137	128	128	126	127	126	119	107	91	70	50	34	25	21	20	14	14	14	18	25	36	52	68	81	94	104	111	116	117	195	182	73	94	84
700	101	92	92	93	94	92	84	73	59	44	32	23	18	15	15	14	13	13	14	18	25	39	52	62	69	77	82	84	85	116	114	50	59	55
850	*****	75	78	81	82	80	72	62	50	38	29	23	19	16	15	14	13	13	17	23	30	39	50	61	69	77	82	84	83	83	82	44	44	44
1000	*****	77	77	80	81	79	72	63	53	44	37	31	25	20	18	16	15	16	19	25	32	41	51	62	72	83	84	85	82	76	74	49	42	46
MEAN	229	183	165	155	147	137	125	110	93	74	56	41	32	28	27	27	26	27	33	45	70	100	130	153	168	181	190	199	207	210	217	86	107	96

DJF 63-73

P (MB)	80S	70S	65S	60S	55S	50S	45S	40S	35S	30S	25S	20S	15S	10S	5S	EQ	5N	10N	15N	20N	25N	30N	35N	40N	45N	50N	55N	60N	65N	70N	80N	SH	NH	GLOBE
50	124	106	107	110	111	108	100	90	81	74	71	68	66	66	68	69	69	69	67	64	63	64	73	88	111	142	175	204	223	231	225	84	105	94
100	98	103	108	117	124	125	118	107	94	81	71	63	58	56	57	58	59	63	64	68	75	82	90	99	113	130	149	168	179	181	170	85	98	92
200	92	98	109	127	144	153	150	137	117	94	73	56	45	39	38	38	40	43	53	68	87	106	122	134	142	148	156	162	163	158	143	89	101	95
300	85	104	120	138	153	159	149	130	104	78	56	39	30	27	27	26	27	28	37	53	76	100	122	141	155	164	171	175	173	166	150	81	98	90
500	65	77	87	98	107	110	103	104	69	49	35	26	22	21	21	20	19	16	23	37	54	68	84	100	112	122	130	133	129	135	136	57	81	71
700	52	58	66	74	81	82	75	64	49	31	26	20	16	14	14	13	13	13	13	24	36	48	60	69	79	89	96	100	100	98	96	42	52	47
850	*****	55	64	71	75	73	65	54	42	31	24	19	15	13	13	12	12	11	11	20	30	41	53	63	71	79	85	88	87	85	84	37	45	41
1000	*****	69	66	71	74	73	66	56	45	35	27	21	17	15	15	14	14	15	15	23	34	47	58	72	84	97	96	96	96	93	83	42	46	43
MEAN	77	81	87	98	106	108	101	88	72	55	43	33	28	26	26	25	25	26	31	40	54	69	84	97	108	119	128	134	135	132	125	60	73	67

JJA 63-73

P (MB)	80S	70S	65S	60S	55S	50S	45S	40S	35S	30S	25S	20S	15S	10S	5S	EQ	5N	10N	15N	20N	25N	30N	35N	40N	45N	50N	55N	60N	65N	70N	80N	SH	NH	GLOBE
50	461	374	302	233	181	144	120	102	87	74	67	64	62	63	64	65	65	65	64	66	69	74	82	89	92	95	97	100	104	108	115	126	80	103
100	165	159	155	148	138	124	109	97	86	76	67	59	54	53	55	57	59	58	58	58	63	73	88	100	104	104	102	101	101	102	102	90	79	84
200	141	151	157	161	158	150	139	128	116	102	84	64	48	38	38	38	40	40	41	45	56	77	106	138	138	130	134	139	141	158	143	98	84	91
300	140	151	159	166	164	168	152	137	119	99	78	57	40	28	27	27	28	28	29	33	43	60	84	105	119	131	137	139	143	143	136	94	75	86
500	107	112	118	122	124	121	111	99	83	67	51	37	26	20	19	18	18	18	20	23	29	38	51	64	76	85	90	93	96	100	98	59	50	59
700	82	86	91	96	97	94	86	74	61	47	35	25	19	15	14	13	13	13	15	19	23	30	32	44	52	60	65	67	69	71	70	44	36	44
850	*****	75	78	83	85	84	68	56	56	43	31	19	15	12	13	10	11	11	11	17	21	26	32	38	44	51	55	57	58	58	57	38	30	38
1000	*****	79	81	83	84	82	77	68	59	49	39	30	21	16	16	16	16	15	15	17	21	26	32	40	48	57	59	61	60	56	57	42	30	42
MEAN	141	130	126	126	123	118	101	95	82	67	54	41	31	25	25	24	25	25	26	29	36	46	60	73	81	88	91	92	94	95	93	72	54	63

TABLE A16. TEMPORAL STANDARD DEVIATION OF SPECIFIC HUMIDITY [$\sigma(q)$] (IN g kg^{-1})

SD-Q   YEAR 63-73   UNITS= 1.E 00 G/KG

P (HB)	80S	70S	65S	60S	55S	50S	45S	40S	35S	30S	25S	20S	15S	10S	5S	EQ	5N	10N	15N	20N	25N	30N	35N	40N	45N	50N	55N	60N	65N	70N	80N	SH	NH	GLOBE
300	0.0	0.0	0.0	0.0	0.1	0.1	0.1	0.1	0.1	0.1	0.1	0.1	0.1	0.1	0.1	0.1	0.1	0.1	0.1	0.1	0.1	0.1	0.1	0.1	0.1	0.1	0.1	0.1	0.1	0.1	0.0	0.10	0.11	0.11
500	0.1	0.2	0.2	0.3	0.3	0.4	0.5	0.6	0.7	0.8	0.9	1.0	1.0	1.0	0.9	0.9	0.9	1.0	1.0	1.0	0.9	0.9	0.8	0.7	0.6	0.6	0.5	0.4	0.4	0.3	0.2	0.67	0.75	0.71
700	0.3	0.4	0.6	0.7	0.7	0.9	1.1	1.2	1.4	1.6	1.8	2.1	2.0	1.9	1.8	1.7	1.8	2.0	2.1	2.0	2.0	1.9	1.8	1.6	1.5	1.4	1.3	1.2	1.1	0.9	0.7	1.52	1.67	1.59
850	*****	0.6	0.7	0.9	1.1	1.4	1.6	1.9	2.1	2.3	2.4	2.4	2.4	2.4	2.1	1.8	2.1	2.4	2.6	2.7	2.7	2.7	2.7	2.6	2.4	2.1	1.9	1.8	1.6	1.4	1.1	1.89	2.25	2.07
900	*****	0.7	0.7	0.9	1.1	1.3	1.5	1.8	2.0	2.2	2.3	2.3	2.1	1.9	1.8	1.7	1.8	2.0	2.3	2.5	2.7	2.7	2.7	2.6	2.4	2.2	1.9	1.8	1.7	1.5	1.1	1.74	2.20	1.97
950	*****	0.7	0.7	0.9	1.1	1.3	1.5	1.8	2.0	2.2	2.3	2.4	2.2	1.9	1.7	1.6	1.7	2.0	2.3	2.7	3.0	3.0	3.0	2.8	2.5	2.3	2.2	2.1	1.9	1.7	1.2	1.76	2.31	2.03
1000	*****	1.0	1.0	1.1	1.2	1.4	1.7	2.0	2.2	2.5	2.5	2.5	2.3	2.0	1.7	1.6	1.6	1.7	2.2	2.7	3.1	3.3	3.2	2.9	2.4	2.0	2.0	1.9	1.8	1.7	1.3	1.89	2.27	2.05
MEAN	0.1	0.2	0.3	0.4	0.5	0.6	0.7	0.8	0.9	1.0	1.1	1.1	1.1	1.0	0.9	0.9	1.0	1.1	1.1	1.2	1.2	1.1	1.0	1.0	0.9	0.8	0.8	0.7	0.6	0.6	0.4	0.83	0.96	0.90

DJF 63-73

P (HB)	80S	70S	65S	60S	55S	50S	45S	40S	35S	30S	25S	20S	15S	10S	5S	EQ	5N	10N	15N	20N	25N	30N	35N	40N	45N	50N	55N	60N	65N	70N	80N	SH	NH	GLOBE
300	0.0	0.0	0.0	0.0	0.1	0.1	0.1	0.1	0.1	0.1	0.1	0.2	0.2	0.1	0.1	0.1	0.1	0.1	0.1	0.1	0.1	0.1	0.1	0.0	0.0	0.0	0.0	0.0	0.0	0.0	0.0	0.11	0.07	0.09
500	0.1	0.2	0.2	0.3	0.4	0.5	0.5	0.6	0.7	0.8	0.9	1.0	1.0	1.0	0.9	0.9	0.9	0.8	0.7	0.6	0.6	0.5	0.4	0.3	0.3	0.2	0.2	0.2	0.2	0.2	0.1	0.71	0.50	0.61
700	0.4	0.4	0.6	0.7	0.7	0.9	1.1	1.2	1.4	1.6	1.8	1.9	2.0	1.9	1.8	1.8	1.9	1.8	1.7	1.5	1.3	1.1	0.9	0.8	0.7	0.6	0.7	0.6	0.6	0.5	0.4	1.53	1.24	1.38
850	*****	0.5	0.6	0.8	1.0	1.3	1.6	1.8	2.0	2.2	2.3	2.2	2.2	2.2	2.0	2.0	2.1	2.0	1.9	1.7	1.6	1.3	1.1	0.9	0.8	0.7	0.8	0.9	0.9	0.8	0.5	1.83	1.57	1.70
900	*****	0.6	0.6	0.8	1.0	1.2	1.5	1.6	1.8	2.0	2.1	2.1	1.9	1.7	1.7	1.7	1.8	1.9	1.9	2.0	1.9	1.6	1.4	1.2	1.0	0.9	1.0	1.0	1.1	1.0	0.6	1.63	1.47	1.55
950	*****	0.6	0.7	0.8	1.0	1.3	1.5	1.7	1.9	2.0	2.0	2.0	1.8	1.6	1.6	1.6	1.7	1.8	1.9	2.0	1.9	1.7	1.4	1.2	1.0	1.0	1.3	1.2	1.2	1.3	0.8	1.57	1.48	1.52
1000	*****	0.7	0.7	0.9	1.0	1.2	1.4	1.5	1.7	1.8	1.9	1.9	1.8	1.6	1.5	1.5	1.6	1.7	1.8	2.0	2.0	1.9	1.7	1.6	1.3	1.1	1.3	1.2	1.2	1.5	0.5	1.63	1.59	1.61
MEAN	0.1	0.2	0.3	0.4	0.5	0.6	0.7	0.8	0.9	1.0	1.1	1.1	1.1	1.0	0.9	0.9	1.0	1.0	1.0	0.9	0.9	0.8	0.7	0.6	0.5	0.4	0.4	0.4	0.4	0.3	0.2	0.81	0.67	0.74

JJA 63-73

P (HB)	80S	70S	65S	60S	55S	50S	45S	40S	35S	30S	25S	20S	15S	10S	5S	EQ	5N	10N	15N	20N	25N	30N	35N	40N	45N	50N	55N	60N	65N	70N	80N	SH	NH	GLOBE
300	0.0	0.0	0.0	0.0	0.1	0.1	0.1	0.1	0.1	0.1	0.1	0.1	0.1	0.1	0.1	0.1	0.1	0.1	0.1	0.2	0.2	0.2	0.1	0.1	0.1	0.1	0.1	0.1	0.1	0.1	0.1	0.07	0.12	0.10
500	0.1	0.1	0.2	0.2	0.3	0.4	0.4	0.5	0.5	0.5	0.6	0.7	0.8	0.8	0.8	0.8	0.9	0.9	0.9	0.9	0.9	0.9	0.9	0.8	0.7	0.7	0.6	0.5	0.5	0.4	0.3	0.50	0.78	0.64
700	0.3	0.4	0.5	0.6	0.7	0.8	0.9	1.1	1.2	1.4	1.6	1.7	1.8	1.8	1.7	1.6	1.6	1.7	1.8	1.7	1.8	1.8	1.8	1.7	1.6	1.5	1.4	1.2	1.1	1.0	0.8	1.27	1.56	1.42
850	*****	0.5	0.6	0.8	1.0	1.2	1.3	1.5	1.6	1.8	1.9	1.9	1.9	1.9	1.8	1.9	1.9	2.0	2.1	2.2	2.2	2.3	2.3	2.3	2.1	1.9	1.7	1.6	1.5	1.4	1.0	1.57	1.98	1.78
900	*****	0.7	0.6	0.8	1.0	1.1	1.2	1.4	1.6	1.8	1.8	1.8	1.8	1.7	1.8	1.6	1.6	1.7	1.8	1.9	2.0	2.1	2.2	2.2	2.1	1.9	1.8	1.6	1.5	1.4	0.9	1.45	1.83	1.64
950	*****	0.6	0.6	0.8	0.9	1.1	1.2	1.4	1.5	1.7	1.7	1.7	1.7	1.7	1.6	1.5	1.5	1.6	1.7	1.8	2.0	2.1	2.3	2.3	2.1	1.9	1.7	1.6	1.4	1.4	0.8	1.41	1.77	1.58
1000	*****	0.7	0.7	0.8	0.9	1.0	1.3	1.5	1.7	1.9	2.0	1.9	2.0	1.7	1.5	1.4	1.4	1.3	1.4	1.5	1.7	2.0	2.3	2.2	1.9	1.6	1.4	1.4	1.4	1.2	0.6	1.50	1.58	1.53
MEAN	0.1	0.2	0.3	0.3	0.4	0.5	0.5	0.6	0.7	0.8	0.9	0.9	0.9	0.9	0.9	0.9	0.9	0.9	0.9	1.0	1.0	1.0	1.0	0.9	0.9	0.8	0.8	0.7	0.6	0.6	0.4	0.68	0.86	0.77

464

TABLE A17. NORTHWARD TRANSPORT OF MOMENTUM BY TRANSIENT EDDIES [v'u'] (IN m² sec⁻²)

$\overline{v'u'}$ YEAR 63-73   UNITS= 1.E 00 M2/S2

P (MB)	80S	70S	65S	60S	55S	50S	45S	40S	35S	30S	25S	20S	15S	10S	5S	EQ	5N	10N	15N	20N	25N	30N	35N	40N	45N	50N	55N	60N	65N	70N	80N	SH	NH	GLOBE
50	-6.9	-12.9	-14.8	-14.4	-11.5	-7.6	-4.5	-2.5	-1.1	-0.4	-0.1	0.0	-0.0	-0.3	-0.5	0.3	0.6	1.0	1.7	2.9	3.7	4.3	4.6	3.8	2.7	1.4	-0.1	-1.9	-9.3	-3.60			1.28	-1.16
100	2.1	-1.3	-6.0	-10.1	-10.9	-10.3	-10.4	-11.0	-10.8	-9.3	-7.5	-6.1	-3.2	-0.7	0.0	-0.0	0.8	3.6	7.8	12.5	14.3	12.5	9.0	7.4	5.4	3.1	1.2	-0.5	-1.7	-1.9	-1.2	-6.02	5.64	-0.19
200	10.1	14.3	12.8	7.5	-2.0	-14.8	-28.6	-38.5	-40.5	-37.4	-30.6	-24.1	-15.1	-6.0	-0.4	5.5	10.3	15.4	29.3	35.7	36.9	31.9	27.6	19.3	7.6	-0.3	-3.7	-4.3	-3.0	0.4	-15.26	17.62	1.18
300	24.2	24.4	18.4	8.3	-4.9	-19.3	-32.7	-42.1	-42.8	-37.2	-27.8	-18.9	-10.9	-4.2	-0.4	2.1	3.7	6.7	11.4	18.1	24.4	26.9	27.1	24.0	17.4	6.8	-2.2	-6.4	-6.1	-3.4	1.2	-14.11	11.72	-1.19
500	8.0	13.6	12.3	7.0	-0.5	-8.0	-13.5	-16.1	-15.5	-13.4	-10.3	-7.0	-4.2	-2.2	-1.0	-0.1	0.9	2.0	3.3	5.6	8.2	8.8	8.4	7.8	6.0	2.2	-1.3	-3.2	-3.2	-1.9	1.7	-4.95	3.58	-0.68
700	4.6	6.6	5.7	3.8	1.0	-2.2	-4.9	-6.1	-6.2	-6.1	-5.3	-3.9	-2.7	-1.9	-1.0	0.1	1.1	1.8	2.1	2.4	3.6	4.2	4.0	3.7	2.6	0.8	-0.9	-1.4	-0.7	0.1	0.9	-2.26	1.86	-0.18
850	5.3	4.8	3.7	1.8	-0.6	-3.0	-3.9	-3.5	-3.0	-2.8	-2.4	-2.5	-2.1	-0.3	1.6	3.0	2.2	1.5	1.7	2.1	2.3	2.2	0.5	-1.3	-1.7	-0.9	-0.7	-0.4	-0.1	-1.50	1.06	-0.22
1000	*****	-0.5	-0.2	0.2	0.3	0.8	1.2	1.1	0.7	-0.4	-0.8	-0.9	-1.7	-2.6	-1.8	0.1	3.5	5.6	3.9	2.4	1.3	0.7	-0.2	-1.6	-2.6	-3.0	-2.0	-1.2	-0.8	-1.9	-0.5	-0.41	1.16	0.28
MEAN	8.7	9.9	7.4	3.5	-1.8	-7.6	-12.8	-16.0	-16.1	-14.4	-11.4	-8.3	-5.2	-2.6	-0.9	0.8	2.4	4.1	5.9	8.5	11.0	12.0	11.3	9.9	6.9	2.4	-1.0	-2.6	-2.6	-1.8	0.2	-5.98	5.24	-0.39

DJF 63-73

P (MB)	80S	70S	65S	60S	55S	50S	45S	40S	35S	30S	25S	20S	15S	10S	5S	EQ	5N	10N	15N	20N	25N	30N	35N	40N	45N	50N	55N	60N	65N	70N	80N	SH	NH	GLOBE
50	-0.8	-0.7	-1.7	-2.9	-3.2	-2.6	-1.8	-0.9	0.1	1.0	1.2	0.6	-0.4	-1.3	-2.4	-3.3	-2.2	-0.0	1.6	2.7	3.9	4.9	5.4	5.6	6.1	6.6	6.8	7.1	6.7	5.1	-11.3	-0.98	2.94	0.98
100	1.1	-0.1	-1.7	-4.2	-6.9	-9.9	-12.8	-14.3	-12.7	-9.0	-5.7	-4.3	-2.6	-1.5	-1.6	-2.7	-2.5	-1.2	1.8	6.3	8.5	9.3	8.6	7.2	5.2	4.4	4.6	4.1	2.5	2.5	3.3	-5.84	3.98	-0.93
200	5.2	8.3	8.3	5.5	-1.3	-13.2	-28.4	-39.0	-38.4	-31.0	-23.9	-16.7	-10.1	-6.6	-6.3	-5.2	0.8	10.6	21.3	29.3	34.2	32.4	23.4	14.6	7.0	2.4	-0.7	-1.9	-1.1	2.9	-14.70	12.16	-1.27
300	17.3	17.0	11.7	1.9	-11.2	-25.5	-35.4	-35.0	-27.6	-18.5	-11.6	-6.9	-3.3	-0.6	1.0	2.7	7.2	13.3	20.0	25.3	29.9	31.3	29.7	20.0	9.7	4.0	-2.1	-4.9	-4.9	-3.0	4.0	-9.93	12.33	1.20
500	8.2	8.8	7.1	2.0	-5.3	-12.2	-15.9	-16.1	-13.0	-9.1	-5.8	-3.5	-2.2	-1.1	-0.3	0.4	1.3	1.8	3.2	6.2	10.1	11.2	10.5	8.6	5.0	0.6	-2.0	-3.1	-2.8	-1.7	3.0	-4.67	4.07	-0.30
700	1.1	4.2	3.3	1.0	-1.8	-4.7	-6.6	-7.3	-6.4	-5.1	-4.1	-3.2	-2.6	-1.9	-0.7	0.0	0.8	1.6	2.3	3.1	4.4	5.1	5.2	4.4	2.7	0.4	-1.5	-1.5	-0.5	0.4	2.1	-2.85	2.16	-0.33
850	*****	3.9	2.5	0.9	-0.8	-2.4	-3.3	-3.2	-2.4	-1.7	-1.9	-2.1	-2.0	-1.9	-1.0	0.2	1.0	1.3	1.4	1.7	1.9	2.1	2.4	2.1	-0.2	-2.6	-2.9	-1.4	-1.1	-0.9	-0.1	-1.43	0.60	-0.41
1000	*****	1.3	0.6	0.3	0.7	1.5	1.8	1.2	0.1	-0.9	-1.1	-0.9	-1.0	-1.1	-1.1	-0.4	0.2	0.8	0.8	0.8	1.0	0.2	-1.1	-2.4	-3.9	-4.2	-4.0	-2.8	-2.4	-1.5	-2.4	-1.1	0.04	-0.81	-0.33
MEAN	6.8	7.3	6.1	3.1	-1.5	-7.1	-12.3	-15.0	-13.9	-10.7	-7.7	-5.3	-3.5	-2.3	-1.4	-0.7	0.2	1.9	4.4	7.7	10.7	12.8	12.7	9.7	5.5	1.4	-0.9	-1.5	-1.4	-0.7	1.5	-5.00	4.75	-0.14

JJA 63-73

P (MB)	80S	70S	65S	60S	55S	50S	45S	40S	35S	30S	25S	20S	15S	10S	5S	EQ	5N	10N	15N	20N	25N	30N	35N	40N	45N	50N	55N	60N	65N	70N	80N	SH	NH	GLOBE
50	-0.4	-5.0	-4.5	-4.0	-4.3	-4.8	-5.2	-4.8	-4.2	-3.3	-2.2	-1.5	-1.1	-0.7	-0.5	-1.1	-1.1	-0.7	-0.5	-0.8	-0.2	-0.0	-0.0	-0.1	0.1	0.5	0.4	0.3	0.2	0.1	-0.0	-0.6	-3.06	-0.14	-1.60
100	-3.3	-2.6	-1.9	-2.2	-3.0	-3.9	-5.9	-7.7	-8.8	-9.4	-7.2	-3.6	-1.0	0.5	1.1	1.9	1.8	1.2	0.7	1.1	1.1	1.9	3.0	3.6	3.6	2.6	1.5	0.6	0.1	-0.3	-0.5	-0.4	-3.71	1.62	-1.05
200	17.3	10.4	4.2	-1.3	-7.0	-14.2	-23.4	-31.8	-38.1	-41.6	-33.6	-21.8	-9.6	1.6	6.1	7.4	6.3	4.9	6.3	10.0	14.6	20.2	24.9	25.5	18.4	9.1	2.6	-1.1	-2.1	-1.8	-0.9	-13.68	10.36	-1.66
300	22.6	19.7	16.3	11.4	1.7	-12.5	-29.6	-43.5	-48.1	-45.0	-34.4	-23.1	-12.5	-3.3	0.4	1.7	1.2	2.8	5.0	8.8	14.0	17.7	19.2	16.5	10.3	2.2	-2.9	-2.3	0.1	1.1	-15.31	6.79	-4.26
500	6.3	11.2	11.9	10.0	5.9	-0.3	-7.6	-13.4	-15.6	-15.5	-12.7	-8.5	-5.0	-2.4	-1.1	-0.4	0.3	0.9	1.5	2.4	4.1	5.0	5.6	6.2	6.0	4.3	2.1	-0.3	-1.2	-0.5	0.2	-4.28	2.58	-0.85
700	7.8	8.1	7.0	5.7	4.0	1.3	-2.1	-4.7	-6.4	-7.5	-6.9	-4.9	-2.7	-1.6	-1.0	-0.3	0.5	1.1	1.2	2.2	3.0	3.0	2.7	1.8	0.8	-0.1	-0.6	-0.2	-0.3	-1.69	1.39	-0.14
850	*****	5.2	4.4	2.9	0.3	-2.7	-5.0	-5.4	-4.7	-4.1	-3.6	-2.6	-1.9	-1.7	-1.2	-0.5	-0.4	-0.2	0.4	1.1	1.0	1.4	1.7	2.1	2.3	1.2	0.0	-0.4	-0.2	-0.2	-2.00	0.76	-0.62
1000	*****	-1.2	-1.1	-0.9	-0.8	0.2	0.9	0.8	0.8	0.3	-0.3	-0.9	-0.8	-0.6	-0.4	-0.2	0.4	1.5	1.3	2.1	2.1	1.5	0.3	-0.5	-0.4	-0.5	-0.7	-1.8	-0.8	-0.2	-0.21	0.71	0.19
MEAN	9.6	8.6	6.7	4.2	0.0	-5.5	-11.3	-15.6	-17.3	-17.2	-13.9	-9.2	-4.9	-1.5	0.0	0.9	1.2	1.5	1.9	2.7	4.4	6.2	7.6	8.0	6.5	3.8	1.1	-0.7	-1.2	-0.7	-0.1	-6.01	3.07	-1.49

TABLE A18. NORTHWARD TRANSPORT OF MOMENTUM BY STATIONARY EDDIES $[\overline{v}^*\overline{u}^*]$ (IN m² sec⁻²)

v*u* YEAR 63-73 UNITS= 1.E 00 M2/S2

P (MB)	80S	70S	65S	60S	55S	50S	45S	40S	35S	30S	25S	20S	15S	10S	5S	EQ	5N	10N	15N	20N	25N	30N	35N	40N	45N	50N	55N	60N	65N	70N	80N	SH	NH	GLOBE
50	-0.1	-2.0	-2.3	-1.8	-1.0	-0.4	0.1	0.3	0.3	0.2	0.1	0.0	-0.1	-0.1	-0.0	0.1	0.1	0.2	0.1	0.0	0.1	0.3	0.3	0.3	0.0	0.0	-0.8	-1.9	-3.2	-3.5	-3.6	-3.5	-0.26	-0.57	-0.42
100	-0.7	-2.3	-2.0	-1.3	-0.5	0.1	0.1	0.5	0.8	0.2	0.4	0.7	1.0	1.0	0.9	0.5	0.6	0.8	1.1	1.3	1.5	0.6	-0.6	-0.8	-1.3	-2.5	-3.9	-4.1	-3.6	-1.7	0.05	-0.27	-0.11
200	0.1	-2.0	-2.8	-2.8	-1.4	0.5	0.8	0.2	-0.6	-0.6	-1.1	-1.2	2.1	5.1	4.3	3.4	1.8	0.4	-0.4	0.6	4.6	7.4	6.0	3.0	2.4	0.2	-3.2	-4.1	-3.7	-0.6	0.61	1.78	1.19
300	0.3	-1.9	-2.5	-2.3	-0.6	1.7	1.5	0.1	-0.8	-0.7	-0.6	-0.9	-0.2	0.5	0.4	0.4	0.3	0.2	0.5	1.4	3.1	4.6	4.6	3.3	4.0	1.1	-3.4	-4.5	-4.3	-0.3	-0.25	1.40	0.58
500	-0.8	-1.1	-1.0	-1.0	-0.6	0.1	-0.2	-0.8	-0.9	-0.6	-0.3	-0.4	-0.4	-0.1	-0.0	-0.2	0.0	-0.2	0.0	0.4	1.0	1.8	1.1	0.7	1.4	1.7	1.7	-2.8	-3.4	-3.1	-0.3	-0.45	0.15	-0.15
700	-1.0	-1.9	-1.4	-0.8	-0.2	0.4	0.3	-0.1	-0.1	0.2	0.2	0.0	-0.4	-0.2	0.1	-0.3	-0.5	-0.8	-0.8	-0.4	0.2	0.7	0.7	0.4	0.7	1.1	0.0	-1.7	-1.8	-1.4	-0.4	-0.18	-0.19	-0.19
850	*****	0.9	-0.2	-0.2	-0.2	-0.0	0.0	0.3	0.2	0.1	-0.1	-0.2	-0.6	-0.5	-0.0	0.0	-0.5	-0.7	-0.9	-0.7	-0.4	-0.6	0.2	1.1	1.3	0.6	0.9	0.3	-0.4	-0.5	-0.4	-0.10	-0.12	-0.11
1000	*****	-0.0	0.0	0.0	-0.1	-0.3	-0.4	-0.2	0.1	0.1	-0.1	0.2	-0.1	0.1	0.1	0.0	0.0	-0.2	0.8	1.7	1.4	0.5	-0.1	-0.2	0.2	0.3	0.3	0.6	0.3	0.1	-0.2	-0.04	0.49	0.19
MEAN	-0.4	-1.3	-1.4	-1.2	-0.5	0.3	0.4	-0.1	-0.4	-0.3	-0.4	0.1	0.7	0.6	0.4	0.1	-0.1	-0.1	0.3	1.1	2.1	1.9	1.2	1.4	-0.1	-1.4	-2.2	-2.8	-2.5	-0.6	-0.13	0.33	0.10

DJF 63-73

P (MB)	80S	70S	65S	60S	55S	50S	45S	40S	35S	30S	25S	20S	15S	10S	5S	EQ	5N	10N	15N	20N	25N	30N	35N	40N	45N	50N	55N	60N	65N	70N	80N	SH	NH	GLOBE
50	0.1	-0.2	-0.3	-0.2	0.0	0.2	0.2	0.1	0.1	0.0	-0.1	-0.1	-0.1	-0.1	-0.2	-0.9	-0.2	0.2	0.4	0.7	1.9	3.8	5.1	3.8	-1.7	-8.4	-14.6	-15.6	-17.1	-19.7	-0.03	-2.16	-1.09
100	0.0	-0.4	-0.4	-0.3	-0.1	0.1	0.1	-0.1	-0.8	-0.3	1.0	2.2	1.2	-0.3	-1.5	-1.1	0.9	3.7	8.6	13.3	9.8	7.0	3.9	-3.2	-9.7	-14.1	-14.5	-12.6	-7.1	0.08	0.64	0.36
200	0.5	-0.9	-1.4	-1.7	-1.5	-0.7	-1.7	-2.4	-3.5	-3.2	0.9	4.5	2.9	2.0	-1.0	-4.0	-0.9	7.0	22.0	36.9	25.3	12.7	9.0	0.6	-7.0	-11.9	-12.4	-10.0	-1.8	-0.36	6.10	2.87
300	0.1	-1.5	-1.4	-1.7	-1.5	-0.0	-1.4	-1.5	-1.5	-2.0	-0.8	0.6	0.5	1.3	4.2	7.7	13.8	22.3	17.8	11.2	7.7	0.8	-6.1	-11.2	-12.2	-10.1	-0.1	-0.39	4.82	2.22
500	-1.1	-0.9	-0.6	-0.8	-0.3	0.5	0.1	-0.5	-0.8	-0.6	-0.4	-0.5	-0.3	-0.1	0.1	0.2	0.3	0.7	1.3	1.6	2.8	5.4	3.1	2.2	2.2	0.2	-4.0	-7.1	-7.4	-6.1	0.3	-0.34	0.45	0.06
700	-1.1	-1.8	-1.3	-0.8	-0.1	0.6	0.5	0.6	0.5	-0.1	-0.4	-0.6	-0.7	-0.3	0.3	0.3	0.1	-0.2	-0.4	-0.1	-0.4	0.6	1.5	0.9	-0.2	0.8	-1.2	-3.4	-3.0	-2.3	0.1	-0.29	-0.22	-0.26
850	0.7	0.1	-0.2	-0.4	0.0	0.5	0.2	-0.0	-0.0	-0.2	-0.4	-0.3	0.4	0.8	-0.1	-0.9	-0.9	-0.2	0.1	-0.0	0.7	0.8	-0.0	-0.3	0.6	0.2	-0.4	-0.1	-0.2	0.1	0.03	-0.12	-0.05
1000	0.1	0.1	-0.1	-0.3	-0.2	-0.7	-0.5	0.2	-0.0	0.2	-0.1	0.2	-0.0	-0.4	-1.8	-2.7	-1.4	1.0	-0.3	-0.4	-0.1	-0.2	-0.3	-0.5	-1.0	-0.6	-0.0	0.1	0.9	0.5	0.6	-0.61	-0.14	-0.40
MEAN	-0.4	-0.8	-0.8	-0.8	-0.4	0.3	0.2	-0.2	-0.7	-0.8	-0.9	0.0	0.7	0.4	0.1	-0.3	-0.3	0.8	2.4	5.7	9.9	7.4	4.4	3.1	0.0	-4.0	-6.9	-7.3	-6.2	-1.7	-0.25	1.30	0.52

JJA 63-73

P (MB)	80S	70S	65S	60S	55S	50S	45S	40S	35S	30S	25S	20S	15S	10S	5S	EQ	5N	10N	15N	20N	25N	30N	35N	40N	45N	50N	55N	60N	65N	70N	80N	SH	NH	GLOBE
50	1.1	-0.8	-1.9	-1.8	-1.1	-0.5	-0.1	0.3	0.3	0.1	0.1	0.0	0.0	0.0	0.3	0.5	0.1	0.4	0.4	0.5	0.5	0.6	0.5	0.3	-0.1	-0.2	-0.1	0.0	-0.0	-0.0	-0.11	0.25	0.07
100	-1.0	-3.0	-2.7	-2.0	-1.0	-0.1	0.4	0.7	0.6	0.4	0.6	0.2	0.0	0.0	0.3	0.8	1.4	3.5	3.6	3.1	4.6	7.1	4.5	0.4	-1.6	-1.3	-0.9	-0.5	-0.3	-0.0	-0.14	2.01	0.93
200	1.2	-2.6	-3.8	-3.4	-1.4	0.9	1.5	0.3	0.4	2.2	3.6	7.5	11.2	9.6	7.7	5.7	5.6	3.7	2.6	5.5	10.1	8.3	2.9	1.0	-0.5	-0.6	-1.0	-0.3	3.17	3.73	3.45
300	2.0	-1.2	-2.4	-2.4	-1.3	0.3	0.3	-1.2	-1.7	-0.4	0.5	0.1	0.2	0.8	0.7	0.8	0.9	1.4	1.5	1.5	2.1	3.7	3.1	1.1	1.9	0.9	0.5	-0.7	-1.0	-1.7	-0.7	-0.21	1.27	0.53
500	-0.8	0.1	-0.5	-0.7	-1.6	-2.3	-2.0	-0.8	-0.2	-0.8	-1.0	-0.0	0.2	-0.2	-0.2	0.6	0.6	0.6	1.1	1.3	0.8	0.5	1.1	0.9	0.3	0.8	0.2	-0.6	-1.1	-1.1	-0.6	-0.78	0.39	-0.20
700	-0.8	-0.7	-0.5	-0.4	-0.3	-0.2	-0.4	-0.7	-0.1	0.6	0.1	-0.5	-0.6	0.2	0.2	-0.4	-0.9	-0.9	-1.2	-1.2	-0.5	0.0	0.0	0.3	0.4	0.3	0.8	0.2	-0.6	-0.6	-0.4	-0.3	-0.20	-0.32	-0.26
850	*****	0.9	-0.4	-0.4	-0.2	-0.2	0.0	0.0	0.4	0.3	-0.3	-0.9	-1.1	-0.9	0.0	-0.6	0.2	0.0	-0.7	-1.3	-0.7	1.0	2.1	1.4	1.2	0.1	0.4	-0.3	-0.2	-0.0	-0.3	-0.34	0.01	-0.17
1000	*****	-0.1	0.2	0.6	0.6	0.5	0.1	-0.0	0.2	0.1	-0.1	0.1	0.5	0.1	1.1	1.1	3.0	9.3	10.4	5.8	1.9	0.9	-0.0	0.1	0.5	0.1	0.2	0.3	0.2	-0.0	-0.1	0.29	3.12	1.52
MEAN	0.2	-0.8	-1.2	-1.1	-0.6	0.0	0.1	-0.4	-0.5	0.0	0.1	-0.4	-0.5	-0.0	0.2	1.1	1.1	2.3	2.1	1.1	1.0	0.6	0.2	-0.5	-0.6	-0.7	-0.4	0.08	0.65	0.46

TABLE A19. NORTHWARD TRANSPORT OF MOMENTUM BY MEAN MERIDIONAL CIRCULATIONS $[\bar{v}]_{\mathrm{ID}}[\bar{u}]$ (IN m² sec⁻²)

V'·U   YEAR 63-73   UNITS= 1.E 00 M2/S2

P (MB)	80S	70S	65S	60S	55S	50S	45S	40S	35S	30S	25S	20S	15S	10S	5S	EQ	5N	10N	15N	20N	25N	30N	35N	40N	45N	50N	55N	60N	65N	70N	80N	SH	NH	GLOBE
50	1.3	1.6	1.2	0.4	-0.2	-0.4	-0.3	-0.2	-0.0	0.0	0.0	-0.0	-0.1	-0.0	0.1	0.1	0.1	0.2	0.3	-0.1	-0.1	-0.0	0.0	-0.0	-0.2	-0.3	-0.2	-0.2	-0.1	-0.1	0.7	0.11	0.02	0.06
100	-0.0	0.9	1.3	1.1	0.5	0.5	0.8	0.8	-0.0	0.0	-0.5	-0.5	-0.8	-0.2	0.2	0.5	0.3	0.6	-1.6	-0.9	-0.7	-0.6	-0.4	-0.1	-0.1	0.3	0.2	0.2	0.23	-0.53	-0.15			
200	-0.9	-0.6	-0.0	1.0	2.7	5.1	6.6	5.5	2.3	-1.1	-3.1	-6.8	-5.1	-1.5	1.0	0.2	-1.4	0.8	4.9	5.0	2.8	-3.3	-5.9	-4.6	-4.5	-3.2	-1.5	-0.3	0.3	0.5	0.1	0.05	-0.50	-0.22
300	-1.6	-0.6	0.3	1.7	3.3	5.3	6.5	5.3	1.8	-1.7	-3.4	-2.8	-0.7	0.0	0.1	-0.2	-0.5	-0.1	1.2	2.5	0.7	-1.1	-2.6	-3.0	-3.4	-3.1	-1.7	-0.4	0.7	0.1	0.68	-0.60	0.04	
500	-0.3	-0.4	-0.1	0.6	1.3	1.8	1.9	1.3	0.5	-0.0	-0.4	-0.2	-0.2	0.5	0.9	1.1	0.5	-0.2	-0.1	0.2	0.2	-0.1	-0.5	-0.5	-0.7	-0.7	-0.4	-0.0	0.2	0.3	0.50	-0.12	0.19	
700	-0.1	-0.1	0.1	0.6	1.3	1.8	1.5	0.7	1.4	1.8	1.5	0.7	1.0	0.9	1.0	0.5	0.9	1.0	0.2	-0.1	-0.2	-0.2	-0.2	-0.1	-0.1	-0.0	0.0	0.1	0.0	0.0	0.0	0.21	0.15	0.18
850	*****	0.0	0.0	0.0	0.2	0.3	0.3	0.2	0.1	0.0	-0.0	-0.0	0.0	-0.9	-1.1	-0.5	0.5	1.5	0.7	0.0	0.0	0.0	0.0	-0.0	-0.1	-0.1	-0.0	0.0	0.0	0.0	0.0	-0.13	0.21	0.04
1000	*****	-0.2	-1.6	-3.0	-4.1	-3.9	-2.0	-0.4	-0.1	-1.3	-2.9	-7.6	-7.2	-4.4	-2.6	-0.9	0.1	1.5	3.1	3.0	5.1	3.1	0.3	-0.4	0.6	1.2	1.6	1.8	0.6	0.0	-0.1	-3.06	1.15	-1.23
MEAN	-0.5	-0.1	0.1	0.4	0.7	1.1	1.5	1.3	0.6	-0.3	-0.9	-1.3	-1.0	-0.5	-0.0	0.1	1.1	1.0	1.1	0.3	-0.8	-1.1	-1.0	-1.0	-0.8	-0.4	-0.1	0.2	0.2	0.1	0.00	-0.06	-0.03	

DJF 63-73

P (MB)	80S	70S	65S	60S	55S	50S	45S	40S	35S	30S	25S	20S	15S	10S	5S	EQ	5N	10N	15N	20N	25N	30N	35N	40N	45N	50N	55N	60N	65N	70N	80N	SH	NH	GLOBE
50	0.0	0.0	0.0	0.0	0.0	-0.0	0.0	0.1	0.1	0.1	-0.4	1.5	0.6	-0.6	-1.0	-0.7	-0.1	0.1	0.0	0.2	0.3	0.1	-0.2	-0.9	-1.3	-0.9	-0.3	0.1	-0.7	3.7	0.08	-0.19	-0.05
100	-0.0	0.1	0.2	0.4	0.7	1.1	1.1	0.6	-0.4	-0.8	-0.3	-0.1	-1.0	-1.1	-1.0	-1.1	1.4	6.1	5.1	4.1	-0.0	-2.5	-2.1	-2.3	-1.6	-0.7	-0.2	0.5	1.0	0.7	-0.14	0.76	0.31
200	-0.3	-0.3	-0.1	0.6	2.2	5.1	6.8	4.7	0.5	-2.1	-3.1	-1.8	-2.2	-1.6	-1.0	-1.1	6.1	22.0	29.6	24.9	21.0	-0.4	-14.2	-9.3	-6.2	-3.8	-1.5	-0.2	0.6	0.9	0.2	-0.33	5.99	2.83
300	-0.6	-0.5	-0.3	0.7	2.4	5.2	6.8	5.0	0.8	-2.1	-1.9	-1.0	-2.0	-2.3	-0.1	3.4	6.6	5.6	3.6	-4.2	-8.1	-7.3	-5.8	-3.5	-1.3	0.1	0.7	0.9	0.1	0.7	0.48	-0.40	0.04
500	-0.1	-0.1	0.1	0.7	1.4	1.8	1.5	0.7	-0.1	-0.4	-0.3	-0.0	0.1	0.4	0.9	1.5	0.9	0.2	0.0	0.6	-1.1	-2.9	-1.5	-1.2	-1.4	-1.0	-0.3	0.1	0.1	0.4	0.45	-0.41	0.02
700	-0.0	0.0	0.0	0.1	0.3	0.4	0.6	0.5	0.2	0.1	-0.0	-0.0	-0.0	-0.0	0.4	0.7	1.4	2.6	2.6	1.4	0.2	-0.1	-0.7	-0.8	-0.3	-0.3	-0.2	-0.1	0.1	0.1	0.45	0.29	0.37
850	*****	-0.0	-0.0	0.1	0.2	0.3	0.4	0.6	0.2	0.1	0.0	-0.0	-0.0	0.0	0.0	0.9	1.5	2.6	5.4	8.2	2.5	0.1	-0.1	-0.1	-0.1	-0.2	-0.1	0.0	0.1	0.1	0.36	1.49	0.92
1000	*****	0.3	-0.0	-1.0	-2.4	-3.7	-3.5	-1.4	-0.1	-0.8	-1.8	-2.5	-3.1	-0.8	-0.1	0.3	5.2	15.6	12.9	5.7	0.9	0.6	2.8	3.5	3.1	2.4	0.5	-0.1	-0.3	-0.1	0.2	-1.55	4.75	1.19
MEAN	-0.2	-0.0	-0.0	0.1	0.5	1.1	1.5	1.2	0.1	0.1	-0.6	-0.6	-0.6	-0.1	-0.1	-0.3	0.2	2.1	4.7	5.2	4.0	2.4	-1.5	-2.9	-2.1	-1.7	-1.1	-0.4	0.0	0.1	0.3	0.11	0.98	0.55

JJA 63-73

P (MB)	80S	70S	65S	60S	55S	50S	45S	40S	35S	30S	25S	20S	15S	10S	5S	EQ	5N	10N	15N	20N	25N	30N	35N	40N	45N	50N	55N	60N	65N	70N	80N	SH	NH	GLOBE
50	0.5	0.5	0.3	0.4	0.4	0.4	0.4	0.1	-0.0	-0.1	-0.1	-0.0	-0.0	0.0	0.2	0.7	1.8	2.3	0.7	-0.2	-0.0	0.0	0.0	0.0	0.0	0.0	-0.0	0.0	0.0	0.0	0.18	0.43	0.31
100	0.6	0.3	0.3	0.3	0.7	1.1	1.1	0.9	-0.4	-1.6	-1.5	-1.2	-0.4	0.9	1.4	2.1	2.8	1.8	-0.2	-0.4	0.0	-0.5	-0.8	-0.4	-0.1	-0.0	-0.0	0.0	0.0	0.0	0.05	0.48	0.26
200	-1.5	0.5	0.9	1.1	1.6	2.9	4.8	6.1	6.3	-0.4	-11.8	-17.3	-17.9	-7.6	10.3	22.3	18.8	9.0	1.1	-1.9	-3.7	-2.2	-0.9	-0.2	0.1	0.1	-1.27	2.78	0.75					
300	-1.9	-0.5	-0.1	0.6	2.2	4.7	6.9	6.9	3.6	-0.3	-4.9	-5.9	-4.1	-1.0	2.5	4.9	3.6	1.5	0.1	-0.2	0.3	0.5	-0.7	-1.0	-1.7	-1.9	-1.2	-0.2	0.2	0.0	0.56	0.32	0.44
500	-0.3	-0.4	-0.3	-0.0	0.6	1.4	2.1	1.9	1.2	0.9	0.3	-0.1	-0.6	0.6	1.4	1.2	0.4	-0.6	-0.4	-0.0	-0.4	-0.3	-0.0	-0.1	-0.1	-0.0	0.1	0.0	0.65	-0.09	0.28			
700	-0.1	-0.4	-0.3	-0.0	0.6	1.4	2.1	1.9	1.2	0.9	0.3	0.0	0.5	0.5	-1.8	-1.6	-0.9	-0.1	0.1	-0.1	-0.1	-0.1	-0.0	-0.0	0.0	0.0	0.0	0.0	-0.08	-0.18	-0.13		
850	*****	-0.0	-0.0	0.1	0.3	0.4	0.3	0.1	0.1	0.0	-0.0	-0.0	-1.2	-5.7	-5.7	-3.4	-1.2	-0.5	-0.2	0.0	-0.0	-0.0	-0.0	-0.0	-0.1	-0.0	-0.0	0.0	-1.21	-0.30	-0.76		
1000	*****	0.5	0.0	-0.9	-2.5	-4.3	-4.7	-3.2	-1.0	-0.0	-1.0	-4.5	-13.7	-17.0	-11.1	-6.1	-1.2	-0.5	0.4	1.3	1.5	0.3	0.1	0.4	0.8	0.9	0.4	0.1	-0.0	-0.0	-5.35	-0.03	-3.04
MEAN	-0.7	0.0	0.0	0.1	0.2	0.4	0.9	1.3	1.5	1.2	0.2	-1.8	-2.9	-3.4	-2.6	-0.2	1.7	1.9	1.0	0.2	-0.0	0.1	0.2	-0.3	-0.5	-0.6	-0.5	-0.3	-0.1	0.0	0.1	0.0	0.24	-0.12

467

TABLE A20. NORTHWARD TRANSPORT OF SENSIBLE HEAT BY TRANSIENT EDDIES [$\overline{v'T'}$] (IN °C m sec^{-1})

$\overline{v'T'}$ YEAR 63-73 UNITS= 1.E 00 M/S °C

P (MB)	80S	70S	65S	60S	55S	50S	45S	40S	35S	30S	25S	20S	15S	10S	5S	EQ	5N	10N	15N	20N	25N	30N	35N	40N	45N	50N	55N	60N	65N	70N	80N	SH	NH	GLOBE
50	13.7	17.9	13.6	6.9	1.2	-2.0	-2.3	-1.7	-1.0	-0.5	-0.3	-0.3	-0.3	-0.5	-0.4	-0.3	-0.0	0.3	0.5	0.6	0.8	1.0	1.5	2.4	3.6	5.0	6.5	7.7	6.6	4.2	7.0	1.44	2.32	1.88
100	-3.5	0.9	2.6	1.8	-1.1	-3.1	-3.1	-2.4	-1.7	-1.0	-0.8	-0.9	-0.7	-0.4	-0.4	-0.1	-0.2	-0.2	0.0	0.4	0.9	1.7	2.3	3.2	4.4	5.6	6.3	6.3	5.1	3.4	1.0	-1.04	2.10	0.53
200	1.2	1.8	-1.5	-5.7	-9.7	-12.2	-12.5	-11.0	-8.0	-4.6	-1.9	-0.3	0.4	0.7	0.7	0.5	-0.1	-1.0	-1.5	-1.4	0.2	3.1	6.2	8.5	9.7	10.2	9.4	7.7	5.5	3.7	0.8	-3.79	3.29	-0.25
300	0.7	1.6	0.8	-0.7	-2.7	-4.6	-5.9	-6.3	-5.7	-4.1	-2.2	-0.6	0.0	0.3	0.3	0.1	-0.2	-0.8	-1.0	-1.0	-0.1	2.6	5.3	6.5	5.8	4.7	3.9	2.9	2.1	1.9	1.0	-1.97	1.88	-0.04
500	-1.7	-2.1	-3.1	-4.5	-6.3	-7.5	-6.2	-4.1	-2.0	-0.6	0.3	0.5	0.5	0.2	0.1	-0.0	-0.2	-0.6	-0.8	-1.0	-0.6	1.2	3.7	6.0	6.7	6.9	7.2	6.6	5.6	5.1	2.8	-2.43	2.44	0.00
700	-4.4	-4.5	-6.3	-8.8	-11.0	-12.7	-11.7	-10.0	-7.0	-3.7	-1.3	0.1	0.2	0.1	-0.1	-0.2	-0.3	-0.6	-0.6	-0.5	0.3	2.5	5.3	6.0	8.2	10.5	11.4	10.9	9.8	8.3	7.4	-4.46	4.04	-0.19
850	-3.9	-6.4	-10.0	-12.7	-13.9	-13.5	-12.3	-10.2	-7.5	-4.5	-2.3	-1.3	-0.6	-0.4	-0.1	0.1	0.1	0.7	1.2	2.8	5.3	8.4	10.7	11.8	12.9	12.4	11.4	10.3	9.0	6.9	6.5	-6.19	5.54	-0.31
1000	-0.8	-1.8	-2.4	-2.8	-3.0	-3.1	-3.1	-2.7	-2.2	-1.6	-1.3	-1.1	-0.7	-0.4	-0.3	-0.0	0.7	1.2	2.2	3.3	5.0	6.0	6.8	6.0	5.7	6.0	5.3	3.5	2.6	-0.5	-1.85	2.84	0.19
MEAN	-0.4	-0.2	-1.9	-4.6	-7.0	-8.6	-8.7	-7.7	-5.9	-3.7	-1.9	-0.7	-0.2	-0.0	-0.0	-0.1	-0.2	-0.4	-0.3	-0.1	0.8	2.8	5.2	7.1	8.0	8.4	8.1	7.4	6.1	4.9	3.1	-3.14	3.20	0.02

DJF 63-73

P (MB)	80S	70S	65S	60S	55S	50S	45S	40S	35S	30S	25S	20S	15S	10S	5S	EQ	5N	10N	15N	20N	25N	30N	35N	40N	45N	50N	55N	60N	65N	70N	80N	SH	NH	GLOBE
50	0.2	-0.4	-0.7	-0.9	-1.0	-1.1	-1.0	-0.8	-0.7	-0.6	-0.6	-0.6	-0.4	-0.6	-0.6	-0.7	-0.7	-0.5	0.1	0.4	0.6	0.8	1.1	1.4	2.0	3.0	4.1	5.2	6.1	6.7	5.3	-0.68	2.20	0.76
100	0.1	-1.0	-2.2	-3.4	-4.3	-4.6	-3.9	-3.1	-2.2	-1.2	-0.8	-0.8	-1.3	-0.8	-0.4	-0.3	-0.6	-0.0	0.0	0.8	1.5	2.1	2.7	3.3	3.7	4.2	4.9	5.1	4.7	4.1	0.0	-1.78	1.80	0.01
200	-1.0	-0.8	-3.3	-7.1	-10.8	-12.6	-11.6	-9.2	-5.8	-2.4	-0.6	0.3	0.4	0.2	0.3	0.2	-0.1	-0.0	-0.2	0.1	1.1	4.0	5.7	7.0	7.7	7.9	7.5	6.4	5.2	5.5	5.2	-3.62	3.11	-0.25
300	3.6	2.4	1.5	0.3	-1.3	-3.2	-4.7	-5.3	-4.7	-3.6	-2.0	-0.7	-0.1	0.2	0.3	0.2	-0.2	-0.5	-0.2	0.2	1.1	3.3	3.9	3.3	2.6	1.9	0.9	0.9	1.4	0.9	0.9	-1.40	1.20	-0.10
500	-0.7	-1.5	-1.8	-2.6	-3.8	-4.7	-4.4	-3.3	-2.1	-1.0	-0.2	0.3	0.3	0.2	0.1	0.1	-0.2	-0.6	-1.1	-1.3	-1.2	0.1	2.5	5.1	6.5	7.1	6.5	5.7	5.2	5.7	6.0	-1.38	2.59	0.60
700	-1.1	-2.4	-4.0	-6.4	-8.7	-9.9	-9.2	-7.2	-4.5	-2.1	-0.6	0.2	0.3	0.2	0.1	0.0	-0.2	-0.5	-0.2	0.2	1.7	5.0	8.7	11.0	12.5	12.4	11.1	9.7	9.4	8.8	6.9	-3.11	5.14	1.04
850	*****	-2.4	-4.0	-6.9	-10.1	-12.2	-12.5	-11.5	-9.3	-5.7	-2.7	-0.9	0.2	0.4	0.1	0.5	1.1	1.3	3.2	5.7	8.8	12.2	15.1	15.4	14.3	13.1	11.9	11.3	9.7	8.6	-4.88	7.16	1.15
1000	*****	-0.4	-1.0	-1.5	-1.8	-2.1	-2.6	-2.8	-2.2	-2.6	-2.2	-1.4	-0.6	-0.3	-0.0	0.0	-0.1	0.1	0.5	1.3	3.2	5.6	6.7	7.5	7.0	6.0	5.2	5.5	5.5	4.7	1.9	-1.15	3.03	0.66
MEAN	-0.4	-0.2	-1.1	-2.4	-4.2	-6.0	-7.2	-7.1	-6.0	-4.4	-2.6	-1.2	-0.2	0.0	0.1	-0.1	-0.1	-0.3	-0.5	-0.3	0.4	2.0	4.3	6.5	7.8	8.2	7.0	6.3	6.1	5.5	4.0	-2.53	3.52	0.48

JJA 63-73

P (MB)	80S	70S	65S	60S	55S	50S	45S	40S	35S	30S	25S	20S	15S	10S	5S	EQ	5N	10N	15N	20N	25N	30N	35N	40N	45N	50N	55N	60N	65N	70N	80N	SH	NH	GLOBE
50	2.0	-5.9	-8.1	-7.9	-6.2	-4.2	-2.9	-1.9	-1.3	-0.9	-0.7	-0.6	-0.6	-0.7	-0.5	-0.1	0.2	0.5	0.6	0.6	0.4	0.2	0.3	0.3	0.5	0.7	0.8	0.7	0.5	0.4	0.2	-2.08	0.45	-0.81
100	2.2	0.3	-0.4	-1.3	-1.9	-2.0	-1.8	-1.7	-1.5	-1.2	-1.0	-0.8	-0.6	-0.2	-0.2	0.6	1.2	1.4	1.1	0.9	0.9	0.6	0.7	0.8	1.4	2.1	2.4	1.9	1.3	1.0	0.2	-0.71	1.22	0.25
200	2.6	4.7	4.0	1.2	-3.0	-6.7	-8.9	-9.3	-7.9	-5.2	-2.4	-0.8	-0.1	0.4	0.4	0.4	0.1	-0.4	-0.4	-0.2	0.0	0.7	1.6	3.4	5.7	7.4	8.4	6.0	3.9	2.5	0.3	-2.33	2.59	0.13
300	3.2	1.9	2.1	1.9	0.7	-1.2	-3.4	-5.1	-5.3	-4.2	-2.4	-1.1	-0.3	0.0	0.0	0.0	-0.2	-0.3	-0.5	-0.4	-0.2	0.0	0.5	2.0	3.5	4.1	4.6	4.2	3.0	2.2	0.3	-1.23	1.51	0.14
500	-2.1	-3.4	-4.2	-5.5	-7.2	-8.6	-9.0	-8.3	-6.4	-4.1	-1.8	-0.1	0.5	0.6	0.3	0.1	0.0	-0.1	-0.1	-0.3	-0.4	-0.5	-0.3	0.2	1.0	2.4	4.0	4.5	3.6	3.3	2.1	-3.36	1.25	-1.06
700	-6.0	-7.4	-8.9	-10.9	-12.4	-13.2	-12.9	-11.8	-9.0	-5.6	-2.5	-0.4	0.3	0.3	0.1	-0.0	-0.1	-0.3	-0.4	-0.5	-0.3	0.3	1.2	2.4	4.2	5.6	5.8	5.3	5.0	4.7	4.2	-5.50	1.75	-1.85
850	*****	-4.1	-7.4	-12.0	-15.4	-16.8	-16.0	-14.4	-11.9	-9.0	-5.8	-3.1	-1.4	-0.3	-0.0	0.1	0.1	-0.6	-0.8	-0.6	0.0	1.0	2.3	3.7	5.4	7.3	7.4	6.6	7.0	7.6	5.5	-7.21	2.45	-2.37
1000	*****	-4.2	-5.0	-5.4	-4.9	-4.3	-4.1	-4.1	-3.7	-2.9	-2.0	-1.0	-0.5	-0.2	0.0	0.0	0.0	-0.2	-0.2	-0.2	-0.1	0.1	0.5	1.5	2.6	3.9	4.8	4.3	3.3	3.5	1.9	-2.62	1.01	-1.04
MEAN	0.2	-1.9	-3.2	-5.1	-6.8	-8.2	-8.7	-8.4	-7.0	-4.9	-2.7	-1.0	-0.3	0.1	0.1	0.2	0.0	-0.2	-0.2	-0.2	-0.1	0.1	0.5	1.5	2.6	3.9	5.0	4.3	3.8	2.8	1.9	-3.54	1.69	-0.93

TABLE A21. NORTHWARD TRANSPORT OF SENSIBLE HEAT BY STATIONARY EDDIES $[\bar{v}^*\bar{T}^*]$ (IN °C m sec⁻¹)

$\bar{v}^*\bar{T}^*$ YEAR 63–73   UNITS= 1.E 00 M/S °C

P (MB)	80S	70S	65S	60S	55S	50S	45S	40S	35S	30S	25S	20S	15S	10S	5S	EQ	5N	10N	15N	20N	25N	30N	35N	40N	45N	50N	55N	60N	65N	70N	80N	SH	NH	GLOBE
50	-1.3	-2.6	-2.5	-2.1	-1.7	-1.1	-0.8	-0.5	-0.3	-0.1	-0.0	0.1	0.1	-0.0	0.0	-0.2	-0.1	0.1	0.2	0.2	0.1	0.2	0.2	0.3	0.9	2.2	4.0	5.7	7.4	7.4	5.7	-0.57	1.65	0.54
100	-1.4	-1.6	-1.3	-1.0	-0.6	-0.3	-0.1	0.1	0.1	0.0	-0.1	0.2	0.3	0.2	0.1	0.3	0.5	1.0	1.4	1.3	0.7	0.2	0.4	0.5	1.1	2.6	4.0	4.3	2.9	0.6	-0.17	1.42	0.62
200	-1.3	-1.9	-1.6	-1.2	-0.7	-0.2	-0.1	0.1	0.1	-0.0	0.1	0.2	0.3	0.2	0.1	0.1	0.1	0.6	1.1	1.1	0.6	2.4	1.6	1.2	1.7	3.1	4.3	3.9	2.2	0.7	-0.3	-0.19	1.53	0.67
300	-0.7	-0.6	-0.4	-0.2	0.1	0.4	0.5	0.4	0.4	0.1	-0.0	0.1	0.2	0.1	-0.0	-0.0	-0.1	0.6	1.1	1.0	-0.0	0.6	0.8	1.4	2.6	3.0	2.3	1.3	0.4	-0.1	0.07	0.86	0.46
500	-1.5	-0.5	-0.2	-0.3	-0.3	-0.2	-0.1	-0.2	-0.3	-0.3	-0.2	-0.1	-0.0	-0.0	-0.2	-0.0	0.3	0.4	0.2	0.7	1.6	2.5	2.7	3.0	2.0	1.2	0.4	0.1	-0.21	0.65	0.22
700	-3.1	-1.5	-0.8	-0.5	-0.5	-0.4	-0.3	-0.2	-0.5	-0.9	-0.7	-0.4	-0.2	-0.2	-0.2	-0.2	0.4	0.9	1.6	2.6	3.2	3.2	2.6	1.6	0.4	0.1	-0.49	0.58	0.05
850	-1.2	-0.5	-0.1	0.1	0.1	0.1	0.1	-0.3	-0.8	-0.6	-0.8	-0.8	-0.5	-0.2	0.2	0.2	-0.1	-1.0	-1.4	-1.6	-0.7	-0.0	0.0	1.1	2.6	3.2	2.9	2.2	0.7	0.3	-0.38	0.24	-0.07
1000	0.1	-0.0	-0.2	-0.2	0.2	0.3	-0.1	-0.6	-1.7	-1.7	-2.0	-1.8	-1.8	-0.9	0.4	1.5	2.1	2.4	1.9	1.1	0.4	0.6	1.5	3.4	5.0	5.1	2.1	-0.1	-0.97	1.14	-0.05
MEAN	-1.4	-1.0	-0.7	-0.5	-0.3	-0.3	-0.1	-0.0	-0.0	-0.2	-0.4	-0.3	-0.2	-0.2	-0.2	-0.1	-0.0	0.0	0.2	0.3	0.4	0.6	1.4	2.6	3.3	3.1	2.2	1.0	0.2	-0.26	0.81	0.27

DJF 63–73

P (MB)	80S	70S	65S	60S	55S	50S	45S	40S	35S	30S	25S	20S	15S	10S	5S	EQ	5N	10N	15N	20N	25N	30N	35N	40N	45N	50N	55N	60N	65N	70N	80N	SH	NH	GLOBE
50	0.2	0.3	0.2	0.2	0.1	0.0	0.1	0.1	0.0	-0.0	-0.0	0.2	0.0	0.1	0.1	-0.1	-0.1	0.0	0.0	0.0	0.0	0.2	1.3	3.4	8.1	14.9	22.4	29.3	27.4	18.3	8.8	0.06	6.10	3.08
100	-0.2	-0.1	0.0	0.1	0.1	0.1	0.2	0.3	0.4	0.4	1.1	1.6	1.9	1.1	0.9	0.3	0.4	0.5	1.5	2.0	1.4	-0.6	-0.5	1.1	2.1	5.1	10.6	15.3	17.0	14.1	8.2	1.3	0.59	3.91	2.25
200	-1.3	-1.1	-0.6	-0.3	-0.2	-0.1	0.2	0.3	0.4	0.4	0.1	0.5	0.4	0.2	0.3	0.1	0.1	-0.3	-0.5	-0.3	-0.6	-0.5	3.0	2.3	2.9	6.3	10.6	12.9	10.8	6.3	2.1	-0.6	-0.03	3.00	1.49
300	-1.1	-0.9	-0.6	-0.4	-0.2	0.1	0.2	0.1	-0.1	-0.1	-0.1	0.2	0.2	0.1	-0.1	-0.3	-0.7	-0.8	-0.3	1.4	2.3	1.1	2.5	5.1	8.2	9.0	6.7	3.1	0.8	-0.2	-0.11	1.98	0.93
500	-1.1	-0.4	-0.3	-0.4	-0.4	-0.4	-0.4	-0.3	-0.2	-0.2	-0.2	-0.0	-0.1	-0.1	-0.1	-0.1	-0.2	-0.4	-0.0	0.4	0.6	1.1	3.4	6.0	8.4	8.6	6.0	3.2	1.5	0.3	-0.21	2.01	0.90
700	-2.3	-0.9	-0.4	-0.4	-0.6	-0.4	-0.1	-0.2	-0.5	-0.2	0.0	0.4	0.0	0.0	-0.0	-0.1	-0.1	-0.1	-0.3	0.1	0.3	2.2	4.9	8.3	11.0	10.8	8.2	4.7	1.5	0.5	-0.31	2.78	1.24
850	-0.2	0.1	0.1	0.1	0.2	0.2	0.3	-0.1	-0.4	-0.0	-0.0	0.1	0.4	-0.0	0.2	0.2	0.2	-0.1	-0.3	-0.3	-0.1	-0.3	0.9	3.4	6.6	10.3	13.3	12.4	9.8	6.3	1.4	0.04	3.39	1.72
1000	0.3	0.2	-0.3	-0.3	0.4	0.7	-0.1	-0.1	-1.3	-2.6	-2.2	-2.8	-2.4	-2.0	-0.7	0.2	0.8	1.1	1.9	2.3	3.1	3.9	6.1	8.0	9.4	10.3	12.1	7.3	1.5	-1.28	2.64	0.42
MEAN	-0.9	-0.5	-0.2	-0.2	-0.2	-0.1	-0.0	-0.0	-0.1	-0.2	0.0	0.2	0.0	-0.0	-0.2	-0.1	-0.1	-0.1	-0.1	0.1	0.5	1.1	1.9	3.8	7.0	10.3	11.1	9.7	6.7	3.0	0.9	-0.11	2.84	1.36

JJA 63–73

P (MB)	80S	70S	65S	60S	55S	50S	45S	40S	35S	30S	25S	20S	15S	10S	5S	EQ	5N	10N	15N	20N	25N	30N	35N	40N	45N	50N	55N	60N	65N	70N	80N	SH	NH	GLOBE
50	-1.1	-3.3	-3.6	-3.0	-2.2	-1.5	-1.3	-1.1	-0.6	-0.3	-0.2	0.0	0.0	-0.0	0.1	0.1	-0.2	-0.1	0.4	0.8	0.7	0.4	0.4	0.4	0.3	0.2	0.1	0.0	-0.0	0.0	0.1	-0.87	0.23	-0.32
100	-0.9	-1.0	-0.8	-0.7	-0.5	-0.2	-0.1	-0.0	0.0	-0.2	-0.9	-0.9	-0.6	-0.4	-0.1	0.2	0.4	1.1	1.5	1.5	1.0	1.1	1.2	1.1	0.6	0.3	0.2	0.1	0.1	0.1	-0.44	0.79	0.17
200	-1.0	-1.9	-1.5	-1.0	-0.7	-0.3	0.0	0.3	0.2	-0.4	-0.7	-0.7	-0.3	0.2	0.1	-0.0	-0.2	-0.8	-0.5	0.7	2.2	1.9	1.7	1.1	0.2	-0.0	0.1	0.4	0.1	0.1	-0.38	0.50	0.06
300	-0.4	-0.3	0.1	0.3	0.6	0.6	0.6	0.3	-0.1	-0.1	0.2	0.2	0.2	0.4	0.3	0.2	0.0	-0.4	-0.7	-0.6	-1.2	-0.9	-0.6	-0.4	-0.1	-0.4	-0.1	0.4	0.5	0.4	0.22	-0.38	-0.08
500	-1.5	-0.5	-0.1	-0.1	-0.3	-0.6	-0.6	-0.8	-1.4	-2.0	-1.7	-1.1	-0.5	-0.4	-0.1	0.1	0.3	0.2	0.0	-0.1	0.0	0.3	0.0	0.1	-0.5	-0.7	-0.8	-0.3	0.1	0.4	-0.37	-0.08	-0.23
700	-3.2	-1.7	-1.0	-0.5	-0.4	-0.5	-0.6	-0.8	-1.4	-2.9	-1.7	-1.5	-1.3	-1.6	-1.7	-1.8	-0.6	-1.3	-1.6	-1.5	-0.9	0.0	0.0	-0.1	-0.5	-0.8	-0.7	-0.3	0.1	0.5	-0.92	-0.57	-0.75
850	-1.8	-1.2	-0.5	-0.2	-0.1	-0.2	-0.4	-1.1	-1.9	-1.8	-2.0	-1.9	-1.6	-1.3	-1.6	-1.7	-1.8	-2.0	-2.2	-4.3	-3.2	-0.9	-1.3	-1.6	-0.5	0.2	1.0	0.2	-1.09	-1.02	-1.05
1000	-0.1	-0.1	-0.1	-0.4	-0.4	-0.2	-0.5	-0.9	-1.8	-2.0	-1.6	-1.3	-1.6	-1.7	-1.8	-1.1	0.0	0.6	2.6	4.9	5.9	5.7	3.5	1.2	-1.5	-0.9	-0.6	0.2	1.0	-1.04	1.57	0.10
MEAN	-1.3	-1.1	-0.7	-0.5	-0.3	-0.3	-0.3	-0.4	-0.7	-0.9	-0.8	-0.6	-0.4	-0.3	-0.1	-0.0	0.0	0.0	-0.2	-0.5	-0.5	-0.1	0.3	-0.1	-0.6	-0.9	-0.7	-0.3	0.2	0.4	-0.51	-0.18	-0.35

TABLE A22. NORTHWARD TRANSPORT OF GEOPOTENTIAL ENERGY BY TRANSIENT EDDIES $[\overline{v'z'}]$ (IN 10^2 gpm m sec^{-1})

$\overline{v'z'}$ YEAR 63-73   UNITS= 1.E 02 M/S GPM

YEAR 63-73

P (MB)	80S	70S	65S	60S	55S	50S	45S	40S	35S	30S	25S	20S	15S	10S	5S	EQ	5N	10N	15N	20N	25N	30N	35N	40N	45N	50N	55N	60N	65N	70N	80N	SH	NH	GLOBE
50	3.6	6.5	6.2	4.5	2.4	0.7	0.0	-0.2	-0.1	-0.0	0.0	0.1	0.0	0.0	-0.1	-0.0	-0.1	-0.0	0.0	0.1	0.0	-0.1	-0.3	-0.4	-0.5	-0.4	-0.2	-0.1	-0.4	-0.8	2.7	0.93	-0.08	0.43
100	-1.2	0.6	1.4	1.6	1.1	0.5	0.2	0.1	0.1	0.0	0.0	0.1	0.1	0.0	-0.0	0.0	-0.1	-0.2	-0.4	-0.5	-0.4	-0.5	-0.5	-0.1	-0.1	-0.1	0.2	0.0	-0.4	-0.4	0.4	0.23	-0.13	0.05
200	-0.2	-0.1	0.1	0.5	0.8	0.9	0.9	0.7	0.6	0.4	0.2	0.5	0.5	0.4	0.1	0.0	0.0	-0.1	-0.5	-0.8	-1.0	-0.8	0.0	0.3	-0.3	-0.8	-0.6	-0.4	-0.3	-0.1	0.1	0.43	-0.35	0.04
300	-1.0	-0.8	-0.7	-0.5	-0.4	-0.4	-0.4	-0.2	-0.0	0.2	0.1	0.2	0.1	0.1	0.0	-0.0	-0.1	-0.2	-0.4	-0.6	-0.2	0.3	0.5	0.1	-0.4	-0.5	-0.5	-0.4	-0.1	-0.4	0.1	-0.12	-0.17	-0.14
500	-0.4	-0.2	-0.1	-0.0	-0.1	-0.2	-0.2	-0.2	-0.1	0.0	0.0	0.0	0.0	-0.0	-0.0	0.0	-0.0	-0.1	-0.1	-0.1	-0.1	0.3	0.1	0.1	-0.0	-0.1	-0.1	-0.0	0.0	-0.1	-0.3	-0.07	-0.03	-0.05
700	0.1	0.1	0.0	-0.0	-0.0	-0.1	-0.2	-0.2	-0.2	-0.1	-0.0	0.0	0.0	0.0	0.0	0.0	-0.0	-0.0	-0.1	-0.1	-0.1	0.0	0.0	0.0	-0.0	-0.1	-0.1	0.0	0.0	-0.2	-0.0	-0.04	-0.02	-0.03
850	*****	0.1	0.2	0.2	0.2	0.1	0.1	0.1	0.1	0.0	0.1	0.1	0.1	0.1	0.0	0.0	0.0	0.1	0.1	0.1	-0.1	0.0	0.0	-0.1	-0.1	-0.1	-0.1	-0.0	-0.1	-0.1	-0.1	0.08	-0.07	0.00
1000	*****	-0.3	-0.3	-0.2	-0.1	0.1	0.2	0.2	0.2	0.2	0.2	0.2	0.1	0.1	0.0	0.1	0.0	-0.1	-0.2	-0.3	-0.3	-0.4	-0.3	-0.3	-0.2	-0.1	0.0	0.0	-0.0	-0.0	0.0	0.08	-0.18	-0.03
MEAN	-0.2	0.2	0.3	0.3	0.2	0.0	-0.0	-0.0	0.0	0.0	0.1	0.1	0.1	0.1	0.1	0.0	0.0	-0.0	-0.1	-0.1	-0.1	-0.2	-0.0	-0.0	-0.2	-0.2	-0.1	-0.2	-0.1	-0.2	0.0	0.08	-0.10	-0.01

DJF 63-73

P (MB)	80S	70S	65S	60S	55S	50S	45S	40S	35S	30S	25S	20S	15S	10S	5S	EQ	5N	10N	15N	20N	25N	30N	35N	40N	45N	50N	55N	60N	65N	70N	80N	SH	NH	GLOBE
50	0.4	0.1	0.2	0.2	0.2	0.2	0.1	-0.0	-0.0	-0.1	-0.1	-0.1	-0.0	0.0	0.0	-0.0	-0.0	0.0	0.0	-0.0	-0.1	-0.1	-0.2	-0.3	-0.4	-0.4	-0.4	-0.6	-0.9	-1.1	-1.3	0.02	-0.26	-0.12
100	-0.1	-0.0	0.3	0.3	0.4	0.4	0.3	0.1	0.0	-0.0	-0.0	-0.1	-0.0	0.0	0.0	0.0	0.1	0.0	0.1	0.1	-0.1	-0.3	-0.4	-0.5	-0.4	-0.4	-0.4	-0.5	-0.4	-0.2	-0.6	0.09	-0.26	-0.09
200	0.0	-0.3	-0.1	0.5	1.1	1.5	1.2	0.8	0.5	0.4	0.3	0.1	0.0	-0.1	-0.1	0.0	-0.1	-0.2	-0.3	-0.5	-0.8	-1.1	-1.2	-1.0	-0.8	-0.7	-0.7	-0.3	-0.1	-0.1	-0.1	0.51	-0.60	-0.05
300	-0.8	-0.3	0.1	0.6	0.8	0.7	0.5	0.3	0.2	0.1	0.1	0.0	-0.1	-0.1	-0.0	-0.1	-0.1	-0.1	-0.2	-0.3	-0.4	-0.5	-0.5	-0.4	-0.4	-0.5	-0.6	-0.6	-0.2	-0.0	0.1	0.16	-0.34	-0.09
500	-0.3	0.0	0.4	0.7	0.8	0.7	0.4	0.1	-0.0	-0.0	0.0	0.0	0.0	-0.0	-0.0	-0.0	-0.0	-0.1	-0.1	-0.1	-0.0	-0.1	-0.0	0.0	0.0	-0.1	-0.2	-0.1	0.0	-0.0	-0.2	0.14	-0.08	0.03
700	-0.0	0.1	0.3	0.4	0.4	0.3	0.1	-0.0	-0.1	-0.1	-0.0	-0.1	-0.0	0.0	0.0	0.0	0.0	0.0	-0.0	-0.1	-0.0	-0.0	-0.1	-0.0	-0.1	-0.1	-0.1	-0.0	-0.1	-0.2	-0.2	0.00	-0.04	-0.02
850	*****	0.1	0.3	0.4	0.4	0.3	0.1	0.1	0.2	0.2	0.2	0.0	0.0	0.0	0.0	0.0	0.0	0.0	0.0	-0.0	-0.1	-0.1	-0.1	-0.2	-0.2	-0.2	-0.1	-0.1	-0.1	-0.1	0.07	-0.10	-0.01
1000	*****	0.2	-0.0	-0.0	0.1	0.1	0.2	0.2	0.2	0.2	0.1	0.1	0.0	0.0	0.0	-0.0	-0.0	-0.0	-0.0	-0.1	-0.1	-0.2	-0.3	-0.4	-0.4	-0.4	-0.2	-0.1	-0.1	-0.1	-0.1	0.08	-0.19	-0.04
MEAN	-0.2	-0.1	0.2	0.4	0.5	0.5	0.3	0.2	0.1	0.0	0.0	-0.0	-0.0	-0.0	-0.0	-0.0	-0.0	-0.0	-0.1	-0.1	-0.2	-0.3	-0.3	-0.3	-0.3	-0.3	-0.3	-0.2	-0.1	-0.1	-0.2	0.12	-0.19	-0.03

JJA 63-73

P (MB)	80S	70S	65S	60S	55S	50S	45S	40S	35S	30S	25S	20S	15S	10S	5S	EQ	5N	10N	15N	20N	25N	30N	35N	40N	45N	50N	55N	60N	65N	70N	80N	SH	NH	GLOBE
50	-2.5	-2.5	-2.2	-2.0	-1.6	-1.2	-0.8	-0.4	-0.2	-0.0	0.0	0.0	0.0	0.0	0.0	-0.0	-0.0	0.0	0.0	0.0	0.0	-0.0	-0.0	0.0	0.0	0.0	0.0	0.0	0.0	0.1	0.2	-0.60	0.00	-0.30
100	0.9	0.1	0.1	0.2	0.3	0.4	0.4	0.3	0.3	0.2	0.1	0.0	0.0	-0.0	-0.0	0.0	0.0	0.0	-0.0	-0.0	-0.0	-0.0	-0.0	-0.1	-0.1	-0.1	-0.0	-0.0	-0.0	-0.0	0.0	0.17	-0.05	0.06
200	0.9	0.3	0.9	1.3	1.4	1.3	1.3	1.5	1.5	1.0	0.6	0.3	0.2	0.0	0.0	0.1	0.0	0.0	0.0	-0.0	-0.0	-0.0	-0.2	-0.3	-0.5	-0.8	-0.8	-0.4	-0.2	-0.0	0.2	0.85	-0.21	0.32
300	-0.1	-1.0	-0.8	-0.3	0.1	0.4	0.6	0.6	0.7	0.5	0.3	0.4	0.2	0.1	0.0	0.0	0.0	0.0	-0.0	-0.0	-0.0	-0.0	-0.1	-0.1	-0.4	-0.5	-0.4	-0.2	-0.2	0.2	0.2	0.13	-0.12	0.00
500	-0.4	-0.5	-0.2	0.0	0.1	0.1	0.1	0.0	0.0	0.1	0.1	0.0	0.1	0.1	0.0	0.0	0.0	0.0	0.0	0.0	0.0	0.0	-0.0	-0.1	-0.1	-0.1	-0.1	-0.0	-0.0	0.2	0.2	-0.02	-0.03	-0.03
700	0.2	-0.5	-0.6	-0.7	-0.7	-0.6	-0.5	-0.3	-0.2	0.0	0.0	0.1	0.1	0.0	0.1	0.0	0.0	0.0	0.0	0.0	0.2	0.0	0.0	0.0	0.0	0.0	0.0	0.0	0.0	0.0	0.0	-0.20	-0.01	-0.10
850	*****	-0.1	-0.1	-0.1	-0.0	-0.0	-0.0	-0.0	0.0	0.1	0.1	0.0	0.1	0.0	0.0	0.0	0.0	0.0	-0.0	-0.1	-0.1	-0.1	-0.1	-0.1	-0.1	-0.1	-0.0	-0.1	-0.1	0.0	0.1	0.01	-0.03	-0.01
1000	*****	0.1	-0.1	-0.1	-0.3	-0.2	-0.1	-0.1	0.1	0.1	0.2	0.1	0.1	0.0	0.0	0.0	0.0	0.0	-0.0	-0.0	-0.0	-0.1	-0.1	-0.1	-0.1	-0.1	-0.0	-0.0	0.0	0.0	0.1	0.07	-0.05	0.02
MEAN	-0.1	-0.4	-0.3	-0.2	-0.1	0.0	0.1	0.1	0.2	0.3	0.2	0.1	0.1	0.0	0.1	-0.0	-0.0	-0.0	-0.1	-0.1	-0.1	-0.1	-0.1	-0.2	-0.2	-0.2	-0.1	-0.0	-0.0	0.1	0.05	-0.06	-0.00

TABLE A23. NORTHWARD TRANSPORT OF GEOPOTENTIAL ENERGY BY STATIONARY EDDIES [$\bar{v}^* \bar{z}^*$] (IN 10^2 gpm m sec^{-1})

v*z* YEAR 63-73 UNITS= 1.E 02 M/S GPM

YEAR 63-73

P (MB)	80S	70S	65S	60S	55S	50S	45S	40S	35S	30S	25S	20S	15S	10S	5S	EQ	5N	10N	15N	20N	25N	30N	35N	40N	45N	50N	55N	60N	65N	70N	80N	SH	NH	GLOBE
50	0.0	-0.6	-0.8	-0.7	-0.4	-0.2	-0.1	-0.1	-0.1	0.0	0.0	0.1	0.1	0.1	0.0	-0.0	0.0	0.1	0.1	0.2	0.3	0.3	0.4	0.5	0.5	0.2	0.4	0.5	0.3	0.1	0.4	-0.18	0.15	-0.01
100	0.1	-0.0	-0.1	-0.2	-0.1	-0.0	-0.0	0.0	0.1	0.1	0.2	0.3	0.3	0.2	0.1	0.0	0.0	0.0	0.0	0.0	0.1	0.2	0.2	0.2	0.2	0.2	0.2	0.2	0.2	0.1	0.1	0.02	0.10	0.06
200	0.0	0.1	-0.0	-0.0	-0.2	-0.2	-0.2	-0.1	-0.1	0.0	0.1	0.2	0.3	0.2	0.1	0.0	0.0	0.0	0.0	0.0	0.0	0.1	0.1	0.1	0.1	0.0	0.0	0.0	0.0	0.0	0.0	0.07	-0.01	0.03
300	0.1	0.1	-0.0	-0.0	-0.2	-0.2	-0.1	-0.1	-0.0	0.0	0.0	0.0	0.0	0.0	0.0	0.0	-0.1	-0.1	-0.1	-0.1	-0.0	0.0	0.1	0.1	0.1	0.1	0.1	0.1	-0.0	0.0	0.0	-0.00	-0.03	-0.01
500	-0.3	-0.1	-0.1	-0.1	-0.1	-0.1	-0.1	-0.1	-0.1	0.0	0.0	0.0	0.0	0.0	0.0	0.0	0.0	-0.0	-0.0	0.0	0.0	0.0	0.0	0.0	0.1	0.1	0.1	0.1	-0.0	-0.0	0.0	-0.06	0.05	-0.00
700	-0.2	-0.1	-0.1	-0.1	-0.1	-0.1	-0.1	-0.1	-0.1	-0.1	-0.1	0.0	0.0	0.0	0.0	0.0	0.0	0.0	0.1	0.1	0.1	0.0	0.0	0.1	0.1	0.1	0.1	0.0	0.0	-0.0	0.0	-0.05	0.05	-0.00
850	*****	-0.0	-0.0	-0.0	-0.0	-0.0	-0.0	-0.0	0.0	0.0	0.0	0.1	0.1	0.0	0.0	0.0	0.0	0.0	0.1	0.1	0.0	0.0	0.0	0.0	0.0	0.1	0.1	0.0	0.0	-0.0	-0.0	-0.01	0.04	0.02
1000	*****	-0.0	0.0	0.0	0.0	-0.0	-0.0	0.0	0.0	0.0	-0.0	-0.1	-0.1	-0.1	0.0	0.1	0.0	0.0	0.0	-0.0	-0.0	-0.0	-0.0	-0.0	0.0	0.0	0.0	0.0	-0.0	-0.0	-0.0	0.04	-0.02	0.01
MEAN	-0.1	-0.1	-0.1	-0.1	-0.1	-0.1	-0.1	-0.1	0.0	0.0	0.0	0.1	0.0	0.0	0.0	0.0	0.0	0.0	0.0	0.1	0.1	0.0	-0.0	0.0	0.0	0.1	0.1	0.1	0.0	0.0	0.0	-0.02	0.03	0.01

DJF 63-73

P (MB)	80S	70S	65S	60S	55S	50S	45S	40S	35S	30S	25S	20S	15S	10S	5S	EQ	5N	10N	15N	20N	25N	30N	35N	40N	45N	50N	55N	60N	65N	70N	80N	SH	NH	GLOBE
50	0.1	-0.0	-0.1	-0.1	-0.1	-0.1	-0.0	-0.0	0.0	-0.0	-0.0	0.0	0.1	0.1	0.0	0.0	0.1	0.1	0.1	0.3	0.3	0.2	0.4	0.1	0.2	0.4	0.9	1.3	1.0	0.0	1.0	-0.04	0.22	0.09
100	0.0	-0.0	-0.1	-0.1	-0.1	-0.2	-0.2	0.2	0.2	0.2	0.2	0.3	0.2	0.1	0.1	0.1	0.0	0.0	0.0	0.0	0.3	0.2	0.4	0.3	0.2	0.2	0.7	0.7	0.5	0.4	0.4	0.12	0.12	0.12
200	-0.0	-0.1	-0.1	-0.2	-0.2	-0.1	-0.0	-0.0	0.1	0.1	0.1	0.0	0.2	0.0	0.0	0.1	-0.1	-0.1	-0.1	-0.0	0.2	-0.4	-1.3	-0.9	-0.6	-0.4	-0.0	-0.1	-0.2	0.0	0.1	0.00	-0.28	-0.14
300	0.1	-0.1	-0.2	-0.2	-0.2	-0.2	-0.1	-0.0	-0.0	-0.0	-0.1	-0.1	-0.0	-0.0	-0.0	0.0	0.0	-0.0	-0.1	-0.0	-0.1	-0.2	-0.6	-0.3	-0.2	-0.3	-0.2	-0.1	-0.2	-0.3	0.1	-0.05	-0.17	-0.11
500	-0.3	-0.1	-0.1	-0.2	-0.2	-0.2	-0.2	-0.1	-0.1	-0.1	-0.1	0.0	0.0	-0.0	-0.0	-0.0	0.0	-0.0	-0.0	0.0	0.0	0.1	-0.0	0.3	0.3	0.2	-0.1	0.2	0.1	-0.1	0.0	-0.06	0.07	0.00
700	-0.1	-0.1	-0.1	-0.1	-0.1	-0.1	-0.1	-0.1	-0.0	-0.0	0.0	0.0	0.0	0.0	0.0	0.0	-0.0	0.0	0.1	0.0	0.0	0.1	0.1	0.2	0.3	0.3	0.1	0.1	0.0	-0.0	0.1	-0.05	0.07	0.01
850	0.0	0.0	0.0	-0.0	-0.0	-0.0	-0.0	0.0	0.1	0.1	0.0	0.0	0.0	-0.0	-0.0	-0.0	-0.0	0.0	0.1	0.0	0.0	-0.0	0.0	0.0	0.2	0.2	0.2	0.1	0.0	-0.0	-0.0	-0.01	0.05	0.02
1000	0.0	0.1	0.1	0.0	0.0	0.0	-0.0	-0.0	0.1	0.2	0.3	0.2	0.0	-0.0	0.2	0.1	-0.1	-0.1	-0.0	-0.0	-0.0	-0.1	-0.1	0.1	0.1	0.1	0.1	0.2	0.1	-0.0	-0.0	0.08	0.01	0.05
MEAN	-0.1	-0.1	-0.1	-0.1	-0.1	-0.1	-0.1	-0.0	-0.0	0.0	0.0	0.1	0.0	0.0	0.0	0.0	-0.0	-0.0	-0.0	-0.0	-0.0	-0.1	-0.3	-0.1	0.0	0.0	0.1	0.1	0.0	0.1	0.1	-0.01	-0.00	-0.01

JJA 63-73

P (MB)	80S	70S	65S	60S	55S	50S	45S	40S	35S	30S	25S	20S	15S	10S	5S	EQ	5N	10N	15N	20N	25N	30N	35N	40N	45N	50N	55N	60N	65N	70N	80N	SH	NH	GLOBE
50	-0.1	-0.6	-0.8	-0.6	-0.4	-0.3	-0.2	-0.2	-0.1	-0.1	-0.1	-0.0	0.0	0.0	-0.0	0.0	0.0	0.1	0.1	0.3	0.4	0.4	0.1	0.0	0.1	-0.1	-0.1	-0.1	-0.0	0.0	0.0	-0.23	0.04	-0.10
100	0.4	0.5	0.3	0.1	-0.0	-0.0	-0.0	-0.1	-0.1	-0.2	-0.2	-0.0	0.1	0.2	0.2	0.0	0.2	0.2	0.2	0.3	0.3	0.1	-0.2	-0.4	-0.2	-0.1	-0.1	-0.0	0.0	0.0	0.0	-0.02	-0.02	-0.02
200	-0.0	0.2	0.2	0.2	-0.0	-0.2	-0.2	-0.1	-0.1	-0.0	0.0	0.1	0.0	0.3	0.2	0.2	0.1	0.0	0.0	0.2	0.1	0.1	-0.5	-0.4	-0.2	-0.0	-0.0	-0.0	0.0	0.0	0.0	0.06	-0.04	0.01
300	-0.1	0.1	0.1	0.1	-0.1	-0.2	-0.2	-0.1	-0.1	-0.1	-0.2	-0.1	0.0	0.1	0.1	0.1	0.1	-0.0	-0.0	0.0	0.0	0.1	-0.2	-0.1	0.0	0.0	-0.0	0.1	0.0	0.0	0.0	-0.02	-0.02	-0.02
500	-0.3	0.0	0.0	-0.1	-0.1	-0.1	-0.1	-0.1	-0.2	-0.3	-0.1	-0.0	0.0	0.0	0.1	0.0	-0.0	-0.0	-0.0	0.0	0.0	0.1	0.0	0.0	0.0	0.0	-0.0	0.1	0.0	-0.0	-0.0	-0.08	0.02	-0.03
700	-0.2	-0.2	-0.1	-0.1	-0.1	-0.1	-0.1	-0.1	-0.1	-0.1	-0.0	-0.0	0.0	0.0	0.0	0.0	-0.0	-0.0	0.0	0.0	0.0	0.1	-0.2	-0.1	0.0	0.0	0.0	0.0	-0.0	-0.0	-0.0	-0.07	0.03	-0.02
850	*****	-0.1	-0.1	-0.1	-0.1	-0.0	-0.0	-0.1	-0.1	-0.1	-0.1	-0.1	-0.0	-0.0	0.0	0.2	0.2	0.2	-0.0	-0.8	-0.3	-0.3	-0.5	-0.3	-0.2	-0.1	0.0	0.2	0.0	0.0	0.0	-0.01	0.04	0.01
1000	*****	-0.0	-0.0	-0.0	-0.0	-0.0	-0.1	-0.0	-0.0	0.0	0.1	0.0	-0.0	0.1	0.0	0.1	0.0	-0.0	-0.0	0.0	0.0	0.0	0.0	-0.1	-0.0	-0.0	-0.0	-0.0	-0.0	-0.0	-0.0	0.03	-0.22	-0.08
MEAN	-0.1	-0.0	-0.0	-0.0	-0.1	-0.1	-0.1	-0.1	-0.1	-0.1	-0.1	-0.0	0.0	0.0	0.1	0.1	0.0	0.0	0.0	0.0	0.0	0.1	0.0	-0.0	-0.0	0.0	0.0	0.0	0.0	0.0	0.0	-0.05	-0.00	-0.02

471

TABLE A24. NORTHWARD TRANSPORT OF HUMIDITY BY TRANSIENT EDDIES $[\overline{v'q'}]$ (IN g kg⁻¹ m sec⁻¹)

$v'q'$ YEAR 63-73   UNITS= 1.E 00 M/S G/KG

P (MB)	80S	70S	65S	60S	55S	50S	45S	40S	35S	30S	25S	20S	15S	10S	5S	EQ	5N	10N	15N	20N	25N	30N	35N	40N	45N	50N	55N	60N	65N	70N	80N	SH	NH	GLOBE
300	-0.1	-0.1	-0.1	-0.1	-0.2	-0.2	-0.2	-0.1	-0.1	-0.0	-0.0	0.0	0.0	0.0	0.0	-0.0	0.0	0.1	0.1	0.1	0.1	0.1	0.1	0.1	0.2	0.1	0.1	0.1	0.1	0.1	0.0	-0.10	0.08	-0.01
500	-0.2	-0.3	-0.4	-0.5	-0.6	-0.8	-1.0	-1.2	-1.2	-1.1	-1.0	-0.8	-0.6	-0.4	-0.1	0.1	0.3	0.4	0.5	0.5	0.7	0.9	1.1	1.2	1.1	1.0	0.9	0.7	0.5	0.4	0.1	-0.67	0.67	-0.00
700	-0.3	-0.7	-1.0	-1.5	-2.0	-2.4	-2.7	-2.9	-2.9	-2.6	-2.0	-1.5	-1.2	-0.8	-0.4	0.1	0.5	0.7	0.8	1.1	1.6	2.2	2.6	2.7	2.6	2.3	1.9	1.4	1.1	0.4		-1.65	1.53	-0.05
850	*****	-1.0	-1.6	-2.3	-2.8	-3.1	-3.0	-2.9	-2.7	-2.4	-1.9	-1.6	-1.4	-1.1	-0.6	-0.0	0.6	1.3	1.5	1.6	2.3	2.9	3.5	3.6	3.2	2.9	2.5	2.1	1.7	1.3	0.6	-1.88	1.97	0.05
900	*****	-0.9	-1.5	-2.3	-3.0	-3.3	-3.0	-3.0	-2.6	-2.3	-1.8	-1.3	-0.9	-1.1	-0.6	0.8	1.3	1.8	2.4	3.0	3.4	3.6	3.4	3.1	2.7	2.1	1.7	1.3	0.6	0.3		-1.95	1.81	-0.08
950	*****	-0.6	-1.2	-1.8	-2.4	-2.8	-2.9	-2.9	-2.6	-2.3	-2.0	-1.8	-1.5	-1.1	-0.6	-0.2	0.4	1.0	1.5	1.8	2.4	3.1	3.7	3.7	3.9	3.2	2.6	1.8	1.5	1.0	0.4	-1.84	1.97	0.02
1000	*****	-0.3	-0.7	-1.0	-1.3	-1.7	-2.2	-2.6	-2.7	-2.4	-1.9	-1.5	-1.3	-0.9	-0.5	-0.4	-0.0	0.4	1.0	1.3	2.1	3.0	4.0	5.3	5.1	3.7	2.9	2.2	1.4	0.8	0.4	-1.52	2.37	0.17
MEAN	-0.1	-0.3	-0.6	-0.9	-1.1	-1.3	-1.4	-1.5	-1.4	-1.3	-1.0	-0.8	-0.7	-0.5	-0.2	0.0	0.2	0.5	0.6	0.8	1.0	1.3	1.5	1.5	1.4	1.3	1.1	0.9	0.7	0.5	0.2	-0.86	0.86	-0.01

DJF 63-73

P (MB)	80S	70S	65S	60S	55S	50S	45S	40S	35S	30S	25S	20S	15S	10S	5S	EQ	5N	10N	15N	20N	25N	30N	35N	40N	45N	50N	55N	60N	65N	70N	80N	SH	NH	GLOBE
300	-0.0	-0.0	-0.0	-0.0	-0.1	-0.1	-0.1	-0.2	-0.2	-0.2	-0.1	-0.1	-0.0	-0.0	0.0	0.1	0.1	0.1	0.1	0.1	0.1	0.1	0.1	0.1	0.1	0.1	0.1	0.1	0.1	0.0	0.0	-0.09	0.09	-0.01
500	-0.2	-0.2	-0.3	-0.4	-0.7	-1.0	-1.2	-1.4	-1.4	-1.3	-1.1	-0.9	-0.7	-0.4	-0.1	0.1	0.4	0.5	0.6	0.6	0.8	1.0	1.0	0.9	0.8	0.6	0.5	0.4	0.3	0.2	0.1	-0.75	0.60	-0.08
700	-0.3	-0.3	-0.6	-1.1	-1.7	-2.3	-2.6	-2.8	-2.7	-2.4	-1.9	-1.4	-1.0	-0.5	-0.0	0.3	0.7	0.9	1.1	1.5	2.0	2.8	3.0	2.6	2.3	1.9	1.6	1.2	0.9	0.7	0.3	-1.44	1.57	0.07
850	*****	-0.8	-1.3	-2.0	-2.6	-3.0	-3.0	-3.0	-2.7	-2.3	-1.8	-1.3	-0.9	-0.4	0.1	0.4	0.5	0.8	1.4	2.6	3.4	3.8	3.8	3.1	2.4	2.2	1.6	1.3	1.2	1.0	0.3	-1.65	1.82	0.09
900	*****	-0.9	-1.6	-2.5	-3.1	-3.5	-3.4	-3.2	-2.5	-2.0	-1.7	-1.4	-0.9	-0.4	-0.1	0.4	0.8	1.3	1.8	2.5	3.4	3.7	3.6	3.0	2.2	2.4	1.6	1.2	0.9	0.3	-1.75	1.65	-0.05
950	*****	-1.0	-1.6	-2.2	-2.7	-3.0	-3.1	-2.9	-2.5	-2.0	-1.4	-0.9	-0.5	-0.2	0.0	0.4	0.5	1.0	1.9	3.1	4.0	4.1	4.1	3.2	2.4	2.4	1.6	1.2	0.9	0.5	0.2	-1.75	1.85	0.10
1000	*****	-0.2	-0.4	-0.6	-1.0	-1.4	-2.1	-2.7	-2.6	-1.8	-0.9	-0.6	-0.3	-0.1	-0.1	0.1	0.3	0.7	1.7	3.5	5.3	4.9	4.6	3.4	2.7	1.7	1.1	0.8	0.5	0.1	-1.07	1.92	0.23
MEAN	-0.1	-0.2	-0.4	-0.6	-1.0	-1.3	-1.4	-1.5	-1.4	-1.2	-1.0	-0.7	-0.5	-0.3	-0.1	0.1	0.3	0.5	0.8	1.2	1.5	1.6	1.5	1.3	1.0	1.0	0.8	0.6	0.5	0.3	0.2	-0.78	0.81	0.01

JJA 63-73

P (MB)	80S	70S	65S	60S	55S	50S	45S	40S	35S	30S	25S	20S	15S	10S	5S	EQ	5N	10N	15N	20N	25N	30N	35N	40N	45N	50N	55N	60N	65N	70N	80N	SH	NH	GLOBE
300	-0.1	-0.1	-0.1	-0.1	-0.1	-0.1	-0.1	-0.1	-0.1	-0.1	-0.0	-0.0	0.0	0.0	-0.0	-0.0	0.0	0.1	0.1	0.1	0.1	0.1	0.2	0.2	0.2	0.2	0.1	0.1	0.0	0.0	-0.09	0.08	-0.01
500	-0.1	-0.3	-0.3	-0.4	-0.6	-0.8	-1.0	-1.1	-1.1	-1.0	-0.9	-0.6	-0.5	-0.5	-0.3	0.0	0.2	0.4	0.4	0.4	0.6	0.9	0.8	1.0	1.2	1.3	1.2	1.0	0.8	0.6	0.2	-0.66	0.60	-0.03
700	-0.4	-0.9	-1.3	-1.8	-2.2	-2.5	-2.8	-3.0	-3.0	-2.6	-2.2	-1.6	-1.1	-0.8	-0.3	0.1	0.2	0.5	0.7	0.8	1.3	1.8	2.2	2.6	2.6	2.6	2.2	1.9	1.6	0.6	-1.73	1.12	-0.30
850	*****	-1.0	-1.8	-2.6	-3.2	-3.4	-3.3	-3.1	-2.9	-2.6	-2.2	-1.8	-1.3	-0.8	-0.4	-0.0	0.2	0.4	0.7	1.0	1.7	2.3	2.6	2.8	2.6	2.3	2.2	2.0	1.1	0.6	-1.98	1.34	-0.32
900	*****	-0.8	-1.4	-2.2	-2.8	-3.1	-3.0	-2.8	-2.5	-2.1	-1.6	-1.2	-0.7	-0.3	0.0	0.2	0.3	0.5	0.7	0.9	1.1	1.6	2.1	2.3	2.4	2.3	2.1	1.8	1.6	0.6	-1.84	1.14	-0.35
950	*****	-0.7	-1.1	-1.6	-2.2	-2.6	-2.9	-3.1	-3.0	-2.5	-2.0	-1.3	-0.8	-0.4	-0.1	0.2	0.4	0.6	0.8	1.1	1.1	1.7	2.2	2.1	1.8	1.7	1.3	1.0	0.6	0.6	-1.90	1.03	-0.47
1000	*****	-1.2	-1.5	-1.7	-1.9	-2.0	-2.4	-2.7	-2.8	-2.6	-1.9	-1.2	-0.6	-0.3	-0.0	-0.1	0.1	0.3	0.3	0.5	1.3	2.4	3.1	2.5	1.7	1.1	1.2	1.3	1.0	0.5	-1.55	0.89	-0.49
MEAN	-0.1	-0.4	-0.7	-0.9	-1.2	-1.3	-1.4	-1.5	-1.5	-1.3	-1.1	-0.8	-0.6	-0.4	-0.2	0.0	0.1	0.3	0.4	0.4	0.9	1.2	1.3	1.2	1.1	1.1	0.9	0.8	0.3	0.3	-0.88	0.59	-0.14

TABLE A25. NORTHWARD TRANSPORT OF HUMIDITY BY STATIONARY EDDIES $[\bar{v}^*\bar{q}^*]$ (IN g kg^{-1} m sec^{-1})

v*Q*   YEAR 63-73   UNITS= 1.E 00 M/S G/KG

P (MB)	80S	70S	65S	60S	55S	50S	45S	40S	35S	30S	25S	20S	15S	10S	5S	EQ	5N	10N	15N	20N	25N	30N	35N	40N	45N	50N	55N	60N	65N	70N	80N	SH	NH	GLOBE
300	-0.0	-0.0	-0.0	0.0	0.0	0.0	0.0	0.0	0.0	0.0	0.1	0.1	0.0	0.0	-0.0	-0.0	-0.0	-0.0	-0.0	0.1	0.1	0.1	0.0	0.0	0.0	0.0	0.0	0.0	-0.0	-0.0	0.0	0.00	0.02	0.01
500	-0.0	-0.0	-0.0	0.0	0.0	0.0	0.1	0.1	0.0	0.0	0.2	0.1	0.1	0.0	-0.0	-0.0	-0.0	0.1	0.1	0.2	0.1	0.1	0.0	0.0	0.0	0.0	0.0	0.0	-0.0	-0.0	0.0	-0.00	0.05	0.02
700	-0.2	-0.1	-0.0	0.1	0.1	0.1	0.1	0.1	0.1	0.0	0.4	0.2	0.0	0.1	-0.1	-0.1	-0.0	0.1	0.1	0.2	0.4	0.3	0.1	0.1	0.0	0.1	0.1	0.1	0.0	-0.0	0.0	0.03	0.08	0.05
850	*****	-0.1	-0.0	-0.0	-0.1	-0.2	-0.4	-0.5	-0.4	-0.3	-0.1	-0.1	-0.1	0.1	0.6	1.4	1.5	1.1	1.0	0.6	0.2	0.2	0.3	0.3	0.2	0.1	-0.0	0.1				-0.17	0.46	0.15
900	*****	-0.1	0.0	-0.0	-0.1	-0.2	-0.8	-0.6	-0.6	-0.7	-0.0	-0.0	-0.2	-0.0	0.5	1.1	1.2	1.1	1.0	0.6	0.2	0.2	0.3	0.4	0.3	0.1	0.1	0.1				-0.31	0.48	0.09
950	*****	0.0	0.0	-0.0	-0.1	-0.3	-1.0	-0.8	-1.0	-0.7	-0.0	0.0	-0.2	0.1	0.6	1.3	1.4	1.2	0.7	0.3	0.3	0.4	0.5	0.4	0.1	0.1						-0.39	0.54	0.07
1000	*****	0.0	-0.0	-0.1	-0.1	-0.5	-1.5	-1.8	-1.4	-1.5	-1.2	-0.5	0.5	1.6	2.1	2.1	1.5	0.8	0.3	0.3	0.5	0.8	0.9	0.3	0.1	0.0	0.0					-0.65	0.75	-0.16
MEAN	-0.0	-0.0	-0.0	0.0	0.0	0.0	-0.0	-0.1	-0.1	-0.1	-0.2	-0.1	-0.1	-0.0	0.2	0.4	0.5	0.3	0.3	0.1	0.1	0.1	0.1	0.1	0.1	0.0	0.0	0.0	0.0	-0.0	0.0	-0.07	0.14	0.03

DJF 63-73

P (MB)	80S	70S	65S	60S	55S	50S	45S	40S	35S	30S	25S	20S	15S	10S	5S	EQ	5N	10N	15N	20N	25N	30N	35N	40N	45N	50N	55N	60N	65N	70N	80N	SH	NH	GLOBE
300	0.0	-0.0	0.0	0.0	0.0	0.0	0.0	0.0	0.0	0.0	0.1	0.1	0.1	0.0	-0.0	-0.0	-0.0	0.0	0.1	0.1	0.2	0.1	0.0	0.0	0.0	0.0	0.2	0.0	0.0	0.0	-0.0	0.01	0.03	0.02
500	-0.0	0.0	0.0	0.0	0.0	0.0	0.0	0.0	0.0	0.0	0.2	0.2	0.1	0.1	0.0	-0.0	0.0	0.1	0.2	0.2	0.2	0.1	0.1	0.2	0.2	0.2	0.1	0.1	0.0	-0.0	0.0	0.03	0.12	0.07
700	-0.2	-0.1	-0.0	0.0	0.0	0.0	0.0	0.1	0.1	0.0	0.8	0.6	0.3	0.1	-0.0	-0.1	-0.0	0.3	0.5	0.7	0.7	0.3	0.3	0.6	0.7	0.6	0.4	0.2	0.0	-0.0	0.1	-0.04	0.26	0.11
850	*****	-0.1	-0.0	-0.0	-0.2	-0.2	-0.4	-0.8	-0.8	-0.9	-0.6	-0.3	-0.2	0.0	0.5	1.0	0.8	0.7	1.0	1.3	1.5	1.6	1.3	0.9	0.6	0.5	0.3	0.1	0.1	-0.0	0.1	-0.35	0.67	0.17
900	*****	-0.1	0.0	0.1	0.1	-0.0	-0.3	-0.6	-0.8	-1.2	-1.1	-0.9	-0.6	-0.3	-0.1	0.3	0.7	0.8	0.9	1.1	1.1	1.1	1.4	1.8	1.9	1.3	0.9	0.6	0.5	0.1	0.1	-0.51	0.67	0.08
950	*****	0.0	0.1	0.1	0.1	-0.0	-0.2	-0.5	-1.1	-1.4	-1.2	-1.0	-0.8	-0.3	-0.2	0.1	0.9	1.0	0.9	1.0	1.2	1.3	1.4	1.8	1.9	1.2	1.0	0.6	0.6	0.1	0.1	-0.60	0.74	0.05
1000	*****	0.1	0.1	-0.1	0.1	-0.0	-0.3	-0.7	-1.2	-2.7	-2.2	-2.2	-1.8	-1.4	-1.1	-0.0	1.4	1.6	1.4	1.4	1.6	1.7	1.8	1.9	1.9	1.6	1.2	0.4	0.2	0.1	-1.31	0.92	-0.34
MEAN	-0.0	-0.0	-0.0	0.0	0.0	0.0	-0.0	-0.1	-0.1	-0.2	-0.2	-0.2	-0.1	-0.0	-0.0	0.2	0.4	0.5	0.6	0.3	0.3	0.3	0.3	0.4	0.5	0.5	0.3	0.2	0.2	0.0	0.0	-0.13	0.23	0.05

JJA 63-73

P (MB)	80S	70S	65S	60S	55S	50S	45S	40S	35S	30S	25S	20S	15S	10S	5S	EQ	5N	10N	15N	20N	25N	30N	35N	40N	45N	50N	55N	60N	65N	70N	80N	SH	NH	GLOBE
300	-0.0	-0.0	-0.0	-0.0	-0.0	-0.0	-0.0	-0.0	-0.0	-0.0	-0.1	-0.1	-0.0	-0.0	-0.0	-0.0	-0.0	0.0	-0.0	0.1	0.1	-0.1	-0.1	-0.0	0.0	0.0	-0.0	-0.0	0.0	0.1	0.0	-0.00	-0.04	-0.02
500	-0.0	-0.1	-0.0	-0.0	-0.0	-0.0	-0.0	-0.0	-0.0	-0.0	-0.2	-0.2	-0.1	-0.2	0.0	0.0	0.1	-0.0	-0.0	0.1	0.2	-0.0	-0.1	-0.1	-0.0	-0.1	-0.0	-0.0	0.1	0.0	-0.07	-0.01	-0.03
700	-0.1	-0.0	-0.0	0.0	0.0	0.0	0.1	0.1	0.1	0.0	0.0	0.0	-0.0	0.0	0.1	0.1	0.0	-0.0	0.5	1.1	1.0	0.6	0.1	-0.3	-0.4	-0.3	-0.0	0.0	0.1	0.1	0.02	0.21	0.11
850	*****	-0.1	-0.0	-0.1	-0.1	-0.1	-0.0	-0.0	0.0	-0.2	-0.3	-0.2	0.0	0.4	0.8	2.6	4.4	4.4	4.8	3.1	2.4	1.0	0.1	-0.5	-0.5	-0.1	0.1	0.2	0.1	-0.09	1.16	0.54
900	*****	-0.0	-0.0	-0.1	-0.1	-0.2	-0.1	-0.2	-0.1	-0.4	-0.7	-0.5	-0.3	0.2	0.9	3.1	4.8	4.1	2.3	1.2	0.2	-0.3	-0.3	-0.2	0.0	0.1	0.0			-0.31	1.23	0.46
950	*****	-0.0	-0.0	-0.1	-0.1	-0.2	-0.2	-0.2	-0.5	-0.8	-0.8	-0.6	-0.3	0.5	1.1	2.9	4.1	4.0	2.4	1.3	0.2	-0.4	-0.4	-0.3	0.0	0.1				-0.32	1.18	0.41
1000	*****	-0.0	-0.1	-0.1	-0.2	-0.1	-0.3	-0.6	-1.4	-1.1	-1.4	-1.1	-0.8	0.9	3.8	5.9	6.3	5.4	3.5	1.6	0.0	-0.3	0.0	0.3	0.4	0.2	0.0			-0.82	2.04	0.42
MEAN	-0.0	-0.0	-0.0	-0.0	-0.0	-0.0	-0.0	-0.0	-0.1	-0.1	-0.2	-0.1	-0.0	0.1	0.3	0.8	1.2	1.1	0.6	0.2	-0.1	-0.2	-0.1	-0.2	-0.1	0.0	0.1	0.0	-0.08	0.31	0.12		

473

TABLE A26. NORTHWARD TRANSPORT OF KINETIC ENERGY BY TRANSIENT EDDIES $[\overline{v'K'}]$ (IN 10^2 m³ sec⁻³)

V·K' YEAR 63-73 UNITS= 1.E 02 M3/S3

YEAR 63-73

P (MB)	80S	70S	65S	60S	55S	50S	45S	40S	35S	30S	25S	20S	15S	10S	5S	EQ	5N	10N	15N	20N	25N	30N	35N	40N	45N	50N	55N	60N	65N	70N	80N	SH	NH	GLOBE
50	-2.3	-5.7	-6.5	-6.0	-4.6	-2.8	-1.4	-0.6	-0.2	-0.0	0.1	0.1	-0.0	-0.0	-0.0	0.1	0.1	-0.0	0.0	0.1	0.3	0.4	0.6	0.8	0.8	0.8	0.7	0.0	-0.7	-3.4		-1.32	0.13	-0.59
100	0.0	-0.8	-1.9	-3.0	-3.2	-2.9	-2.7	-2.5	-2.2	-1.6	-1.1	-0.7	-0.4	-0.3	-0.2	0.1	0.2	-0.0	0.7	1.8	2.3	1.6	1.2	0.9	0.6	0.3	-0.1	-0.1	-0.1	-0.1		-1.37	0.67	-0.35
200	-2.2	1.4	2.8	1.3	-1.5	-5.1	-7.9	-8.5	-7.7	-5.6	-3.6	-1.8	-0.6	-0.4	-0.0	0.6	0.6	2.2	4.5	7.4	8.7	6.5	5.8	4.4	1.8	-0.9	-2.0	-1.7	-1.1	0.1	0.8	-2.84	2.36	-0.24
300	1.9	2.4	1.8	0.4	-1.8	-5.3	-8.3	-9.1	-7.9	-5.1	-2.1	-2.6	-0.8	-0.3	-0.1	0.1	0.3	1.2	2.8	5.0	6.1	1.5	1.4	1.1	0.8	0.6	0.2	-0.2	-0.3	-0.1	0.2	-2.54	1.90	-0.32
500	0.5	0.1	0.2	0.0	-0.3	-0.7	-1.3	-1.8	-2.1	-1.5	-0.8	-0.3	-0.1	0.0	0.0	0.0	0.2	0.6	1.2	1.5	1.4	1.1	0.8	0.6	0.2	-0.3	-0.1	-0.1	0.2			-0.75	0.52	-0.11
700	-0.7	-0.5	-0.5	-0.6	-0.7	-0.8	-0.9	-1.0	-0.9	-0.7	-0.4	-0.2	-0.1	-0.0	-0.0	0.0	0.2	0.4	0.5	0.6	0.6	0.7	0.8	0.8	0.3	0.1	-0.1	-0.1	0.1			-0.54	0.33	-0.10
850	-1.1	-1.4	-1.5	-1.4	-1.2	-1.1	-0.9	-0.7	-0.5	-0.4	-0.6	-0.7	-0.8	-0.7	0.4	0.6	0.6	0.7	0.8	0.8	0.5	0.3	0.2	-0.1	0.2							-0.55	0.26	-0.14
1000	-0.0	-0.4	-0.6	-0.7	-0.6	-0.5	-0.4	-0.2	-0.0	0.0	0.1	-0.1	-0.1	-0.0	-0.1	0.1	0.1	0.1	0.1	0.0	-0.1	-0.2	-0.1	0.2								-0.18	-0.03	-0.12
MEAN	-0.0	0.1	-0.0	-0.0	-0.4	-0.9	-1.6	-2.4	-3.0	-3.1	-2.7	-1.9	-1.0	-0.4	-0.1	-0.0	0.1	0.4	1.0	1.9	2.5	2.2	1.7	1.1	0.4	0.0	-0.3	-0.4	-0.3			-1.21	0.77	-0.22

DJF 63-73

P (MB)	80S	70S	65S	60S	55S	50S	45S	40S	35S	30S	25S	20S	15S	10S	5S	EQ	5N	10N	15N	20N	25N	30N	35N	40N	45N	50N	55N	60N	65N	70N	80N	SH	NH	GLOBE
50	0.1	-0.3	-0.9	-1.5	-1.6	-1.3	-0.8	-0.3	0.0	0.2	0.2	0.2	0.2	0.1	0.4	0.2	0.1	0.1	0.1	0.3	0.6	0.9	1.1	1.3	1.6	1.8	1.6	0.1	-7.8			-0.21	0.42	0.11
100	-0.1	-0.5	-1.2	-2.2	-3.3	-4.1	-4.7	-4.8	-4.0	-2.6	-1.3	-0.5	-0.2	-0.1	-0.1	0.1	1.0	1.2	1.7	1.9	1.8	1.3	1.0	0.9	0.9	0.7	0.3	0.4	0.9			-1.78	1.13	-0.32
200	0.7	1.1	1.3	0.6	-1.5	-5.1	-7.8	-7.3	-5.0	-2.4	0.1	1.2	1.6	1.9	3.7	6.4	8.6	8.7	6.4	2.4	-0.0	-1.3	-1.9	-3.0	-2.2	-2.5	-1.9	0.6				-1.33	2.80	0.73
300	1.3	2.4	3.2	3.4	2.9	1.4	-1.9	-5.0	-5.6	-4.0	-1.7	-0.2	0.4	0.6	1.1	2.5	4.4	6.3	7.7	7.4	3.9	1.1	-0.3	-0.7	-1.9	-3.0	-3.1	2.5				-0.56	2.43	0.94
500	0.3	0.5	0.7	0.5	-0.1	-0.1	-0.4	-0.5	-0.2	-0.1	0.0	0.0	0.3	1.0	2.0	2.4	1.9	1.0	0.5	0.3	0.0	-0.4	-0.7	-0.4								-0.40	0.65	0.13
700	-1.4	-0.4	-0.1	-0.1	-0.4	-0.7	-0.9	-1.1	-1.1	-0.9	-0.7	-0.4	-0.2	-0.2	-0.1	0.0	0.1	0.3	0.6	0.8	0.9	1.0	0.9	1.1	1.0	0.9	0.5	0.3	0.1	-0.1		-0.48	0.46	-0.00
850	-0.2	-0.4	-0.7	-0.9	-0.9	-0.7	-0.5	-0.4	-0.3	-0.2	-0.2	-0.2	-0.1	-0.0	-0.0	0.2	0.6	0.6	0.9	1.0	1.0	0.6	0.5	0.3	-0.1							-0.35	0.34	-0.01
1000	0.4	0.1	-0.1	-0.2	-0.3	-0.2	-0.1	-0.1	0.1	0.1	0.0	-0.0	-0.0	-0.0	-0.1	-0.1	-0.1	-0.2	-0.1	-0.0	-0.1	-0.2	-0.0	0.2								0.00	-0.08	-0.03
MEAN	0.3	0.6	0.6	0.3	-0.3	-1.0	-1.9	-2.5	-2.3	-1.5	-0.7	-0.1	0.1	0.1	0.2	0.4	0.8	1.6	2.5	3.0	2.7	1.5	0.7	0.5	0.1	-0.3	-0.6	-0.7	0.2			-0.60	1.00	0.20

JJA 63-73

P (MB)	80S	70S	65S	60S	55S	50S	45S	40S	35S	30S	25S	20S	15S	10S	5S	EQ	5N	10N	15N	20N	25N	30N	35N	40N	45N	50N	55N	60N	65N	70N	80N	SH	NH	GLOBE
50	-2.0	-7.1	-7.3	-6.0	-4.2	-2.5	-1.5	-1.0	-0.6	-0.3	-0.2	-0.1	-0.1	-0.1	-0.1	-0.1	-0.1	-0.1	-0.1	-0.1	0.2	0.6	1.1	1.3	1.6	1.9	-0.0	-0.0	-0.1			-1.51	-0.06	-0.78
100	-1.1	-0.1	0.3	0.1	-0.4	-0.8	-1.3	-1.7	-1.7	-1.6	-1.2	-0.7	-0.4	-0.4	-0.3	-0.3	-0.3	-0.1	-0.1	-0.1	0.2	0.2	0.1	0.2	0.1	0.1	0.0	0.0	0.0			-1.78	-0.09	-0.44
200	-7.0	-1.3	1.3	3.0	2.6	0.8	-2.1	-4.7	-6.8	-8.5	-7.1	-4.8	-3.0	-2.0	-1.5	-1.1	-0.5	-0.2	0.3	1.6	3.4	4.0	2.1	-0.0	-0.4	-0.2	-0.1	0.0	-0.3			-3.15	0.36	-1.39
300	-3.2	1.9	5.4	7.9	7.1	3.1	-3.5	-9.5	-11.2	-7.7	-4.2	-1.8	-0.5	-0.1	-0.1	-0.4	-0.3	-0.1	0.1	0.4	1.4	2.7	3.7	3.1	1.7	1.1	1.0	0.6	1.9			-2.71	0.97	-0.87
500	0.2	0.4	1.1	2.0	2.2	1.7	0.3	-1.2	-2.3	-2.9	-2.3	-1.2	-0.5	-0.1	0.0	0.1	0.0	0.1	0.2	0.5	0.9	1.2	1.1	0.8	0.6	0.3	0.1					-0.42	0.39	-0.02
700	0.1	-0.6	-0.7	-0.8	-0.7	-0.7	-0.9	-1.1	-1.3	-0.9	-0.5	-0.2	-0.1	0.0	0.0	-0.0	0.1	0.2	0.4	0.5	0.6	0.6	0.4	0.2	0.1	-0.1	-0.1					-0.64	0.20	-0.22
850	-1.6	-1.9	-2.3	-2.4	-2.3	-2.1	-1.7	-1.3	-0.9	-0.5	-0.3	-0.2	0.0	0.0	0.1	0.1	0.2	0.3	0.5	0.5	0.3	0.2	0.1	-0.0								-0.94	0.19	-0.37
1000	-0.2	-0.4	-0.5	-0.6	-0.6	-0.5	-0.3	-0.1	-0.0	0.1	0.1	0.1	0.1	0.0	0.1	0.2	0.2	0.3	-0.0	-0.1	-0.1	0.1										-0.18	0.07	-0.07
MEAN	-1.4	-0.4	0.2	0.6	0.4	-0.5	-1.8	-2.9	-3.5	-3.6	-2.7	-1.5	-0.7	-0.3	-0.2	-0.2	-0.1	-0.0	0.2	0.6	1.1	1.3	1.0	0.6	0.4	0.3	0.1	0.2	-1.32	0.31	-0.51			

474

TABLE A27. NORTHWARD TRANSPORT OF KINETIC ENERGY BY STATIONARY EDDIES $[\bar{v}^{*}\bar{K}^{*}]$ (IN 10^{2} m³ sec⁻³)

V*K* YEAR 63-73   UNITS= 1.E 02 M3/S3

YEAR 63-73

p (HB)	80S	70S	65S	60S	55S	50S	45S	40S	35S	30S	25S	20S	15S	10S	5S	EQ	5N	10N	15N	20N	25N	30N	35N	40N	45N	50N	55N	60N	65N	70N	80N	SH	NH	GLOBE
50	-0.1	-0.4	-0.4	-0.2	-0.1	-0.0	0.0	0.0	0.0	0.0	0.0	0.0	0.0	-0.0	-0.0	0.0	0.0	0.0	0.0	-0.0	-0.0	0.0	0.0	-0.0	-0.1	-0.1	-0.2	0.0	0.2	0.2	-0.1	-0.04	0.01	-0.02
100	-0.1	-0.3	-0.3	-0.2	-0.0	0.1	0.1	-0.0	-0.1	-0.1	-0.1	-0.0	0.0	0.0	-0.0	-0.0	0.0	0.0	0.1	0.3	0.2	0.4	0.2	0.5	0.7	0.9	-0.2	-0.2	-0.2	-0.0	-0.1	-0.03	0.02	-0.01
200	0.1	-0.1	-0.1	-0.2	-0.3	-0.1	0.1	0.1	0.0	0.0	0.1	0.3	0.2	-0.1	0.0	-0.1	0.0	0.1	0.4	0.9	1.5	1.0	1.0	0.8	0.9	0.6	0.6	-0.2	-0.3	-0.3	-0.1	0.04	0.45	0.25
300	0.1	-0.1	-0.2	-0.3	-0.1	0.5	0.5	0.1	-0.0	-0.1	-0.1	0.1	0.5	0.3	-0.0	-0.0	0.0	0.1	0.4	1.2	1.6	1.0	0.8	1.2	1.3	0.4	0.4	0.0	-0.2	-0.3	-0.0	0.02	0.50	0.26
500	-0.1	-0.1	-0.1	-0.2	-0.1	0.0	-0.0	-0.1	-0.1	-0.0	0.0	0.0	0.0	0.0	0.0	0.0	0.0	0.0	0.1	0.2	0.2	0.2	0.2	0.4	0.6	0.3	0.2	0.1	-0.1	-0.1	-0.0	-0.05	0.13	0.04
700	-0.1	-0.1	-0.0	-0.0	-0.0	0.0	0.0	0.0	-0.0	-0.1	-0.1	-0.0	0.0	0.0	0.0	0.0	0.0	0.0	0.1	0.1	0.1	0.1	0.1	0.1	0.2	0.2	0.1	0.1	-0.0	-0.0	0.0	0.00	0.05	0.03
850	.....	-0.0	-0.0	0.0	0.1	0.0	0.1	0.0	-0.0	-0.0	-0.0	0.0	0.0	0.0	0.0	0.0	0.0	0.0	0.0	0.0	0.0	0.1	0.1	0.1	0.1	0.1	0.1	0.0	0.0	0.0	0.0	0.01	0.05	0.03
1000	.....	0.0	0.0	-0.0	0.0	-0.0	0.0	-0.0	-0.0	-0.0	0.0	0.0	0.0	0.0	0.0	0.0	0.0	0.0	-0.0	-0.0	0.0	0.0	0.0	0.1	0.1	0.1	0.1	0.0	0.0	0.0	-0.0	-0.01	0.02	0.00
MEAN	-0.0	-0.1	-0.1	-0.1	-0.1	0.1	0.1	-0.0	-0.0	-0.0	0.0	0.1	0.1	0.0	0.0	-0.0	0.0	0.0	0.1	0.2	0.4	0.3	0.2	0.4	0.5	0.4	0.1	0.1	-0.0	-0.1	-0.0	-0.01	0.17	0.08

DJF 63-73

p (HB)	80S	70S	65S	60S	55S	50S	45S	40S	35S	30S	25S	20S	15S	10S	5S	EQ	5N	10N	15N	20N	25N	30N	35N	40N	45N	50N	55N	60N	65N	70N	80N	SH	NH	GLOBE
50	0.0	-0.0	-0.0	-0.0	-0.0	-0.0	-0.0	-0.0	-0.0	0.1	0.1	0.0	0.0	-0.0	0.0	0.1	0.1	0.1	0.0	0.0	0.1	0.2	0.5	0.6	0.5	-0.1	-0.3	-0.1	0.2	-0.2	-0.8	-0.01	0.09	0.04
100	-0.0	-0.0	-0.0	-0.0	-0.0	-0.1	-0.0	-0.1	-0.2	-0.2	-0.1	0.0	0.0	-0.1	0.0	0.0	0.1	0.2	0.3	0.7	1.9	3.2	2.4	1.5	0.7	-0.2	-0.6	-0.8	-0.7	-0.3	-0.2	-0.02	0.68	0.33
200	0.0	-0.1	-0.2	-0.3	-0.2	-0.1	-0.1	-0.1	-0.2	-0.2	-0.1	0.3	0.5	0.3	0.1	0.1	0.1	0.4	0.9	1.9	3.2	6.3	13.1	8.1	3.6	2.3	0.8	0.2	-0.4	-0.6	-0.8	0.01	2.45	1.23
300	0.0	-0.1	-0.2	-0.3	-0.2	0.6	0.4	-0.1	-0.3	-0.1	-0.1	0.3	0.5	0.0	-0.1	0.0	0.1	0.3	0.8	2.0	3.3	6.4	4.4	4.6	3.2	2.6	2.0	1.5	0.8	0.2	-0.3	-0.03	1.75	0.86
500	-0.0	-0.1	-0.1	-0.1	-0.1	0.1	0.4	-0.1	-0.3	-0.1	-0.1	0.0	0.0	-0.1	-0.0	0.0	0.1	0.1	0.3	0.8	3.3	0.8	0.5	0.7	1.0	1.5	0.8	0.4	0.1	-0.1	-0.1	-0.03	0.35	0.16
700	-0.0	0.0	-0.0	-0.1	0.1	0.1	0.0	-0.0	-0.1	-0.0	0.0	0.0	-0.0	-0.0	0.0	0.0	0.0	0.1	0.1	0.1	0.0	0.1	0.1	0.1	0.4	0.6	0.5	0.2	0.1	-0.1	-0.0	0.01	0.12	0.07
850	.....	0.0	0.1	0.0	0.1	0.0	0.0	-0.0	-0.0	0.0	0.0	-0.0	-0.0	-0.0	0.0	0.0	0.0	0.0	0.0	0.0	0.0	0.1	0.1	0.1	0.2	0.4	0.4	0.2	0.1	-0.0	-0.0	-0.00	0.09	0.05
1000	.....	-0.0	-0.0	0.0	0.0	0.0	0.0	-0.0	-0.0	-0.0	-0.0	-0.0	-0.0	-0.0	-0.0	0.0	0.0	0.0	0.0	0.0	0.0	0.0	0.1	0.1	0.1	0.1	0.0	0.0	0.1	0.1	0.1	0.01	0.03	0.01
MEAN	-0.0	-0.0	-0.1	-0.1	-0.1	0.1	0.1	-0.1	-0.1	-0.1	-0.1	0.1	0.1	0.0	0.0	0.0	0.0	0.1	0.3	1.3	2.7	1.9	1.2	1.0	0.8	0.5	0.2	-0.0	-0.2	-0.1	-0.2	-0.02	0.68	0.33

JJA 63-73

p (HB)	80S	70S	65S	60S	55S	50S	45S	40S	35S	30S	25S	20S	15S	10S	5S	EQ	5N	10N	15N	20N	25N	30N	35N	40N	45N	50N	55N	60N	65N	70N	80N	SH	NH	GLOBE
50	0.1	-0.6	-0.8	-0.4	-0.2	-0.1	-0.0	-0.1	-0.1	0.0	0.0	0.0	0.0	-0.0	-0.0	-0.0	-0.1	-0.1	-0.1	-0.1	-0.1	-0.0	-0.0	0.0	0.0	-0.0	-0.0	-0.0	0.0	0.0	0.0	-0.07	-0.03	-0.05
100	0.0	-0.4	-0.4	-0.3	-0.1	-0.1	0.1	0.1	0.0	0.0	0.0	0.1	0.0	-0.0	-0.0	-0.0	-0.2	-0.6	-0.6	-0.3	-0.1	0.3	0.4	0.2	-0.2	-0.2	-0.1	-0.0	0.0	0.0	0.0	-0.03	-0.12	-0.07
200	0.3	-0.2	-0.4	-0.4	-0.1	0.3	0.5	0.5	0.1	0.0	0.0	0.9	1.1	0.7	-0.1	-0.0	-0.6	-0.6	-0.5	-0.3	0.1	0.5	0.5	-0.3	-0.4	-0.2	-0.0	0.0	0.0	0.0	0.0	0.19	-0.11	0.04
300	0.3	-0.1	-0.3	-0.3	-0.1	0.2	0.1	-0.3	-0.4	-0.1	0.1	0.2	0.1	0.0	-0.0	-0.0	-0.1	-0.1	-0.1	-0.0	0.0	0.3	0.4	-0.0	0.2	0.2	0.3	0.1	0.1	0.0	0.0	-0.05	0.10	0.03
500	-0.2	-0.1	-0.1	-0.1	-0.1	-0.1	-0.1	-0.2	-0.3	-0.0	0.0	-0.0	-0.0	-0.0	-0.0	0.0	-0.1	-0.0	-0.0	-0.0	0.0	0.0	-0.0	0.1	0.1	0.1	0.0	-0.0	-0.0	-0.0	-0.0	-0.11	0.02	-0.05
700	-0.2	-0.1	-0.1	-0.1	-0.1	-0.1	-0.1	-0.1	-0.1	-0.0	0.0	0.0	-0.0	-0.0	-0.0	0.0	-0.1	-0.0	-0.0	-0.0	-0.0	-0.0	-0.0	0.0	0.1	0.0	0.0	-0.0	-0.0	-0.0	-0.0	-0.04	0.01	-0.01
850	.....	-0.1	-0.0	0.0	0.3	0.5	0.5	0.1	-0.0	-0.0	-0.0	-0.0	0.0	0.0	0.0	0.0	0.0	0.0	0.0	0.0	0.0	0.0	0.0	0.0	0.0	0.0	0.0	0.0	0.0	0.0	0.0	0.02	0.03	0.02
1000	.....	0.0	0.0	0.0	0.1	0.2	0.1	-0.0	-0.0	-0.0	-0.0	-0.0	0.0	0.0	0.0	0.0	0.0	0.0	0.0	0.0	0.0	0.1	0.1	0.1	0.1	0.0	0.0	0.0	0.0	0.0	-0.0	-0.01	0.02	0.00
MEAN	0.0	-0.2	-0.2	-0.2	-0.1	0.0	0.0	-0.0	-0.1	-0.0	0.0	0.1	0.1	0.1	-0.0	-0.1	-0.1	-0.1	-0.1	-0.1	0.0	0.1	0.2	0.0	0.0	0.0	0.0	0.0	0.0	0.0	0.0	-0.02	0.00	-0.01

Table A28. Northward Transport of Total Energy by Transient Eddies $[\overline{v'E'}]$ (in °C m sec⁻¹)

$v'E'$ YEAR 63-73   UNITS = 1.E 00 M/S °C

P (HB)	80S	70S	65S	60S	55S	50S	45S	40S	35S	30S	25S	20S	15S	10S	5S	EQ	5N	10N	15N	20N	25N	30N	35N	40N	45N	50N	55N	60N	65N	70N	80N	SH	NH	GLOBE
50	17.4	24.4	19.8	11.4	3.5	-1.2	-2.3	-1.9	-1.2	-0.6	-0.3	-0.3	-0.5	-0.5	-0.4	-0.1	0.2	0.5	0.7	0.8	0.8	0.8	1.2	1.9	3.1	4.6	6.3	7.7	6.2	3.4	9.7	2.37	2.24	2.31
100	-4.8	1.5	4.0	3.4	-0.0	-2.6	-2.9	-2.3	-1.6	-1.0	-0.8	-0.8	-0.7	-0.4	-0.1	-0.2	0.0	0.3	1.4	2.4	3.4	4.4	5.5	6.2	6.2	4.7	3.0	1.4				-0.80	1.96	0.58
200	1.0	1.6	-1.4	-5.2	-8.9	-11.3	-11.6	-10.3	-7.5	-4.2	-1.4	0.2	0.8	0.9	0.8	0.6	-0.1	-1.1	-2.0	-2.2	-0.8	2.3	6.2	8.8	9.4	9.4	8.8	7.3	5.2	3.6	0.9	-3.35	2.94	-0.21
300	-0.4	0.7	-0.0	-1.4	-3.3	-5.2	-6.5	-7.0	-6.2	-4.4	-2.3	-0.7	-0.0	0.3	0.3	0.2	-0.2	-0.8	-1.1	-1.3	-0.6	2.6	5.9	7.4	6.3	4.7	3.7	2.6	1.9	0.7	0.7	-2.33	1.92	-0.21
500	-2.5	-3.0	-4.1	-5.7	-8.0	-9.9	-10.3	-9.3	-7.2	-4.7	-2.9	-1.7	-1.0	-0.4	-0.1	0.2	0.4	0.3	0.1	1.0	3.5	6.6	9.2	9.6	9.4	8.3	6.9	9.4	8.3	2.9	-4.18	-4.18	4.08	-0.05
700	-5.2	-6.1	-8.9	-12.7	-16.2	-18.7	-17.5	-14.3	-10.2	-6.3	-3.6	-2.7	-1.8	-1.0	-0.0	0.8	1.1	1.5	2.3	4.1	7.9	12.7	14.8	17.2	17.8	16.7	14.4	11.9	10.0	5.4	-8.62	-8.62	7.84	-0.35
850	*****	-6.4	-10.3	-15.6	-19.6	-21.4	-21.0	-19.6	-16.9	-13.4	-9.3	-6.2	-4.7	-3.2	-1.8	-0.2	1.5	3.2	4.3	5.1	8.5	12.4	16.9	19.5	19.6	19.9	18.5	16.7	14.4	12.2	8.4	-10.83	10.39	-0.19
1000	*****	-1.8	-3.7	-5.0	-6.2	-7.3	-8.4	-9.4	-9.2	-8.0	-6.2	-4.9	-4.0	-2.8	-1.5	-1.3	-0.1	3.9	6.2	9.3	13.1	18.0	18.8	19.2	14.9	12.9	11.3	8.9	5.6	3.6	-0.6	-5.58	8.59	0.58
MEAN	-1.0	-0.8	-3.1	-6.3	-9.6	-11.8	-12.2	-11.4	-9.4	-6.7	-4.3	-2.6	-1.8	-1.1	-0.6	-0.0	0.4	0.7	1.0	1.5	3.0	5.8	9.0	11.0	11.5	11.4	10.7	9.5	7.6	6.0	3.6	-5.22	5.24	-0.01

DJF 63-73

P (HB)	80S	70S	65S	60S	55S	50S	45S	40S	35S	30S	25S	20S	15S	10S	5S	EQ	5N	10N	15N	20N	25N	30N	35N	40N	45N	50N	55N	60N	65N	70N	80N	SH	NH	GLOBE
50	0.6	-0.2	-0.5	-0.7	-0.8	-0.9	-0.8	-0.8	-0.7	-0.6	-0.7	-0.7	-0.7	-0.5	-0.5	0.0	0.5	0.6	0.6	0.8	1.0	1.3	1.9	2.7	3.7	4.8	5.7	6.1	4.4	1.6	1.1	-0.66	1.94	0.64
100	-0.0	-1.0	-2.1	-3.1	-3.9	-4.2	-3.7	-2.7	-1.8	-0.9	-0.4	-0.3	-0.7	-1.2	-1.5	-1.1	-0.7	-0.9	-0.9	-0.5	1.1	1.2	1.7	2.2	2.9	3.3	3.8	4.4	4.6	3.9	-0.6	-1.69	1.55	-0.07
200	-0.9	-1.1	-3.4	-6.7	-9.7	-11.1	-10.0	-8.0	-5.0	-1.9	-0.2	0.6	0.6	0.6	0.5	-0.1	-0.5	-0.9	-0.9	-0.5	1.1	2.9	4.5	5.8	6.7	7.1	6.7	5.8	4.6	3.6	0.5	-3.10	2.51	-0.30
300	2.7	2.1	1.5	0.8	-0.6	-2.6	-4.4	-5.3	-5.0	-4.1	-2.5	-1.1	-0.4	0.1	0.2	0.1	0.3	0.0	-0.0	0.2	1.8	2.9	3.6	3.1	2.4	1.6	0.5	-0.0	0.9	1.6	1.1	-1.49	1.09	-0.20
500	-1.5	-2.0	-2.1	-3.0	-4.7	-6.4	-7.0	-6.7	-5.5	-4.2	-2.9	-1.9	-1.3	-0.7	-0.2	0.1	0.3	0.0	-0.0	0.2	3.5	6.8	7.6	9.1	9.1	8.0	6.8	6.1	6.6	5.0	-3.11	-3.11	4.00	0.45
700	-1.8	-3.1	-5.3	-8.8	-13.0	-15.7	-15.8	-14.4	-11.8	-8.1	-5.4	-3.3	-2.1	-1.0	0.0	0.8	1.5	1.7	2.3	3.5	6.8	11.8	16.1	17.4	18.3	17.1	14.9	12.6	11.7	7.5	5.0	-6.70	9.01	1.20
850	*****	-4.4	-7.0	-11.6	-16.3	-19.4	-19.6	-18.9	-16.1	-12.2	-7.6	-4.0	-2.0	-0.9	0.4	1.2	1.1	2.9	6.6	12.1	17.2	22.1	24.3	23.0	20.2	17.8	15.6	14.3	11.9	9.7	7.5	-8.94	11.60	1.35
1000	*****	-0.8	-2.0	-3.1	-4.1	-5.6	-7.7	-9.4	-8.5	-5.7	-2.9	-1.6	-0.9	-0.2	-0.2	-0.3	0.1	1.0	2.1	5.5	11.7	18.3	18.5	18.5	15.1	12.5	9.2	8.3	7.5	5.9	2.0	-3.76	7.64	1.20
MEAN	-0.6	-1.9	-3.5	-5.7	-8.1	-10.0	-10.3	-9.6	-7.8	-5.6	-3.5	-2.0	-1.2	-0.5	-0.2	0.1	0.4	0.4	0.9	2.1	4.6	7.7	10.1	11.2	11.1	10.1	8.7	7.5	7.2	6.3	4.2	-4.36	5.36	0.49

JJA 63-73

P (HB)	80S	70S	65S	60S	55S	50S	45S	40S	35S	30S	25S	20S	15S	10S	5S	EQ	5N	10N	15N	20N	25N	30N	35N	40N	45N	50N	55N	60N	65N	70N	80N	SH	NH	GLOBE
50	-0.5	-8.4	-10.3	-9.9	-7.8	-5.4	-3.6	-2.3	-1.5	-0.9	-0.7	-0.6	-0.6	-0.7	-0.5	-0.2	0.2	0.5	0.6	0.6	0.3	0.2	0.3	0.5	0.7	0.8	0.8	0.7	0.5	0.4	0.4	-2.67	0.46	-1.11
100	3.1	0.4	-0.3	-1.1	-1.5	-1.6	-1.4	-1.2	-1.0	-0.9	-0.8	-0.6	0.6	1.3	1.0	0.6	1.3	1.0	0.9	0.9	0.6	0.6	0.8	1.3	1.9	2.3	2.2	1.8	1.3	0.9	0.1	-0.54	1.17	0.32
200	3.5	5.0	4.9	2.5	-1.6	-4.7	-7.6	-8.0	-6.4	-3.8	-1.3	0.5	0.5	0.1	-0.4	-0.4	-0.4	-0.1	0.1	0.6	0.7	1.3	3.1	5.2	6.6	7.7	7.2	5.6	3.7	2.4	0.1	-1.49	2.38	0.45
300	2.8	0.8	1.0	1.0	0.2	-1.3	-3.2	-4.9	-5.0	-3.8	-2.2	-1.0	-0.3	-0.0	-0.1	-0.3	-0.4	-0.5	-0.4	0.1	0.6	2.2	3.8	4.3	4.7	4.7	4.2	3.1	2.3	2.4	-1.33	-1.33	1.59	0.13
500	-2.8	-4.5	-5.3	-6.6	-8.5	-10.5	-11.4	-11.0	-9.2	-6.7	-4.1	-2.0	-1.0	-0.5	-0.4	-0.2	0.1	0.5	0.8	0.7	1.3	2.1	3.4	5.3	7.1	7.7	7.7	7.5	6.5	5.1	4.1	-5.05	2.71	-1.17
700	-6.7	-10.0	-12.8	-16.1	-18.6	-20.3	-19.6	-16.8	-12.7	-8.0	-4.4	-2.4	-1.1	-0.5	-0.2	0.0	0.3	0.8	1.1	1.5	2.5	4.5	6.6	9.8	12.0	12.1	10.8	9.7	8.7	5.8	-10.02	-10.02	4.53	-2.70
850	*****	-6.6	-11.9	-18.6	-23.4	-25.2	-24.2	-22.2	-19.1	-15.5	-11.4	-7.6	-4.6	-2.3	-1.0	-0.0	0.3	1.0	1.5	2.4	3.6	6.6	9.5	11.8	14.2	13.9	12.4	12.7	8.2	-12.16	5.77	-3.17
1000	*****	-7.1	-8.8	-10.0	-9.9	-9.4	-9.9	-10.6	-10.5	-9.0	-6.6	-3.8	-2.0	-0.9	-0.1	-0.1	-0.3	0.3	0.7	1.4	4.0	7.8	10.5	8.6	6.0	4.4	5.5	6.4	6.0	3.1	-6.41	3.18	-2.24
MEAN	-0.1	-3.3	-5.2	-7.6	-9.8	-11.4	-12.2	-12.0	-10.5	-8.0	-5.3	-3.0	-1.7	-0.8	-0.3	0.0	0.1	0.2	0.5	0.7	1.1	1.8	3.4	5.1	6.7	8.1	8.1	7.3	6.6	5.8	3.7	-5.68	3.12	-1.30

476

TABLE A29. NORTHWARD TRANSPORT OF TOTAL ENERGY BY STATIONARY EDDIES $[\bar{v}^*\bar{E}^*]$ (IN °C m sec⁻¹)

v*E* YEAR 63-73 UNITS= 1.E 00 M/S °C

p (MB)	80S	70S	65S	60S	55S	50S	45S	40S	35S	30S	25S	20S	15S	10S	5S	EQ	5N	10N	15N	20N	25N	30N	35N	40N	45N	50N	55N	60N	65N	70N	80N	SH	NH	GLOBE
50	-0.8	-2.4	-2.6	-2.3	-1.8	-1.2	-0.9	-0.6	-0.4	-0.3	-0.2	-0.0	0.1	0.0	-0.1	-0.1	0.0	-0.1	0.2	0.3	0.2	0.4	1.0	2.3	4.0	5.6	7.2	7.1	5.2	2.4		-0.64	1.65	0.51
100	-1.0	-1.2	-1.1	-0.9	-0.7	-0.3	-0.1	0.0	0.1	0.0	-0.1	0.1	0.3	0.3	0.2	0.3	0.5	1.1	1.5	1.5	0.7	0.2	0.4	0.5	1.2	2.6	4.1	5.0	4.3	2.8	0.5	-0.12	1.45	0.67
200	-1.0	-1.4	-1.3	-1.1	-0.8	-0.2	0.1	0.1	0.2	0.0	-0.0	0.2	0.4	0.4	0.3	0.2	0.6	1.6	2.2	1.8	0.8	0.8	0.6	1.4	2.9	4.0	3.8	2.2	0.8	-0.1	-0.07	1.34	0.63
300	-0.4	-0.3	-0.3	-0.3	-0.3	-0.1	0.3	0.4	0.3	0.0	-0.1	0.1	0.2	0.2	0.1	0.0	0.2	0.7	1.1	0.7	1.1	0.7	-0.1	0.4	1.2	2.2	2.7	2.1	0.5	0.1	0.07	0.72	0.39
500	-1.0	-0.3	-0.2	-0.3	-0.3	-0.2	-0.2	-0.3	-0.4	-0.3	-0.1	-0.1	-0.1	-0.1	-0.0	0.0	0.2	0.3	0.5	0.6	0.5	0.1	0.6	0.7	1.4	2.1	2.3	1.8	1.1	0.4	0.2	-0.22	0.68	0.23
700	-2.0	-1.1	-0.6	-0.4	-0.4	-0.4	-0.3	-0.2	-0.3	-0.6	-0.7	-0.6	-0.2	-0.0	-0.2	-0.2	-0.4	-0.3	0.1	0.7	0.7	0.1	0.1	1.0	1.5	2.3	2.9	2.6	1.6	0.3	0.1	-0.36	0.62	0.13
850	*****	-1.2	-0.6	-0.1	0.1	0.1	0.0	0.0	-0.3	-0.6	-1.6	-1.8	-1.7	-1.3	-0.5	-0.2	-0.3	0.3	1.5	1.6	1.3	1.3	0.9	1.4	2.8	3.6	3.4	2.5	0.7	0.3	-0.70	1.14	0.22
1000	*****	0.1	-0.1	-0.4	-0.3	0.2	0.4	-0.2	-1.6	-4.6	-5.6	-5.2	-4.9	-6.2	-5.5	-4.8	-2.3	1.3	4.8	6.5	6.8	5.1	2.7	1.0	1.3	2.5	5.2	6.8	5.7	2.1	-0.8	-2.85	2.53	-0.51
MEAN	-0.9	-0.8	-0.6	-0.5	-0.3	-0.1	-0.1	-0.1	-0.1	-0.3	-0.6	-0.7	-0.6	-0.4	-0.3	-0.2	-0.2	-0.2	-0.1	0.1	0.5	0.1	1.1	1.3	0.5	1.4	2.5	3.2	2.3	1.0	-0.3	-0.40	1.01	0.31

DJF 63-73

p (MB)	80S	70S	65S	60S	55S	50S	45S	40S	35S	30S	25S	20S	15S	10S	5S	EQ	5N	10N	15N	20N	25N	30N	35N	40N	45N	50N	55N	60N	65N	70N	80N	SH	NH	GLOBE
50	0.2	0.2	0.1	0.0	-0.1	-0.1	-0.1	-0.1	-0.1	-0.1	-0.1	-0.0	0.0	0.1	0.0	-0.1	-0.1	0.0	0.0	0.1	-0.1	0.2	1.2	3.2	7.7	14.3	21.8	29.1	27.7	18.0	9.3	-0.03	6.01	2.99
100	-0.2	-0.1	-0.0	0.0	0.0	0.1	0.3	0.4	0.5	0.7	1.4	2.3	1.8	1.0	0.4	0.4	0.4	0.6	1.6	2.1	1.4	-0.6	-0.9	0.4	1.6	4.7	10.2	15.3	17.6	15.1	2.2	0.72	3.92	2.32
200	-1.2	-1.0	-0.5	-0.3	-0.2	-0.0	-0.1	0.3	0.4	0.5	0.7	0.1	0.1	0.6	0.4	0.3	0.2	0.3	0.1	-0.4	-0.7	-0.4	0.7	1.8	1.9	5.5	9.7	12.1	10.5	6.5	3.0	-0.01	2.54	1.28
300	-0.7	-0.7	-0.6	-0.5	-0.2	0.2	0.3	0.2	-0.1	-0.2	-0.1	-0.3	-0.6	-0.6	-0.2	-0.1	-0.3	-0.6	-0.2	0.4	1.2	1.7	0.7	2.6	4.9	7.3	8.2	6.2	3.4	1.6	0.4	-0.07	1.89	0.91
500	-0.9	-0.3	-0.2	-0.3	-0.4	-0.2	-0.0	-0.0	-0.1	-0.0	0.0	0.2	0.3	0.2	-0.0	-0.1	-0.0	0.2	0.4	0.8	0.7	0.6	1.2	1.7	3.5	5.6	7.2	7.6	5.7	3.2	1.7	-0.06	2.05	0.99
700	-1.7	-0.7	-0.3	-0.3	-0.3	-0.4	-0.4	-0.2	0.1	-0.1	-0.4	-0.3	-0.2	-0.1	-0.1	-0.2	-0.6	-0.3	-0.2	0.0	0.5	0.7	0.6	2.5	5.6	8.8	10.9	10.6	8.4	4.9	1.4	-0.27	2.94	1.34
850	*****	-0.4	-0.1	0.0	0.0	0.2	0.3	0.2	0.2	-0.8	-1.9	-1.8	-1.2	-1.1	-0.7	-0.2	-0.4	-0.5	-0.2	0.4	1.4	0.9	0.9	4.9	8.7	12.5	15.2	13.9	11.0	6.7	1.5	-0.60	4.31	1.86
1000	*****	0.5	0.6	-0.0	-0.0	-0.2	0.8	1.3	-0.3	-3.9	-8.3	-6.9	-7.1	-9.4	-7.5	-4.8	-1.2	0.7	1.7	2.5	3.8	4.5	6.4	7.8	10.0	11.5	12.6	10.9	9.6	7.0	0.0	-4.11	3.84	-0.66
MEAN	-0.7	-0.4	-0.2	-0.2	-0.1	0.1	0.1	-0.0	-0.0	-0.4	-0.7	-0.5	-0.3	-0.4	-0.5	-0.4	-0.3	-0.1	-0.1	0.2	0.6	0.8	1.1	2.0	4.3	7.5	10.3	11.1	9.9	6.8	3.3	-0.31	3.03	1.35

JJA 63-73

p (MB)	80S	70S	65S	60S	55S	50S	45S	40S	35S	30S	25S	20S	15S	10S	5S	EQ	5N	10N	15N	20N	25N	30N	35N	40N	45N	50N	55N	60N	65N	70N	80N	SH	NH	GLOBE
50	-1.1	-3.8	-4.4	-3.7	-2.7	-1.9	-1.5	-1.1	-0.7	-0.5	-0.4	-0.2	-0.1	-0.1	-0.1	0.1	0.1	0.0	0.0	0.6	1.1	0.7	0.3	0.2	0.2	0.1	0.0	-0.0	-0.0	0.1	-1.08	0.26	-0.41
100	-0.3	-0.6	-0.6	-0.7	-0.6	-0.3	-0.2	-0.0	-0.1	-0.6	-0.7	-0.9	-0.6	-0.4	-0.1	0.2	0.5	1.1	1.5	1.6	1.2	1.2	1.1	0.9	0.4	0.2	0.1	0.1	0.1	0.1	-0.43	0.78	0.18
200	-0.7	-1.5	-1.2	-1.0	-0.9	-0.6	-0.3	-0.2	0.0	-0.1	-0.6	-0.7	-0.6	-0.2	0.3	0.2	0.4	0.6	0.6	0.6	1.4	0.6	0.5	-0.4	-0.8	-0.4	-0.2	0.0	0.0	0.1	-0.40	-0.10	-0.25
300	-0.1	-0.1	0.1	0.2	0.3	0.3	0.1	-0.3	-0.2	-0.1	-0.1	-0.3	-0.2	-0.1	-0.1	0.1	0.3	0.1	-0.1	-0.6	-0.6	-0.1	1.4	-2.6	-2.4	-1.5	-1.3	-0.8	-0.4	0.4	0.13	-0.87	-0.37
500	-1.0	-0.4	-0.1	-0.1	-0.2	-0.4	-0.8	-1.2	-1.5	-1.1	-1.0	-0.5	-0.4	-0.2	-0.0	0.0	0.1	0.2	0.1	0.2	0.0	0.0	-0.6	-0.6	-0.9	-1.3	-1.4	-1.1	-0.7	-0.3	0.2	-0.57	-0.28	-0.42
700	-1.5	-1.1	-0.8	-0.5	-0.4	-0.4	-0.6	-0.4	-0.6	-1.2	-2.1	-1.8	-1.3	-0.6	-0.4	0.1	0.5	1.4	1.1	2.9	5.8	6.4	4.0	0.6	-1.3	-2.7	-2.8	-1.5	-0.1	0.3	0.1	-0.76	-0.34	-0.55
850	*****	-1.6	-1.1	-0.4	-0.1	-0.2	-0.4	-0.6	-1.2	-2.1	-4.6	-4.1	-3.7	-4.6	-4.1	-4.9	-5.1	-3.1	2.7	11.0	17.8	19.8	18.0	10.7	3.8	-1.5	-1.5	-1.3	-0.8	0.8	0.1	-1.03	1.32	0.15
1000	*****	-0.1	-0.3	-0.6	-0.8	-0.7	-0.2	-0.8	-1.8	-4.0	-4.6	-4.1	-3.7	-4.6	-4.9	-5.1	-3.1	2.7	11.0	17.8	19.8	18.0	10.7	10.7	3.8	-1.1	-0.1	1.4	0.4	-0.1	-2.60	5.99	1.13
MEAN	-0.8	-0.9	-0.7	-0.5	-0.5	-0.4	-0.5	-0.6	-0.9	-0.6	-0.5	-0.3	-0.1	0.0	0.2	0.4	1.4	2.2	1.9	0.8	-0.4	-1.1	-1.5	-1.1	-1.3	-0.8	-0.2	0.2	0.1	-0.65	0.26	-0.20

TABLE A30. NORTHWARD TRANSPORT OF TOTAL ENERGY BY MEAN MERIDIONAL CIRCULATIONS $[\bar{v}]_{lp}[\bar{E}]$ (IN °C m sec⁻¹)

V'·E   YEAR 63-73   UNITS= 1.E 00 M/S °C

P (MB)	80S	70S	65S	60S	55S	50S	45S	40S	35S	30S	25S	20S	15S	10S	5S	EQ	5N	10N	15N	20N	25N	30N	35N	40N	45N	50N	55N	60N	65N	70N	80N	SH	NH	GLOBE
50	10.0	10.1	7.3	2.8	-1.5	-3.5	-3.3	-2.7	-1.9	-0.8	-0.3	0.1	1.7	0.6	-3.9	-3.1	-2.5	-3.3	-4.1	1.7	2.3	2.5	0.7	-0.7	-2.8	-3.9	-3.3	-2.2	-1.0	-0.8	15.7	0.06	-0.81	-0.37
100	-0.0	2.6	3.7	3.0	1.4	1.2	2.1	2.0	0.4	-1.3	-1.6	-3.8	-2.6	-6.7	-6.6	-2.3	4.3	13.6	12.8	5.2	-0.7	-4.8	-4.5	-2.9	-2.5	-2.4	-1.9	-0.8	0.9	2.1	2.8	-1.21	1.85	0.32
200	0.8	0.2	0.0	0.0	0.4	1.2	1.8	1.9	0.9	-0.5	-1.9	-6.0	-8.9	-12.3	-8.7	-0.9	9.4	18.0	11.7	5.7	2.1	-1.7	-2.5	-1.8	-1.5	-0.8	-0.2	0.3	0.5	0.2	-2.99	3.34	0.23
300	8.9	2.3	-0.7	-2.6	-3.3	-3.5	-2.9	-1.4	-0.2	-0.2	-1.1	-1.9	-1.0	0.1	-0.3	0.5	2.0	3.4	3.0	2.3	0.3	-0.2	0.0	0.7	1.5	2.2	1.8	0.3	-1.1	-1.7	-0.4	-0.68	1.18	0.25
500	4.6	3.7	0.6	-1.8	-2.6	-2.7	-2.2	-1.3	-0.4	0.0	0.1	0.0	-0.8	-0.9	-0.9	-0.4	0.1	0.2	0.4	0.5	1.1	1.5	1.2	0.1	-1.1	-1.1	-1.9	-0.2	-0.45	0.13	-0.16			
700	2.5	2.0	0.8	-0.2	-0.8	-1.2	-1.4	-1.0	-0.7	-0.4	-0.1	-0.0	0.5	0.1	-0.0	-0.1	-0.2	0.9	-0.7	-1.6	-0.3	0.0	0.1	-0.1	0.0	0.1	0.3	-0.9	-0.0	-0.0	-0.15	0.11	-0.02	
850	.....	0.9	-0.7	-0.8	-1.0	-1.1	-1.1	-0.8	-0.4	-0.4	-0.4	-0.5	-0.2		1.0	0.2	-0.4	-0.7	-0.3	-0.9	-0.9	-0.0	-0.0	-0.15	-0.4	-1.17	-0.17	-0.67						
1000	......	-23.7	2.2	14.4	20.6	24.3	22.3	13.1	4.0	-0.5	-0.2	3.2	16.3	25.4	23.2	21.3	8.9	-18.9	-19.5	-8.1	-0.6	-0.8	-5.6	-11.0	-19.0	-22.7	-16.0	-2.2	12.8	20.1	10.6	11.98	-4.61	4.77
MEAN	-2.4	0.1	1.2	1.6	1.7	2.1	2.2	1.5	0.4	-0.4	-0.4	-0.8	-1.2	-0.5	-0.6	-0.6	0.3	1.3	1.8	1.3	1.1	0.4	-0.7	-1.4	-1.5	-2.1	-2.3	-1.7	-0.5	0.8	1.5	2.1	0.20	0.06	0.13

DJF 63-73

P (MB)	80S	70S	65S	60S	55S	50S	45S	40S	35S	30S	25S	20S	15S	10S	5S	EQ	5N	10N	15N	20N	25N	30N	35N	40N	45N	50N	55N	60N	65N	70N	80N	SH	NH	GLOBE
50	1.0	1.8	3.0	2.6	0.7	-0.6	-1.2	-1.7	-2.1	-1.5	0.8	3.0	-10.5	-5.3	7.7	20.2	19.9	4.7	-5.1	4.9	4.3	3.7	1.4	-2.0	-7.1	-9.2	-6.1	-1.6	0.4	-4.0	34.1	0.25	2.92	1.58
100	-0.2	1.1	2.4	3.1	3.6	4.3	3.9	1.9	-1.3	-3.2	-2.1	-3.4	5.0	7.2	7.9	11.2	22.7	36.6	28.5	12.3	7.3	-0.1	-4.6	-4.6	-6.4	-5.2	-2.6	-0.7	1.6	3.5	3.8	2.48	8.45	5.46
200	-1.1	-0.6	-0.1	0.6	1.5	2.8	3.4	2.4	0.3	-1.4	-2.4	-6.4	4.8	21.2	36.5	52.0	57.8	55.6	31.2	14.8	8.4	-3.5	-2.2	-1.5	-1.0	-0.4	-0.0	0.0	-0.0	-0.1	-0.2	7.31	15.98	11.64
300	3.3	2.1	0.6	-0.8	-1.5	-1.9	-1.2	-0.4	0.0	0.1	-1.0	-1.8	-2.0	2.5	8.2	9.1	10.4	10.5	10.1	6.2	2.3	0.6	0.1	1.7	3.2	4.2	4.1	2.2	-0.3	-2.2	-3.5	-0.4	0.1	15.98
500	2.2	1.8	-0.3	-1.9	-2.3	-1.9	-1.2	-0.8	-0.3	0.0	0.1	-0.1	-0.6	-1.0	-1.3	-1.5	-0.9	-0.3	-0.0	-0.0	-0.1	0.4	1.7	1.4	1.6	2.6	2.7	1.3	-0.8	-2.4	-3.4	-0.56	0.30	-0.13
700	0.6	1.2	-0.1	-0.7	-0.9	-0.9	-1.0	-0.8	-0.3	-0.0	0.0	0.1	-0.1	-0.6	-1.0	-1.3	-0.3	0.5	0.6	0.3	1.4	1.7	1.7	0.8	1.0	1.1	0.8	-1.6	-1.6	0.5	-0.46	0.43	-0.01	
850	.....	0.2	-1.7	-1.5	-1.0	-0.4	-0.4	-0.5	-0.3	-0.1	-0.0	-0.0	-1.9	-3.6	-4.2	-5.5	-5.4	-1.8	1.6	0.4	1.3	0.8	0.5	1.1	1.0	-0.4	-1.2	-1.5	-0.6	-0.7	-0.4	-1.99	-0.60	-1.29
1000	......	-20.6	0.5	11.3	17.6	21.3	18.5	8.6	0.7	0.4	3.2	6.8	13.2	7.0	1.3	-2.6	-26.7	-48.9	-27.4	-6.4	0.8	-8.1	-21.8	-30.0	-37.6	-34.5	-17.7	5.9	26.8	36.9	21.7	6.85	-16.15	-3.15
MEAN	-2.0	-0.8	0.3	1.2	1.9	2.6	2.5	1.3	-0.1	-0.8	-0.5	-0.6	0.9	2.6	4.2	6.0	6.9	6.9	4.9	3.5	2.7	-0.5	-3.0	-4.2	-4.0	-2.2	-0.1	1.7	2.5	4.5	1.23	1.64	1.43	

JJA 63-73

P (MB)	80S	70S	65S	60S	55S	50S	45S	40S	35S	30S	25S	20S	15S	10S	5S	EQ	5N	10N	15N	20N	25N	30N	35N	40N	45N	50N	55N	60N	65N	70N	80N	SH	NH	GLOBE
50	1.8	1.7	1.1	1.1	1.4	1.9	2.0	1.0	-0.0	-0.8	-2.3	-2.5	9.2	1.9	14.3	-18.1	-19.7	-15.2	-4.0	0.9	0.3	0.0	-0.2	-0.1	-0.5	-0.3	-0.1	-0.0	-0.2	-0.4	0.7	-0.79	-4.13	-2.46
100	0.6	0.3	0.3	0.4	0.5	1.3	2.2	2.3	1.9	-0.8	-3.8	-4.5	-7.3	-14.5	-12.5	-8.5	-8.3	-8.1	-5.3	0.6	3.1	0.5	-3.7	-4.1	-2.0	-0.7	-0.2	-0.0	0.1	0.2	0.6	-3.40	-2.58	-2.99
200	4.1	-0.8	-1.0	-0.7	-0.5	-0.2	0.4	1.1	1.5	-0.1	-4.4	-9.5	-19.3	-36.7	-43.0	-42.4	-31.3	-16.5	-2.9	2.3	3.3	2.0	-1.8	-2.5	-2.4	-1.6	-0.9	-0.3	0.1	0.3	0.4	-11.24	-6.13	-8.69
300	10.9	1.8	0.3	-1.4	-3.6	-5.4	-5.6	-3.7	-1.1	0.0	-1.7	-3.1	-5.0	-6.9	-7.3	-5.3	-2.7	-0.3	1.1	1.4	0.7	-0.4	-0.2	0.0	0.0	0.5	0.7	0.2	-0.3	-0.0	-0.0	-2.50	-0.76	-1.63
500	4.4	4.2	2.0	0.0	-1.6	-3.2	-3.7	-2.6	-1.3	-0.7	-0.2	0.1	0.1	-0.0	-0.4	-0.2	0.3	0.3	0.5	0.6	0.4	-0.1	-0.5	-0.1	-0.3	0.0	0.3	0.5	0.6	0.4	0.2	-0.43	0.09	-0.17
700	5.1	2.4	1.1	0.5	-0.3	-1.4	-2.0	-1.7	-1.4	-1.0	-0.2	0.4	-1.3	-1.6	-0.6	-0.0	0.0	0.3	0.2	0.1	0.0	0.0	-0.1	0.0	0.1	0.3	0.4	0.2	-0.1	-0.4	-0.0	-0.59	0.03	-0.28
850	.....	0.5	-0.9	-1.6	-2.1	-2.0	-1.2	-0.4	-0.3	-0.8	-0.8	-0.3	-2.2	-4.0	-0.3	2.9	3.3	2.4	0.6	-0.0	-0.0	-0.0	-0.1	-0.1	0.1	0.1	0.4	0.3	0.2	-0.1	-0.1	-0.3	-2.24	0.72	-0.75
1000	......	-18.1	-0.2	8.8	18.5	27.3	28.3	20.0	10.6	2.8	-3.3	-2.3	10.2	29.4	33.7	37.4	39.0	14.5	-2.3	-5.5	-5.6	-1.4	0.5	-2.6	-8.3	-11.9	-9.4	-3.7	2.4	5.6	2.2	14.12	5.55	10.40
MEAN	-3.2	-0.1	0.4	0.8	1.4	2.1	2.4	2.1	1.3	-0.1	-1.9	-2.8	-2.9	-5.1	-5.7	-4.7	-3.3	-2.2	-0.8	0.1	0.5	0.3	-0.7	-1.0	-1.1	-1.1	-0.7	-0.2	0.2	0.4	0.3	-1.23	-0.93	-1.08

APPENDIX B. THE MEAN MERIDIONAL CIRCULATION

In an earlier section on the vertical structure of the circulation (Section 3.1) we mentioned the indirect method used in this study to derive estimates of the mean meridional circulation. Because of the crucial role played by mean meridional circulations in the general circulation, we will give here a fairly detailed discussion of the actual method used to derive $[\bar{v}]$.

As we have seen before, the $[\bar{v}]$ values are very hard to measure directly in middle latitudes because they constitute the small average of large $\bar{v}$ values along the latitude circle with alternating positive and negative sign. Especially in the Southern Hemisphere, there is no hope of finding reasonable estimates of the strength of the Ferrel cell directly from the $\bar{v}$ distributions. It would lead to a Ferrel cell with meridional velocities up to 1.5 m sec^{-1} (see Table A3 and Fig. B1), implying clearly erroneous equatorward energy fluxes (see Oort, 1978). Of necessity, one must use an indirect method, such as the one described next, of calculating $[\bar{v}]$ in the Southern Hemisphere extratropics.

B.1. Extratropics

Taking the zonal average of the balance equation for angular momentum and neglecting the (small) time rate of change term gives

$$0 = -\partial[\overline{Mv}] \cos \phi/R \cos \phi\ \partial\phi - \partial[\overline{M\omega}]/\partial p + R \cos \phi\ [\bar{F}_\lambda] \quad (B.1)$$

where

$$M = \Omega R^2 \cos^2 \phi + uR \cos \phi \quad (B.2)$$

After expanding the flux terms in eddy and mean contributions, using the advective form for the mean terms, and dividing the entire equation by $R \cos \phi$, one obtains

$$0 = -(D_1 + D_2) + f[\bar{v}] - (D_3 + D_4) + [\bar{F}_\lambda] \quad (B.3)$$

where

$$D_1 = \partial([\overline{u'v'}] + [\bar{u}^*\bar{v}^*]R \cos^2 \phi)/R^2 \cos^2 \phi\ \partial\phi \quad (B.4)$$

$$D_2 = \partial([\overline{u'\omega'}] + [\bar{u}^*\bar{\omega}^*])/\partial p \quad (B.5)$$

$$D_3 = [\bar{v}]\ \partial[\bar{u}]R \cos \phi/R^2 \cos \phi\ \partial\phi \quad (B.6)$$

$$D_4 = [\bar{\omega}]\ \partial[\bar{u}]/\partial p \quad (B.7)$$

Equation (B.3) was used with certain simplifications first by Kuo (1956) and later by Gilman (1965), Holopainen (1967), Newell *et al.* (1972), and others to compute $[\bar{v}]$.

Above the surface boundary layer frictional effects are generally small, and presumably vertical eddy effects can be neglected. Thus Eq. (B.3) can be simplified to

$$[\bar{v}] = (D_1 + D_3 + D_4)/f \tag{B.8}$$

Given the zonal wind $[\bar{u}]$ and the eddy momentum fluxes $[\overline{u'v'}]$ and $[\bar{u}^*\bar{v}^*]$, the only unknown quantities in Eq. (B.8) are $[\bar{v}]$ and $[\bar{\omega}]$. With the aid of the equation for the conservation of mass in the zonal mean form

$$\partial[\bar{v}] \cos \phi/R \cos \phi \, \partial\phi + \partial[\bar{\omega}]/\partial p = 0 \tag{B.9}$$

one can solve for $[\bar{v}]$ and $[\bar{\omega}]$ by an iterative method. Thus, the initial guess field for $[\bar{v}]$ is obtained from Eq. (B.8) by putting $D_3 = 0$ and $D_4 = 0$, and for $[\bar{\omega}]$ from Eq. (B.9) by integrating downward starting with $[\bar{\omega}] = 0$ at $p = 25$ mbar. Next, these initial values for $[\bar{v}]$ and $[\bar{\omega}]$ are used to estimate the mean advection terms D_3 and D_4, and one can compute a new $[\bar{v}]$ value using all terms in Eq. (B.8). Then a new $[\bar{\omega}]$ value is calculated from Eq. (B.9). After a few iterations no appreciable change occurs in the $[\bar{v}]$ and $[\bar{\omega}]$ values.

One obvious difficulty is the area close to the equator, where the Coriolis parameter becomes very small. Problems were avoided by interpolating the values of $[\bar{v}]$ between 6° S and 6° N rather than using Eq. (B.8).

B.2. Surface Boundary Layer

In the surface boundary layer friction and small-scale vertical eddy fluxes are the most important terms balancing the Coriolis term, so that Eq. (B.3) can be simplified to

$$[\bar{v}] = (D_3 + D_4)/f + [\bar{F}_\lambda]/f \tag{B.10}$$

Assuming a vertical profile of $[\bar{F}_\lambda]$, one can in principle compute $[\bar{v}]$ from Eqs. (B.10) and (B.9) again in an iterative fashion. However, we have chosen a simpler method, using the fact that the net mass flow across each latitude circle must (practically) vanish:

$$\int_0^{p_0} [\bar{v}] \, dp/g = 0 \tag{B.11}$$

Thus, integrating down to the top of the surface boundary layer, one knows how large the return flow should be in the boundary layer itself.

Following Holopainen (1967), we used the simplest assumption of a return flow that does not vary with height. North of 74° S the boundary layer was assumed to lie between 875 mbar and the latitudinal mean surface pressure. When there are no mountains, this means a thickness of the boundary layer of 1012.5 − 875 = 137.5 mbar. Not the actual, observed surface pressure was used in our calculations but a climatological value of 1012.5 mbar at sea level reduced by 1 mbar for each 8 m of mountain height. South of 84° S over Antarctica the boundary layer was assumed to lie between 650 mbar and the surface pressure, whereas between 84° S and 74° S the top of the boundary layer was assumed to decrease gradually to 875 mbar. Inspection of Fig. B1, showing a comparison at 50° S and 50° N of the direct and indirect $[\bar{v}]$, clearly indicates the effect of our assumption of a uniform return flow in the surface boundary layer. It further shows the unreliability of the direct method in the Southern Hemisphere.

B.3. Tropics

As was noticed before by Newell et al. (1972, pp. 13, 251, 252), the indirect method breaks down in the Hadley cells in the inner tropics, giving only a weak Hadley circulation. Here direct estimates show a large, more zonally uniform, meridional circulation with meridional velocities on the order of 2–3 m sec^{-1}. Obviously vertical eddy fluxes (i.e., term D_2) must be important in the tropics, and Eq. (B.8) is not valid. Fortunately the rawinsonde network appears to be able to measure the essential features of the Hadley cells (Oort, 1978). After some testing we decided to use the indirect method poleward of 20° latitude, the direct method equatorward of 10°, and a weighted average of the direct and indirect values between 10° and 20° latitude.

B.4. Final Comments

We should mention the studies by Starr et al. (1970) and Hantel and Hacker (1978), who compared the directly and indirectly evaluated meridional velocities in the Northern Hemisphere, and interpreted the differences in terms of vertical eddy fluxes of momentum. The residual calculations show that in the tropics, in the surface boundary layer, and near the jet core vertical eddies are important. The first two regions are probably taken care of reasonably well in our mixed method. In the jet region our indirect method might give a somewhat too broad (not sharp enough in the vertical) picture of the flow, since Starr et al. (1970) found a strengthening ($D_2 < 0$) effect by the vertical eddies. However their results are too

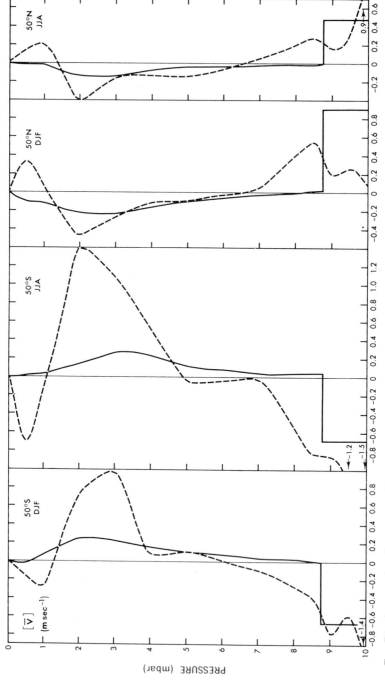

Fig. B1. A comparison between vertical profiles of the directly (---) and indirectly (——) computed mean meridional velocities at 50° S and 50° N for the DJF and JJA seasons.

tentative to allow a correction to our calculation of the Ferrel cell in middle latitudes.

In the Southern Hemisphere extratropics there is no choice but to use the indirect method because of the sparse rawinsonde network, which makes direct evaluation of $[\bar{v}]$ impossible. A comparison between the results obtained by using the direct and indirect methods to calculate the transports of total atmospheric energy by the mean meridional circulations is presented in Fig. B2. Obviously the direct values are useless south of about 15° S. Here we have to rely on the accuracy of measuring the eddy momentum flux. The standing eddy component $[\bar{v}^*\bar{u}^*]$ cannot be

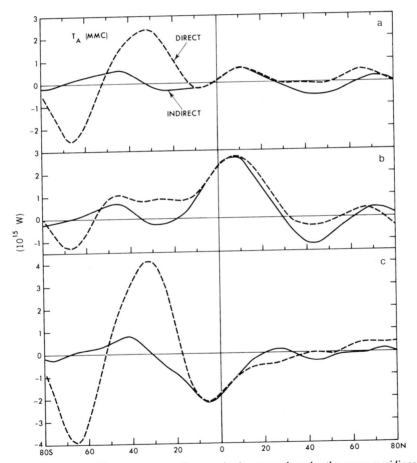

FIG. B2. The meridional transport of energy in the atmosphere by the mean meridional circulations, based on both directly (----) and indirectly (——) computed mean meridional velocities, for the year (a) and for the DJF (b) and JJA (c) seasons.

measured but is thought to be quite small due to the more zonally uniform surface conditions in the Southern Hemisphere (see, e.g., Obasi, 1963; Van Loon *et al.*, 1973). On similar grounds one may argue that the $\overline{v'u'}$ pattern should also be zonally uniform and that a few stations along a latitude circle in the Southern Hemisphere may be sufficient to derive a representative value for $[\overline{v'u'}]$. Some diagnostic model calculations (Oort, 1978) suggest fairly large errors of about 50% in the transient eddy momentum flux near 60° S, where there is a 5° latitude zone with no rawinsonde station at all.

For a thorough diagnostic study of how eddy momentum convergence, diabatic heating, and frictional processes act to maintain the mean meridional circulation in the Northern Hemisphere, the reader is referred to a recent article by Pfeffer (1981).

LIST OF SYMBOLS AND DEFINITIONS

$B(A)$ Boundary flux of A

$\mathbf{c}$ Velocity vector in (x, y, z) system $[= (u, v, w)]$

c_D Drag coefficient

c_p, c_v Specific heat at constant pressure and volume, respectively

$C(A, B)$ Conversion rate from A into B

$D(A)$ Dissipation rate of A

D_1–D_4 Momentum flux divergence and advection terms

DJF   Time period from December 1 to February 28/29

div   Three-dimensional divergence $[= (\partial/R \cos \phi \, \partial\lambda, \partial(\cos \phi)/R \cos \phi \, \partial\phi, \partial/\partial z)]$

div_2 Two-dimensional divergence $[= (\partial/R \cos \phi \, \partial\lambda, \partial(\cos \phi)/R \cos \phi \, \partial\phi)]$

dm Mass element $(= dx \, dy \, dp/g = R \cos \phi \, d\lambda \, R \, d\phi \, dp/g)$

dV Volume element $(= dx \, dy \, dz)$

E Total energy per unit mass $[= c_v T + gz + Lq + \frac{1}{2}(u^2 + v^2)]$

f Coriolis parameter $(= 2\Omega \sin \phi)$

F Frictional force $[= (F_\lambda, F_\phi)]$

F_BA Flux of energy at earth's surface (positive if downward)

F_TA Net flux of radiation at top of the atmosphere (positive if downward)

g Acceleration due to gravity

$G(A)$ Generation rate of A

IE   Internal energy per unit mass

$\mathbf{J}_\mathrm{q}$ Energy flux by radiation, conduction, pressure work and subgrid-scale terms $= \mathbf{F}_\mathrm{rad} + \mathbf{F}_\mathrm{con} + p\mathbf{c} + \boldsymbol{\tau} \cdot \mathbf{c}$

JJA   Time period from June 1 to August 31

K Horizontal kinetic energy $(= K_\mathrm{TE} + K_\mathrm{SE} + K_\mathrm{M} = K_\mathrm{E} + K_\mathrm{M})$

$K_\mathrm{M}, K_\mathrm{TE}, K_\mathrm{SE}$ Horizontal kinetic energy associated with zonal mean circulations, transient eddies, and stationary eddies, respectively

L Heat of condensation

M Absolute angular momentum $(= M_\Omega + M_\mathrm{r})$

M_Ω, M_r Ω and relative angular momentum, respectively

p Pressure

p_0 Pressure at ground level (where there are no mountains $p_0 = 1012.5$ mbar)

p_t Top level of vertical integration $(= 25$ mbar)

P Available potential energy $(= P_\mathrm{TE} + P_\mathrm{SE} + P_\mathrm{M} = P_\mathrm{E} + P_\mathrm{M})$

P_M, P_{TE}, P_{SE} Available potential energy associated with zonal mean conditions, transient eddies, and stationary eddies, respectively

PE   Potential energy per unit mass

q Specific humidity

R Mean radius of the earth (= 6371 km) or gas constant for dry air

S_A, S_O Rate of storage of energy in atmosphere and oceans, respectively

t Time

T Temperature

T_A, T_O Total meridional energy transport in atmosphere and oceans, respectively

u Zonal wind component (positive if eastward)

v Meridional wind component (positive if northward)

v_g, v_{ag} Geostrophic and ageostrophic part of meridional wind component, respectively

$\mathbf{v}$ Two-dimensional wind vector $[= (u, v)]$

$[\bar{v}]_{ID}$ Indirectly derived mean meridional velocity

w Vertical wind component

z Geopotential height

z_E, z_W Geopotential height at east and west sides of a mountain range, respectively

z_{SA} Geopotential height in NMC standard atmosphere

γ Measure of static stability in atmosphere $[= -(\theta/T)(R/c_p p)(\partial\bar{\theta}/\partial p)^{-1}]$

θ Potential temperature

λ Geographic longitude

ρ Density

$\sigma(A)$ Standard deviation of A

τ Three-dimensional stress tensor

$\tau_{E\phi}$, τ_{Ez} Meridional and vertical fluxes, respectively, used in defining ψ_E

τ_ϕ, τ_p Total stress components in the λ direction due to all scales of motion on ϕ = constant and p = constant surfaces, respectively

$\tau_{F\phi}$, τ_{Fp} Stress components in the λ direction due to friction and subgrid-scale processes on ϕ = constant and p = constant surfaces, respectively

τ_0 Surface friction stress in the λ direction (positive if atmosphere gains westerly momentum) $(= -\tau_{Fp}/g$ at $p = p_0)$

ϕ Geographic latitude

ψ Stream function for mass in (ϕ, p) plane

ψ_E Stream function for energy in (ϕ, p) plane

ψ_M Stream function for angular momentum in (ϕ, p) plane

ω "Vertical" pressure velocity (positive if downward) $(= dp/dt \approx -\rho gw)$

Ω Angular velocity of the earth

Mathematical Operators

$\bar{A} = (t_2 - t_1)^{-1} \int_{t_1}^{t_2} A\, dt$ Time average of A

$A' = A - \bar{A}$ Departure from time average of A

$[A] = (2\pi)^{-1} \int_0^{2\pi} A\, d\lambda$ Zonal average of A

$A^* = A - [A]$ Departure from zonal average of A

$\hat{A} = (p_0 - p_t)^{-1} \int_{p_t}^{p_0} A\, dp$ Mass-weighted "vertical" average of A

$\tilde{A} = \int_{-\pi/2}^{\pi/2} [A] \cos \phi\, d\phi / 2$ Global average of A

Example of Nomenclature

$$\widehat{[vA]} = \widehat{[v'A']} + \widehat{[\bar{v}^*\bar{A}^*]} + \widehat{[\bar{v}][\bar{A}]}$$

where

$\widehat{[v'A']}$ Vertical average of meridional transport of A resulting from transient eddies

$\widehat{[\bar{v}^*\bar{A}^*]}$ Vertical average of meridional transport of A resulting from standing eddies

$\widehat{[\bar{v}][\bar{A}]}$ Vertical average of meridional transport of A resulting from mean meridional circulations

ACKNOWLEDGMENTS

The authors would like to thank the members of the Academy of Sciences of Lisbon for the opportunity to participate in the international symposium "Advances in the Theory of Climate," where excerpts of this article were presented, Dr. J. Smagorinsky for his generous overall support of our work, Dr. B. Saltzman for his encouragement and editorial help, Messrs. J. Welsh and H. M. Frazier for their advice in handling the huge data sets, Mr. M. Rosenstein and Ms. M. A. Maher for their invaluable help in the data reduction and analysis, Mr. S. Hellerman, Mr. S. Levitus, Dr. G. G. Campbell, and Dr. T. H. Vonder Haar for the use of their data before publication, Messrs. P. Tunison, W. H. Ellis, and M. E. Zadworny for the excellent drafting of the figures, Drs. K. Bryan, N.-C. Lau, and S. Manabe for their critical review of the manuscript, and finally Ms. J. Kennedy for her outstanding typing of the paper. The visits of J. P. Peixóto to the Geophysical Fluid Dynamics Laboratory were supported through NOAA Grant 04-7-022-44017.

REFERENCES

Bjerknes, J. (1953). The maintenance of the zonal circulation of the atmosphere. *P.-V. Seances Assoc. Meteorol., 1951* Presidential Address, pp. 1–21.

Bjerknes, J. (1966). A possible response of the atmospheric Hadley circulation to equatorial anomalies of ocean temperature. *Tellus* **18,** 820–829.

Bowman, K. P. (1982). Sensitivity of an annual mean diffusive energy balance model with an ice sheet. *JGR, J. Geophys. Res.* **87,** No. C11, 9667–9674.

Bryan, K., and Lewis, L. J. (1979). A water mass model of the world ocean. *JGR, J. Geophys. Res.* **84,** No. C5, 2503–2517.

Budyko, M. I. (1963). "Atlas of the Heat Balance of the Earth." Globnaia Geofiz. Observ., Moscow (in Russian).

Bunker, A. F. (1976). Computations of surface energy flux and annual air-sea interaction cycles of the North Atlantic Ocean. *Mon. Weather Rev.* **104,** 1122–1140.

Campbell, G. G., and Vonder Haar, T. H. (1980). "Climatology of Radiation Budget Measurements from Satellites," Atmos. Sci. Pap. No. 323. Dept. Atmos. Sci., Colorado State University, Fort Collins.

Campbell, G. G., and Vonder Haar, T. H. (1982). Latitude average radiation budget over land and ocean from satellite observations and some implications for energy transport and climate modeling. *Mon. Weather Rev.* **110** (in press).

Crutcher, H. L. (1961). "Meridional Cross-Sections, Upper Winds over the Northern Hemisphere," Tech. Pap. No. 41. Weather Bureau, Washington, D.C.

Ellis, J. S., Vonder Haar, T. H., Levitus, S., and Oort, A. H. (1978). The annual variation in the global heat balance of the earth. *JGR, J. Geophys. Res.* **83,** No. C4, 1958–1962.

Gilman, P. A. (1965). The mean meridional circulation of the Southern Hemisphere inferred from momentum and mass balance. *Tellus* **17,** 277–284.

Hall, M. M., and Bryden, H. L. (1982). Direct estimates and mechanisms of ocean heat transport. *Deep-Sea Res.* **29,** 339–359.

Hantel, M., and Hacker, J. M. (1978). On the vertical eddy transports in the northern atmosphere. 2. Vertical eddy momentum transport for summer and winter. *JGR, J. Geophys. Res.* **83,** No. C3, 1305–1318.

Hastenrath, S. (1982). On meridional heat transports in the World Ocean. *J. Phys. Oceanogr.* **12,** 922–927.

Hellerman, S. (1967). An updated estimate of the wind stress on the world ocean. *Mon. Weather Rev.* **95,** 607–626 (see also corrections **96,** No. 1, pp. 63–74).

Hellerman, S. (1982). Wind stress on the world ocean with error estimates. *J. Phys. Oceanogr.* (in preparation).

Hide, R., Birch, N. T., Morrison, L. V., Shea, D. J., and White, A. A. (1980). Atmospheric angular momentum fluctuations and changes in the length of the day. *Nature (London)* **286,** 114–117.

Holopainen, E. O. (1967). On the mean meridional circulation and the flux of angular momentum over the Northern Hemisphere. *Tellus* **19,** 1–13.

Holopainen, E. O. (1969). On the maintenance of the atmosphere's kinetic energy over the Northern Hemisphere in winter. *Pure Appl. Geophys.* **77,** 104–121.

Holopainen, E. O. (1982). Long-term budget of zonal momentum in the free atmosphere over Europe in winter. *Q. J. R. Meteorol. Soc.* **108,** 95–102.

Holopainen, E. O., and Oort, A. H. (1981). Mean surface stress curl over the oceans as determined from the vorticity budget of the atmosphere. *J. Atmos. Sci.* **38,** 262–280.

Kung, E. C. (1969). Further study on the kinetic energy balance. *Mon. Weather Rev.* **97,** 573–581.

Kuo, H. L. (1956). Forced and free meridional circulation in the atmosphere. *J. Meteorol.* **13,** 561–568.

Lambeck, K. (1981). "The Earth's Variable Rotation." Press Syndicate of the University of Cambridge, Cambridge, England.

Lau, N.-C., and Oort, A. H. (1981). A comparative study of observed Northern Hemisphere circulation statistics based on GFDL and NMC analyses. Part I. *Mon. Weather Rev.* **109,** 1380–1403.

Levitus, S. (1982). "Climatological Atlas of the World Ocean," NOAA Prof. Pap. No. 13. U.S. Gov. Printing Office, Washington, D.C. (in press).

Lorenz, E. N. (1955). Available potential energy and the maintenance of the general circulation. *Tellus* **7,** 157–167.

Lorenz, E. N. (1963). Deterministic nonperiodic flow. *J. Atmos. Sci.* **20**, 130–141.

Lorenz, E. N. (1967). The nature and theory of the general circulation of the atmosphere. *WMO [Publ.]* **218**, TP115, 1–161.

Mak, M.-K. (1969). Laterally driven stochastic motions in the tropics. *J. Atmos. Sci.* **26**, 41–64.

Mak, M.-K. (1978). The observed momentum flux by standing eddies. *J. Atmos. Sci.* **35**, 340–346.

Min, K. D., and Horn, L. H. (1982). Available potential energy in the Northern Hemisphere during the FGGE year. *Tellus* **34**, 526–539.

Mintz, Y. (1954). The observed zonal circulation of the atmosphere. *Bull. Am. Meteorol. Soc.* **35**, 208–214.

Namias, J. (1979). "Short Period Climatic Variations, Collected Works of J. Namias 1934 through 1974." University of California, San Diego.

Newell, R. E., Kidson, J. W., Vincent, D. G., and Boer, G. J. (1972). "The General Circulation of the Tropical Atmosphere," Vol. 1. MIT Press, Cambridge, Massachusetts.

Newell, R. E., Kidson, J. W., Vincent, D. G., and Boer, G. J. (1974). "The General Circulation of the Tropical Atmosphere," Vol. 2. MIT Press, Cambridge, Massachusetts.

Newton, C. W. (1971a). Mountain torques in the global angular momentum balance. *J. Atmos. Sci.* **28**, 623–628.

Newton, C. W. (1971b). Global angular momentum balance: Earth torques and atmospheric fluxes. *J. Atmos. Sci.* **28**, 1329–1341.

Obasi, G. O. P. (1963). Poleward flux of atmospheric angular momentum in the Southern Hemisphere. *J. Atmos. Sci.* **20**, 516–528.

Oort, A. H. (1964). On estimates of the atmospheric energy cycle. *Mon. Weather Rev.* **92**, 483–493.

Oort, A. H. (1971). The observed annual cycle in the meridional transport of atmospheric energy. *J. Atmos. Sci.* **28**, 325–339.

Oort, A. H. (1977). "The Interannual Variability of Atmospheric Circulation Statistics," NOAA Prof. Pap. No. 8. U.S. Gov. Printing Office, Washington, D.C.

Oort, A. H. (1978). On the adequacy of the rawinsonde network for global circulation studies tested through numerical model output. *Mon. Weather Rev.* **106**, 174–195.

Oort, A. H. (1983). "Global Atmospheric Circulation Statistics, 1958–1973," NOAA Prof. Pap. No. 14. U.S. Gov. Printing Office, Washington, D.C. (in press).

Oort, A. H., and Bowman, H. D. (1974). A study of the mountain torque and its interannual variations in the Northern Hemisphere. *J. Atmos. Sci.* **31**, 1974–1982.

Oort, A. H., and Peixóto, J. P. (1974). The annual cycle of the energetics of the atmosphere on a planetary scale. *JGR, J. Geophys. Res.* **79**, 2705–2719.

Oort, A. H., and Rasmusson, E. M. (1971). "Atmospheric Circulation Statistics," NOAA Prof. Pap. No. 5. U.S. Gov. Printing Office, Washington, D.C.

Oort, A. H., and Vonder Haar, T. H. (1976). On the observed annual cycle in the ocean-atmosphere heat balance over the Northern Hemisphere. *J. Phys. Oceanogr.* **6**, 781–800.

Oort, A. H., Vonder Haar, T. H., and Levitus, S. (1983). On the observed annual cycle in the global heat balance. *J. Phys. Oceanogr.* (in preparation).

Palmén, E., Riehl, H., and Vuorela, L. A. (1958). On the meridional circulation and release of kinetic energy in the tropics. *J. Meteorol.* **15**, 271–277.

Pearce, R. P. (1978). On the concept of available potential energy. *Q. J. R. Meteorol. Soc.* **104**, 737–755.

Peixóto, J. P. (1958). "Hemispheric Humidity Conditions During the Year 1950," Sci. Rep. No. 3 (Gen. Circ. Proj.). Massachusetts Institute of Technology, Cambridge.

Peixóto, J. P. (1960). "Hemispheric Temperature Conditions During the Year 1950," Sci. Rep. No. 4 (Planet. Circ. Proj.). Massachusetts Institute of Technology, Cambridge.

Peixóto, J. P. (1965). On the role of water vapor in the energetics of the general circulation of the atmosphere. *Port. Phys.* **4**, 135–170.

Peixóto, J. P. (1973). "Atmospheric Vapour Flux Computations for Hydrological Purposes," Rep. No. 20. WMO, Geneva, Switzerland.

Peixóto, J. P. (1974). Enthalpy distribution in the atmosphere over the Southern Hemisphere. *Riv. Ital. Geofis.* **23**, No. 3/4, 223–242.

Peixóto, J. P., and Corte-Real, J. (1982). The energetics of the general circulation of the atmosphere in the Southern Hemisphere during the IGY. Part I. Distribution of atmospheric energy. Part II. The cycle of energetics of the atmosphere in the Southern Hemisphere. *Arch. Meteorol., Geophys. Bioklimatol., Ser. A* **31** (in press).

Peixóto, J. P., and Oort, A. H. (1974). The annual distribution of atmospheric energy on a planetary scale. *JGR, J. Geophys. Res.* **79**, 2149–2159.

Peixóto, J. P., and Oort, A. H. (1983). The atmospheric branch of the hydrological cycle and climate. *In* "Variations of the Global Water Budget," pp. 5–65. Reidel, London.

Peixóto, J. P., Rosen, R. D., and Wu, M.-F. (1976). Seasonal variability in the pole-to-pole water vapor balance during the IGY. *Nord. Hydrol.* **7**, 95–114.

Peixóto, J. P., Rosen, R. D., and Salstein, D. A. (1978). Seasonal variability in the pole-to-pole modes of water vapor transport during the IGY. *Arch. Meteorol. Geophys. Bioklimatol., Ser. A* **27**, 233–255.

Pfeffer, R. A. (1981). Wave-mean flow interactions in the atmosphere. *J. Atmos. Sci.* **38**, 1340–1359.

Phillips, N. A. (1956). The general circulation of the atmosphere: A numerical experiment. *Q. J. R. Meteorol. Soc.* **82**, 123–164.

Priestley, C. H. B. (1951). A survey of the stress between the ocean and atmosphere. *Aust. J. Sci. Res., Ser. A* **4**, 315–328.

Rabiner, L. R., Sambur, M. R., and Schmidt, C. E. (1975). Applications of a non-linear smoothing algorithm to speech processing. *IEEE Trans. Acoust., Speech, Signal Process.* **ASSP-23**, No. 6, 552–557.

Rosen, R. D. (1976). The flux of mass across latitude walls in the atmosphere. *JGR, J. Geophys. Res.* **81**, 2001–2002.

Rosen, R. D., and Salstein, D. A. (1980). A comparison between circulation statistics computed from conventional data and NMC Hough analyses. *Mon. Weather Rev.* **108**, No. 8, 1226–1247.

Saltzman, B. (1957). Equations governing the energetics of the larger scales of atmosphere turbulence in the domain of wave number. *J. Meteorol.* **14**, 513–523.

Starr, V. P. (1948). An essay on the general circulation of the earth's atmosphere. *J. Meteorol.* **5**, 39–43.

Starr, V. P. (1951). Applications of energy principles to the general circulation. *In* "Compendium of Meteorology," pp. 568–574. Am. Meteorol. Soc., Boston, Massachusetts.

Starr, V. P. (1953). Note concerning the nature of the large-scale eddies in the atmosphere. *Tellus* **5**, 494–498.

Starr, V. P. (1968). "Physics of Negative Viscosity Phenomena." McGraw-Hill, New York.

Starr, V. P., and Peixóto, J. P. (1971). Pole-to-pole eddy transport of water vapor in the atmosphere during the IGY. *Arch. Meteorol. Geophys. Bioklimatol., Ser. A* **20,** 85–114.
Starr, V. P., and Wallace, J. M. (1964). Mechanics of eddy processes in the tropical troposphere. *Pure Appl. Geophys.* **58,** 138–144.
Starr, V. P., and White, R. M. (1951). A hemispherical study of the atmospheric angular-momentum balance. *Q. J. R. Meteorol. Soc.* **77,** 215–225.
Starr, V. P., and White, R. M. (1952). Meridional flux of angular momentum in the tropics. *Tellus* **4,** 118–125.
Starr, V. P., and White, R. M. (1954). Balance requirements of the general circulation. Air Force Cambridge Research Directorate, *Geophys. Res. Pap.* **35,** 1–57.
Starr, V. P., Peixóto, J. P., and McKean, R. G. (1969). Pole-to-pole moisture conditions for the IGY. *Pure Appl. Geophys.* **75,** 300–331.
Starr, V. P., Peixóto, J. P., and Gaut, N. E. (1970). Momentum and zonal kinetic energy balance of the atmosphere from five years of hemispheric data. *Tellus* **22,** 251–274.
Stephens, G. L., Campbell, G. G., and Vonder Haar, T. H. (1981). Earth radiation budgets. *JGR, J. Geophys. Res.* **86,** 9739–9760.
Swanson, G. S., and Trenberth, K. E. (1981). Trends in the Southern Hemisphere tropospheric circulation. *Mon. Weather Rev.* **109,** No. 9, 1879–1889.
Trenberth, K. E. (1979). Mean annual poleward energy transports by the oceans in the Southern Hemisphere. *Dyn. Atmos. Oceans* **4,** 57–64.
Trenberth, K. E., and Van Loon, H. (1981). Comment on "Impact of FGGE buoy data on Southern Hemisphere analyses." *Bull. Am. Meteorol. Soc.* **62,** 1486–1488.
Van Loon, H., Taljaard, J. J., Jenne, R. L., and Crutcher, H. L. (1971). Zonal geostrophic winds. *In* "Climate of the Upper Air: Southern Hemisphere," Vol. 2, NAVAIR 50-1C-56 and NCAR TN/STR-57, pp. 1–42.
Van Loon, H., Jenne, R. L., and Labitzke, K. (1973). Zonal harmonic standing waves. *JGR, J. Geophys. Res.* **78,** 4463–4471.
White, R. M. (1949). The role of the mountains in the angular momentum balance of the atmosphere. *J. Meteorol.* **6,** 353–355.

INDEX